Studies in Lo

Mathematical Logic and Foundations

Volume 56

Dualities for Structures of Applied Logics

Volume 45
Errors of Reasoning. Naturalizing the Logic of Inference
John Woods

Volume 46
Questions, Inferences, and Scenarios
Andrzej Wiśniewski

Volume 47
Logic Across the University: Foundations and Applications. Proceedings of the Tsinghua Logic Conference, Beijing, 2013
Johan van Benthem and Fenrong Liu, eds.

Volume 48
Trends in Belief Revision and Argumentation Dynamics
Eduardo L. Fermé, Dov M. Gabbay, and Guillermo R. Simari

Volume 49
Introduction to Propositional Satisfiability
Victor Marek

Volume 50
Intuitionistic Set Theory
John L. Bell

Volume 51
Metalogical Contributions to the Nonmonotonic Theory of Abstract Argumentation
Ringo Baumann

Volume 52
Inconsistency Robustness
Carl Hewitt and John Woods, eds.

Volume 53
Aristotle's Earlier Logic
John Woods

Volume 54
Proof Theory of N4-related Paraconsistent Logics
Norihiro Kamide and Heinrich Wansing

Volume 55
All about Proofs, Proofs for All
Bruno Woltzenlogel Paleo and David Delahaye, eds

Volume 56
Dualities for Structures of Applied Logics
Ewa Orłowska, Anna Maria Radzikowska and Ingrid Rewitzky

Studies in Logic Series Editor
Dov Gabbay dov.gabbay@kcl.ac.uk

Dualities for Structures of Applied Logics

Ewa Orłowska,
Anna Maria Radzikowska
and
Ingrid Rewitzky

ISBN 978-1-84890-181-0

College Publications
Scientific Director: Dov Gabbay
Managing Director: Jane Spurr

http://www.collegepublications.co.uk

Original cover design by Orchid Creative www.orchidcreative.co.uk
Printed by Lightning Source, Milton Keynes, UK

Preface

A mathematical structure arises whenever one attempts to describe and study an activity or entity or an idea using mathematical concepts and tools, so that the structure would grasp the essential features of the phenomena. The structures considered in the book are algebras and relational systems used in studying applied logics via their semantics. In particular, semantics structures for philosophical logics and computer science logics arising when modelling uncertainty and incompleteness of information are among the main themes of the book. Throughout the book, according to a logical tradition, relational systems will be referred to as frames.

There is a diversity about opinions of an appropriate formalism for mathematical structures. David Hilbert [128] said 'No one shall drive us from the paradise which Cantor created for us'. Some 'attacked' set theory. For example, Wittgenstein replied 'If one person can see it as a paradise of mathematicians, why should not another see it as a joke?'. Others like Mac Lane felt that this 'is a mistakenly one-sided view of mathematics' and his letter [157] may be interpreted as saying that 'In spite of the fundamental achievements of set theory, the perfect paradise is still to be found'. With few exceptions, each structure considered in the book is presented axiomatically in a first order language, with the underlying universe of the structure providing the domain over which the quantified variables of the first order language range.

In many mathematical contexts, duality is used as the main principle for describing relationships between different classes of mathematical structures. This is formulated in terms of two classes of mathematical structures, a translation from each class to the other, and a property preserved under the translations. For example: in *projective duality* points and lines are dual with respect to the property of incidence; in *order duality* a lattice and its opposite lattice are dual with respect to the property of the underlying orders being converses of each other; in order duality ideals and filters are dual concepts because a filter (resp. ideal) of a lattice is the ideal (resp. filter) of its opposite lattice; in *contravariant dualities*, such as Stone duality and Priestley duality, a topological space of entities and a lattice of properties are dual with respect to the property of being isomorphic; in *modal logic* the possibility operator is dual to the necessity operator because one is definable in terms of the other and the negation. The notion of duality considered in this book is formulated in terms of two classes of structures being dual with respect to the property of being embeddable.

In each chapter of the three parts of the book a starting point is a pair of such structures – a class $\mathcal{A}lg$ of lattices, and a class $\mathcal{F}rm$ of frames. Next, from each frame in the class $\mathcal{F}rm$ a concrete algebra in the class $\mathcal{A}lg$ is defined as the closure system with respect to some closure operator on the given frame. On the other hand, from each algebra in the class $\mathcal{A}lg$ a concrete frame in the class $\mathcal{F}rm$ is defined over certain subsets of the given algebra. Then we approach the problem of providing discrete representation theorems and, if possible, a discrete duality by determining whether each algebra in the class $\mathcal{A}lg$ and each frame in the class $\mathcal{F}rm$ can be defined concretely in this way. The term discrete duality is used for a system ($\mathcal{A}lg$, $\mathcal{F}rm$, $\mathcal{C}m$, $\mathcal{C}s$) where

> $\mathcal{C}m : \mathcal{F}rm \to \mathcal{A}lg$ is a mapping assigning to every frame $X \in \mathcal{F}rm$ its complex algebra $\mathcal{C}m(X)$ in such a way that $\mathcal{C}m(X) \in \mathcal{A}lg$,
>
> $\mathcal{C}s : \mathcal{A}lg \to \mathcal{F}rm$ is a mapping assigning to every algebra $L \in \mathcal{A}lg$ its canonical frame $\mathcal{C}s(L)$ in such a way that $\mathcal{C}s(L) \in \mathcal{F}rm$,

and the following two representation theorems hold.

Theorem 1 (Representation theorem for algebras)
Every algebra $L \in \mathcal{A}lg$ is embeddable into $\mathcal{C}m(\mathcal{C}s(L))$ which is the complex algebra of its canonical frame.

Theorem 2 (Representation theorem for frames)
Every frame $X \in \mathcal{F}rm$ is embeddable into $\mathcal{C}s(\mathcal{C}m(X))$ which is the canonical frame of its complex algebra.

Then we say that the system $(\mathcal{A}lg, \mathcal{F}rm, \mathcal{C}m, \mathcal{C}s)$ is a *discrete duality* between the classes $\mathcal{A}lg$ and $\mathcal{F}rm$. Frames, being semantic structures of formal languages do not involve topology and therefore this duality is said to be discrete. A source of inspiration for discarding topology in the class $\mathcal{F}rm$ can be traced to the papers by Viorica Sofronie-Stokkermans [222, 224]. Motivated by efficiency issues in automated theorem proving she showed that in some logics deduction procedures based on a frame semantics are more efficient than those based on an algebraic semantics.

The complex algebras are set algebras but not necessarily the powerset algebras, except in the case of some structures considered in Part 1. Similarly, the canonical frames are not always defined on the families of prime filters of the lattice in question, as is done in classical Stone duality.

We emphasize that the representation structures in Theorem 1 and Theorem 2 are always constructed from two structures one of which is of an algebraic origin and the other of a logical origin. Our view is that employing the concepts and structures from these two sources makes the developments transparent, easy to comprehend, and enables their interdisciplinary applications.

In the three parts of the book we consider representations and dualities for various classes of Boolean lattices, distributive lattices, and general, that is not necessarily distributive, lattices, respectively. The results for Boolean algebras and distributive lattices could be obtained as instances of the results for general lattices. However, we prefer to describe them directly and independently, which usually makes the presentation simpler and more intuitive. Moreover, those interested in a particular class of structures do not need to study first a more general class. In Section 2.8 and Section 18.7 the consequences of distributivity of lattices are discussed enabling a derivation of representation theorems for distributive lattices from those for general lattices.

Among the classes of lattices investigated in the book there are lattices with monotone operators, antitone operators, operators of both of these types, the operators which are neither monotone nor antitone, and various axiomatic extensions of these classes. The operators are most often unary such as, for example, modal monotone or antitone operators, or binary such as, for example, adjoint pairs of operators in residuated lattices. We do not consider n-ary generalizations of the operators. Among the axiomatic extensions of lattices with operators considered in the book there are few classes for which the corresponding classes of frames are known from the correspondence theory, but many of the classes of frames are novel, in particular, those associated to some lattices with operators which require ternary relations in the frames, with operators characterized by second order axioms, or non-canonical classes of lattices.

Relationships between a discrete representation and a discrete duality considered in the book and the other types of representations and dualities known in the literature are discussed in Chapter 2 and Chapter 18. We have attempted to make the book self-contained and complete in that the proofs are provided in a sufficient detail enabling an easy verification. For simplicity in the table of contents we list the names of classes of algebras only, although the sections also deal with the classes of corresponding frames.

The book might be of interest for the readers interested in semantics of philosophical logics including modal, intuitionistic and relevant logics, computational logics including spatial and temporal logics, and logics for reasoning with uncertainty including fuzzy logics and rough set logics. Algebraic and frame semantics for a variety of those logics are considered and their relationships are discussed within the framework of what is called a discrete duality and a duality via truth.

The concept of discrete duality and duality via truth proposed in the book enable to apply the tools of universal algebra for discussing those two types of semantics directly with a purely algebraic method without involving a topology.

Ewa Orłowska gratefully acknowledges the financial support from the National Science Centre project DEC-2011/02/A/HS1/00395. Ingrid Rewitzky gratefully acknowledges the financial support from the South African National Research Foundation and the research leave opportunities afforded her by Stellenbosch University. Moreover, we would like to thank our publishers for the way they have handled the whole process since signing of the contract, through to being patient enough to wait for the long overdue delivery of the typescript, and finally going through copy-editing with such a great sense of professionalism.

Ewa Orłowska (Warsaw, National Institute for Telecommunications)
Ania Radzikowska (Warsaw University of Technology)
Ingrid Rewitzky (Stellenbosch University)
October 2015

Contents

Chapter 1

Preliminaries

1.1 Introduction

In this chapter we establish the notation and terminology to be used in the book. In addition, we provide an overview of various structures and their properties which play an important role in the discrete duality framework developed in the book, as well as an overview of the operators and relations with which these structures will be enriched for further cases studies of the framework.

In Section 1.2 the standard set-theoretical notation is fixed. Then, in Section 1.5 and Section 1.6, the definitions and basic properties of ordered sets and lattices, in particular distributive lattices, residuated lattices, and Boolean algebras, are outlined. Standard references on the lattice theory include [14, 33, 116]. The next three sections, namely, Section 1.3, Section 1.4, and Section 1.8, are relevant for the operators and/or relations with which these algebras are to be enriched in the subsequent chapters of the book. The notion of frame is recalled in Section 1.9, and then related to the notion of coalgebra in Section 1.13. In Section 1.11 there is a brief overview of topological spaces, in particular, the Alexandroff topology for further consideration in Section 2.5. For the purposes of relating discrete duality to other dualities, we recall the notion of natural transformation and categorical equivalence in Section 1.12.

1.2 Set-theoretical notations

We assume the usual notation for the operations defined on sets. For sets A and B contained in some non-empty set X,

- the *union* $A \cup B \stackrel{\text{def}}{=} \{x \in X : x \in A \text{ or } x \in B\}$;
- the *intersection* $A \cap B \stackrel{\text{def}}{=} \{x \in X : x \in A \ \&\ x \in B\}$;
- the *difference* $A - B \stackrel{\text{def}}{=} \{x \in X : x \in A \ \&\ x \notin B\}$;
- the *complement* $-A \stackrel{\text{def}}{=} = \{x \in X : x \notin A\}$;
- the *Cartesian product* $\mathrm{A} \times \mathrm{B} \stackrel{\text{def}}{=} \{(\mathrm{x},\mathrm{y}) : \mathrm{x} \in \mathrm{A} \ \&\ \mathrm{y} \in \mathrm{B}\}$.

A *cofinite set* A is a subset of a set X whose complement $-A$ is a finite set. Two sets are said to be *disjoint* if their intersection is the empty set, $\emptyset$.

The relationship of *set inclusion* is denoted by $A \subseteq B$. Two sets A and B are *equal*, if both $A \subseteq B$ and $B \subseteq A$. Set equality is denoted by $A = B$ and set *inequality*by $A \neq B$. If $A \subseteq B$ and $A \neq B$, then A is a *proper subset* of B. Proper inclusion is denoted by $A \subset B$. The set of all subsets of a set X is called the *powerset* of X and denoted by 2^X.

Let I be any set and $\{A_i : i \in I\}$ be a family of subsets of a set X indexed by I. Then the union, the intersection, the Cartesian product, and the disjoint union of these sets are defined respectively by

$$\bigcup_{i \in I} A_i \stackrel{\text{def}}{=} \{x \in X : x \in A_i \text{ for some } i \in I\}$$

$$\bigcap_{i \in I} A_i \stackrel{\text{def}}{=} \{x \in X : x \in A_i \text{ for every } i \in I\}$$

$$\prod_{i \in I} A_i \stackrel{\text{def}}{=} \{f : I \to \bigcup_{i \in I} A_i : f(i) \in A_i \text{ for every } \ i \in I\}$$

$$\dot{\bigcup_{i \in I}} A_i \stackrel{\text{def}}{=} \bigcup_{i \in I} \{(x, i) \in X \times I : x \in A_i\}.$$

If $I = \emptyset$ then $\bigcup_{i \in I} A_i = \emptyset$ and $\bigcap_{i \in I} A_i = X$. The Cartesian product of finite $n \geq 1$ copies of a set A is denoted by A^n. For each $i \in I$, the function

$$\prod_i : \prod_{i \in I} A_i \to A_i$$

defined by $\prod_i(f) = f(i)$ is called the i-th projection map.

1.3 Binary relations

A *binary relation* from a set X to a set Y is a subset of the Cartesian product $X \times Y$, that is, a set of ordered pairs (x, y) with $x \in X$ and $y \in Y$. Mostly we will deal with the case where $X = Y$. That is, we will be concerned with a binary relation $R \subseteq X^2$, and we say that R is a binary relation on set X. In the rest of the book (unless otherwise stated) we denote binary relations by the symbols R, S, T and we often write 'xRy' instead of '$(x, y) \in R$'.

Binary relations on X, being sets, are partially ordered by inclusion, and allow the set-theoretical operations of intersection $\cap$, union $\cup$, and complement $(-)$. The smallest relation with respect to this ordering is the empty subset of X^2, called the *empty relation*, and the largest relation is X^2 itself, called the *universal relation*. In addition, for any two binary relations $R, S \subseteq X^2$, their *relational product* or *composition* $R\,;S \subseteq X^2$ is given by

$$R\,; S \stackrel{\text{def}}{=} \{(x, y) : \exists z \in X,\ xRz \ \&\ zRy\}.$$

We sometimes write 'R^2' for the product $R\,; R$, and in general 'R^{n+1}' for $R^n\,; R$, for any natural number $n \geq 0$. The *converse* of a binary relation R is the relation R^{-1} given by

$$R^{-1} \stackrel{\text{def}}{=} \{(x, y) : yRx\}.$$

The *identity* relation Id_X on a set X is given by

$$Id_X \stackrel{\text{def}}{=} \{(x, x) : x \in X\}.$$

If X is known within a certain context, we often omit the subscript. By convention we put $R^0 \stackrel{\text{def}}{=} Id_X$ for any relation $R \subseteq X^2$.

There are many interesting and useful special binary relations characterised in terms of additional algebraic conditions not satisfied by all binary relations. In particular, a binary relation $R \subseteq X^2$ is

- *reflexive* if $Id_X \subseteq R$;
- *weakly reflexive* if $(R; R^{-1}) \cap Id_X = R$;
- *irreflexive* if, $R \cap Id_X = \emptyset$;
- *serial* if, for every $x \in X$, there is some $y \in X$, xRy;
- *transitive* if $R ; R \subseteq R$;
- *n-transitive* if $R^n \subseteq R$, for natural number $n \geq 3$;
- *symmetric* if $R \subseteq R^{-1}$;
- *antisymmetric* if $R \cap R^{-1} \subseteq Id_X$;
- a *partial order* if it is reflexive, transitive, and antisymmetric;
- an *equivalence relation* if it is reflexive, transitive, and symmetric.

For a given property α of a binary relation R by the *co-property*(co-α) we mean the property α of the complement $X^2 - R$ of R.

For the algebraic properties of relational operations the original paper [233] is still a good reference.

Lemma 1.3.1 *For any binary relations* $R, S, T \subseteq X^2$,

(a) $Id_X; R = R = R; Id_X$
(b) $(R; S); T = R ; (S; T)$
(c) $R ; (S \cup T) = (R; S) \cup (R; T)$ *and* $(R \cup S); T = (R; T) \cup (S; T)$
(d) $(R^{-1})^{-1} = R$
(e) $(R \cup S)^{-1} = R^{-1} \cup S^{-1}$
(f) $(R; S)^{-1} = S^{-1}; R^{-1}$.

For every binary relation $R \subseteq X^2$ and for all $x, y \in X$, we define

- the *domain* of R by $dom(R) \stackrel{\text{def}}{=} \{x : \exists y \in X,\ xRy\}$;
- the *range* of R by $ran(R) \stackrel{\text{def}}{=} \{y : \exists x \in X,\ xRy\}$;
- the *image set* of x under R by $R(x) \stackrel{\text{def}}{=} \{y : xRy\}$;
- the *inverse image* set of y under R by $R^{-1}(y) \stackrel{\text{def}}{=} \{x \in X : xRy\}$.

1.4 Functions

Of the binary relations from a set X to a set Y the most widely used are the *functions.*

A binary relation $R \subseteq X \times Y$ is

- a *partial function* if, xRy and xRz imply $y = z$ for every $x \in dom(R)$ and all $y, z \in Y$;
- a *function* (or *unary operator*) if, for every $x \in X$ there is a unique $y \in Y$ such that xRy.

We use the symbols f, g, h instead of the relational symbols to denote functions and write $f : X \to Y$'. The notation '$y = f(x)$' is used instead of 'xfy'. A function f with domain X and range Y is said to be a function from X *onto* Y; if in addition $Y \subseteq Z$, then f is a function from X *into* Z; if $X = Y$ then f is a function *on* X.

For functions $f : X \to Z$ and $g \ : Z \to Y$ there is a natural mapping, their *composition* $g \circ f : X \to Y$ given by

$$g \circ f \stackrel{\text{def}}{=} f\,;g.$$

Note that $g \circ f : X \to Y$ is a function since both f and g are and it is given by $(g \circ f)(x) = g(f(x))$, for every $x \in X$. The *inverse* of a function $f : X \to Y$ is a mapping $f^{-1} : Y \to X$ given, for every $x \in X$ and for every $y \in Y$, by

$$f^{-1}(y) = x \stackrel{\text{def}}{\Leftrightarrow} f(x) = y.$$

Note that f^{-1} is not necessarily a function since there may be more than one x such that $f(x) = y$. The identity relation over a set X is a total function so the *identity* function on X is the mapping $\mathrm{id}_X : X \to X$ given by $\mathrm{id}_X(x) = x$.

A function $f : X \to Y$ is

- *injective*(or *one-one*)if, $f(x) = f(y)$ implies $x = y$ for all $x, y \in X$;
- *surjective* (or *onto*) if, for every $y \in Y$ there is some $x \in X$ such that $f(x) = y$.

An injective and surjective function $f : X \to Y$ is called a *bijection*, or a *bijective correspondence* between X and Y. Note that the inverse f^{-1} of a function f is a function if f is one-one.

For functions we can extend the notions of image and inverse image set from objects to sets of objects. For any function $f : X \to Y$ and for any sets $A \subseteq X$ and $B \subseteq Y$, we define

- the *image* of A under f by

$$f(A) \stackrel{\text{def}}{=} \{y \in Y : \exists x \in A,\ y = f(x)\};$$

- the *inverse image* of B under f by

$$f^{-1}(B) \stackrel{\text{def}}{=} \{x \in X : \exists y \in B,\ y = f(x)\}.$$

Lemma 1.4.1 *For every function $f : X \to Y$, for all sets $A \subseteq X$ and $B \subseteq Y$,*

(a) $A \subseteq f^{-1}(f(A))$;
$A = f^{-1}(f(A))$, *if f is injective;*

(b) $f(f^{-1}(B)) \subseteq B$;
$f(f^{-1}(B)) = B$, *if f is surjective;*

(c) $f(A) \subseteq B \Leftrightarrow A \subseteq f^{-1}(B)$;

(d) $f^{-1}(Y - B) = X - f^{-1}(B)$;

(e) *If $X = Y$ and $A \subseteq B$, then $f(A) \subseteq f(B)$ and $f^{-1}(A) \subseteq f^{-1}(B)$.*

1.5 Ordered sets

Let X be a non-empty set. A *pre-order* on X is a subset of $X \times X$ which is reflexive and transitive. A non-empty set together with a pre-order $\leqslant$, written $(X, \leqslant)$, is called a *pre-ordered set.* A pre-order $\leqslant$ on X which is antisymmetric is a *partial order*, and $(X, \leqslant)$ is called a *partially ordered set* or *poset.* A poset is *linearly ordered* whenever, for all $x, y \in X$, either xRy or yRx. A non-empty subset C of a poset X is a *chain* if it is linearly ordered. Note that $(X, \leqslant)$ is a poset if $(X, \geqslant)$ is a poset. This means that if a statement on posets is dualised (that is, $\leqslant$ replaced by $\geqslant$), then if the statement is true, so is its dual.

When considering more than one pre-ordered set, a subscript on each order is used to indicate the set over which it is defined – for example, $(X, \leqslant_X)$ and $(Y, \leqslant_Y)$.

Let $(X, \leqslant)$ be a poset. For any subset $A \subseteq X$,

- A is $\leqslant$*-increasing* if, for all $x, y \in X$, $x \in A$ and $x \leqslant y$ imply $y \in A$;
- A is $\leqslant$*-decreasing* if, for all $x, y \in X$, $y \in A$ and $x \leqslant y$ imply $x \in A$;
- the *up-closure* of A is $\uparrow_{\leqslant} A \stackrel{\text{def}}{=} \{x \in X : \exists a \in A,\ a \leqslant x\}$;
- the *down-closure* of A is $\downarrow_{\leqslant} A \stackrel{\text{def}}{=} \{x \in X : \exists a \in A,\ x \leqslant a\}$.

For singletons we simplify the notation by writing $\uparrow_{\leqslant} x$ instead of $\uparrow_{\leqslant} \{x\}$ and $\downarrow_{\leqslant} y$ instead of $\downarrow_{\leqslant} \{y\}$. A subset $A \subseteq X$ is called an *up-closed set* if $A = \uparrow_{\leqslant} A$, that is, if $x \in A$ and $x \leqslant y$ together imply $y \in A$. Dually, a subset $A \subseteq X$ is called a *down-closed set* if $A = \downarrow_{\leqslant} A$, that is, if $x \in A$ and $y \leqslant x$ together imply $y \in A$. Up-closed sets and down-closed sets are also related by set-theoretical complementation.

Lemma 1.5.1 *Let $(X, \leqslant)$ be a poset. A subset $A \subseteq X$ is an up-closed set if and only if its set complement $-A$ is a down-closed set.*

Up-closed sets determine the order $\leqslant$ on a poset $(X, \leqslant)$.

Lemma 1.5.2 *Let $(X,\leqslant)$ be a poset. For any $x,y\in X$,*

$$x\leqslant y \Leftrightarrow \uparrow_{\leqslant} x \supseteq \uparrow_{\leqslant} y \Leftrightarrow \text{ for every up-closed set } A\subseteq X,\ x\in A \text{ implies } y\in A.$$

This theorem can be dualised to show that the down-closed sets of a poset determine the order on its dual.

Many properties of posets are expressed in terms of least upper bounds and greatest lower bounds. Let A be a subset of a poset $(X,\leqslant)$. Then $x\in X$ is an *upper bound* for A if for all $a\in A$, $a\leqslant x$. An upper bound x of A is the *least upper bound* (abbreviated as 'lub') if $x\leqslant y$ for any other upper bound y of A. Dually, we may define the concepts *lower bound* and *greatest lower bound* (abbreviated as 'glb'). The lub of a set A is denoted $\bigvee A$. The glb is denoted $\bigwedge A$. If A is a two-element set, say $\{x,y\}$, then we write the lub and glb as $x\vee y$ and $x\wedge y$, and call these the *join* and *meet* of x and y, respectively. The same notation and terminology is often used when A is any finite set. An element $y\in X$ is called the *least element* or *bottom* of X if for all $x\in X$, $y\leqslant x$. Depending on the context, the least element of $(X,\leqslant)$ is sometimes denoted 0_X, or 0 when no confusion arises. Dually, the *greatest element* or *top* of X is an element $z\in X$ such that for all $x\in X$, $x\leqslant z$, and this greatest element is often denoted by 1_X, or 1 when no confusion arises.

The next lemma contains some useful information for working with least upper bounds and greatest lower bounds.

Lemma 1.5.3 *Let $(X,\leqslant)$ be a poset and let $A,B\subseteq X$. Assume $\bigwedge A,\bigwedge B,\bigvee A$ and $\bigvee B$ exist in X.*

(a) *For all $x\in A$, $x\leqslant \bigvee A$ and $x\geqslant \bigwedge A$;*
(b) *If $A\subseteq B$, then $\bigvee A\leqslant \bigvee B$ and $\bigwedge B\leqslant \bigwedge A$;*
(c) *For any $x\in X$, $x\leqslant \bigwedge A$ if and only if $x\leqslant y$ for every $y\in A$;*
(d) *For any $x\in X$, $\bigvee A\leqslant x$ if and only if $y\leqslant x$ for every $y\in A$.*

The natural structure-preserving maps between posets are those which preserve the order. Let $(X,\leqslant_X)$ and $(Y,\leqslant_Y)$ be posets. A map $f: X\to Y$ is

- *monotone* (or *order-preserving*) if, $x_1\leqslant_X x_2$ implies $f(x_1)\leqslant_Y f(x_2)$;
- *antitone* (or *order-reversing*) if, $x_1\leqslant_X x_2$ implies $f(x_2)\leqslant_Y f(x_1)$;
- an *order-embedding* if, $x_1\leqslant_X x_2$ if and only if $f(x_1)\leqslant_Y f(x_2)$;
- an *order-isomorphism* if it is a surjective order-embedding.

The composition of a finite number of order-preserving mappings is order-preserving. If $f: X\to Y$ is an order-isomorphism, then we say that X and Y are *order-isomorphic*.

If f is a function on a poset $(X,\leqslant)$, then

- f is *expanding* if and only if $x\leqslant f(x)$ for every $x\in X$;
- f is *contracting* if and only if $f(x)\leqslant x$ for every $x\in X$;
- f is *idempotent* if and only if $f(x)=f(f(x))$ for every $x\in X$;
- f is an *interior operator* if and only if f is monotone, contracting, and idempotent;
- f is a *closure operator* if and only if f is monotone, expanding, and idempotent.

1.6 Lattices

A *lattice* is a non-empty poset $(L, \leq)$ in which every two-element subset $\{x, y\}$ has both a lub and a glb. That is, $x \wedge y$ and $x \vee y$ exist for all $x, y \in L$. Alternatively, a lattice can be defined algebraically to be a set L together with two binary operations $\vee$ and $\wedge$ satisfying, for all $x, y, z \in L$,

- *idempotent laws:* $x \vee x = x, \quad x \wedge x = x;$
- *commutative laws:* $x \vee y = y \vee x, \quad x \wedge y = y \wedge x;$
- *associative laws:* $x \vee (y \vee z) = (x \vee y) \vee z \quad x \wedge (y \wedge z) = (x \wedge y) \wedge z;$
- *absorption laws:* $x \vee (x \wedge y) = x \quad x \wedge (x \vee y) = x.$

In this book a lattice will be denoted by $(L, \vee, \wedge)$ or simply by L, and elements of L will be denoted by a, b, c etc.

The equivalence of the two characterizations is shown using the following connection between $\leq$, $\wedge$, and $\vee$

$$a \leq b \Leftrightarrow a \wedge b = a \Leftrightarrow a \vee b = b, \text{ for all } a, b \in L.$$

So a lattice can be thought of either as a special kind of poset $(L, \leq)$ or as an algebraic structure $(L, \vee, \wedge)$.

The dual lattice L^{op} of L is a lattice with the same elements as $(L, \vee, \wedge)$, but the reverse underlying partial ordering, that is, $a \leq b$ in L^{op} if and only if $b \leq a$ in L. Joins and meets in L^{op} are respectively meets and joins in L.

For a family $\{L_i\}_{i \in I}$ of lattices we may define the *direct product* $\Pi_{i \in I} L_i$ where the underlying set is the Cartesian product of lattices L_i, for $i \in I$, and the operations of $\vee$ and $\wedge$ on the tuples in $\Pi_{i \in I} L_i$ are defined componentwise. A *subdirect product* of a family $\{L_i\}_{i \in I}$ of lattices is a sublattice of $\Pi_{i \in I} L_i$ such that each i-th projection map $\prod_i$ is surjective.

A *sublattice* of a lattice $(L, \vee, \wedge)$ is a non-empty subset A of L that is closed under meets and joins, that is, $a, b \in A$ implies $a \vee b$, $a \wedge b \in A$. Note that a subset A of a lattice $(L, \vee, \wedge)$ such that $(A, \vee, \wedge)$ is itself a lattice is not necessarily a sublattice of $(L, \vee, \wedge)$.

In any lattice $(L, \vee, \wedge)$, $\bigvee A$ and $\bigwedge A$ exist for all finite non-empty subsets A of L.

A lattice $(L, \leq)$ is

- *bounded* if, it has the least element 0 $(= \bigvee \emptyset = \bigwedge L)$ and the greatest element 1 $(= \bigvee L = \bigwedge \emptyset)$ satisfying $0 \leq x \leq 1$ for all $x \in L$;
- *complete* if, $\bigvee A$ and $\bigwedge A$ exist for all subsets A of L;
- *distributive* if, for all $a, b, c \in L$,
  $a \vee (b \wedge c) = (a \vee b) \wedge (a \vee c)$ and $a \wedge (b \vee c) = (a \wedge b) \vee (a \wedge c)$;
- *complemented* if, it is bounded and every $a \in L$ has a complement $-a$ defined as being both the greatest element in the set $\{c \in L : a \wedge c = 0\}$ and the least element in the set $\{c \in L : a \vee c = 1\}$. It follows that $a \wedge -a = 0$ and $a \vee -a = 1$.

Any complete lattice is a bounded lattice. In a distributive complemented lattice each element has a unique complement. A poset $(L, \leq)$ in which every two-element subset $\{x, y\}$ has a lub (but not necessarily a glb) is called a *join-semilattice.* A poset $(L, \leq)$ in which every two-element subset $\{x, y\}$ has a glb (but not necessarily a lub) is called a *meet-semilattice.*

The natural structure-preserving maps between lattices are those which preserve joins and meets. Let $(L, \vee, \wedge)$ and $(K, \vee, \wedge)$ be lattices. A map $f : L \to K$ is

- a *meet-homomorphism* if, $f(a \wedge b) = f(a) \wedge f(b)$ for all $a, b \in L$;
- a *join-homomorphism* if, $f(a \vee b) = f(a) \vee f(b)$ for all $a, b \in L$;
- a *homomorphism* (or, for emphasis, a lattice homomorphism) if it is a meet- and join-homomorphism.

Note that a mapping is a meet-homomorphism over a lattice L if it is a join-homomorphism over L^{op}. A bijective (lattice) homomorphism is a (*lattice*) *isomorphism.* If $f : L \to K$ is a one-one homomorphism, then the sublattice $f(L)$ of K is isomorphic to (or an isomorphic copy of) L, and f is called an *embedding* (of L into K). Since lattices are ordered sets, order-preserving maps are also structure-preserving maps between lattices. The relationship between homomorphisms and order-preserving maps between lattices is given in the following lemma.

Lemma 1.6.1 *Let $(L, \vee, \wedge)$ and $(K, \vee, \wedge)$ be lattices and $f : L \to K$ a map. Then*

(a) *If f is a homomorphism, then f is order-preserving (but not necessarily vice versa);*

(b) *f is a lattice isomorphism if and only if f is an order-isomorphism if and only if f is a bijective order-embedding.*

In lattice theory it is useful to consider basic elements of the lattice from which others may be built up using joins and meets.

Let $(L, \vee, \wedge)$ be a lattice. An element $a \in L$ is

- *join-irreducible* if for all $b, c \in L$, $a = b \vee c$ implies $a = b$ or $a = c$;
- *meet-irreducible* if for all $b, c \in L$, $a = b \wedge c$ implies $a = b$ or $a = c$.

If L has the least element 0, then an element $a \in L - \{0\}$ is an *atom* if for every $e \in L$, $e \leq a$ implies $e = 0$ or $e = a$. A distributive lattice L, with the least element 0, is said to be *atomic* if for any $b \in L$ with $b \neq 0$, there is an atom a such that $a \leq b$.

Lemma 1.6.2 *Let $(L, \vee, \wedge)$ be a distributive lattice. Then*

(a) *An element $a \in L$ is join-irreducible if for all $b, c \in L$, $a \leq b \vee c$ implies $a \leq b$ or $a \leq c$;*

(b) *If L has the least element then every atom of L is join-irreducible.*

Every non-zero element in a finite bounded distributive lattice is a join of join-irreducible elements. This is essentially Birkhoff's representation theorem [14]. However, in an infinite distributive lattice there may be no join-irreducible elements. For example, in the lattice of cofinite subsets of a given set L, every cofinite set $A \subseteq L$, may be written as $A = (A - \{x\}) \cup (A - \{y\})$, for distinct $x, y \in A$. We need a generalization of the notion of join-irreducible element. This is provided by the notion of *prime filter*, defined below.

Let $(L, \vee, \wedge)$ be a lattice.

- A non-empty subset $A \subseteq L$ is *meet-closed*, if $a, b \in A$ implies $a \wedge b \in A$ for all $a, b \in A$;
- A non-empty subset $F \subseteq L$ is called a *filter* if it is a meet-closed up-closed set (equivalently, $a, b \in F$ if and only if $a \wedge b \in F$);
  if a lattice has the greatest element 1, then $F \subseteq L$ is a filter if, $1 \in F$ and, for all $a, b \in L$, $a, b \in F$ implies $a \wedge b \in F$;
- A filter F of L is said to be *proper* if F is a proper subset of L, that is $F \neq L$; if a lattice has the least element 0, then F is proper if $0 \notin F$;
- A proper filter is called a *prime filter* if for all $a, b \in L$, $a \vee b \in F$ implies $a \in F$ or $b \in F$;
- A *principal filter* is a filter F arising from a single element a by up-closure, that is, $F = \uparrow_{\leq} a$;
- The filter F is generated by a subset $A \subseteq L$ if

$$F = \{a \in L : \exists n \geq 1,\ \exists a_1, \ldots, a_n \in A,\ a_1 \wedge \ldots \wedge a_n \leq a\};$$

- A filter is *maximal*, or an *ultrafilter*, if the only filter properly containing it is the improper filter L itself (equivalently, it is a maximal element with respect to set inclusion in the family of all proper filters of L).

The lattice-theoretic duals of filters are *ideals*, defined below.

Let $(L, \vee, \wedge)$ be a lattice.

- A non-empty subset $A \subseteq L$ is *join-closed* if for all $a, b \in L$, $a, b \in A$ implies $a \vee b \in A$;
- A non-empty subset $I \subseteq L$ is called an *ideal* if it is a join-closed down-closed set; (equivalently, $a, b \in I$ if and only if $a \vee b \in I$);
  if a lattice has the least element 0, then $I \subseteq L$ is an ideal if, $0 \in F$ and, for all $a, b \in L$, $a, b \in I$ implies $a \vee b \in I$;
- An ideal I of L is *proper* if $I \neq L$; if a lattice has the greatest element 1, then I is proper if $1 \notin I$;
- A proper ideal is called a *prime ideal* if for all $a, b \in L$, $a \wedge b \in I$ implies $a \in I$ or $b \in I$;
- A *principal ideal* I is an ideal arising from a single element a by down-closure, that is, $I = \downarrow_{\leq} a$;
- The ideal I is generated by a subset $A \subseteq L$ if

$$I = \{a \in L : \exists n \geq 1,\ \exists a_1, \ldots, a_n \in A,\ a \leq a_1 \vee \ldots \vee a_n\};$$

- An ideal is *maximal* if the only ideal properly containing it is the improper ideal L itself.

Theorem 1.6.3 *Let L be a lattice. Then*

(a) *The union of any chain of filters (resp. ideals) of L is a filter (resp. ideal) of L;*

(b) *If L has the least (resp. greatest) element, then the union of any chain of proper filters (resp. ideals) of L is a proper filter (resp. ideal) of L;*

(c) *If L has the least (resp. greatest) element, then every proper filter (resp. ideal) of L is included in a maximal filter (resp. ideal);*

(d) *If L has the least element 0 (resp. the greatest element 1), then for every element $a \neq 0$ (resp. $a \neq 1$) in L there exists a maximal filter F (resp. ideal I) of L such that $a \in F$ (resp. $a \in I$).*

The next lemma shows that the notion of a prime filter is indeed a generalization of join-irreducible, and dually for prime ideal and meet-irreducible.

Lemma 1.6.4 *Let $(L, \vee, \wedge)$ be a finite distributive lattice and let $a \in L$. Then*

(a) *a is join-irreducible if the principal filter $\uparrow_{\leq} a$ is prime;*

(b) *a is meet-irreducible if the principal ideal $\downarrow_{\leq} a$ is prime.*

In a lattice the notions of prime filter and prime ideal are related by set-complementation as stated in the following lemma.

Lemma 1.6.5 *Let L be a lattice. If A and B are mutually exclusive and jointly exhaustive subsets of L, then A is a prime filter of L if and only if B is a prime ideal of L.*

Lemma 1.6.6 *For every lattice L,*

(a) *The complement of a prime ideal (resp. minimal prime ideal) of L is a prime filter (resp. maximal prime filter) of L;*

(b) *Each prime ideal of L is of the form $L - F$ for some prime filter F of L;*

(c) *An ideal (resp. a prime ideal, a minimal ideal) of L^{op} is a filter (resp. a prime filter, a minimal filter) of L.*

Consequently, a proof of a result about ideals of a lattice L may be obtained by applying the proof of the dual result about filters to the dual lattice of L.

Similarly, the complement of a prime filter (resp. a maximal prime filter) of L is a prime ideal (resp. a minimal prime ideal) of L, every prime filter of L is of the form $-I$ for some prime ideal I of L, and a filter (resp. prime filter, maximal filter) of L^{op} is an ideal (resp. prime ideal, maximal ideal) of L.

Lemma 1.6.7 *If L is a bounded distributive lattice, then every maximal filter (ideal) of L is prime (but not necessarily conversely).*

Representation results for bounded distributive lattices crucially depend on the existence of prime filters as established in the following *prime filter theorem for bounded distributive lattices.*

Theorem 1.6.8 *Let L be a bounded distributive lattice. If I is an ideal of L and F' is a filter of L such that $I \cap F' = \emptyset$, then there is a prime filter F of L such that $F' \subseteq F$ and $I \cap F = \emptyset$.*

As a consequence of this existence theorem we have the following corollary.

Corollary 1.6.9 *Let L be a bounded distributive lattice. Then*

(a) *If F' is a filter of L and $a \notin F'$, then there is a prime filter F of L with $F' \subseteq F$ and $a \notin F$;*
(b) *If $a \not\leq b$, then there is a prime filter F of L such that $a \in F$ and $b \notin F$;*
(c) *If a and b are distinct elements of L, then there is a prime filter F of L containing precisely one of a and b;*
(d) *Every filter in L is the intersection of all prime filters containing it.*

The prime ideal theorem analogous to Theorem 1.6.8 and its corollary analogous to Corollary 1.6.9 can be formulated dually.

A *monoid* is an algebra $(L, \odot, 1')$ such that $\odot$ is a binary associative operation on L and $1' \in L$ is a unit element of $\odot$, that is, for any $a \in L$,

$$a \odot 1' = a = 1' \odot a.$$

A lattice-ordered monoid is a monoid such that L is a lattice.

A *residuated lattice*, or a *residuated lattice-ordered monoid*, is an algebra $(L, \vee, \wedge, \odot, \rightarrow, \leftarrow, 1')$ such that $(L, \vee, \wedge)$ is a lattice, $(L, \odot, 1')$ is a lattice-ordered monoid, and for all $a, b, c, \in L$,

$$a \odot b \leq c \Leftrightarrow b \leq a \rightarrow c \Leftrightarrow a \leq b \leftarrow c.$$

The operators $\rightarrow$ and $\leftarrow$ are called the right and the left residual of $\odot$, respectively. In residuated lattices product $\odot$ distributes over join $\vee$ in both arguments. If a residuated lattice has the least element 0, then it also has the least element 1. In a bounded residuated lattice L, for every $a \in L$,

- $a \odot 0 = 0 \odot a = 0$;
- $0 \leftarrow a = 0 \rightarrow a = 1$;
- $a \leftarrow 1 = a \rightarrow 1 = 1$.

Many properties of residuated lattices can be found in [16, 15].

A filter (resp. ideal) of a residuated lattice is that of its lattice reduct.

Let $(L, \odot, \rightarrow, \leftarrow, 1')$ be a residuated lattice and let $A, B \subseteq L$. We extend the product $\odot$ to a set operator by

$$A \odot B \stackrel{\text{def}}{=} \{c \in L : \exists a \in A\ \exists b \in B\ a \odot b \leq c\}.$$

Lemma 1.6.10 *Let L be a residuated lattice. If F and G are filters of L, then so is $F \odot G$.*

Proof:
Clearly, $F \odot G$ is non-empty and increasing. Let $c, c' \in F \odot G$. Then there exist $a, a' \in F$ and $b, b' \in G$ such that $a \odot b \leq c$ and $a' \odot b' \leq c'$. Furthermore, $a \wedge a' \in F$, $b \wedge b' \in G$, and $(a \wedge a') \odot (b \wedge b') \in F \odot G$. Since $(a \wedge a') \odot (b \wedge b') \leq (a \odot b) \wedge (a' \odot b') \leq c \wedge c'$, $c \wedge c' \in F \odot G$. Hence $F \odot G$ is closed on $\wedge$. □

For distributive residuated lattices the following counterpart to the prime filter theorem for distributive lattices is proved in [237]. Its proof presented below is from [93].

Theorem 1.6.11 *Let L be a distributive residuated lattice. Let F, G, H be filters of L such that H is prime and $F \odot G \subseteq H$. Then there exist prime filters F' and G' of L such that $F \subseteq F'$, $G \subseteq G'$, and $F' \odot G' \subseteq H$.*

Proof:
Consider the set

$$I = \{a \in L : \{a\} \odot G \not\subseteq H\}.$$

We show that I is an ideal of L. Let $a \in I$ and $b \leq a$. Suppose $b \notin I$, then $\{b\} \odot G \subseteq H$. Thus for every $c \in G$, $b \odot c \in H$. Since $b \odot c \leq a \odot c$ and H is a filter, for every $c \in G$, $a \odot c \in H$. Hence $\{a\} \odot G \subseteq H$. Hence $a \notin I$, a contradiction. Furthermore, let $a_1, a_2 \in I$. Then $\{a_1\} \odot G \not\subseteq H$ and $\{a_2\} \odot G \not\subseteq H$. This means that for some $b_1, b_2 \in G$, $a_1 \odot b_1 \notin H$ and $a_2 \odot b_2 \notin H$. Since H is a prime filter, $(a_1 \odot b_1) \vee (a_2 \odot b_2) \notin H$. Hence we have

$$(a_1 \vee a_2) \odot (b_1 \wedge b_2) = a_1 \odot (b_1 \wedge b_2) \vee a_2 \odot (b_1 \wedge b_2) \leq (a_1 \odot b_1) \vee (a_2 \odot b_2) \notin H.$$

Thus $(a_1 \vee a_2) \odot (b_1 \wedge b_2) \notin H$, which yields $\{a_1 \vee a_2\} \odot G \not\subseteq H$. It follows that $a_1 \vee a_2 \in I$, and hence I is an ideal.

Now, $F \cap I = \emptyset$ because $\{a\} \odot G \subseteq H$ for every $a \in F$. By the prime filter theorem for distributive lattices there is a prime filter F' of L such that $F \subseteq F'$ and $F' \cap I = \emptyset$. Thus for every $a \in F'$, $a \notin I$. Hence, for every $a \in F'$, $\{a\} \odot G \subseteq H$. It follows that $F' \odot G \subseteq H$.

The existence of a prime filter G' such that $G \subseteq G'$ and $F \odot G' \subseteq H$ can be proved in a similar way. □

A *Boolean algebra* $(B, \vee, \wedge, -, 0, 1)$ is a complemented bounded distributive lattice. A *subalgebra* of a Boolean algebra B is a sublattice of the lattice $(B, \vee, \wedge, 0, 1)$ which is closed under complementation. A Boolean algebra is *complete* if it is complete as a lattice, and *atomic* if it is atomic as a lattice.

The simplest non-trivial Boolean algebra is the two-element set $\{0, 1\}$ with the partial order $\leq$ defined by $0 \leq 0$, $0 \leq 1$, $1 \leq 1$. It is also a complete lattice. The standard example of a Boolean algebra is the collection of all subsets of a set X, that is, $(2^X, \cup, \cap, -, \emptyset, X)$. We call such a Boolean algebra a *powerset Boolean algebra*. The atoms are the singleton subsets of B.

Let $(B_1, \vee, \wedge, -, 0, 1)$ and $(B_2, \vee, \wedge, -, 0, 1)$ be Boolean algebras. Then a mapping $f : B_1 \to B_2$ is a *Boolean algebra homomorphism* if it is a lattice homomorphism which also preserves complements, that is $f(-a) = -f(a)$. A Boolean algebra homomorphism f which is also injective and surjective is a *(Boolean algebra) isomorphism.*

For Boolean algebras the notions of prime filter and maximal filter coincide, although this is not true for lattices in general. For this reason, for any Boolean algebra the prime filters are called *ultrafilters.*

Lemma 1.6.12 *Let B be a Boolean algebra and let F be a filter of B. Then the following conditions are equivalent:*

(a) *The filter F is prime;*

(b) *The filter F is an ultrafilter;*

(c) *The filter F is proper and, for every $a \in B$, either $a \in F$ or $-a \in F$ (but not both).*

In view of Theorem 1.6.3 and the above lemma we have the following Boolean algebraic version of Theorem 1.6.8, the prime filter theorem for lattices. It plays an important role in the representation theorems for Boolean algebras considered in Part I of the book.

Theorem 1.6.13 *Let B be a Boolean algebra. For every proper filter F' of B there exists an ultrafilter F of B such that $F' \subseteq F$.*

Clearly, Theorem 1.6.8 and Corollary 1.6.9 carry over to Boolean algebras.

1.7 Galois connections

A Galois connection is a pair of maps from a poset to another poset which captures a relationship between the posets that respects the orders in a sense and is weaker than an isomorphism. This concept was introduced in Galois [95], for a modern presentation see Birkhoff [14], Edwards [73], and Erné et al. [75]. In Dunn and Hardegree [46] some other connections are discussed and referred to as a dual Galois connection, residuation and dual residuation. Among a variety of examples of maps which form these connections, the modal order-preserving operators of necessity and possibility, and the modal order-reversing operators of sufficiency and dual sufficiency play an important role, see Section 1.8, Section 8.9, Section 8.10, and Section 8.11.

Let $(X, \leqslant_X)$ and $(Y, \leqslant_Y)$ be posets and let $f : X \to Y$ and $g : Y \to X$ be maps. We consider the following Galois-style connections.

(f, g) is called a *Galois pair* (or a *Galois connection*) if

(G) $y \leqslant_Y f(x) \Leftrightarrow x \leqslant_X g(y)$ for all $x \in X, y \in Y$, or equivalently,

(G') f, g are antitone and $\mathrm{id}_Y \leqslant_Y f \circ g$ and $\mathrm{id}_X \leqslant_X g \circ f$.

(f, g) is called a *dual Galois pair* if

(dG) $f(x) \leqslant_Y y \Leftrightarrow g(y) \leq_X x$ for all $x \in X, y \in Y$, or equivalently,

(dG') f, g are antitone and $f \circ g \leqslant_Y \mathrm{id}_Y$ and $g \circ f \leqslant_X \mathrm{id}_X$.

(f, g) is called a *residuated pair* if

(R) $f(x) \leqslant_Y y \Leftrightarrow x \leqslant_X g(y)$ for all $x \in X, y \in Y$, or equivalently,

(R') f, g are monotone and $f \circ g \leqslant_Y \mathrm{id}_Y$ and $\mathrm{id}_X \leqslant_X g \circ f$.

(f, g) is called a *dual residuated pair* if

(dR) $y \leqslant_Y f(x) \Leftrightarrow g(y) \leqslant_X x$ for all $x \in X, y \in Y$, or equivalently,

(dR') f, g are isotone and $\mathrm{id}_Y \leqslant_Y f \circ g$ and $g \circ f \leqslant_X \mathrm{id}_X$.

Note that for posets $(X, \leqslant_X)$ and $(Y, \leqslant_Y)$ and maps $f : X \to Y$ and $g : Y \to X$,

- (f, g) is a residuated pair with respect to the posets $(X, \leqslant_X)$ and $(Y, \leqslant_Y)$ if (f, g) is a Galois pair with respect to the posets $(X, \leqslant_X)$ and $(Y, \geqslant_Y)$, where $\geqslant_Y$ is $\leqslant_Y^{-1}$.
- (f, g) is a residuated pair if and only if (g, f) is a dual residuated pair.

1.8 Operators determined by binary relations

For any non-empty set X and any binary relation $R \subseteq X \times X$,

- the *necessity* operator $[R] : 2^X \to 2^X$ is defined, for every $A \subseteq X$, as

$$[R]A \stackrel{\text{def}}{=} \{x \in X : \forall y \in X,\ xRy \Rightarrow y \in A\};$$

- the *possibility* operator $\langle R \rangle : 2^X \to 2^X$ is defined, for every $A \subseteq X$, as

$$\langle R \rangle A \stackrel{\text{def}}{=} \{x \in X : \exists y \in X,\ xRy \ \&\ y \in A\};$$

- the *sufficiency* operator $[\![R]\!] : 2^X \to 2^X$ is defined, for every $A \subseteq X$, as

$$[\![R]\!]A \stackrel{\text{def}}{=} \{x \in X : \forall y \in X,\ y \in A \Rightarrow xRy\};$$

- the *dual sufficiency* operator $\langle\!\langle R \rangle\!\rangle : 2^X \to 2^X$ is defined, for every $A \subseteq X$, as

$$\langle\!\langle R \rangle\!\rangle A \stackrel{\text{def}}{=} \{x \in X : \exists y \in X,\ \text{not } xRy \ \&\ y \notin A\}.$$

Equivalently,

$$[R]A = \{x \in X : R(x) \subseteq A\}$$
$$\langle R\rangle A = \{x \in X : A \cap R(x) \neq \emptyset\}$$
$$[\![R]\!]A = \{x \in X : A \subseteq R(x)\}$$
$$\langle\!\langle R\rangle\!\rangle A = \{x \in X : (-A) \cap (-R(x)) \neq \emptyset\}.$$

Collectively, we refer to these operators as modal operators. Note that if relation R on a set X is a partial order, then for every subset A of X, $[\![R]\!]A$ and $[\![R^{-1}]\!]A$ are the collections of all the lower bounds of A and upper bounds of A, respectively. The well-known relationships between these operators are as follows.

Lemma 1.8.1 *For all $R, S \subseteq X \times X$ and for every $A \subseteq X$,*

(a) $\langle R\rangle A = -[R](-A)$;
(b) $\langle\!\langle R\rangle\!\rangle A = -[\![R]\!](-A)$;
(c) $[\![R]\!]A = [-R](-A)$;
(d) $\langle\!\langle R\rangle\!\rangle A = \langle -R\rangle(-A)$.

Further relationships between these operators may described in terms of Galois-style connections.

Lemma 1.8.2 *For any binary relation $R \subseteq X \times X$, for any $A \subseteq X$,*

(a) $\langle R^{-1}\rangle[R]A \subseteq A$;
(b) $A \subseteq [R^{-1}]\langle R\rangle A$;
(c) $A \subseteq [\![R]\!][\![R^{-1}]\!]A$;
(d) $\langle\!\langle R\rangle\!\rangle\langle\!\langle R^{-1}\rangle\!\rangle A \subseteq A$;
(a') $A \subseteq [R]\langle R^{-1}\rangle A$;
(b') $\langle R\rangle[R^{-1}]A \subseteq A$;
(c') $A \subseteq [\![R^{-1}]\!][\![R]\!]A$;
(d') $\langle\!\langle R^{-1}\rangle\!\rangle\langle\!\langle R\rangle\!\rangle A \subseteq A$.

As a consequence we have the following theorem.

Theorem 1.8.3 *For any binary relation $R \subseteq X \times X$,*

(a) $([\![R]\!], [\![R^{-1}]\!])$ *and* $([R]-, [R^{-1}]-)$ *are Galois pairs;*
(b) $(\langle\!\langle R\rangle\!\rangle, \langle\!\langle R^{-1}\rangle\!\rangle)$ *and* $(\langle R\rangle-, \langle R^{-1}\rangle-)$ *are dual Galois pairs.*
(c) $(\langle R\rangle, [R^{-1}])$ *and* $(\langle R^{-1}\rangle, [R])$ *are residuated pairs;*
(d) $([R], \langle R^{-1}\rangle)$ *and* $([R^{-1}], \langle R\rangle)$ *are dual residuated pairs;*

Here are some useful properties of these modal operators. We list many of them although not all of them are independent.

Lemma 1.8.4 *For all $R, S \subseteq X \times X$ and for all $A, B \subseteq X$,*

(a) $[R](A \cap B) = [R]A \cap [R]B$
$[R]X = X$;
(b) $\langle R\rangle(A \cup B) = \langle R\rangle A \cup \langle R\rangle B$
$\langle R\rangle\emptyset = \emptyset$;

(c) $[\![R]\!](A \cup B) = [\![R]\!]A \cap [\![R]\!]B$
$[\![R]\!]\emptyset = X$;

(d) $\langle\!\langle R\rangle\!\rangle(A \cap B) = \langle\!\langle R\rangle\!\rangle A) \cup \langle\!\langle R\rangle\!\rangle B$
$\langle\!\langle R\rangle\!\rangle X = \emptyset$;

(e) $[R]A \cup [R]B \subseteq [R](A \cup B)$;

(f) $\langle R\rangle(A \cap B) \subseteq \langle R\rangle A \cap \langle R\rangle B$;

(g) $[\![R]\!]A \cup [\![R]\!]B \subseteq [\![R]\!](A \cap B)$;

(h) $\langle\!\langle R\rangle\!\rangle(A \cup B) \subseteq \langle\!\langle R\rangle\!\rangle A \cap \langle\!\langle R\rangle\!\rangle B$;

(i) $R \subseteq S$ *implies* $[S]A \subseteq [R]A$ *and* $\langle R\rangle A \subseteq \langle S\rangle A$;

(j) $R \subseteq S$ *implies* $[\![R]\!]A \subseteq [\![S]\!]A$ *and* $\langle\!\langle S\rangle\!\rangle A \subseteq \langle\!\langle R\rangle\!\rangle A$;

(k) $[R]$ *and* $\langle R\rangle$ *are monotone;*

(l) $[\![R]\!]$ *and* $\langle\!\langle R\rangle\!\rangle$ *are antitone;*

(m) $[R\,;S]A = [R][S]A$;

(n) $\langle R\,;S\rangle A = \langle R\rangle\langle S\rangle A$;

(o) $\langle Id_X\rangle A = A$ *and* $\langle R\rangle\emptyset = \emptyset$.

Combining these properties for each operator, we obtain the following characterisations.

Lemma 1.8.5 *For every* $R \subseteq X \times X$,

(a) $[R]$, $\langle R^{-1}\rangle[R]$, *and* $\langle\!\langle R^{-1}\rangle\!\rangle\langle\!\langle R\rangle\!\rangle$ *are interior operators;*

(b) $\langle R\rangle$, $[R]\langle R^{-1}\rangle$, *and* $[\![R]\!][\![R^{-1}]\!]$ *are closure operators.*

Lemma 1.8.6 *Let* $(X, \leqslant)$ *be a poset. Then for every* $A \subseteq X$,

(a) $\uparrow_{\leqslant} A = \langle\geqslant\rangle A$;

(b) $\downarrow_{\leqslant} A = \langle\leqslant\rangle A$;

(c) $A = \langle\geqslant\rangle A \Leftrightarrow A = [\leqslant]A$;

(d) $A = \langle\leqslant\rangle A \Leftrightarrow A = [\geqslant]A$.

In view of this lemma throughout the book we will use general modal operators instead of arrows $\uparrow$ and $\downarrow$. We need such operators determined by various binary relations, not necessarily by an ordering relation. The only exceptions are the closure operators acting on singleton sets. In these cases we write $\uparrow_{\leq} a$ or $\downarrow_{\leq} a$ for an element a of a lattice ordered by $\leq$ instead of $\langle\geq\rangle\{a\}$ and $\langle\leq\rangle\{a\}$, respectively. Similarly, we write $\uparrow_{\leqslant} x$ and $\downarrow_{\leqslant} x$ for an element x of a poset $(X, \leqslant)$.

Lemma 1.8.7 *Let* $(X, \leqslant)$ *be a poset and define* $\Diamond \stackrel{\text{def}}{=} \langle\leqslant\rangle$ *and* $\Box \stackrel{\text{def}}{=} [\leqslant]$. *Then*

(a) $\Diamond$, $\Box$, $\Box\Diamond$, $\Diamond\Box$, $\Box\Diamond\Box$, *and* $\Diamond\Box\Diamond$ *are pairwise different operators;*

(b) *Any other operator defined as a composition of a finite number of* $\Diamond$ *and/or* $\Box$ *equals one of those from* (a).

1.9 Frames

A *frame* is a relational system $(X, \{R_i : i \in I\})$ where X is a non-empty set and each R_i is a relation on X. In the book mostly we deal with frames with a finite index set I. Arities of relations are determined by an arity function $\rho_I : I \to \omega$ such that for every $i \in I$, $0 < \rho_I(i) \in \omega$ where ω is the set of natural numbers. The signature of a frame is the string $(\rho_I(1), \ldots, \rho_I(n))$ where n is the cardinality of I. Frames $(X_1, \{R_i : i \in I_1\})$ and $(X_2, \{S_i : i \in I_2\})$ are similar whenever $I_1 = I_2$ and $\rho_{I_1} = \rho_{I_2}$.

Let $(X_1, \{R_i : i \in I\})$ and $(X_2, \{S_i : i \in I\})$ be similar frames with an arity function ρ_I. A function $f : X_1 \to X_2$ is

- a *homomorphism* from X_1 to X_2 if it preserves the relations, that is for all $x_0, \ldots, x_{\rho_I(i)} \in X_1$,

$$R_i(x_0, \ldots, x_{\rho_I(i)}) \text{ if and only if } S_i(f(x_0), \ldots, f(x_{\rho_I(i)}));$$

- an *embedding* from X_1 to X_2 if f is an injective homomorphism;
- an *isomorphism* from X_1 to X_2 if it is a bijective homomorphism;
- a *bounded morphism* from X_1 to X_2 if it is a homomorphism such that, for every $i \in I$, every $x_0 \in X_1$, and all $y_1, \ldots y_{\rho_I(i)} \in X_2$,
  $S_i(f(x_0), y_1, \ldots y_{\rho_I(i)})$
  implies
  there exist $x_1, \ldots x_{\rho_i(i)}$ such that
  $thegreatest R_i(x_0, x_1, \ldots x_{\rho_I(i)})$ and, for every i, $f(x_i) = y_i$.

Generalising the notion of bounded morphism to a relation we obtain the notion of a bisimulation. Let $(X_1, \{R_i : i \in I\})$ and $(X_2, \{S_i : i \in I\})$ be similar frames with an arity function ρ_I. A relation $N \subseteq X_1 \times X_2$ is a *bisimulation* from X_1 to X_2 if it satisfies, for every $i \in I$,

(forth condition) for all $s_1, t_1 \in X_1, s_2 \in X_2$,
$(s_1, s_2) \in N$ and $(s_1, t_1) \in R_i$ imply $\exists t_2 \in X_2, (s_2, t_2) \in S_i$ and $(t_1, t_2) \in N$

(back condition) for all $s_2, t_2 \in X_2, s_1 \in X_1$,
$(s_1, s_2) \in N$ and $(s_2, t_2) \in S_i$ imply $\exists t_1 \in X_1, (s_1, t_1) \in R_i$ and $(t_1, t_2) \in N$.

By a *reduct of a frame* $(X, \{R_i : i \in I\})$ with an arity function ρ_I we mean a frame $(X, \{R_i : i \in J\})$ where $J \subseteq I$ and the arity function is $\rho_J = \rho_I|_J$. Throughout the book we often refer to the lattice frame reduct of the frames associated to lattices with operators.

1.10 Correspondence Theory

In this section we briefly mention some tools which may be helpful in providing definitions of classes of frames associated with classes of algebras considered in the book. The tools were developed within the framework of correspondence theory. Given a propositional language, correspondence theory provides theorems which relate syntactically characterized classes of propositional formulas with the first order definable classes of frames in terms of validity. Given a propositional language whose semantics is defined with a class $\mathcal{F}rm$ of frames, validity of a formula of the language in the class means that the formula is true in every frame of $\mathcal{F}rm$. In general, the statement of its validity is a sentence of monadic second order logic. However, for some formulas there is a first order sentence which defines a class of frames such that the validity of the formula in that class is guaranteed. The correspondence theory is well developed for modal logics which require binary relations in their semantic structures, although it is undecidable whether for a given modal formula there is a first order definable class of frames which validate it, see [23]. The Sahlqvist theorem provides a syntactic characterization of such a class of modal formulas, see [144]. However, this is only an existential non-constructive statement, and a concrete condition defining the appropriate class of frames must be discovered. For that purpose a computer system SQEMA [32] was developed which, if terminates, generates first order conditions describing properties of binary relations in term of which validity of modal formulas is defined. The system is available at www.fmi.uni-sofia.bg/fmi/logic/sqema/. A systematic modern study of modal correspondence theory originated in [243]. A list of some modal correspondences used in the book is given in the following lemma.

Lemma 1.10.1 *For all binary relations R and S on a set X the following conditions are satisfied*

(a) *R is serial* $\Leftrightarrow [R]A \subseteq \langle R\rangle A$ *for every* $A \subseteq X$;

(b) *R is reflexive* $\Leftrightarrow [R]A \subseteq A$ *for every* $A \subseteq X$;

(c) *R is irreflexive* $\Leftrightarrow [\![R]\!]A \subseteq -A$ *for every* $A \subseteq X$;

(d) *R is weakly reflexive* $\Leftrightarrow \langle R\rangle X \cap [R]A \subseteq A$ *for every* $A \subseteq X$;

(e) *R is co-weakly reflexive* $\Leftrightarrow A \cap [\![R]\!]A \subseteq [\![R]\!]X$ *for every* $A \subseteq X$;

(f) *R is symmetric* $\Leftrightarrow A \subseteq [R]\langle R\rangle A$ *for every* $A \subseteq X$
$\Leftrightarrow A \subseteq [\![R]\!][\![R]\!]A$ *for every* $A \subseteq X$;

(g) *R is transitive* $\Leftrightarrow [R]A \subseteq [R][R]A$ *for every* $A \subseteq X$;

(h) *R is co-transitive* $\Leftrightarrow [\![R]\!]A \subseteq [\![R]\!](-[\![R]\!]A)$ *for every* $A \subseteq X$;

(i) $R^{-1} \subseteq S \Leftrightarrow A \subseteq [R]\langle S\rangle A$ *for every* $A \subseteq X$;
$\Leftrightarrow A \subseteq [\![S]\!][\![R]\!]A$ *for every* $A \subseteq X$;

(j) *if S is reflexive, then $R \subseteq (R;S)$ and $R \subseteq (S;R)$.*

For the correspondences in logics whose operators require ternary relations in the frames less is known. One of the important developments in this direction is the system SCAN which is based on a method of elimination of second order quantifiers from formulas of the monadic second order logic. The elimination method was developed in [2] and then described and studied in [231, 232] and [90], see also [169], [91]. The foundations of the system can be found in [19]. Given a formula of the monadic second order logic, the algorithm computes, provided that it terminates, an equivalent first order formula. In the literature there are also some results on Sahlqvist-style characterization of classes of formulas of the logics which require ternary relations in their semantic structures, see for example [106, 230].

1.11 Topological spaces

This section provides a summary of the notions and results from topology referred to in Chapter 2. Proofs and further motivation can be found in the standard references on topology, such as [148, 44, 255], and in the [221] survey paper on topology.

Let X be a non-empty set. A collection Ω of subsets of X is called a *topology* on X if Ω is closed under the formation of arbitrary unions and finite intersections. A *topological space* consists of a non-empty set X and a topology Ω on X. The sets in Ω are the *open sets* of the topological space (X, Ω); the elements of X are called the *points*. A *closed set* in a topological space is a set whose (set-theoretic) complement in X is an open set. The collection of closed sets is closed under arbitrary intersections and finite unions. Sets that are both open and closed are called *clopen sets*. So any topological space has at least two clopen sets – the empty set and the full space X (since they are respectively the union and intersection of the empty class of sets). A sufficient condition for the existence of other clopen sets is provided by a notion of *total disconnectedness*. That is,

A topological space (X, Ω) is *totally disconnected* if for any two distinct elements $x, y \in X$, there exists a clopen subset of X containing x but not y.

There are various separation conditions which ensure that a topology on a set X contains enough open sets to distinguish between the points of X, in some way.

A topological space (X, Ω) is

- T_0 if, for any two elements $x, y \in X$, there exists either an open set U such that $x \in U$ and $y \notin U$ or an open set V such that $x \notin V$ and $y \in V$;

- T_2 (or *Hausdorff*) if, for any two elements $x, y \in X$, there exist open sets U_1, U_2 such that $x \in U_1$ and $y \in U_2$ and $U_1 \cap U_2 = \emptyset$.

Note that every T_2-space is T_0, and every totally disconnected space is T_2.

A topology Ω on a non-empty set X can be specified without describing every open set. For example, if a family $\mathcal{S}$ of subsets of X is closed under finite intersections, then the unions of sets in $\mathcal{S}$ form a topology Ω on X, and $\mathcal{S}$ is called a *basis* for Ω. Likewise, if we take any family $\mathcal{S}$ of subsets of X, first form all finite intersections of sets in $\mathcal{S}$, and then form all arbitrary unions of the resulting set, this will again yield a topology Ω, and $\mathcal{S}$ is then called a *subbasis* for Ω.

Any topological space (X, Ω) has a pre-order induced by the topology, namely its *specialisation order*, defined for all $x, y \in X$ by

$$x \leqslant_\Omega y \quad \text{iff} \quad \text{for all open sets } O \in \Omega, x \in O \text{ implies } y \in O.$$

In general this is a pre-order, but for a T_0-space it is a partial order. There are natural structure-preserving maps between topological spaces which preserve both the topological and set-theoretic structure.

Theorem 1.11.1 *For any topological spaces (X, Ω_X) and (Y, Ω_Y) and any map $f : X \to Y$ the following conditions are equivalent.*

(a) *$f^{-1}(U)$ is open in X whenever U is open in Y;*

(b) *$f^{-1}(U)$ is closed in X whenever U is closed in Y;*

(c) *$f^{-1}(U)$ is open in X for all $U \in \mathcal{S}$, where $\mathcal{S}$ is a basis or a subbasis for Ω_Y.*

A map $f : X \to Y$ between topological spaces (X, Ω_X) and (Y, Ω_Y) is called *continuous* if it satisfies any of these conditions. The map $f : X \to Y$ is called a *homeomorphism* if f is one-one, onto and continuous and f^{-1} is also continuous. In this case, X and Y are said to be *homeomorphic*. So a homeomorphism is the appropriate topological notion of isomorphism.

Any ordered set $(X, \leqslant)$, where $\leqslant$ is reflexive and transitive, induces a natural topology $\Omega_\leqslant$ on X by taking the collection $\{\uparrow_\leqslant x : x \in X\}$ as a basis. The resulting topology, $\Omega_\leqslant$ is called the Alexandroff topology, first introduced in [4]. If the order $\leqslant$ is also antisymmetric then the Alexandroff topology $\Omega_\leqslant$ is T_0. The T_0-Alexandroff topological spaces are interesting since we may characterise topological properties entirely in terms of the specialisation order $\leqslant_{\Omega_\leqslant}$ which is exactly the order $\leqslant$ of the poset.

Lemma 1.11.2 *Let $(X, \leqslant)$ be any poset and let $\Omega_\leqslant$ be the T_0-Alexandroff topology on X. For any $x, y \in X$ and for every $A \subseteq X$,*

(a) *$x \leqslant_{\Omega_\leqslant} y$ iff $x \leqslant y$*

(b) *A is open iff $A = \uparrow_\leqslant A$;*

(c) *A is closed iff $A = \downarrow_\leqslant A$;*

(d) *the interior $int(A) = \{x \in A : \uparrow_\leqslant x \subseteq A\}$;*

(e) *the closure $cl(A) = \bigcup\{\downarrow_\leqslant x : x \in A\}$.*

There is also a connection between the structure-preserving maps.

Lemma 1.11.3 *Let $(X, \leqslant_X)$ and $(Y, \leqslant_Y)$ be posets. For any map $f : X \to Y$, f is order-preserving iff f is continuous between $(X, \Omega_{\leqslant_X})$ and $(Y, \Omega_{\leqslant_Y})$.*

For any poset $(X, \leqslant)$, we will consider in subsequent chapters the collection

$$L_X = \{A \subseteq X : A = \uparrow_\leqslant A\}.$$

By Lemma 1.11.2(a), this is the collection of open sets of the Alexandroff topology on $(X, \leqslant)$ and hence is closed under arbitrary unions and arbitrary intersections, and also contains the empty set $\emptyset$ and the full set X. The collection L_X is sometimes called the *interior algebra* of X, and provides an algebraic characterisation of the Alexandroff topological space.

1.12   Categories

This section provides some category-theoretic notions that will be referred to in Chapter 2. Our main reference for Category Theory is [158].

A *category* $\mathcal{C}$ consists of a class of objects and has, for any two objects A, B, a set $\mathcal{C}(A, B)$ of morphisms from A to B. Morphisms $f : A \to B$ and $g : B \to C$ have a composition $g \circ f$ and for each object A there is an identity morphism id_A. Standard examples of categories are the category SET with sets as objects and functions as morphisms, and the category BA with Boolean algebras as objects and homomorphisms as morphisms.

Let $\mathcal{C}$ and $\mathcal{D}$ be two categories. A *functor* F from $\mathcal{C}$ to $\mathcal{D}$ is a mapping such that

- for each object C in $\mathcal{C}$, $F(C)$ is an object in $\mathcal{D}$.

- for each morphism $f : C \to B$ in $\mathcal{C}$, $F(f) : F(C) \to F(B)$ is a morphism in $\mathcal{D}$ such that $F(\mathrm{id}_C) = \mathrm{id}_{F(C)}$, for every object C in $\mathcal{C}$, and $F(g \circ f) = F(g) \circ F(f)$, for all morphisms $f : C \to B$ and $g : B \to A$ in $\mathcal{C}$.

An *endofunctor* is a functor that maps a category to itself. A *covariant functor* between $\mathcal{C}$ and $\mathcal{D}$, denoted $F : \mathcal{C} \to \mathcal{D}$, maps objects of $\mathcal{C}$ to objects of $\mathcal{D}$ and morphisms $f : A \to B$ in $\mathcal{C}$ to morphisms $Ff : FA \to FB$ in $\mathcal{D}$, preserving identities and composition. For example, in SET, the covariant *power set functor* $\mathcal{P}$ maps any set X to its power set 2^X and a function $f : X \to Y$ to the image map $\mathcal{P}(f)$ given, for any $A \subseteq X$, by $\mathcal{P}(f)(A) = f(A) = \{f(a) \in a \in A\}$.

A *contravariant functor* $F : \mathcal{C} \to \mathcal{D}$ is a covariant function $\mathcal{C}^{op} \to \mathcal{D}$, or equivalently, $\mathcal{C} \to \mathcal{D}^{op}$, that is, it reverses the direction of the arrows. The functors that arise in duality theory are mostly contravariant functors. For example, in SET, the contravariant *power set functor* $\mathcal{P}$ maps any set X to its power set 2^X and a function $f : X \to Y$ to the inverse image map $\mathcal{P}(f)$ given, for any $B \subseteq Y$, by $\mathcal{P}(f)(B) = f^{-1}(B) = \{x \in X : \exists y \in B, y = f(x)\}$.

For each category $\mathcal{C}$, the *dual category*, denoted by $\mathcal{C}^{op}$, is obtained by taking objects of $\mathcal{C}$ as its objects and by reversing the morphisms, that is, for each morphism $f : X \to Y$ in $\mathcal{C}$, $f^{op} : Y \to X$ is a morphism in $\mathcal{C}^{op}$. Composition of morphisms $f^{op} : Y \to X$ and $g^{op} : Z \to Y$ in $\mathcal{C}^{op}$ is defined by $f^{op} \circ g^{op} = (g \circ f)^{op}$. The identity morphism in $\mathcal{C}^{op}$ is the identity morphism in $\mathcal{C}$.

Given functors F, G from the category $\mathcal{C}$ to the category $\mathcal{D}$, a *natural transformation* $h : F \to G$ consists of a family of maps $h_C : FC \to GC$, for every object C in $\mathcal{C}$, such that for all $f : C \to B$ we have $G(f) \circ h_C = h_B \circ G(f)$. That is, the following diagram commutes

$$
\begin{array}{ccc}
F(C) & \xrightarrow{\;\;h_C\;\;} & G(C) \\
{\scriptstyle F(f)}\downarrow & & \downarrow{\scriptstyle G(f)} \\
F(B) & \xrightarrow{\;\;h_B\;\;} & G(B)
\end{array}
$$

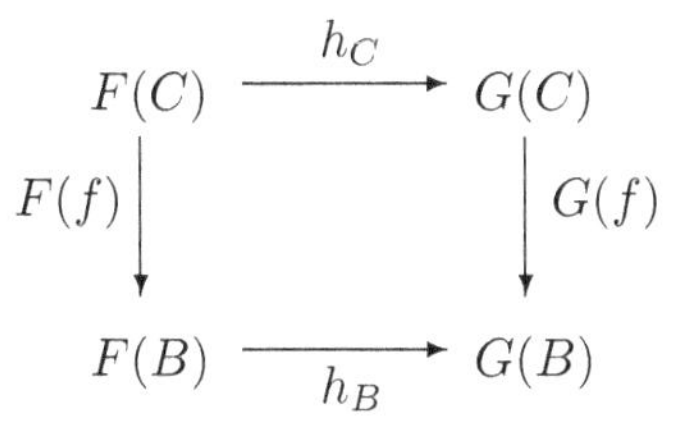

An *isomorphism* is a morphism $f : A \to B$ such that there is a morphism $g : B \to A$ with $f \circ g = \mathrm{id}_B$ and $g \circ f = \mathrm{id}_A$. If for every object A in $\mathcal{C}$, the morphism h_A is an isomorphism in $\mathcal{D}$, then h is called a *natural isomorphism* and the functors F and G are said to *naturally isomorphic*.

Given two functions $F : \mathcal{C} \to \mathcal{D}$ and $G : \mathcal{D} \to \mathcal{C}$, the categories $\mathcal{C}$ and $\mathcal{D}$ are called *equivalent* if there are natural isomorphisms $h_{\mathcal{C}} : \mathcal{C} \to GFC$ and $k_{\mathcal{D}} : \mathcal{D} \to FGD$. Categories $\mathcal{C}$ and $\mathcal{D}$ are *dually equivalent* if $\mathcal{C}^{op}$ and $\mathcal{D}$ are equivalent.

1.13 Coalgebras

This section provides some coalgebraic notions that will be referred to in Chapter 2. Further details and motivations may be found in [216].

Given a category $\mathcal{C}$ and an endofunctor $T : \mathcal{C} \to \mathcal{C}$, a T-coalgebra in $\mathcal{C}$ is a pair (A, f_A) where A is an object of $\mathcal{C}$, called the *carrier* of the coalgebra, and $f_A : A \to T(A)$ is a morphism in $\mathcal{C}$, called the *transition map* of the coalgebra. A T-morphism of coalgebras (A, f_A) and (B, f_B) is a morphism $n : A \to B$ in $\mathcal{C}$ such that $f_B \circ n = T(n) \circ f_A$. That is, the following diagram commutes

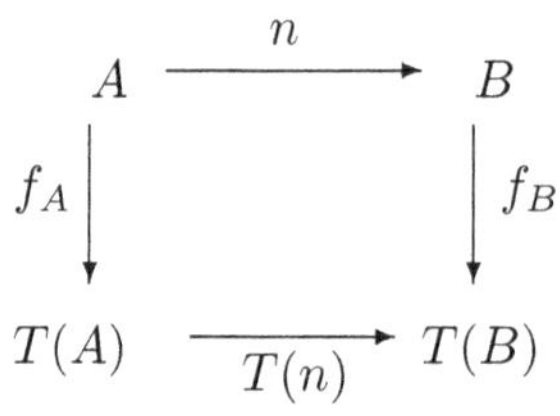

$$
\begin{array}{ccc}
A & \xrightarrow{\;\;n\;\;} & B \\
{\scriptstyle f_A}\big\downarrow & & \big\downarrow{\scriptstyle f_B} \\
T(A) & \xrightarrow[T(n)]{} & T(B)
\end{array}
$$

Given a category $\mathcal{C}$ and an endofunctor $T : \mathcal{C} \to \mathcal{C}$, a T-algebra in $\mathcal{C}$ is a pair (A, g_A) where A is an object of $\mathcal{C}$ and $g_A : T(A) \to A$ is a morphism in $\mathcal{C}$. A T-morphism of algebras (A, g_A) and (B, g_B) is a morphism $n : A \to B$ in $\mathcal{C}$ such that $n \circ g_A = g_B \circ T(n)$.

The obvious similarities between the notions of algebra and coalgebra are made precise by the algebra-coalgebra duality: given any category $\mathcal{C}$ and any endofunctor T on $\mathcal{C}$, the category of T-coalgebras is dually isomorphic to the category of algebras over the functor T^{op}, which acts like T on objects and morphism and is an endofunctor on $\mathcal{C}^{op}$.

For our purposes, the following observations will be useful. Any frame (X, R), where $R \subseteq X^2$, may be viewed as a coalgebra for the (covariant) power set functor $\mathcal{P}$ over the category SET as follows. The binary relation $R \subseteq X^2$ may be viewed as a map $f_R : X \to 2^X$ defined, for every $x \in X$, by $f_R(x) = \{y \in X : xRy\}$. This connection may be attributed to [1].

Let (X_1, R_1) and (X_2, R_2) be two frames. The respective coalgebraic representations are (X_1, f_{R_1}) and (X_2, f_{R_2}). Consider any relation $n : X_1 \to X_2$. Recalling, from Section 1.9, the notion of a bounded morphism between frames it is easy to show that, for every $x \in X_1$,

$$\forall y \in X_1, (x,y) \in R_1 \Rightarrow (n(x), n(y)) \in R_2$$

iff

$$\mathcal{P}(n) \circ f_{R_1}(x) \subseteq f_{R_2} \circ n(x),$$

$$\forall y_2 \in X_2, (n(x), y_2) \in R_2 \Rightarrow \exists y_1 \in X_1, (x, y_1) \in R_1 \ \& \ n(y_1) = y_2$$

iff

$$\mathcal{P}(n) \circ f_{R_1}(x) \supseteq f_{R_2} \circ n(x).$$

Hence, n is a bisimulation from (X_1, R_1) and (X_2, R_2) iff it is a coalgebraic morphism from (X_1, f_{R_1}) to (X_2, f_{R_2}). Formally, we may say that the category of frames with bounded morphisms is isomorphic to the category of $\mathcal{P}$-coalgebras and $\mathcal{P}$-coalgebra morphisms.

A natural question is which dualities are instances of the algebra-coalgebra duality. q

Chapter 2

Foundations

2.1 Introduction

In this book we consider classes of algebras which are signature and/or axiomatic extensions of lattices and classes of frames which are signature and/or axiomatic extensions of posets. The only exception is the case of Boolean algebras, where the corresponding frames are just sets which, however, may be viewed as structures having the empty family of relations. The inspiration for the developments in the book are the well known classical topological dualities and representations: Stone duality for Boolean algebras [227], Priestley duality for distributive lattices [202, 203], and Urquhart representation for bounded, not necessarily distributive lattices [235]. Since our intention is to treat algebras and frames as semantic structures of formal languages, we need a framework which allows for presentation of algebras and frames and for proving relationships between them, in particular, the equivalence of semantics which these structures determine for formal languages. For that purpose a discrete framework is needed rather than a topological framework. We apply the ideas of Jónsson and Tarski [140] and Goldblatt [113] in such a way that topology is not involved in any of the underlying structures. The resulting concept is referred to as discrete duality. Furthermore, a concept of duality via truth is formulated for expressing that a class of algebras and a class of frames provide the equivalent semantics for a formal language understood as validation of the same formulas. In other words, duality via truth states that the algebras and frames under consideration determine the equivalent notions of truth.

In Section 2.2 the notion of discrete duality considered in the book is presented. Establishing the discrete duality between a class of algebras and a class of frames results, among others, in holding of representation theorems for them such that a topology is not involved in the construction of the representation structures into which the given structures are embeddable. In Section 2.3 and Section 2.5 the dualities for Boolean algebras and bounded distributive lattices, respectively, are presented within the framework of discrete duality. In Section 2.8 a discrete Urquhart-style representation for bounded lattices, not necessarily distributive, is described. In Section 2.4, Section 2.7, and Section 2.9 dualities via truth are shown for the algebras and frames considered in Section 2.3, Section 2.5, and Section 2.8, respectively.

Duality theory for distributive lattices started with the work of Marshall Stone [228], where he extended his fundamental duality between Boolean algebras to the distributive lattice setting. The spaces used in this duality later became known as spectral spaces. Since then a variety of dualities for distributive lattices have evolved, for example, [124]. Many of these are in terms of ordered topological spaces following the ideas of Hilary Priestley [202, 203, 76]. One of the advantages of Priestley duality is the resulting correspondence between many algebraic concepts arising in the study of distributive lattices and topological concepts. However, shortcomings of this approach have been encountered when investigating duality for bounded (not necessarily distributive) lattices as, for example, in [235]. This has inspired new approaches. There are those based on purely topological spaces, defined in terms of stably compact spaces, as shown in [143, 141]. There are others based on bitopological spaces, defined in terms of concepts introduced by Sergio Salbany in [218]. Examples include [142, 11]. Two interesting questions stimulating much discussion and research are: *What are the connections between these various dualities*? and *What distinctive features of distributive lattices is each of these dualities capturing*?

2.2 Discrete duality and duality via truth

Discrete representations of classes of algebras (respectively, frames) are the representation theorems such that a topological structure is not involved in the construction of the representation algebras (respectively, frames) in which the given algebras (respectively, frames) are embeddable.

Discrete representations and duality via truth are obtained by the following framework. Given a class $\mathcal{A}lg$ of algebras (respectively, a class $\mathcal{F}rm$ of frames) define a class $\mathcal{F}rm$ of frames (respectively, a class $\mathcal{A}lg$ of algebras) in such a way that the theorems listed in the framework hold.

Step 1
With every algebra L from $\mathcal{A}lg$ associate a canonical frame X_L of the algebra and show that it belongs to class $\mathcal{F}rm$.

Proposition 2.2.1 $X_L \in \mathcal{F}rm$.

In classical Stone-like or Priestley-like dualities X_L would be additionally equipped with a topology and called a dual space.

Step 2
With every frame X from $\mathcal{F}rm$ associate a complex algebra L_X of the frame and show that it belongs to class $\mathcal{A}lg$:

Proposition 2.2.2 $L_X \in \mathcal{A}lg$.

Usually, the universe of L_X is a family of subsets of X or a family of pairs of subsets of X satisfying some closure property.

Step 3
Prove a *discrete representation theorem* for algebras:

Theorem 2.2.3 Representation theorem for algebras
Every algebra L of $\mathcal{A}lg$ is embeddable in the complex algebra of its canonical frame, L_{X_L}.

Step 4
Prove a *discrete representation theorem* for frames:

Theorem 2.2.4 Representation theorem for frames
Every frame X of $\mathcal{F}rm$ is embeddable in the canonical frame of its complex algebra, X_{L_X}.

Step 5
Define a propositional language $\mathcal{L}an_{\mathcal{A}lg}$ from a countably infinite set Var of propositional variables and propositional connectives which correspond to the operations of the algebras from $\mathcal{A}lg$. The formulas of the language are generated from propositional variables with the propositional connectives in the usual way. If the algebras from $\mathcal{A}lg$ are based on lattices with the greatest element 1, then most often the concept of truth is defined as follows. For $L \in \mathcal{A}lg$ we say that a formula α is true in L whenever for every assignment $v : Var \to L$ extended homomorphically to all the formulas of $\mathcal{L}an_{\mathcal{A}lg}$, $v(\alpha) = 1$. We slightly abuse the notation and denote the universe of an algebra and the algebra with the same symbol. A formula α is $\mathcal{A}lg$-valid (or an $\mathcal{A}lg$-tautology) if α is true in every algebra from $\mathcal{A}lg$. Formulas $\alpha_1, \ldots, \alpha_n$, $n \geqslant 1$, entail a formula α in $\mathcal{A}lg$ whenever for every $L \in \mathcal{A}lg$, truth of all α_i, $i = 1, \ldots, n$, in $\mathcal{A}lg$ implies truth of α in $\mathcal{A}lg$. If the algebras from $\mathcal{A}lg$ do not have a designated element 1 or there are no $\mathcal{A}lg$-tautologies in the language, then usually the notion of truth applies to sequents which are expressions of the form $\alpha \vdash \beta$, where $\alpha, \beta \in \mathcal{L}an_{\mathcal{A}lg}$. A sequent $\alpha \vdash \beta$ is true in algebra L whenever for every assignment v, $v(\alpha) \leq v(\beta)$. The notions of $\mathcal{A}lg$-validity of a sequent and entailment of sequents in $\mathcal{A}lg$ are analogous to those for formulas. Truth determined by $\mathcal{F}rm$-semantics is stated in terms of models $M = (X, m)$, where $X \in \mathcal{F}rm$ and $m : Var \to 2^X$ is a meaning function. Again, we slightly abuse the notation and denote the universe of a frame and the frame with same symbol. Extend m to all formulas in such a way that for every model $M = (X, m)$ the meaning function m is a valuation in the complex algebra L_X. Again we denote a frame and its universe with the same symbol. A formula α is true in M whenever $m(\alpha) = X$. A sequent $\alpha \vdash \beta$ is true in M whenever $m(\alpha) \subseteq m(\beta)$. The notions of $\mathcal{F}rm$-validity of formulas and sequents and entailment in $\mathcal{F}rm$ are analogous to those for formulas.

Step 6
Prove the following theorem:

Theorem 2.2.5 (Complex algebra theorem) *For every frame $X \in \mathcal{F}rm$ and for every formula α of $\mathcal{L}an_{\mathcal{A}lg}$, α is true in all models (X, m) if and only if α is true in the complex algebra of X.*

Steps 1, 2, 3, and 4 lead to what we call a discrete duality between the classes $\mathcal{A}lg$ and $\mathcal{F}rm$. Thus the discrete duality may be seen as a system

$(\mathcal{A}lg, \mathcal{F}rm, \mathcal{C}m, \mathcal{C}s)$ such that $\mathcal{C}m : X \to L_X$ is a mapping that assigns a complex algebra to every frame $X \in \mathcal{F}rm$, $\mathcal{C}s : L \to X_L$ is a mapping that assigns a canonical frame to every algebra $L \in \mathcal{A}lg$ and Propositions 2.2.1 and 2.2.2 and Theorems 2.2.3 and 2.2.4 hold.

Steps 1, 2, 3, 5, and 6 enable us to establish *duality via truth* between the classes $\mathcal{A}lg$ and $\mathcal{F}rm$.

Theorem 2.2.6 (Duality via truth) *For every formula α of $\mathcal{L}an_{\mathcal{A}lg}$ the following conditions are equivalent:*

(a) *α is true in all algebras $L \in \mathcal{A}lg$;*

(b) *α is true in all models (X, m) for $X \in \mathcal{F}rm$.*

Proof:
(a)$\Rightarrow$(b) If (a) holds, then in particular α is true in the complex algebras L_X of the frames $X \in \mathcal{F}rm$. From theorem of Step 6 condition (b) holds.

(b)$\Rightarrow$(a) Consider the following two cases.
Case 1. Assume that the algebras from $\mathcal{A}lg$ are based on lattices with the greatest element 1. Suppose that for some algebra $L \in \mathcal{A}lg$ and for some assignment v in L we have $v(\alpha) \neq 1$. Consider the canonical frame X_L of algebra L. The representation theorem of Step 3 says that there is an embedding $h : L \to L_{X_L}$. It follows that $h(v(\alpha)) \neq X_L$. Consider a model $M = (X_L, m)$ such that $m = h \circ v$. Since both v and h are homomorphisms, for every formula α we have $m(\alpha) = h(v(\alpha))$. Since the canonical frame X_L is in the class $\mathcal{F}rm$, by assumption α is true in every model based on X_L, hence $m(\alpha) = X_L$, a contradiction.
Case 2. Now assume that the algebras from $\mathcal{A}lg$ do not have a designated element 1 and suppose that for some algebra $L \in \mathcal{A}lg$ and for some assignment v in L we have not $v(\alpha) \leq v(\beta)$. Considering a canonical frame of L and a model (X_L, m) defined above we get not $m(\alpha) \subseteq m(\beta)$, a contradiction. □

Discrete duality contributes also to a completeness result once a deductive system for the language $\mathcal{L}an_{\mathcal{A}lg}$ is given. Assume that an algebraic semantics of $\mathcal{L}an_{\mathcal{A}lg}$ is given in terms of a class $\mathcal{A}lg$ of algebras and a relational semantics in terms of a class $\mathcal{F}rm$ of frames such that a discrete duality holds between these two classes. We assume that the algebras from $\mathcal{A}lg$ are based on bounded lattices. To prove completeness we define a binary relation $\approx$ in the set of formulas of $\mathcal{L}an_{\mathcal{A}lg}$ in terms of provability of double implication, if it is among the propositional operations of $\mathcal{L}an_{\mathcal{A}lg}$, or otherwise in terms of provability of a sequent built with a pair of formulas. Next we show that this relation is an equivalence relation and a congruence with respect to all the propositional operations admitted in $\mathcal{L}an_{\mathcal{A}lg}$. Then we form the Lindenbaum algebra $A_{\approx}$ of $\mathcal{L}an_{\mathcal{A}lg}$. Its universe consists of equivalence classes $|\phi|$ (with respect to relation $\approx$) of formulas. Then we show that the algebra $A_{\approx}$ belongs to the class $\mathcal{A}lg$ of algebras. Now, depending on whether we are interested in completeness with respect to the algebraic or relational semantics we proceed as follows.

To prove completeness of the deduction system with respect to the relational semantics we consider the canonical frame $X_{A_{\approx}}$ of the Lindenbaum algebra. Its universe consists of prime filters of $A_{\approx}$. Then we form a model $M_{\approx}$ based on this frame. Preservation of operations by the mapping that provides an embedding of $A_{\approx}$ into $L_{X_{A_{\approx}}}$ guaranteed by Theorem 2.2.3 enables us to prove the truth lemma saying that satisfaction of a formula ϕ in model $M_{\approx}$ by a filter F is equivalent to $|\phi| \in F$. From this lemma the completeness follows in the usual way.

To prove completeness of the deduction system with respect to the algebraic semantics we define a valuation of atomic formulas of $\mathcal{L}an_{\mathcal{A}lg}$ in $\mathcal{A}_{\approx}$ as $v(p) = |p|$ and we prove that it extends to all the formulas so that $v(\phi) = |\phi|$. Then we show that provability of a formula ϕ is equivalent to $v(\phi) = 1_{\approx}$, where $1_{\approx}$ is the unit element of the lattice reduct of $A_{\approx}$. Then the completeness follows.

Discrete duality is also relevant for the correspondence theory which aims at finding relationships between truth of formulas in a frame an properties of relations in the frame. Typically, a correspondence has the following form:

Theorem 2.2.7 *A formula $\phi \in \mathcal{L}an_{\mathcal{A}lg}$ is true in a frame X iff the relations of the frame have a certain property.*

Given the classes $\mathcal{A}lg$ and $\mathcal{F}rm$ for which a discrete duality and duality via truth theorem with respect to a language $\mathcal{L}an_{\mathcal{A}lg}$ hold, we may consider the following correspondences:

Theorem 2.2.8 *The relations of a frame $X \in \mathcal{F}rm$ have a certain property iff a formula $\phi \in \mathcal{L}an_{\mathcal{A}lg}$ is true in the complex algebra L_X.*

Theorem 2.2.9 *A formula $\phi \in \mathcal{L}an_{\mathcal{A}lg}$ is true in an algebra $L \in \mathcal{A}lg$ iff the relations of the canonical frame X_L have a certain property.*

It is known that these corespondences are related to the classical correspondence Theorem 2.2.7. The left-to-right implication of Theorem 2.2.8 and the right-to-left implication of Theorem 2.2.5 imply the right-to-left implication of Theorem 2.2.7 . The right-to-left implication of Theorem 2.2.8 and the left-to-right implication of Theorem 2.2.5 imply left-to-right implication of Theorem 2.2.7. Examples of the correspondences of these types can be found in [136]. In the book we present several examples of discrete dualities and duality via truth.

2.3 Discrete duality for Boolean algebras

In this section we recall Stone duality for Boolean algebras [227] within the framework of discrete duality as it is understood in this book.

A *Boolean frame* is any non-empty set X. The *complex algebra of a Boolean frame* X, B_X, is the powerset Boolean algebra $(2^X, \cup, \cap, -, \emptyset, X)$. Clearly,

Proposition 2.3.1 *The complex algebra of a Boolean frame is a Boolean algebra.*

Given a Boolean algebra B, its *canonical frame*, X_B, is the set of all ultrafilters of B.

The Stone mapping $h : B \to B_{X_B}$ is defined, for every $a \in B$, by

$$h(a) \stackrel{\text{def}}{=} \{F \in X_B : a \in F\}.$$

Since $B_{X_B} = 2^{X_B}$, h is well defined.

Lemma 2.3.2 *The mapping h is an embedding.*

Proof:
First, we prove that the mapping h preserves the operations $\vee$ and $\wedge$. We show $h(a \vee b) = h(a) \cup h(b)$. Let $F \in h(a \vee b)$. Then $a \vee b \in F$. Since F is a prime filter, $a \in F$ or $b \in F$. Thus $F \in h(a) \cup h(b)$. Conversely, let $a \in F$ or $b \in F$ and let $\leq$ be the natural ordering on B defined as in Section 1.6. Since $a \leq a \vee b$ and $b \leq a \vee b$, in both cases $a \vee b \in F$. The preservation of meet can be proved in the similar way. Since F is an ultrafilter, $-a \in F$ iff $a \notin F$, hence $h(-a) = -h(a)$. Also, $h(0) = \emptyset$ and $h(1) = X_B$.

Second, we show that h is injective. Take $a, b \in B$ such that $a \neq b$. Then, by antisymmetry of $\leq$, either $a \not\leq b$ or $b \not\leq a$. Without loss of generality, assume that $a \not\leq b$. Then by Corollary 1.6.9(b) there is an ultrafilter F such that $a \in F$ and $b \notin F$. Hence $h(a) \neq h(b)$. □

Now, let X be a Boolean frame. Consider a mapping $k : X \to X_{2^X}$ defined, for every $x \in X$,

$$k(x) \stackrel{\text{def}}{=} \{A \in 2^X : x \in A\}.$$

Since X_{2^X} is the set of ultrafilters of 2^X and $k(x)$ is a principal filter of 2^X generated by $\{x\}$, k is well defined.

Lemma 2.3.3 *The mapping k is an embedding.*

Proof:
To show that k is injective, take $x, y \in X$ such that $k(x) = k(y)$. This means that for every $A \subseteq X$, $A \in k(x)$ if and only if $A \in k(y)$, that is $x \in A$ if and only if $y \in A$. In particular, since $x \in \{x\}$, $y \in \{x\}$, that is $y = x$, as required. □

We now obtain discrete representations for Boolean algebras and Boolean frames.

Theorem 2.3.4

(a) *Every Boolean algebra is embeddable into the complex algebra of its canonical frame;*

(b) *Every Boolean frame is embeddable into the canonical frame of its complex algebra.*

By Theorem 2.3.4, together with Proposition 2.3.1, we establish a discrete duality, for Boolean algebras and Boolean frames.

2.4 Duality via truth for Boolean algebras

Let $\mathcal{L}an_B$ be the language of the classical propositional calculus. Its formulas are built from propositional variables taken from a countable infinite set Var with the classical propositional operators of negation ($\neg$), disjunction ($\vee$), conjunction ($\wedge$), and implication ($\rightarrow$). We slightly abuse the language by denoting the operations of join and meet in Boolean algebras and the propositional operations of disjunction and conjunction in $\mathcal{L}an_B$ with the same symbols.

The class $\mathcal{A}lg_B$ of Boolean algebras is now shown to provide an algebraic semantics for $\mathcal{L}an_B$. Let B be a Boolean algebra. A valuation in B is a function $v : Var \rightarrow B$ which assigns elements of B to propositional variables and extends homomorphically to all the formulas of $\mathcal{L}an_B$, that is

$$\begin{aligned} v(\neg\alpha) &\stackrel{\text{def}}{=} -v(\alpha) \\ v(\alpha\vee\beta) &\stackrel{\text{def}}{=} v(\alpha)\vee v(\beta) \\ v(\alpha\wedge\beta) &\stackrel{\text{def}}{=} v(\alpha)\wedge v(\beta) \\ v(\alpha\rightarrow\beta) &\stackrel{\text{def}}{=} -v(\alpha)\vee v(\beta). \end{aligned}$$

A formula α in $\mathcal{L}an_B$ is true in a Boolean algebra B whenever $v(\alpha)=1$ for every v on B. A formula α is true in the class $\mathcal{A}lg_B$ if and only if it is true in every algebra $B\in\mathcal{A}lg_B$.

The class $\mathcal{F}rm_B$ of Boolean frames is now shown to provide a set semantics for $\mathcal{L}an_B$. Let X be a Boolean frame. A model based on X is a system $\mathcal{M}=(X,m)$ where $m : Var\rightarrow 2^X$ is a meaning function. The satisfaction relation $\models$ is defined inductively, namely, we say that in a model M a state $x\in X$ satisfies a formula α whenever the following conditions hold:

$$\begin{aligned} \mathcal{M},x\models p &\stackrel{\text{def}}{\Leftrightarrow} x\in m(p) \text{ for every } p\in Var \\ \mathcal{M},x\models\neg\alpha &\stackrel{\text{def}}{\Leftrightarrow} \mathcal{M},x\not\models\alpha \\ \mathcal{M},x\models\alpha\vee\beta &\stackrel{\text{def}}{\Leftrightarrow} \mathcal{M},x\models\alpha \text{ or } \mathcal{M},x\models\beta \\ \mathcal{M},x\models\alpha\wedge\beta &\stackrel{\text{def}}{\Leftrightarrow} \mathcal{M},x\models\alpha \ \&\ \mathcal{M},x\models\beta \\ \mathcal{M},x\models\alpha\rightarrow\beta &\stackrel{\text{def}}{\Leftrightarrow} \mathcal{M},x\models\alpha \text{ implies } \mathcal{M},x\models\beta. \end{aligned}$$

A notion of truth determined by this frame semantics is defined as follows. A formula $\alpha\in\mathcal{L}an_B$ is true in a model $\mathcal{M}=(X,m)$ whenever for every $x\in X$ we have $\mathcal{M},x\models\alpha$. A formula α is true in a frame X if and only if it is true in every model based on this frame. And finally a formula α is true in the class $\mathcal{F}rm_B$ of frames whenever it is true in every frame $X\in\mathcal{F}rm_B$.

It is easy to see that the complex algebra theorem holds.

Theorem 2.4.1 (Complex algebra theorem) *For every frame $X\in\mathcal{F}rm_B$ and for every formula $\alpha\in\mathcal{L}an_B$ the folowing conditions are equivalent:*

(a) *α is true in every model based on X;*

(b) *α is true in the complex algebra B_X of X.*

Proof:
For every model (X, m) we extend the meaning function m to all the formulas as follows: $m(\neg\alpha) = X - m(\alpha)$, $m(\alpha \vee \beta) = m(\alpha) \cup m(\beta)$, $m(\alpha \wedge \beta) = m(\alpha) \cap m(\beta)$, $m(\alpha \rightarrow \beta) = -m(\alpha) \cup m(\beta)$. Then α is true in a model (X, m) whenever $m(\alpha) = X$. In this way every function m may be seen as a valuation in the complex algebra B_X of the frame X. Consequently, truth of a formula in every model based on X is equivalent to its truth in the algebra B_X. □

Finally, we prove the duality via truth theorem for Boolean algebras and frames.

Theorem 2.4.2 (Duality via truth) *For every formula $\alpha \in \mathcal{L}an_B$ the following conditions are equivalent:*

(a) *α is true in every Boolean algebra in $\mathcal{A}lg_B$.*

(b) *α is true in every frame in $\mathcal{F}rm_B$.*

Proof:
(a)⇒(b) This part follows directly from the complex algebra theorem.

(b)⇒(a) Assume that for every frame X and for every meaning function m on X, $m(\alpha) = X$ and suppose that α is not true in some Boolean algebra B. It follows that for some valuation v on B, $v(\alpha) \neq 1$. By assumption α is true in the canonical model X_B. Consider a meaning function on X_B defined as $m(p) = h(v(p))$, where h is the Boolean embedding providing the representation stated in Theorem 2.3.4. Since both h and v are homomorphisms, for every formula β, $m(\beta) = h(v(\beta))$. Since h is injective, $m(\alpha) \neq \mathit{Ult}(B)$, a contradiction. □

2.5 Discrete duality for bounded distributive lattices

A *bounded distributive lattice frame* is a structure $(X, \leqslant)$ such that X is a non-empty set and $\leqslant$ is a partial order on X.

The *complex algebra of a bounded distributive lattice frame* $(X, \leqslant)$ is an algebraic structure $(L_X, \cup, \cap, 0_X, 1_X)$ where

$$L_X \stackrel{\text{def}}{=} \{A \subseteq X : [\leqslant]A = A\},$$

and $0_X = \emptyset$ and $1_X = X$.

Proposition 2.5.1 *The complex algebra of a bounded distributive lattice frame is a bounded distributive lattice.*

Proof:
L_X is closed on the operations, that is, for any $A, B \in L_X$, $A \cup B = [\leqslant](A \cup B)$ and $A \cap B = [\leqslant](A \cap B)$.

The case of $\cap$ follows from Lemma 1.8.4(b) which says that $[\leqslant]$ distributes over $\cap$. For $\cup$ the inclusion $\supseteq$ follows from the reflexivity of $\leqslant$. For the other inclusion, assume $x \in A \cup B = [\leqslant]A \cup [\leqslant]B$. Suppose $x \notin [\leqslant](A \cup B)$. Then, for some x', $x \leqslant x'$ and $x' \notin A \cup B$. Also, from the assumption, for every $z \in X$ $x \leqslant z$ implies $z \in A$ and for every $w \in X$, $x \leqslant w$ implies $w \in B$. Hence $x' \in A$ and $x' \in B$, which gives the required contradiction.
Note that $[\leqslant]X = X$ and $[\leqslant]\emptyset = \emptyset$, so the complex algebra of a distributive lattice frame is a distributive bounded lattice. □

In fact, the lattice L_X is always a complete lattice. Note that in the case of Boolean algebras, which are particular bounded distributive lattices with a complementation operator, the only prime filters are the maximal filters and the ordering collapses to equality, the discrete order. This is why in context of Boolean algebras prime filters are called ultrafilters. However, in the more general setting of bounded distributive lattices there may be non-maximal filters which are prime, and hence the collection of prime filters of a bounded distributive lattice is therefore naturally ordered by inclusion.

The *canonical frame of a bounded distributive lattice* L is a structure $(X_L, \subseteq)$ where X_L is the set of prime filters of L ordered by set inclusion.

Proposition 2.5.2 *The canonical frame of a bounded distributive lattice is a bounded distributive lattice frame.*

The Stone mapping $h : L \to L_{X_L}$ is defined, for every $a \in L$, by

$$h(a) \stackrel{\text{def}}{=} \{F \in X_L : a \in F\}.$$

Lemma 2.5.3 *The mapping h is an embedding.*

Proof:
The map h is well-defined in the sense that, for any $a \in L$, $h(a) = [\subseteq]h(a)$. Also h is a lattice embedding as shown in Lemma 2.3.2. □

Now, let X be a bounded distributive lattice frame. Consider the mapping $k : X \to X_{L_X}$ defined, for every $x \in X$,

$$k(x) \stackrel{\text{def}}{=} \{A \in L_X : x \in A\}.$$

Since X_{2^X} is the set of prime filters of 2^X and $k(x)$ is a principal filter of 2^X generated by $\{x\}$, k is well defined.

Lemma 2.5.4 *The mapping k is an embedding.*

Proof:
For every $x \in X$, $k(x)$ is a prime filter of L_X. To see that k is injective, note

that $\uparrow_{\leqslant} x, \uparrow_{\leqslant} y \in L_X$ and $\uparrow_{\leqslant} x \in k(x)$ and $\uparrow_{\leqslant} y \in k(y)$. Thus if $x \neq y$, then $k(x) \neq k(y)$. All that remains is to show that k is order preserving, that is, for all $x, y \in X$, $x \leqslant y$ if and only if $k(x) \subseteq k(y)$. Take any $x, y \in X$ such that $x \leqslant y$ and $A \in k(x)$. Then $x \in A$. So, since $A = [\leqslant]A$, $y \in A$, that is, $A \in k(y)$. On the other hand, take any $x, y \in X$ such that $k(x) \subseteq k(y)$. Then, since $x \in \uparrow_{\leqslant} x$, $\uparrow_{\leqslant} x \in k(x)$ and hence $\uparrow_{\leqslant} x \in k(y)$. Thus $y \in \uparrow_{\leqslant} x$ and hence $x \leqslant y$, as required. □

We now obtain the discrete representations for bounded distributive lattices and bounded distributive lattice frames.

Theorem 2.5.5

(a) *Every bounded distributive lattice is embeddable into the complex algebra of its canonical frame;*

(b) *Every bounded distributive lattice frame is embeddable into the canonical frame of its complex algebra.*

Theorem 2.5.5, together with Proposition 2.5.1 and Proposition 2.5.2, establish a discrete duality for bounded distributive lattices and bounded distributive lattice frames.

The representation in Theorem 2.5.5 (a) may be viewed as a topological representation since, in view of the discussion at the end of Section 1.11, any bounded distributive lattice frame $(X, \leqslant)$ may be viewed as a topological space with the Alexandroff topology $\Omega_{\leqslant}$. With this perspective, it is easy to see that the lattice L_{X_L} is always a complete lattice. So since L may not be a complete lattice, the embedding h need not be surjective. Hence, we have that any bounded distributive lattice L can be embedded in the (complete) lattice of open subsets of $(X_L, \Omega_{\leqslant})$. In general, L may not be isomorphic to the lattice L_{X_L}.

In fact there are several topological representation theorems for bounded distributive lattices.

Priestley [202] showed that any bounded distributive lattice is isomorphic to the lattice of clopen upsets of the Priestley space $(X_L, \subseteq, \Omega_P)$, where the Priestley topology Ω_P is generated by the basis $\mathcal{B} = \{h(a) - h(b) : a, b \in L\}$. This can be viewed as another topological representation for bounded distributive lattices: any bounded distributive lattice L can be embedded in the lattice of open subsets of (X_L, Ω_P).

Stone [228] showed that any bounded distributive lattice is isomorphic to the lattice of open subsets of the spectral space (X_L, Ω_S), where the spectral topology Ω_S is generated by the subbasis $\mathcal{S} = \{h(a) : a \in L\}$. This can be viewed as another topological representation for bounded distributive lattices: any bounded distributive lattice L can be embedded in the lattice of open subsets of (X_L, Ω_S).

The spectral topology on X_L consists of the open upsets of the Priestley space $(X_L, \subseteq, \Omega_P)$. Hence, $\Omega_S \subseteq \Omega_P$. On the other hand, the Priestley topology Ω_P is the patch topology of Ω_S in the sense that Ω_P is generated by the set $\{U \cap V : U \in \mathcal{S}$ and $X_L - V \in \mathcal{S}\}$. Also the specialisation order of the spectral topology Ω_S is the set-theoretic inclusion $\subseteq$. Hence, the Priestley space $(X_L, \subseteq, \Omega_P)$ may be recovered from the spectral space (X_L, Ω_S)

Consequently, we obtain three representations for bounded distributive lattices: the discrete representation by means of order, the spectral representation by means of topology only, and the Priestley representation by means of topology and order.

In a sense, the spectral representation is the "most efficient" since the spectral topology is the intersection of the Alexandroff topology and Priestley topology. It is tradeoff to decide which representation is the most appropriate for a particular context.

2.6 Discrete duality for bounded distributive lattice homomorphisms

In this section we establish a correspondence between bounded distributive lattice homomorphisms and maps between bounded distributive lattice frames.

A *bounded morphism* between bounded distributive lattice frames $(X_1, \leqslant_1)$ and $(X_2, \leqslant_2)$ is a map $n : X_1 \to X_2$ satisfying, for all $x, y \in X_1$, $x \leqslant_1 y$ implies $n(x) \leqslant_2 n(y)$, and if $n(x) \leqslant_2 y_2$ then for some $y_1 \in X_1$, $x \leqslant_1 y_1$ and $f(y_1) = y_2$.

Theorem 2.6.1 *Let L_1 and L_2 be bounded distributive lattices and let $f : L_1 \to L_2$ be a lattice homomorphism between them. Let $(X_1, \leqslant_1)$ and $(X_2, \leqslant_2$) be bounded distributive lattice frames and let $n : X_1 \to X_2$ be a bounded morphism between them. Then inverse image map $f^{-1} : X_{L_2} \to X_{L_1}$ is a bounded morphism, and inverse image map $n^{-1} : L_{X_2} \to L_{X_1}$, is a lattice homomorphism.*

It is not difficult to extend Theorem 2.6.1 to show that injective/surjective homomorphisms correspond to surjective/injective bounded morphisms and vice versa. We can also show that the embeddings h and k used to prove Theorem 2.5.5 "commute" with lattice homomorphisms and bounded morphisms in the sense of the diagrams in Theorem 2.6.2.

Theorem 2.6.2 *Let f and n be as in* Theorem 2.6.1. *Suppose that the maps $h_{L_1} : L_1 \to L_{X_{L_1}}$ and $h_{L_2} : L_2 \to L_{X_{L_2}}$, and $k_{X_1} : X_1 \to X_{L_{X_1}}$ and $k_{X_2} : X_2 \to X_{L_{X_2}}$ are the embeddings used in Theorem 2.5.5. Then*

$$(f^{-1})^{-1} \circ h_{L_1} = h_{L_2} \circ f \quad \textit{and} \quad (n^{-1})^{-1} \circ k_{X_1} = k_{X_2} \circ n.$$

That is, the following diagrams commute.

$$\begin{array}{ccc} L_1 & \xrightarrow{h_{L_1}} & L_{X_{L_1}} \\ \downarrow f & & \downarrow (f^{-1})^{-1} \\ L_2 & \xrightarrow{h_{L_2}} & L_{X_{L_2}} \end{array} \qquad \begin{array}{ccc} X_1 & \xrightarrow{k_{X_1}} & X_{L_{X_1}} \\ \downarrow n & & \downarrow (n^{-1})^{-1} \\ X_2 & \xrightarrow{k_{X_2}} & X_{L_{X_2}} \end{array}$$

Proof:
(a) For every $F \in X_{L_2}$, $f^{-1}(F) \in X_{L_1}$ and

$$\begin{aligned} F \in (f^{-1})^{-1}(h_{L_1}(a)) &\Leftrightarrow f^{-1}(F) \in h_{L_1}(a) \\ &\Leftrightarrow a \in f^{-1}(F) \\ &\Leftrightarrow f(a) \in F \\ &\Leftrightarrow F \in h_{L_2}(f(a)). \end{aligned}$$

(b) For every $B \subseteq Y$,

$$\begin{aligned} B \in (n^{-1})^{-1}(k_{X_1}(x)) &\Leftrightarrow n^{-1}(B) \in k_{X_1}(x) \\ &\Leftrightarrow x \in n^{-1}(B) \\ &\Leftrightarrow n(x) \in B \\ &\Leftrightarrow B \in k_{X_2}(n(x)). \end{aligned}$$

□

These properties of the embeddings h and k are expressed categorically by saying that they are natural transformations, as defined in Section 1.12. In particular, let Dlat denote the category with bounded distributive lattices as objects and lattice homomorphisms as morphisms, and let Pos denote the category with partially ordered sets as objects and bounded morphisms as morphisms.

We can define a functor D from Dlat to Pos which takes any bounded distributive lattice L of Dlat to $D(L)$ defined to be its canonical frame $(X_L, \subseteq)$, and which takes any lattice homormorphism $f : L_1 \to L_2$ to the morphism $D(f) : D(L_2) \to D(L_1)$ defined, for any $G \in D(L_2)$, by $D(f)(G) = f^{-1}(G) = \{x \in L_1 : f(x) \in G\}$.

We can define a functor E from Pos to Dlat which takes any lattice frame $(X, \leqslant)$ to $E(X)$ defined to be its complex algebra L_X, and which takes any bounded morphism $n : X_1 \to X_2$ to the morphism $E(n) : E(X_2) \to E(X_1)$ defined, for any $B \in E(X_2)$, by $E(n)(B) = n^{-1}(B) = \{x \in X_1 : n(x) \in B\}$.

It is easy to show that D and E are well-defined functors. Also, by Theorem 2.6.2, the functors $\mathrm{id}_{\mathsf{Dlat}}$ and ED from Dlat to Dlat, and the functors $\mathrm{id}_{\mathsf{Pos}}$ and DE from Pos to Pos satisfy, for every lattice homormorphism $f : L_1 \to L_2$ and for every bounded morphism $n : X_1 \to X_2$, the following commuting diagrams

$$\begin{array}{ccc} \mathrm{id}_{\mathsf{Dlat}}(L_1) & \xrightarrow{h_{L_1}} & ED(L_1) \\ {\scriptstyle \mathrm{id}_{\mathsf{Dlat}}(f)} \downarrow & & \downarrow {\scriptstyle ED(f)} \\ \mathrm{id}_{\mathsf{Dlat}}(L_2) & \xrightarrow{h_{L_2}} & ED(L_2) \end{array} \qquad \begin{array}{ccc} \mathrm{id}_{\mathsf{Pos}}(X_1) & \xrightarrow{k_{X_1}} & DE(X_1) \\ {\scriptstyle \mathrm{id}_{\mathsf{Pos}}(n)} \downarrow & & \downarrow {\scriptstyle DE(n)} \\ \mathrm{id}_{\mathsf{Pos}}(X_2) & \xrightarrow{k_{X_2}} & DE(X_2) \end{array}$$

Hence, the family of maps $h_L : \mathrm{id}_{\mathsf{Dlat}}(L) \to ED(L)$, for every bounded distributive lattice L in Dlat, denoted by $h : \mathrm{id}_{\mathsf{Dlat}} \to ED$, is a natural transformation. Similarly, the family of maps $k_X : \mathrm{id}_{\mathsf{Poset}}(X) \to DE(X)$, for every lattice frame X in Poset, denoted by $k : \mathrm{id}_{\mathsf{Poset}} \to DE$, is a natural transformation.

Since the mappings h_L, for every bounded distributive lattice L in Dlat, and k_X, for every lattice frame X in Poset, are not isomorphisms we do not have a categorical equivalence or categorical dual equivalence.

2.7 Duality via truth for bounded distributive lattices

Given a lattice $(L, \vee, \wedge)$, formulas of the language $\mathcal{L}an_L$ are built up from propositional variables of a countable infinite set Var using the propositional connectives $\vee$ and $\wedge$ of disjunction and conjunction, respectively. For a bounded distributive lattice $(L, \vee, \wedge, 0, 1)$ we endow the language $\mathcal{L}an_L$ with propositional constants T and F. By a sequent we mean an expression $\alpha \vdash \beta$ where α and β are formulas of $\mathcal{L}an_L$.

We slightly abuse the language by denoting the operations in bounded distributive lattices and the propositional connectives of $\mathcal{L}an_L$ with the same symbols.

Algebraic semantics of the language is determined by the class $\mathcal{A}lg_{dL}$ of bounded distributive lattices.

Let L be a bounded distributive lattice and let $\leq$ stands for its natural ordering. By a valuation in L we mean a function $v : Var \to L$. We extend valuation v homomorphically to all the formulas of $\mathcal{L}an_L$ in the usual way:

$$v(\alpha \vee \beta) = v(\alpha) \vee v(\beta)$$
$$v(\alpha \wedge \beta) = v(\alpha) \wedge v(\beta).$$

For bounded distributive lattices we define $v(T) = 1$ and $v(F) = 0$.

We say that a sequent $\alpha \vdash \beta$ is satisfied by a valuation v whenever $v(\alpha) \leqslant v(\beta)$ and $\alpha \vdash \beta$ is true in a lattice L if and only if it is satisfied by every valuation in L.

Let $\mathcal{F}rm_{dL}$ be the class of bounded distributive lattice frames presented in Section 2.5. A model based on a lattice frame $(X, \leqslant)$ is a system $\mathcal{M} = (X, \leqslant, m)$ where $m : Var \to L_X$ is a meaning function. If $(X, \leqslant)$ is a bounded distributive lattice frame, then $m(T) = X$ and $m(F) = \emptyset$. The satisfaction relation $\models$ is defined for all formulas α and β of $\mathcal{L}an_L$ by

$$\mathcal{M}, x \models p \overset{\text{def}}{\Leftrightarrow} x \in m(p) \text{ for every } p \in Var$$
$$\mathcal{M}, x \models \alpha \wedge \beta \overset{\text{def}}{\Leftrightarrow} \mathcal{M}, x \models \alpha \;\&\; \mathcal{M}, x \models \beta$$
$$M, x \models \alpha \vee \beta \overset{\text{def}}{\Leftrightarrow} \mathcal{M}, x \models \alpha \text{ or } \mathcal{M}, x \models \beta.$$

Whenever $\mathcal{M}$ is based on a bounded distributive lattice frame then, for every $x \in X$, $\mathcal{M}, x \models T$ and $\mathcal{M}, x \not\models F$.

We define $m(\alpha) = \{x \in X : \mathcal{M}, x \models \alpha\}$. One can easily verify that

$$m(\alpha \wedge \beta) = m(\alpha) \cap m(\beta)$$
$$m(\alpha \vee \beta) = m(\alpha) \cup m(\beta).$$

We say that a sequent $\alpha \vdash \beta$ is true in a model $\mathcal{M} = (X, \leqslant, m)$ whenever $m(\alpha) \subseteq m(\beta)$ and $\alpha \vdash \beta$ is true in a bounded distributive lattice frame X if and only if it is true in every model based on X.

Theorem 2.7.1 (Complex algebra theorem) *For all formulas α and β of $\mathcal{L}an_L$ and for every bounded distributive lattice frame $X \in \mathcal{F}rm_{dL}$ the following conditions are equivalent:*

(a) *A sequent $\alpha \vdash \beta$ is true a distributive lattice frame $X \in \mathcal{F}rm_{dL}$;*

(b) *A sequent $\alpha \vdash \beta$ is true in the complex algebra L_X of X.*

Proof:
The proof follows from the fact that meaning functions of models based on a lattice frame X can be viewed as valuations in the complex algebra L_X of X. □

Finally, we prove the *duality via truth theorem* for bounded distributive lattices and bounded distributive lattice frames.

Theorem 2.7.2 (Duality via truth) *For all formulas α and β of $\mathcal{L}an_L$ the following conditions are equivalent:*

(a) *A sequent $\alpha \vdash \beta$ is true in every bounded distributive lattice in $\mathcal{A}lg_{dL}$;*

(b) *A sequent $\alpha \vdash \beta$ is true in every model based on a bounded distributive lattice frame in $\mathcal{F}rm_{dL}$.*

Proof:
Let α and β be any formulas of $\mathcal{L}an_L$.

(a)⇒(b) Assume that a sequent $\alpha \vdash \beta$ is not true in some model $\mathcal{M} = (X, \leqslant, m)$ based on a bounded distributive lattice frame X. By Theorem 2.7.1, it is not true in the complex algebra of X. Hence, by Proposition 2.5.1, $\alpha \vdash \beta$ is not true in some bounded distributive lattice L.

(b)⇒(a) Assume that a sequent $a \vdash \beta$ is not true in some bounded distributive lattice L. Then there exists a valuation v in L such that $v(\alpha) \not\leq v(\beta)$. Consider a model $\mathcal{M} = (X_L, \subseteq, m)$ based on the canonical frame X_L of L with the meaning function $m = h \circ v$, where h is the embedding defined in Section 2.5. Since $v(\alpha) \not\leq v(\beta)$, by Corollary 1.6.9(b) there is some prime filter of L, say F, such that $v(\alpha) \in F$ and $v(\beta) \notin F$. Hence $h(v(\alpha)) \not\subseteq h(v(\beta))$, so $m(\alpha) \not\subseteq m(\beta)$. Therefore the sequent $\alpha \vdash \beta$ is not true in the model $\mathcal{M}$. Since by Proposition 2.5.2 X_L is a bounded distributive lattice frame, the sequent $\alpha \vdash \beta$ is not true is some model based on that bounded distributive lattice frame. □

2.8 Urquhart-style discrete representation for bounded lattices

In this section we give the formulation of the Urquhart-style (see [235]) discrete representation for bounded, not necessarily distributive, lattices, henceforth referred to as *general lattices.* It is discrete in the sense that it is given in terms of doubly ordered sets and does not assume any topological structure.

A *general lattice frame* is a structure $(X, \leqslant_1, \leqslant_2)$ where X is a non-empty set and $\leqslant_1$ and $\leqslant_2$ are two pre-orders such that, for any $x, y \in X$,

$$x \leqslant_1 y \ \& \ x \leqslant_2 y \ \Rightarrow \ x = y.$$

Each of these pre-orders, viewed as a binary relation over X, induces antitone unary operations l and r over 2^X defined, for every $A \subseteq X$, by

$$l(A) \stackrel{\text{def}}{=} \{x \in X : \forall y,\ x \leqslant_1 y \Rightarrow y \notin A\} = [\leqslant_1](-A)$$
$$r(A) \stackrel{\text{def}}{=} \{x \in X : \forall y,\ x \leqslant_2 y \Rightarrow y \notin A\} = [\leqslant_2](-A).$$

Note that, for any $A \subseteq X$,

$$lr(A) = [\leqslant_1] - [\leqslant_2](-A) = [\leqslant_1]\langle\leqslant_2\rangle(A).$$

Following the notational convention described in Section 1.9 for the remainder of this chapter we will denote a lattice frame by the underlying set.

In Lemma 2.8.1 we state some properties of general lattice frames needed for the subsequent results.

Lemma 2.8.1 *Let X be a general lattice frame. For all $A, B \subseteq X$,*

(a) *$l(A)$ is $\leqslant_1$-increasing and $r(A)$ is $\leqslant_2$-increasing;*
(b) *$l(A) \subseteq -A$ and $r(A) \subseteq -A$;*
(c) *l and r are antitone.*

Lemma 2.8.2 *Let X be a general lattice frame. The antitone maps l and r form a Galois connection between the lattice of $\leqslant_1$-increasing subsets of X and the lattice of $\leqslant_2$-increasing subsets of X.*

Note that Lemma 2.8.2 states that for every $\leqslant_1$-increasing $A \subseteq X$ and for every $\leqslant_2$-increasing $B \subseteq X$,

$$A \subseteq l(B) \ \Leftrightarrow \ B \subseteq r(A).$$

A subset A of X will be called *l-stable* (resp. *r-stable*) if $lr(A) = A$ (resp. $rl(A) = A$).

Lemma 2.8.3 *Let X be a general lattice frame. For every subset $A \subseteq X$,*

(a) *If A is $\leqslant_1$-increasing, then $r(A)$ is r-stable;*

(b) *If A is $\leqslant_2$-increasing, then $l(A)$ is l-stable;*

(c) *lr and rl are closure operators on 2^X.*

Proof:
By way of example we show (a). Let A be $\leqslant_1$-increasing. We have to show that $r(A) = rlr(A)$. By Lemma 2.8.1(c) it follows that $r(A) \subseteq rlr(A)$, so it suffices to show that $rlr(A) \subseteq r(A)$. Take $x \in rlr(A)$ and suppose that $x \notin r(A)$. Thus there is some $y \in X$ such that $x \leqslant_2 y$ and $y \in A$. From $x \in rlr(A)$ and $x \leqslant_2 y$ we get $y \notin lr(A)$, so there is some $z \in X$ such that $y \leqslant_1 z$ and $z \in r(A)$. Hence $z \notin A$, since $\leqslant_2$ is reflexive. But A is $\leqslant_1$-increasing, so $y \in A$ and $y \leqslant_1 z$ imply $z \in A$, a contradiction. □

The *complex algebra of a general lattice frame* $(X, \leqslant_1, \leqslant_2)$ is a structure $(L_X, \vee_X, \wedge_X, \emptyset, X)$ where $L_X = \{A \subseteq X : lr(A) = A\}$ and for any $A, B \subseteq X$,

$$A \vee_X B \overset{\text{def}}{=} l(r(A) \cap r(B))$$
$$A \wedge_X B \overset{\text{def}}{=} A \cap B$$

Lemma 2.8.4 (a) *$\emptyset$ and X are l-stable;*

(b) *For all $A, B \subseteq X$, $A \vee_X B$ is l-stable;*

(c) *If $A, B \subseteq X$ are l-stable, then so is $A \wedge_X B$.*

Proof:
We show (a). The proofs of (b) and (c) follow easily from Lemma 2.8.3. Note that $lr(\emptyset) = [\leqslant_1]\langle\leqslant_2\rangle\emptyset$. By Lemma 1.8.4(o), $\langle\leqslant_2\rangle\emptyset = \emptyset$. Hence $lr(\emptyset) = [\leqslant_1]\emptyset$, and by reflexivity of $\leqslant_1$, $lr(\emptyset) = \emptyset$. Next, $lr(X) = [\leqslant_1]\langle\leqslant_2\rangle X = [\leqslant_1]X$ by reflexivity of $\leqslant_2$, so by Lemma 1.8.4(a), $lr(X) = X$. □

From Lemma 2.8.4 the following fact easily follows.

Proposition 2.8.5 *The complex algebra of a general lattice frame is a general lattice.*

From any general lattice L, we can construct a general lattice frame by taking *filter-ideal pairs* (F_1, F_2) where F_1 is a filter of L and F_2 is an ideal of L such that $F_1 \cap F_2 = \emptyset$. Two pre-orders $\leqslant_{1L}$ and $\leqslant_{2L}$ may be defined on such pairs in terms of set inclusion on filters and set inclusion on ideals. That is, for two filter-ideal pairs $F = (F_1, F_2)$ and $G = (G_1, G_2)$,

$$F \leqslant_{1L} G \overset{\text{def}}{\Leftrightarrow} F_1 \subseteq G_1$$
$$F \leqslant_{2L} G \overset{\text{def}}{\Leftrightarrow} F_2 \subseteq G_2.$$

Then

$$F \leqslant_L G \overset{\text{def}}{\Leftrightarrow} F \leqslant_{1L} G \ \&\ F \leqslant_{2L} G.$$

A filter-ideal pair is *maximal* whenever it is maximal with respect to $\leqslant_L$. It was shown in [235] that any filter-ideal pair can be extended to a maximal pair.

Theorem 2.8.6 *Let L be a general lattice. For any filter-ideal pair (F_1, F_2) there is a maximal filter-ideal pair (G_1, G_2) with $F_1 \subseteq G_1$ and $F_2 \subseteq G_2$.*

Proof:
If (F_1, F_2) is a filter-ideal pair, then the set of all filters containing F_1 and disjoint from F_2 is not empty and closed under unions of chains, and thus, has a maximal element, say G_1. Similarly, the set of all ideals containing F_2 and disjoint from G_1 is not empty and closed under union of chains. If G_2 is maximal with those properties, then (G_1, G_2) is a maximal filter-ideal pair. $\square$

The following lemma will be useful later.

Lemma 2.8.7 *Let L be a general lattice and let $a, b \in L$ be such that $a \not\leqslant b$. Then there is some maximal filter-ideal pair (F_1, F_2) such that $a \in F_1$ and $b \notin F_1$.*

Proof:
Assume $a \not\leqslant b$. Then $\uparrow_{\leq} a \cap \downarrow_{\leq} b = \emptyset$, for otherwise there is some $c \in L$ such that $c \in \uparrow_{\leq} a$, that is $a \leqslant c$, and $c \in \downarrow_{\leq} b$, that is $c \leqslant b$, thus $a \leqslant b$, a contradiction. Clearly, $\uparrow_{\leq} a$ is a filter of L and $\downarrow_{\leq} b$ is an ideal of L. Then $(\uparrow_{\leq} a, \downarrow_{\leq} b)$ is a filter-ideal pair and by Lemma 2.8.6 it can be extended to a maximal filter-ideal pair, say (F_1, F_2). Then $\uparrow_{\leq} a \subseteq F_1$, hence $a \in F_1$, and $\downarrow_{\leq} b \subseteq F_2$, so $b \in F_2$. But $F_1 \cap F_2 = \emptyset$, so $b \notin F_1$. $\square$

The *canonical frame of a general lattice* L is a structure $(X_L, \leqslant_{1L}, \leqslant_{2L})$ where X_L is the family of all maximal filter-ideal pairs of lattice L and the relations $\leqslant_{1L}$ and $\leqslant_{2L}$ are defined as above.

Proposition 2.8.8 *The canonical frame of a general lattice is a general lattice frame.*

Thus every general lattice gives rise to a general lattice frame, and every general lattice frame gives rise to a general lattice. We now show that any general lattice can be embedded into the general lattice obtained from its general lattice frame.

Let L be a general lattice. Consider the Stone-style mapping $h : L \rightarrow L_{X_L}$ defined, for every $a \in L$, by

$$h(a) \stackrel{\text{def}}{=} \{F \in X_L : a \in F_1\}.$$

Lemma 2.8.9 *Let L be a general lattice. For every $a \in L$,*

(a) $rh(a) = \{F \in X_L : a \in F_2\}$

(b) *$h(a)$ is an l-stable set.*

Proof:
(a) For the left-to-right inclusion, take $F \in X_L$ and $a \in L$ such that $a \notin F_2$. Suppose $\uparrow_{\leq} a \cap F_2 \neq \emptyset$. Then there is some $b \in F_2$ such that $a \leqslant b$. Since F_2 is an ideal of L, we get $a \in F_2$, a contradiction. Therefore $(\uparrow_{\leq} a, F_2)$ is a filter-ideal

pair. By Lemma 2.8.6 there is some $G=(G_1,G_2)\in X_L$ such that $\uparrow_\leq a\subseteq G_1$ and $F_2\subseteq G_2$. Clearly, $a\in G_1$, which means $G\in h(a)$. Then we have $G\in X_L$ such that $F\leqslant_{2L} G$ and $G\in h(a)$, thus $F\notin rh(a)$. For the right-to-left inclusion, take $F\in X_L$ and $a\in L$ such that $F\notin rh(a)$. Then there exists $G\in X_L$ such that $F_2\subseteq G_2$ and $a\in G_1$. Hence $a\notin G_2$, thus $a\notin F_2$.

(b) We have to show that for every $a\in L$, $h(a)=lrh(a)$. Let $F\in h(a)$ and let $G\in X_L$ be such that $F_1\subseteq G_1$. We have to show that there is $H\in X_L$ such that $G_2\subseteq H_2$ and $a\in H_1$. Taking G for H yields $F\in lr(h(a))$. For the other inclusion assume $F\notin h(a)$. Then $a\notin F_1$, so $F_1\cap\downarrow_\leq a=\emptyset$. Thus $(F_1,\downarrow_\leq a)$ is a filter-ideal pair and it can be extended to a maximal filter-ideal pair, say G. Then $F_1\subseteq G_1$ and $a\in G_2$. By (a) we get $G\in rh(a)$. Hence $F\notin lrh(a)$. □

We now obtain the following discrete representation for general lattices. Other representations will be considered in Chapter 18.

Theorem 2.8.10 *Every general lattice is embeddable into the complex algebra of its canonical frame.*

Proof:
We prove that the mapping h is a lattice embedding. First, we show that h is injective. Assume that for some $a,b\in L$, $a\neq b$. Without loss of generality we may assume that $a\nleq b$. Then $\uparrow_\leq a\cap\downarrow_\leq b=\emptyset$ so by Lemma 2.8.6, there is some $F\in X_L$ such that $a\in F_1$ and $b\in F_2$. Thus, $h(a)\nsubseteq h(b)$.
Now we show that h preserves lattice operations. Observe that for any $a,b\in L$,

$h(a)\vee_{X_L} h(b)$	
$=l(r(\{F\in X_L : a\in F_1\})\cap r(\{F\in X_L : b\in F_1\}))$	definition of $\vee_X$
$=l(\{F\in X_L : a\in F_2\}\cap\{F\in X_L : b\in F_2\})$	Lemma 2.8.9(a)
$=l(\{F\in X_L : a\vee b\in F_2\})$	F_2 is an ideal of L
$=l(r(\{F\in X_L : a\vee b\in F_1\})$	Lemma 2.8.9(a)
$=lr(h(a\vee b))$	definition of h
$=h(a\vee b)$	Lemma 2.8.3(b).

$h(a)\wedge_{X_L} h(b)$	
$=\{F\in X_L : a\in F_1\}\cap\{F\in X_L : b\in F_1\}$	definition of $\wedge_{X_L}$
$=\{F\in X_L : a\wedge b\in F_1\}$	F_1 is a filter of L
$=h(a\wedge b)$	definition of h

which completes the proof. □

Unfortunately, we do not have a representation theorem for general lattice frames, and hence we have not established a discrete duality.

Note that, as a consequence of Lemma 2.8.11 and Lemma 2.8.12, the discrete representation for bounded distributive lattices obtained in Theorem 2.5.5 may be obtained as a special case of Theorem 2.8.10.

Lemma 2.8.11 *Let $(X,\leqslant_1,\leqslant_2)$ be a general lattice frame such that $\leqslant_2=\leqslant_1^{-1}$. Then for every $A\subseteq X$, $A=lr(A)$ if and only if $A=[\leqslant_1]A$.*

Proof:
Assume $A = lr(A)$ and suppose $A \not\subseteq [\leqslant_1]A$. Then there is $x \in A$ such that, for some $y \in X_1$, $x \leqslant_1 y$ and $y \notin A$. Since A is $\leqslant_1$-increasing, by Lemma 2.8.1(a), we obtain a contradiction. Since $\leqslant_1$ is reflexive, by Lemma 1.10.1(b), we have $A = [\leqslant_1]A$.

For the reverse implication note that, by Lemma 2.8.3(c), $A \subseteq lr(A)$. Let $A = [\leqslant_1]A$ and suppose $lr(A) \not\subseteq A$. Then there is $x \in lr(A)$ such that $x \notin A$. Since $\leqslant_1$ is reflexive, by Lemma 1.10.1(b), $x \in \langle\leqslant_1^{-1}\rangle A$. Thus, for some $y \in X$, $y \leqslant_1 x$ and $x \in A$, a contradiction. Hence, $A = lr(A)$. □

The following Lemma was proved in [46].

Lemma 2.8.12 *If L is a distributive lattice and (F, I) is a maximal disjoint filter-ideal pair of L then $F \cup I = L$.*

Proof:
Suppose that there is some $d \in L$ such that $d \notin F \cup I$. Let G be a filter generated by $F \cup \{d\}$ and let J be an ideal generated by $I \cup \{d\}$. Then $G \cap I \neq \emptyset$ and $F \cap J \neq \emptyset$. Thus there are $c \in G \cap I$ and $c' \in F \cap J$ such that for some $a_1, \ldots, a_n \in F$ and $a'_1, \ldots, a'_{n'} \in I$, $a_1 \wedge \ldots \wedge a_n \wedge d \leq c \leq a'_1 \vee \ldots \vee a'_{n'}$ and for some $b_1, \ldots b_m \in F$ and $b'_1, \ldots b'_{m'} \in I$, $b_1 \wedge \ldots \wedge b_m \leq c' \leq d \vee b'_1 \vee \ldots \vee b'_{m'}$. Let $a = a_1 \wedge \ldots \wedge a_n \wedge b_1 \wedge \ldots \wedge b_m$ and let $a' = a'_1 \vee \ldots \vee a'_{n'} \vee b'_1 \vee \ldots \vee b'_{m'}$. Thus $a \in F$ and $a' \in I$. Then $a \wedge d \leq a'$ and $a \leq d \vee a'$. Since L is distributive, $a \leq a \wedge (d \vee a') \leq (a \wedge d) \vee a' \leq a'$. Hence, since $a \in F$, $a' \in F$. It follows that $F \cap I \neq \emptyset$ which is a contradiction because (F, I) is a disjoint filter-ideal pair. □

The only operation in L_X which depends on $\leqslant_2$ is the join $\vee_X$ defined as $A \vee_X B = l(r(A) \cap r(B))$. It is easy to see that that $A \vee_X B = lr(A \cup B)$. By Lemma 2.8.11, $A \vee_X B = [\leqslant_1](A \cup B)$. It follows that L_X, the complex algebra of X, is a distributive lattice and $(X, \leqslant_1, \leqslant_2)$ may be viewed as a distributive lattice frame $(X, \leqslant_1)$.

By Lemma 2.8.12, if L is distributive then for every maximal disjoint filter-ideal pair (F, I), $I = -F$. It follows that F is a prime filter. Conversely, if F is a prime filter, then $(F, -F)$ is a maximal disjoint pair of L. Thus X_L, the canonical frame of L, may be viewed as a structure $(\{F : F \text{ is a prime filter of } L\}, \subseteq)$ which is a distributive lattice frame.

2.9 Duality via truth for general lattices

In this section we extend the discrete representation of general lattices to a duality via truth. The lattice language $\mathcal{L}an_L$ is the same as the language defined in Section 2.7 with two propositional constants T and F.

Algebraic semantics of this language, determined by the class $\mathcal{A}lg_L$ of general lattices, is defined similarly to the algebraic semantics presented in Section 2.7 with a general lattice instead of a distributive lattice as an interpretation structure. The same concerns the semantics of sequents.

The class $\mathcal{F}rm_L$ of lattice frames provides a Kripke-style semantics for $\mathcal{L}an_L$. A model based on a general lattice frame $(X, \leqslant_1, \leqslant_2)$ is a system $\mathcal{M} = (X, \leqslant_1, \leqslant_2, m)$ where $m : Var \cup \{T, F\} \to L_X$ is a meaning function such that $m(T) = X$ and $m(F) = \emptyset$. The satisfaction relation $\models$ is defined for all formulas α and β of $\mathcal{L}an$ by:

$$\mathcal{M}, x \models T$$

$$\mathcal{M}, x \not\models F$$

$$\mathcal{M}, x \models p \overset{\text{def}}{\Leftrightarrow} x \in m(p) \text{ for every } p \in Var$$

$$\mathcal{M}, x \models \alpha \wedge \beta \overset{\text{def}}{\Leftrightarrow} \mathcal{M}, x \models \alpha \;\&\; \mathcal{M}, x \models \beta$$

$$M, x \models \alpha \vee \beta \overset{\text{def}}{\Leftrightarrow} \forall y,\ x \leqslant_1 y \Rightarrow \exists z,\ y \leqslant_2 z \;\&\; (\mathcal{M}, z \models \alpha \text{ or } \mathcal{M}, z \models \beta).$$

We define $m(\alpha) = \{x \in X : \mathcal{M}, x \models \alpha\}$. One can easily verify the next lemma.

Lemma 2.9.1

(a) $m(T) = 1,\ m(F) = 0;$

(b) $m(\alpha \wedge \beta) = m(\alpha) \cap m(\beta);$

(c) $m(\alpha \vee \beta) = l(r(m(\alpha)) \cap r(m(\beta))).$

It follows that we have the following fact.

Lemma 2.9.2 *For every formula α of $\mathcal{L}an$, $m(\alpha)$ is an l-stable set.*

The truth of a sequent in a model and the truth of a sequent in a frame are defined as in the case of distributive lattices in Section 2.7.

Theorem 2.9.3 (Complex algebra theorem) *For all formulas α and β of $\mathcal{L}an_L$ and for every general lattice frame $X \in \mathcal{F}rm_L$ the following conditions are equivalent:*

(a) *A sequent $\alpha \vdash \beta$ is true in X;*

(b) *A sequent $\alpha \vdash \beta$ is true in the complex algebra L_X of X.*

Proof:
The proof follows from the fact that meaning functions of a model based on a general lattice frame X can be viewed as valuations in the complex algebra L_X of X. □

Finally, we prove the *duality via truth theorem* for general lattices and general lattice frames.

Theorem 2.9.4 (Duality via truth) *For all formulas α and β of $\mathcal{L}an_L$ the following conditions are equivalent:*

(a) *A sequent $\alpha \vdash \beta$ is true in every general lattice in $\mathcal{A}lg_L$;*

(b) *A sequent $\alpha \vdash \beta$ is true in every model based on a general lattice frame in $\mathcal{F}rm_L$.*

Proof:
(a)⇒(b) Assume that a sequent $\alpha \vdash \beta$ is not true in some model $\mathcal{M} = (X, m)$ based on a general lattice frame X. By Theorem 2.9.3, it is not true is the complex algebra of X. Hence, by Proposition 2.8.5, $\alpha \vdash \beta$ is not true in some general lattice L.

(b)⇒(a) Assume that a sequent $a \vdash \beta$ is not true in some general lattice L. Then there exists a valuation v in L such that $v(\alpha) \nleq v(\beta)$. Consider a model $\mathcal{M} = (X_L, m)$ based on the canonical frame X_L of L with the meaning function $m = h \circ v$, where h is the embedding quaranteed by the representation Theorem 2.8.10. By Lemma 2.8.7, $v(\alpha) \nleq v(\beta)$ implies that there is some maximal filter-ideal pair $(F_1, F_2) \in X_L$ such that $v(\alpha) \in F_1$ and $v(\beta) \notin F_1$. Hence $h(v(\alpha)) \nsubseteq h(v(\beta))$, so $m(\alpha) \nsubseteq m(\beta)$. Therefore the sequent $\alpha \vdash \beta$ is not true in the model $\mathcal{M}$. Since by Proposition 2.8.8, X_L is a general lattice frame, the sequent $\alpha \vdash \beta$ is not true in some model based on that general lattice frame. □

Part I

Boolean algebras with operators

Chapter 3

Boolean algebras with modal operators

3.1 Introduction

In this chapter discrete duality for Boolean algebras with two groups of operators is discussed: normal (transforming 0 to 0) and additive (transforming join to join) and co-normal (transforming 0 to 1) and co-additive (transforming join to meet). Following the logical tradition we refer to these operators as possibility and sufficiency, respectively. The first class of these algebras is an instance of the class of Boolean algebras with unary operators of Jónsson and Tarski [139]. The logics based on the algebras of this class are the traditional modal logics. The second class of algebras, referred to as sufficiency algebras, was introduced in [55] as an algebraic counterpart to the logic with a sufficiency operator originated in [98], see also [97]. In [111] the operator of sufficiency is considered as a negation in an orthologic, and in [242] it plays the role of an operator of obligation. An operator similar to sufficiency was axiomatized in [132]. Possibility operators are monotone and sufficiency operators are antitone. The dual operators of necessity and dual sufficiency are definable in terms of possibility and sufficiency, respectively, with the Boolean negation. In the book we refer to all these operators collectively as modal operators.

In Section 3.2 discrete duality for Boolean algebras with a possibility operator and the corresponding frames is presented. It is reconstructed from the Jónsson-Tarski duality [140] in the style of Goldblatt [113, 109] where the role of frames is made explicit. The section concludes with a presentation of duality via truth for semantic structures of classical propositional calculus with a possibility operator. In Section 3.3 discrete duality for Boolean algebras with a sufficiency operator and the corresponding frames is presented following [183]. The section is concluded with a brief outline of duality via truth for the algebraic and frame semantic structures of classical propositional calculus with a sufficiency operator. In Section 3.4 we consider the structures which are related to formal concept analysis originated in Wille [256], also Ganter and Wille [96]. The Boolean algebras are endowed with two sufficiency operators which are interpreted as intension and extension of a concept. The two relations in the corresponding frames are the converses of each other. Following [180] we present discrete duality for these algebras and frames and develop a duality via truth for the algebraic and frame semantic structures of a context logic. In the following three sections we extend the developments of Section 3.2 and Section 3.3 to Boolean algebras with both possibility and sufficiency operators. Invoking the results of Goranko [114] in Section 3.5 the frames corresponding to these algebras are endowed with two relations such that one is the complement of the other. In the algebras a specific axiom is postulated which says how the possibility and sufficiency operators are related to each other. In Section 3.6 a more involved case of mixed algebras is considered based on [55, 59]. The mixed algebras require a second order axiom. The frames associated to the mixed algebras have a single relation which serves as a basis for the definitions of possibility and sufficiency operators in their complex algebras. The complex algebras of the frames are not necessarily mixed algebras, but the complex algebras of the canonical frames are. Therefore, although the discrete representation theorems for algebras and frames hold, we do not have a discrete duality for them. In Section 3.7 we present an axiomatic extension of mixed algebras for which a discrete duality can be proved. In Section 3.8 mixed algebras provide a basis for an algebraic approach to preference. Preference structures are a subject of investigations in a variety of fields, for example in theory of social choice, economics, game theory [215, 36, 190]. Following [57] we present preference algebras and frames and discrete representation theorems for them.

Finally, in Section 3.9 we consider a class of algebras corresponding to event structures. Event structures introduced in [257] are models of concurrency which is a prominent field of computer science. For a survey see [258] and for some further developments on event structures see [145].

Apart from their role in modal algebras, the modal operators dealt with in this chapter are extensively used throughout the book in presentation of complex algebras of frames. In particular, in Chapter 18 the sufficiency operators determined by binary relations on a set are extended to the operators determined by binary heterogenous relations being subsets of a Cartesian product of two, possibly different, sets. Then they coincide with Birkhoff polarities and are a means for presentation of complex algebras of two-sorted frames which are used in the constructions of representation algebras of general lattices such as MacNeille and Wille representations, see Sections 18.5–18.10.

3.2 Possibility Boolean algebras

In this section we review Jónsson-Tarski duality for Boolean algebras with modal operators of possibility within the framework of discrete duality. This is then used as a case study for illustrating how the duality via truth approach extends this duality with dual notions of truth of formulas of a propositional language.

A *possibility Boolean algebra*, a *P-Boolean algebra* for short, is a structure $(B, \Diamond)$ where B is a Boolean algebra and $\Diamond$ is a unary operator on B such that, for all $a, b \in B$,

(P1) $\Diamond(a \vee b) = \Diamond a \vee \Diamond b$ additive

(P2) $\Diamond 0 = 0$ normal.

The operator of *necessity* dual to the possibility operator is defined, for every $a \in B$, by

$$\Box a \stackrel{\text{def}}{=} -\Diamond -a.$$

Using the definitions in Section 1.4 the image and inverse image of any set $A \subseteq B$ under the possibility operator $\Diamond$ are given by

$$\Diamond(A) = \{\Diamond a \in B : a \in A\} \quad \text{and} \quad \Diamond^{-1}(A) = \{a \in B : \Diamond a \in A\}.$$

Lemma 3.2.1 *Let $(B, \Diamond)$ be a P-Boolean algebra. Then for all $a, b \in B$ and for all $A, A_1, A_2 \subseteq B$,*

(a) *If $a \leq b$, then $\Diamond a \leq \Diamond b$;*

(b) *If $A_1 \subseteq A_2$, then $\Diamond(A_1) \subseteq \Diamond(A_2)$ and $\Diamond^{-1}(A_1) \subseteq \Diamond^{-1}(A_2)$;*

(c) *If A is an ideal of B, then so is $\Diamond^{-1}(A)$;*

(d) *If A is an ultrafilter of B, then $-\Diamond^{-1}(A)$ is an ideal of B.*

Proof:
By way of example we prove (c) and (d).

(c) Let A be an ideal of B. Since $\Diamond 0 = 0 \in A$, $\Diamond^{-1}(A) \neq \emptyset$. Take $a, b \in B$ such that $a \in \Diamond^{-1}(A)$ and $b \in \Diamond^{-1}(A)$. Then $\Diamond a \in A$ and $\Diamond b \in A$, so $\Diamond a \vee \Diamond b \in A$. By axiom (P1), $\Diamond(a \vee b) \in A$, whence $a \vee b \in \Diamond^{-1}(A)$. Furthermore, let $a, b \in B$ be such that $a \in \Diamond^{-1}(A)$ and $b \leq a$. Then $\Diamond a \in A$ and, by (a), $\Diamond b \leq \Diamond a$. Hence $\Diamond b \in A$, that is $b \in \Diamond^{-1}(A)$.

(d) Let A be a ultrafilter of B. Since $\Diamond 0 = 0 \notin A$, $-\Diamond^{-1}(A) \neq \emptyset$. Take $a, b \in B$ such that $a \in -\Diamond^{-1}(A)$ and $b \in -\Diamond^{-1}(A)$. Then $\Diamond a \notin A$ and $\Diamond b \notin A$, so by primeness of A, $\Diamond a \vee \Diamond b \notin A$. Hence, by axiom (P1), $\Diamond(a \vee b) \notin A$, so $a \vee b \in -\Diamond^{-1}(A)$. Next, let $a, b \in B$ be such that $a \in -\Diamond^{-1}(A)$ and $b \leq a$. Then $\Diamond a \notin A$. Moreover, by (a), $\Diamond b \leq \Diamond a$. Hence $\Diamond b \notin A$, that is $b \in -\Diamond^{-1}(A)$. □

Observe that this lemma holds in a more general setting of bounded distributive lattices with the operator satisfying (P1) and (P2), as in Section 8.2.

A *possibility Boolean frame*, or a *P-Boolean frame*, is a structure (X, R) where X is a non-empty set and R is a binary relation on X.

The *complex algebra of a P-Boolean frame* (X, R) is a structure $(2^X, \Diamond_R)$ where 2^X is the complex algebra of its Boolean frame reduct and $\Diamond_R : 2^X \to 2^X$ is defined, for every $A \subseteq X$, by

$$\Diamond_R(A) \stackrel{\text{def}}{=} \langle R \rangle A = \{x \in X : R(x) \cap A \neq \emptyset\}$$

and its dual, namely, $\Box_R : 2^X \to 2^X$ is defined, for every $A \subseteq X$, by

$$\Box_R(A) \stackrel{\text{def}}{=} [R]A = -\Diamond_R(-A).$$

Based on the properties of modal operators listed in Section 1.8 it is easy to show that the operator $\Diamond_R$ is normal and additive, and its dual $\Box_R$ satisfies $\Box 1 = 1$, and is multiplicative, that is, $\Box(a \wedge b) = \Box a \wedge \Box b$. Thus we have the following fact.

Proposition 3.2.2 *The complex algebra of a P-Boolean frame is a P-Boolean algebra.*

The *canonical frame of a P-Boolean algebra* $(B, \Diamond)$ is a structure $(X_B, R_\Diamond)$ where X_B is the canonical frame of its Boolean reduct and $R_\Diamond$ is a binary relation on X_B defined, for all $F, G \in X_B$, by

$$F R_\Diamond G \stackrel{\text{def}}{\Leftrightarrow} G \subseteq \Diamond^{-1}(F).$$

Clearly, we have the following fact.

Proposition 3.2.3 *The canonical frame of a P-Boolean algebra is a P-Boolean frame.*

Lemma 3.2.4 *For any P-Boolean frame* (X, R) *and for all* $F, G \in X_B$,

$$G \subseteq \Diamond^{-1}(F) \Leftrightarrow \Box^{-1}(F) \subseteq G.$$

We now show how any P-Boolean algebra can be embedded in the P-Boolean algebra it gives rise to. That is, if we start with a P-Boolean algebra $(B, \Diamond)$, form its canonical frame $(X_B, R_\Diamond)$ and form the complex algebra $(2^{X_B}, \Diamond_{R_\Diamond})$ of that, then the original P-Boolean algebra can be embedded in it. For this it suffices to show that the Stone mapping $h : B \to 2^{X_B}$, used in Section 2.3, preserves the possibility operator on B.

Lemma 3.2.5 *For any P-Boolean algebra* $(B, \Diamond)$ *and for every* $a \in B$,

$$h(\Diamond a) = \Diamond_{R_\Diamond}(h(a)).$$

Proof:
For this we show that for every $F \in X_B$ and for every $a \in B$,

$$\Diamond a \in F \Leftrightarrow \exists G \in X_B,\ G \subseteq \Diamond^{-1}(F) \ \&\ a \in G.$$

Clearly, if $a \in G$ and $G \subseteq \Diamond^{-1}(F)$, then $\Diamond a \in F$. On the other hand, consider the set $Z_\Box = \{b \in B : \Box b \in F\}$. Let F' be the filter generated by $Z_\Box \cup \{a\}$, that is, $F' = \{b \in B : \exists a_1, \ldots, a_n \in Z_\Box,\ a_1 \wedge \ldots \wedge a_n \wedge a \leq b\}$. Then F' is proper. For suppose otherwise, then for some $a_1, \ldots, a_n \in Z_\Box$, $a_1 \wedge \ldots \wedge a_n \wedge a = 0$, that is, $a_1 \wedge \ldots \wedge a_n \leq -a$. Since $\Box$ is monotone, $\Box(a_1 \wedge \ldots \wedge a_n) \leq \Box(-a)$, that is, $\Box a_1 \wedge \ldots \wedge \Box a_n \leq \Box(-a)$. By definition of $Z_\Box$ we have $\Box a_1, \ldots, \Box a_n \in F$ so, since F is a filter, $\Box a_1 \wedge \ldots \wedge \Box a_n \in F$ and hence $\Box(-a) \in F$. Thus $-a \in Z_\Box$

which is a contradiction. Thus by Theorem 1.6.13 there is an ultrafilter G containing F'. Since $a \in F'$, $a \in G$ and hence $G \in h(a)$. Also $G \subseteq \Diamond^{-1}(F)$ since if $b \notin \Diamond^{-1}(F)$, then $\Diamond b \notin F$, that is, $\Box(-b) \in F$, so $-b \in F' \subseteq G$ and hence $b \notin G$. □

On the other hand, we show how any P-Boolean frame can be embedded in the P-Boolean frame it gives rise to. That is, if we start with a P-Boolean frame (X, R), form its complex algebra $(2^X, \Diamond_R)$ and form the canonical frame $(X_{2^X}, R_{\Diamond_R})$ of that, then the original P-Boolean frame can be embedded in it. For this we show that the mapping $k : X \to X_{2^X}$, used in Section 2.3, preserves the relation R.

Lemma 3.2.6 *Let (X, R) be a P-Boolean frame. For any $x, y \in X$,*

$$xRy \Leftrightarrow k(x)R_{\Diamond_R}k(y).$$

Proof:
Note, for any $x, y \in X$, that

$$\begin{aligned} k(x)R_{\Diamond_R}k(y) &\Leftrightarrow k(y) \subseteq \Diamond_R^{-1}(k(x)) \\ &\Leftrightarrow \forall A \subseteq X,\ A \in k(y) \Rightarrow A \in \Diamond_R^{-1}(k(x)) \\ &\Leftrightarrow \forall A \subseteq X,\ y \in A \Rightarrow x \in \langle R \rangle A. \end{aligned}$$

Assume xRy. Take any $A \subseteq X$ with $y \in A$. Hence $A \in \{B : x \in \Diamond_R(A)\}$. On the other hand, assume $k(x)R_{\Diamond_R}k(y)$. Since $y \in \{y\}$, by the above, $x \in \Diamond_R(\{y\})$, that is, xRy. □

As a consequence of the above theorems we obtain discrete representations, and hence a Jónsson-Tarski-style duality, between P-Boolean algebras and P-Boolean frames.

Theorem 3.2.7

(a) *Every P-Boolean algebra is embeddable into the complex algebra of its canonical frame;*

(b) *Every P-Boolean frame (X, R) is embeddable into the canonical frame of its complex algebra.*

Proof:
As shown in Section 2.3 the mapping h is a Boolean embedding and the mapping k is a Boolean frame embedding. Combining these with Lemma 3.2.5 and Lemma 3.2.6 the theorem follows. □

In order to extend the discrete duality in Theorem 3.2.7 to a duality via truth, we build a modal language, which we denote by $\mathcal{L}an_P$, from $\mathcal{L}an_B$ and the possibility operator ($\Diamond$).

The class $\mathcal{A}lg_P$ of P-Boolean algebras is now shown to provide an algebraic semantics for $\mathcal{L}an_P$. Let $(B, \Diamond)$ be a P-Boolean algebra. The Boolean valuations on formulas in $\mathcal{L}an_B$, defined in Section 2.4, can be extended inductively to all formulas in $\mathcal{L}an_P$ by

$$v(\Diamond\alpha) = \Diamond(v(\alpha)).$$

The notion of truth determined by this semantics is defined as in Section 2.4.

The class $\mathcal{F}rm_P$ of P-Boolean frames is now shown to provide a well-known frame semantics for $\mathcal{L}an_P$. A model based on a frame (X, R) is a structure $M = (X, R, m)$ where $m : Var \to 2^X$ is a meaning function. The satisfaction relation $\models$, defined in Section 2.4, may be extended inductively to all formulas in $\mathcal{L}an_P$ by

$$M, x \models \Diamond\alpha \Leftrightarrow \exists y,\ M, y \models \alpha \ \&\ xRy.$$

The notion of truth determined by this semantics is defined as in Section 2.4.

It is easy to see that the complex algebra theorem holds.

Theorem 3.2.8 *A formula $\alpha \in \mathcal{L}an_P$ is true in every model based on a P-Boolean frame (X, R) if and only if α is true in the complex algebra $(2^X, \Diamond_R)$ of that frame.*

Proof:
Let (X, R) be any P-Boolean frame. For every model (X, R, m) we extend the meaning function m to all the formulas by $m(\alpha) = \{x \in X : M, x \models \alpha\}$. Then it is easy to see that $m(\Diamond\alpha) = \langle R \rangle m(\alpha)$. From this and Theorem 2.4.1, it follows that the meaning function m of any model (X, R, m) based on the P-Boolean frame (X, R) coincides with the valuation function on the complex P-Boolean algebra $(2^X, \Diamond_R)$ of (X, R). □

Finally, we prove the duality via truth theorem for P-Boolean algebras and P-Boolean frames.

Theorem 3.2.9 *A formula $\alpha \in \mathcal{L}an_P$ is true in every algebra of $\mathcal{A}lg_P$ if and only if α is true in every frame of $\mathcal{F}rm_P$.*

Proof:
($\Rightarrow$) Let (X, R) be a P-Boolean frame. By Proposition 3.2.2 the complex algebra of (X, R) is a P-Boolean algebra. So, by the assumption, α is true in the complex algebra $(2^X, \Diamond_R)$ of (X, R). By Theorem 3.2.8, α is true in every model based on (X, R), thus it is true in (X, R).

($\Leftarrow$) Assume that α is not true in some P-Boolean algebra $(B, \Diamond)$. Then there is a valuation v in B such that $v(\alpha) \neq 1$. Let $(X_B, R_\Diamond)$ be the canonical frame of that algebra. By Proposition 3.2.3, it is a P-Boolean frame. Consider a model (X_B, m) based on X_B such that $m = h \circ v$. Since h is an embedding, $m(\alpha) \neq X_B$. Hence α is not true in $(X_B, R_\Diamond)$. □

3.3 Sufficiency Boolean algebras

A *sufficiency Boolean algebra, S-Boolean algebra*, is a structure $(B, \boxbox)$ where B is a Boolean algebra and $\boxbox$ is a unary operator on B such that, for all $a, b \in B$,

(S1) $\boxbox(a \vee b) = \boxbox a \wedge \boxbox b$ co-additive

(S2) $\boxbox 0 = 1$ co-normal.

The operator of dual sufficiency is defined, for every $a \in B$, by

$$\diamonddiamond a \stackrel{\text{def}}{=} -\boxbox(-a).$$

Lemma 3.3.1 *Let B be an S-Boolean algebra.*

(a) *For any $a, b \in B$, $a \leq b$ implies $\boxbox b \leq \boxbox a$;*

(b) *If $F \subseteq B$ is an ultrafilter of B, then $\boxbox^{-1}(F)$ is an ideal of B.*

Proof:
The proof of (a) is straightforward. For (b), assume that F is a ultrafilter of B and take $a, b \in B$ such that $a \in \boxbox^{-1}(F)$ and $b \leq a$. By (a), $\boxbox$ is antitone, so $\boxbox a \leq \boxbox b$. Since $\boxbox a \in F$, $\boxbox b \in F$, that is, $b \in \boxbox^{-1}(F)$. Now, let $a, b \in \boxbox^{-1}(F)$. Thus $\boxbox a \in F$ and $\boxbox b \in F$, so $\boxbox a \wedge \boxbox b \in F$. By (S1), $\boxbox(a \vee b) \in F$, hence $a \vee b \in \boxbox^{-1}(F)$. □

A *sufficiency Boolean frame*, or an *S-Boolean frame*, is a structure (X, R) where X is a non-empty set and R is a binary relation on X.

Given a S-Boolean frame (X, R), its *complex algebra* is a structure $(2^X, \boxbox_R)$ where 2^X is the complex algebra of its Boolean-frame reduct and $\boxbox_R : 2^X \to 2^X$ is defined, for every $A \subseteq X$, by

$$\boxbox_R A \stackrel{\text{def}}{=} [\![R]\!] A = \{x \in X : A \subseteq R(x)\}$$

and its dual $\diamonddiamond_R : 2^X \to 2^X$ is defined, for every $A \subseteq X$, by

$$\diamonddiamond_R A \stackrel{\text{def}}{=} \langle\!\langle R \rangle\!\rangle A = \{x \in X : R(x) \cup A \neq X\}.$$

Proposition 3.3.2 *The complex algebra of a S-Boolean frame is a S-Boolean algebra.*

Proof:
Observing that the operators $\boxbox_R$ and $\diamonddiamond_R$ may be defined in terms of the operators in Section 3.2 by $\boxbox_R A = \Box_{-R}(-A)$ it follows that $\boxbox_R$ is co-normal and co-additive. □

Let $(B, \boxbox)$ be a S-Boolean algebra. Its *canonical frame* is a structure $(X_B, R_{\boxbox})$ where X_B is the canonical frame of its Boolean reduct and $R_{\boxbox}$ is a binary relation on X_B defined, for all $F, G \in X_B$, by

$$F R_{\boxbox} G \stackrel{\text{def}}{\Leftrightarrow} \boxbox(G) \cap F \neq \emptyset.$$

Proposition 3.3.3 *The canonical frame of a S-Boolean algebra is a S-Boolean frame.*

We now show that any S-Boolean algebra can be embedded in the S-Boolean algebra it gives rise to. That is, if we start with an S-Boolean algebra $(B, \boxbar)$, form its canonical frame $(X_B, R_{\boxbar})$ and form the complex algebra $(2^{X_B}, \boxbar_{R_{\boxbar}})$ of that, then the original S-Boolean algebra can be embedded in it. For this it suffices to show that the Stone mapping $h : B \rightarrow 2^{X_B}$, used in Section 2.3, preserves the operator $\boxbar$ on B.

Lemma 3.3.4 *For any S-Boolean algebra $(B, \boxbar)$ and for any $a \in B$,*

$$h(\boxbar a) = \boxbar_{R_{\boxbar}} h(a).$$

Proof:
For any $a \in B$,

$$\begin{aligned}\boxbar_{R_{\boxbar}} h(a) &= \{F \in X_B : h(a) \subseteq R_{\boxbar}(F)\} \\ &= \{F \in X_B : \forall G \in X_B,\ a \in G \Rightarrow \boxbar(G) \cap F \neq \emptyset\}.\end{aligned}$$

To show that this is equal to $h(\boxbar a) = \{F \in X_B : \boxbar a \in F\}$ we show that

$$\boxbar a \notin F \Leftrightarrow \exists G \in X_B,\ a \in G \ \&\ \boxbar(G) \cap F = \emptyset.$$

The right-to-left direction is easy, because if $a \in G$ and $\boxbar(G) \cap F = \emptyset$, then $\boxbar a \in \boxbar(G)$ so $\boxbar a \notin F$.
For the left-to-right direction, assume $\boxbar a \in F$. Consider the set

$$Z_{\boxbar} = \{b \in B : \Diamond b \notin F\}.$$

Let F' be the filter generated by $Z_{\boxbar} \cup \{a\}$, that is,

$$F' = \{b \in B : \exists a_1, \ldots, a_n \in Z_{\boxbar},\ a_1 \wedge \ldots \wedge a_n \wedge a \leq b\}.$$

Suppose that $0 \in F'$. Then for some $a_1, \ldots, a_n \in Z_{\boxbar}$, $a_1 \wedge \ldots \wedge a_n \wedge a = 0$, that is, $a \leq -(a_1 \wedge \ldots \wedge a_n) = -a_1 \vee \ldots \vee -a_n$. Since $\boxbar$ is antitone,

$$\boxbar(-a_1 \vee \ldots \vee -a_n) \leq \boxbar(a).$$

Thus $\boxbar(-a_1) \wedge \ldots \wedge \boxbar(-a_n) \leq \boxbar(a)$, that is, $-\Diamond(a_1) \wedge \ldots \wedge -\Diamond(a_n) \leq \boxbar(a)$. By definition of $Z_{\boxbar}$ we have $\Diamond(a_1), \ldots, \Diamond(a_n) \notin F$ and hence we have

$$-\Diamond a_1, \ldots, -\Diamond a_n \in F.$$

Since F is a filter of B, $-\Diamond a_1 \wedge \ldots \wedge -\Diamond a_n \in F$ and hence $\boxbar a \in F$ which contradicts the original assumption. Thus by Theorem 1.6.13 there is an ultrafilter G containing F'. Since $a \in F'$, $a \in G$ and hence $G \in h(a)$. Suppose there is some $b \in G$ such that $\boxbar b \in F$. Then $\Diamond -b \notin F$ hence $-b \in Z_{\boxbar} \subseteq F' \subseteq G$ and thus $b \notin G$, a contradiction. Hence $\boxbar(G) \cap F = \emptyset$. $\square$

On the other hand any S-Boolean frame can be embedded in the S-Boolean frame it gives rise to. That is, if we start with a S-Boolean frame (X, R), form its complex algebra $(2^X, \boxbar_R)$ and form the canonical frame $(X_{2^X}, R_{\boxbar_R})$ of that, then the original S-Boolean frame can be embedded in it. For this we show that the mapping $k : X \rightarrow X_{2^X}$, defined in Section 2.3, preserves relation R.

Lemma 3.3.5 *For any S-Boolean frame* (X, R) *and for any* $x, y \in X$,

$$xRy \Leftrightarrow k(x)R_{\boxminus_R}k(y).$$

Proof:
Note, for any $x, y \in X$,

$$k(x)R_{\boxminus_R}k(y) \Leftrightarrow \boxminus_R(k(y)) \cap k(x) \neq \emptyset \Leftrightarrow \{\boxminus_R(A) : y \in A\} \cap \{B : x \in B\} \neq \emptyset.$$

We now prove the desired double implication. For the left-to-right direction, assume xRy. Then $\{y\} \subseteq R(x)$, so $x \in \boxminus_R(\{y\})$. Hence $\boxminus_R\{y\} \in \boxminus_R(k(y)) \cap k(x)$. Thus $k(x)R_{\boxminus_R}k(y)$. For the right-to-left direction, assume $k(x)R_{\boxminus_R}k(y)$. Since $y \in \{y\}$, by the above, $\boxminus_R\{y\} \in \{B : x \in B\}$. Thus $x \in \boxminus_R\{y\}$, that is, $\{y\} \subseteq R(x)$, and hence xRy. □

Therefore we have discrete representations for S-Boolean algebras and S-frames.

Theorem 3.3.6

(a) *Every S-Boolean algebra is embeddable into the complex algebra of its canonical frame;*

(b) *Every S-Boolean frame is embeddable into the canonical frame of its complex algebra.*

We now extend the discrete duality obtained in Theorem 3.3.6 to a duality via truth. Consider a modal-type language, denoted by $\mathcal{L}an_S$, whose formulas are built from $\mathcal{L}an_B$ and the sufficiency operator $\boxminus$. Let $\mathcal{A}lg_S$ denote the class of S-Boolean algebras, and $\mathcal{F}rm_S$ denote the class of S-Boolean frames.

The algebraic semantics for $\mathcal{L}an_S$ is defined, as in Section 2.4, for $\mathcal{L}an_B$ with the Boolean valuations extended inductively to all formulas in $\mathcal{L}an_P$ by

$$v(\boxminus\alpha) = \boxminus(v(\alpha)).$$

The frame semantics for $\mathcal{L}an_S$ is defined in a similar way as for the language $\mathcal{L}an_P$. Let (X, R) be a frame and let $M = (X, R, m)$ be a model based on that frame. The satisfaction relation extends to the formulas with the sufficiency operator as follows

$$M, x \models \boxminus\alpha \Leftrightarrow \forall y,\ M, y \models \alpha \Rightarrow xRy.$$

The notions of truth of a formula in the algebraic semantics and in the frame semantics are defined as in Section 2.4. Using analogous reasoning to that in Theorem 3.2.8 and Theorem 3.2.9, we obtain the following results.

Theorem 3.3.7 *A formula* $\alpha \in \mathcal{L}an_S$ *is true in every model based on a frame* (X, R) *if and only if* α *is true in the complex algebra* $(2^X, \boxminus_R)$ *of that frame.*

Theorem 3.3.8 *A formula* $\alpha \in \mathcal{L}an_S$ *is true in every algebra of* $\mathcal{A}lg_S$ *if and only if* α *is true in every frame of* $\mathcal{F}rm_S$.

3.4 Context algebras in formal concept analysis

Central in formal concept analysis is the notion of a Galois connection, defined in Section 1.7, between two types of entities. Algebraically this may be captured by two maps e and i, and relationally by a relation between the two types of entities. In this section we formalise this in the notions of context algebra and context frame introduced in [180], and present a discrete duality between these structures.

A *context algebra*, a *Ctx-algebra* for short, is a structure (B,e,i) where B is a Boolean algebra and e,i are sufficiency operators on B satisfying for all $a,b\in B$,

(Ctx1) $e(a\vee b)=e(a)\wedge e(b)$ and $i(a\vee b)=i(a)\wedge i(b)$

(Ctx2) $e(0)=1$ and $i(0)=1$

(Ctx3) $a\leq e(i(a))$

(Ctx4) $a\leq i(e(a))$.

It follows that the operators e and i are antitone and form a Galois connection, that is, for all $a,b\in B$,

$$a\leq i(b) \Leftrightarrow b\leq e(a).$$

From this Galois connection we can separate the two types of entities and the relation between them thereby deriving a formal context arising in formal concept analysis. From any Ctx-algebra (B,e,i) we may define the formal context (G_B,M_B,I_B), where

$$\begin{aligned} G_B &\stackrel{\text{def}}{=} \{o\in B:\exists a, o\leq e(a)\}\\ M_B &\stackrel{\text{def}}{=} \{a\in B:\exists o, a\leq i(o))\}\\ I_B &\stackrel{\text{def}}{=} \{(o,a):o\leq e(a)\}=\{(o,a):a\leq i(o)\}. \end{aligned}$$

Then $G_B=dom(I_B)$ and $M_B=ran(I_B)$.

Lemma 3.4.1 *Let (B,e,i) be any Ctx-algebra. The sufficiency operators e and i are the mappings of extent and intent determined by the corresponding formal context (G_B,M_B,I_B). That is, for any $a,o\in B$,*

$$\begin{aligned} o\leq e(a) &\Leftrightarrow o\in\{o:oI_Ba\}\\ a\leq i(o) &\Leftrightarrow a\in\{a:oI_Ba\}. \end{aligned}$$

On the other hand, given any formal context (G,M,I), we may define a Ctx-algebra $(2^{G\cup M},e_I,i_I)$ where$e_I=[\![I]\!]$, and $i_I=[\![I^{-1}]\!]$, and, for any $A\in 2^{G\cup M}$ and $T\in\{I,I^{-1}\}$, $[\![T]\!](A)=\{x\in G\cup M:\forall y,\ y\in A\Rightarrow xTy\}$.

Lemma 3.4.2 *The mappings of extent and intent determined by a formal context (G,M,I) are the sufficiency operators e_I and i_I, respectively, of the corresponding Ctx-algebra $(2^{G\cup M},e_I,i_I)$.*

Theorem 3.4.3

(a) *Any context* (G, M, I), *where* $G = dom(I)$ *and* $M = ran(I)$, *coincides with the formal context from the context algebra* $(2^{G\cup M}, e_I, i_I)$;

(b) *Any complete and atomic Ctx-algebra* $(2^X, e, i)$, *where*

$$X = \{o : o \in e(\{a\})\} \cup \{a : a \in i(\{o\})\},$$

coincides with the Ctx-algebra from the formal context (G_B, M_B, I_B).

A *context frame*, *Ctx-frame* for short, is a structure (X, R, S) where X is a non-empty set that is a Boolean frame and R and S are binary relations on X such that $S = R^{-1}$. Although relation S is definable from R, the setting with two relations enables us to avoid any relation-algebraic structure (in this case the operation of converse of a relation) in the language of a context logic. In this way the intended object language is singular, the required constraint is formulated only in the definition of its semantics, that is, in the metalanguage.

From a Ctx-frame (X, R, S) we may define the formal context (G_X, M_X, I_X) where $G_X = dom(R)$, $M_X = ran(R)$, and $I_X = R$.

Lemma 3.4.4 *Let* (X, R, S) *be a Ctx-frame. The sufficiency operators determined by* R *and* S *are the mappings of extent and intent determined by a formal context* (G_X, M_X, I_X).

On the other hand, given a formal context (G, M, I), we may define a Ctx-frame $(G \cup M, I, I^{-1})$.

Lemma 3.4.5 *The mappings of extent and intent determined by the formal context* (G, M, I) *are the sufficiency operators* $[\![I]\!]$ *and* $[\![I^{-1}]\!]$, *respectively, determined by the corresponding Ctx-frame* $(G \cup M, I, I^{-1})$.

Theorem 3.4.6

(a) *Any context* (G, M, I), *where* $G = \mathrm{dom}(I)$ *and* $M = \mathrm{ran}(I)$, *coincides with the formal context corresponding to the Ctx-frame* $(G \cup M, I, I^{-1})$;

(b) *Any Ctx-frame* (X, R, S), *where* $X = \mathrm{dom}(R) \cup \mathrm{ran}(R)$, *coincides with the Ctx-frame corresponding to the formal context* (G_X, M_X, I_X).

We now establish a discrete duality between Ctx-algebras and Ctx-frames. First, we show that from any Ctx- frame we can define a Ctx-algebra.

The *complex algebra* of a Ctx-frame (X, R, S) is a structure $(2^X, e_R, i_S)$ where 2^X is the complex algebra of the Boolean frame X and e_R and i_S are defined, for every $A \subseteq X$, by

$$e_R(A) \stackrel{\text{def}}{=} [\![R]\!](A) = \{x \in X : \forall y \in X,\ y \in A \Rightarrow xRy\}$$
$$i_S(A) \stackrel{\text{def}}{=} [\![S]\!](A) = \{x \in X : \forall y \in X,\ y \in A \Rightarrow xSy\}.$$

Proposition 3.4.7 *The complex algebra of a Ctx-frame is a Ctx-algebra.*

Proof:
The operators e_R and i_S are sufficiency operators as shown in [183]. By way of example we show that (Ctx3) is satisfied. Let $A \subseteq X$. We show that $A \subseteq e_R(i_S(A))$. Let $x \in X$ and suppose that $x \notin e_R(i_S(A))$. It follows that there is $y \in i_S(A)$ such that $(x,y) \notin R$. By definition of i_S, for every $z \in A$, ySz. In particular, taking z to be x, we have ySx. Since $S = R^{-1}$, we have xRy, a contradiction.
The proof of (Ctx4) is similar. □

The *canonical frame* of a Ctx-algebra (B,e,i) is the relational structure (X_B, R_e, S_i) where X_B is the canonical frame of its Boolean reduct and R_e and S_i are defined, for every $F, G \in X_B$, by

$$FR_eG \stackrel{\text{def}}{\Leftrightarrow} e(G) \cap F \neq \emptyset$$

$$FS_iG \stackrel{\text{def}}{\Leftrightarrow} i(G) \cap F \neq \emptyset$$

where for $A \subseteq B$ and $g \in \{e,i\}$, $g(A) = \{g(a) : a \in A\}$.

Proposition 3.4.8 *The canonical frame of a Ctx-algebra is a Ctx-frame.*

Proof:
We show that $R_e^{-1} \subseteq S_i$. Let $(F,G) \in R_e^{-1}$. Then $(G,F) \in R_e$, that is $e(F) \cap G \neq \emptyset$. It follows that there is $a \in B$ such that $e(a) \in G$ and $a \in F$. Suppose that $(F,G) \notin S_i$. It follows that $i(G) \cap F = \emptyset$ so for every $b \in B$, $b \notin G$ or $i(b) \notin F$. In particular, taking b to be $e(a)$, we have $i(e(a)) \notin F$ or $e(a) \notin G$. By (Ctx4) and since F is up-closed, $a \notin F$. Hence in both cases we obtain a contradiction.
The proof of the other inclusion is similar. □

We now show that any Ctx-algebra (B,e,i) can be embedded in the Ctx-algebra it gives rise to. For this it suffices to show that the Stone embedding $h : B \to 2^{X_B}$, used in Section 2.3, preserves the operators i and e over B. Since they are sufficiency operators, the proofs are as in Section 3.3.

Lemma 3.4.9 *For any Ctx-algebra (B,e,i) and for any $a \in B$,*

(a) $h(e(a)) = e_{R_e}(h(a))$;
(b) $h(i(a)) = i_{S_i}(h(a))$. □

On the other hand, any Ctx-frame (X,R,S) can be embedded in the Ctx-frame it gives rise to. For this it suffices to show that the embedding $k : X \to X_{2^X}$, used in Section 2.3, preserves the relations R and S. Since these are relations defined from sufficiency operators, the reasoning is as in Section 3.3.

Lemma 3.4.10 *For any Ctx-frame (X,R,S) and for any $x,y \in X$,*

(a) $xRy \Leftrightarrow k(x)R_{e_R}k(y)$;
(b) $xSy \Leftrightarrow k(x)S_{i_S}k(y)$. □

Therefore we have the following discrete representations for Ctx-algebras and Ctx-frames.

Theorem 3.4.11

(a) *Every Ctx-algebra* (B, e, i) *is embeddable into the complex algebra of its canonical frame;*

(b) *Every Ctx-frame* (X, R, S) *is embeddable into the canonical frame of its complex algebra.*

In order to extend the duality established in Theorem 3.4.11 to a duality via truth, we need a logical language. Let $\mathcal{L}an_{Ctx}$ be a modal language whose formulas are built from $\mathcal{L}an_B$, defined in Section 2.4, and two unary operators $\boxminus_1$ and $\boxminus_2$. We may define $\Diamond_i$, by

$$\Diamond_i\alpha \stackrel{\text{def}}{=} \neg\boxminus_i\neg\alpha, \quad i=1,2.$$

Let $\mathcal{A}lg_{Ctx}$ denote the class of context algebras and $\mathcal{F}rm_{Ctx}$ denote the class of context frames. Then $\mathcal{A}lg_{Ctx}$ and $\mathcal{F}rm_{Ctx}$ may be shown to provide an algebraic semantics and frame semantics, respectively, for Ctx-logic using analogous reasoning to that in Section 3.2. Also we have the complex algebra and duality via truth theorems.

Theorem 3.4.12 *A Ctx-formula* α *is true in every model based on a Ctx-frame* (X, R, S) *if and only if* α *is true in the complex algebra* $(2^X, e_R, i_S)$ *of that frame.*

Theorem 3.4.13 *A formula* α *is true in every algebra of* $\mathcal{A}lg_{Ctx}$ *if and only if* α *is true in every frame of* $\mathcal{F}rm_{Ctx}$.

3.5 Boolean algebras with possibility and sufficiency operators

In this section we consider Boolean algebras endowed with two modal operators of possibility and sufficiency.

A *PS-Boolean algebra* is an algebraic structure $(B, \Diamond, \boxminus)$ where $(B, \Diamond)$ is a P-Boolean algebra and $(B, \boxminus)$ is an S-Boolean algebra. A *PS-Boolean frame* is a structure (X, R, S) where (X, R) is a P-Boolean frame, (X, S) is an S-Boolean frame.

Given a PS-Boolean frame (X, R, S), its *complex algebra* is a structure $(2^X, \Diamond_R, \boxminus_S)$ where 2^X is the complex algebra of its Boolean-frame reduct and the operators $\Diamond_R$ and $\boxminus_S$ are defined as in Sections 3.2 and 3.3, respectively. On the other hand, given a PS-Boolean algebra $(B, \Diamond, \boxminus)$, its canonical frame is a structure $(X_B, R_\Diamond, S_\boxminus)$ where X_B is the canonical frame of its Boolean reduct and $R_\Diamond$ and $S_\boxminus$ are binary relations on X_B defined as in Sections 3.2 and 3.3, respectively.

It is easy to see that the complex algebra of any PS-frame is a PS-algebra and the canonical frame of any PS-algebra is a PS-frame. Furthermore, the usual discrete representations theorems for PS-algebras and PS-frames hold thus providing a discrete duality between these structures.

The original motivation for introducing an inaccessibility operator in [132], and a sufficiency operator in [98], was a need for a modal characterisation of the complemented relations along the lines of a Kripke-style semantics. For that purpose we consider PS$^+$-Boolean algebras obtained from PS-Boolean algebras by adding an axiom reflecting that one operator may be defined in terms of the other, namely, for every $a \in L$,

(PS) $\Diamond a = -\boxminus a$

and PS$^+$-frames obtained from PS-frames by adding

(FPS) $S = -R$.

Proposition 3.5.1

(a) *The complex algebra of a PS$^+$-Boolean frame is a PS$^+$-Boolean algebra.*
(b) *The canonical frame of a PS$^+$-Boolean algebra is a PS$^+$-Boolean frame.*

Proof:
(a) Note that by Lemma 1.8.1, $[\![S]\!]A = -\langle -S \rangle A$ for every $A \subseteq X$. Since by (FPS) $R = -S$, we get $\Diamond_R(A) = -\boxminus_S(A)$, as required.

(b) We have to show that $R_\Diamond = -S_\boxminus$, that is, $G \subseteq \Diamond^{-1}(F) \Leftrightarrow F \cap \boxminus(G) = \emptyset$ for all $F, G \in X_B$. Assume $G \subseteq \Diamond^{-1}(F)$ and $a \in F \cap \boxminus(G)$. Then $a \in F$ and $a = \boxminus b$ for some $b \in G$. By the assumption, $\Diamond b \in F$. So, by axiom (PS) and Lemma 1.6.12, $a = \boxminus b \notin F$, which is a contradiction. On the other hand, assume that $F \cap \boxminus(G) = \emptyset$ and let $a \in G$. Then $\boxminus a \in \boxminus(G)$, so $\boxminus a \notin F$. Thus $-\boxminus a \in F$, which by axiom (PS) means $\Diamond a \in F$, that is, $a \in \Diamond^{-1}(F)$. Hence $G \subseteq \Diamond^{-1}(F)$. □

Now, by Theorem 3.2.7 and Theorem 3.3.6 we obtain the discrete representations between PS$^+$-Boolean algebras and PS$^+$-Boolean frames.

Theorem 3.5.2

(a) *Every PS$^+$-Boolean algebra is embeddable into the complex algebra of its canonical frame.*
(b) *Every PS$^+$-Boolean frame is embeddable into the canonical frame of its complex algebra.*

Invoking the results of Sections 3.3 and 3.5, we may extend the discrete duality in Theorem 3.5.2 to a duality via truth.

3.6 Mixed algebras

Mixed algebras are PS-algebras such that their corresponding frames are of the form (X, R, R), written (X, R). In mixed algebras we are able to express properties of both a binary relation and its complement without explicitly introducing the complemented relation in the frames. The tradeoff is that a second order axiom is postulated in the definition of a mixed algebra.

Let $(B, \vee, \wedge, -, 0, 1)$ be a Boolean algebra and X_B be the set of all ultrafilters of B. The Boolean algebra $(2^{X_B}, \cup, \cap, -, \emptyset, X_B)$ is referred to as the *canonical extension of* B ([139]) or the *canonical embedding algebra* ([113, 109]). Recall the Stone embedding $h : B \to 2^{X_B}$ defined, for any $a \in B$, by $h(a) = \{F \in X_B : a \in F\}$.

For each unary operator $f : B \to B$, we may define two kinds of mappings over 2^{X_B}, namely, $f^\sigma : 2^{X_B} \to 2^{X_B}$ and $f^\pi : 2^{X_B} \to 2^{X_B}$ defined, for every $\mathsf{A} \subseteq X_B$, by

$$(\sigma)\ f^\sigma(\mathsf{A}) \stackrel{\text{def}}{=} \bigcup\{\bigcap\{h(f(a)) : a \in F\} : F \in \mathsf{A}\}$$

$$(\pi)\ f^\pi(\mathsf{A}) \stackrel{\text{def}}{=} \bigcap\{\bigcup\{h(f(a)) : a \in F\} : F \in \mathsf{A}\}.$$

In particular, if $F \in X_B$, then we have

$$(\sigma')\ f^\sigma(\{F\}) = \bigcap\{h(f(a)) : a \in F\}$$

$$(\pi')\ f^\pi(\{F\}) = \bigcup\{h(f(a)) : a \in F\}.$$

If f is a modal possibility operator $\Diamond$, then $\Diamond^\sigma$ is a complete possibility operator and if f is a sufficiency operator $\boxbar$, then $\boxbar^\pi$ is a complete sufficiency operator.

Following the developments of Section 3.2 we show the following lemma.

Lemma 3.6.1 *If $(B, \Diamond)$ is a Boolean algebra with a possibility operator, then its canonical extension $(2^{X_B}, \Diamond^\sigma)$ coincides with the complex algebra of the canonical frame of $(B, \Diamond)$.*

Proof:
Let $\mathsf{A} \subseteq X_B$. Then we have

$$\begin{aligned} F \in \Diamond^\sigma(\mathsf{A}) &\Leftrightarrow \exists G \in \mathsf{A},\ \forall a \in B,\ a \in G \Rightarrow \Diamond a \in F \\ &\Leftrightarrow \exists G \in \mathsf{A},\ \Diamond(G) \subseteq F \\ &\Leftrightarrow \exists G \in \mathsf{A},\ F\, R_\Diamond\, G \\ &\Leftrightarrow F \in \langle R_\Diamond \rangle \mathsf{A} \\ &\Leftrightarrow F \in \Diamond_{R_\Diamond}(\mathsf{A}), \end{aligned}$$

which completes the proof. □

Similarly, the developments of Section 3.3 lead to the following fact.

Lemma 3.6.2 *If $(B, \boxbar)$ is a Boolean algebra with a sufficiency operator, then the canonical extension $(2^{X_B}, \boxbar^\pi)$ coincides with the complex algebra of the canonical frame of $(B, \boxbar)$.*

A *mixed algebra* is a structure $(B, \Diamond, \boxbar)$ such that B is a Boolean algebra, $\Diamond$ is a possibility operator, $\boxbar$ is a sufficiency operator and, for every ultrafilter $F \in X_B$,

(Mix) $\Diamond^\sigma(\{F\}) = \boxed{\Box}^\pi(\{F\})$

where $\Diamond^\sigma$ and $\boxed{\Box}^\pi$ are defined as in (σ') and (π'), respectively.

Lemma 3.6.3 *Let $(B, \Diamond, \boxed{\Box})$ be a mixed algebra. Then, for every atom a of B, $\Diamond a = \boxed{\Box} a$.*

Proof:
Observe that if a is an atom of B, $\leq$ is the natural ordering on B, and h is a Stone mapping, then $h(a) = \{\uparrow_\leq a\}$ and it is an atom of 2^{X_B}. Hence, by (Mix), $\Diamond^\sigma(h(a)) = \boxed{\Box}^\pi(h(a))$. Applying Lemma 3.6.1, Lemma 3.6.2, Lemma 3.2.5, and Lemma 3.3.4 we get $h(\Diamond a) = h(\boxed{\Box} a)$. Since h is injective, $\Diamond a = \boxed{\Box} a$. ☐

Lemma 3.6.4 *Let $(B, \Diamond, \boxed{\Box})$ be a mixed algebra. Then for every $F \in X_B$ the following are equivalent*

(a) $\Diamond^\sigma(\{F\}) = \boxed{\Box}^\pi(\{F\})$;
(b) $\forall G \in X_B$, $\Diamond(G) \subseteq F \Leftrightarrow \boxed{\Box}(G) \cap F \neq \emptyset$.

Proof:
Note that

$$\begin{aligned}
&\Diamond^\sigma(\{F\}) = \boxed{\Box}^\pi(\{F\}\\
&\quad\Leftrightarrow \bigcap\{\{G \in X_B : \Diamond a \in G\} : a \in F\} = \bigcup\{\{G \in X_B : \boxed{\Box} a \in G\} : a \in F\}\\
&\quad\Leftrightarrow \{G : \forall a,\ a \in F \Rightarrow \Diamond a \in G\} = \{G : \exists a,\ a \in F \ \&\ \boxed{\Box} a \in G\}\\
&\quad\Leftrightarrow \forall G,\ [(\forall a,\ a \in F \Rightarrow \Diamond a \in G)) \Leftrightarrow (\exists a,\ a \in F \ \&\ \boxed{\Box} a \in G)]\\
&\quad\Leftrightarrow \forall G,\ (\Diamond(G) \subseteq F) \Leftrightarrow (\boxed{\Box}(G) \cap F \neq \emptyset),
\end{aligned}$$

which completes the proof. ☐

The purpose of the second order axiom (Mix) is to enable a single relation in the frames adequate for mixed algebras.

A *mixed frame* is a structure (X, R) where X is a non-empty set and R is a binary relation on X.

The *complex algebra of a mixed frame* (X, R) is a structure $(2^X, \Diamond_R, \boxed{\Box}_R)$ where 2^X is the complex algebra of its Boolean-frame reduct and, for every $A \subseteq X$,

$$\Diamond_R(A) \stackrel{\text{def}}{=} \langle R\rangle A$$

$$\boxed{\Box}_R(A) \stackrel{\text{def}}{=} [\![R]\!]A.$$

The complex algebra of a mixed frame is not necessarily a mixed algebra. It satisfies $\Diamond_R(\{x\}) = \boxed{\Box}_R(\{x\})$, but the axiom (Mix) is not necessarily satisfied. More precisely, $\Diamond^\sigma_R(\{F\})$ is not necessarily included in $\boxed{\Box}^\pi_R(\{F\})$, we only have $\boxed{\Box}^\pi_R(\{F\}) \subseteq \Diamond^\sigma_R(\{F\})$ for any ultrafilter F of 2^X. However, the complex algebra of the canonical frame of a mixed algebra is a mixed algebra due to the axiom (Mix) and Lemma 3.6.4. Hence, we have a discrete representation theorem for mixed algebras.

Lemma 3.6.5 *Let* $(2^X, \langle R\rangle, [\![R]\!])$ *be the complex algebra of a mixed frame. Then, for every* $F \in X_{2^X}$*, we have* $[\![R]\!]^\pi(\{F\}) \subseteq \langle R\rangle^\sigma(\{F\})$.

Proof:
Applying Lemma 3.6.4 it is sufficient to show that for every $G \in X_{2^X}$

$$\exists B \subseteq X,\ (B \in G\ \&\ [\![R]\!]B \in F) \Rightarrow \forall A \subseteq X,\ (A \in G \Rightarrow \langle R\rangle A \in F).$$

Take $A \in G$. Since $B \in G$ and G is a filter, $A \cap B \in G$. Since G is an ultrafilter, $A \cap B \neq \emptyset$. Hence for some $x \in X$, $\{x\} \subseteq A \cap B$. Then we have $[\![R]\!]B \subseteq [\![R]\!]\{x\} = \langle R\rangle\{x\} \subseteq \langle R\rangle A$. Since $[\![R]\!]B \in F$, $\langle R\rangle A \in F$, as required. □

The following example shows that the converse implication does not necessarily hold.

Example 3.6.6 *Consider a mixed frame* (X, R) *such that* $R = (X \times X) - Id_X$. *Then for every* $A \subseteq X$ *such that the cardinality of* A *is greater than 1 we have* $\langle R\rangle A = X$. *Thus* $\langle R\rangle G \subseteq F$ *for every* $F \in X_{2^X}$. *In particular,* $\langle R\rangle G \subseteq G$. *On the other hand,* $[\![R]\!]A = X - A$ *for every* $A \subseteq X$*, so that* $G \cap [\![R]\!]G = \emptyset$.

Thus the complex algebra of a mixed frame is not necessarily a mixed algebra. As a consequence, if we define a logic with possibility and sufficiency operators where frame semantics is determined by mixed frames, we will not be able to prove the complex algebra theorem of the form presented in Theorem 2.2.5.

The *canonical frame of a mixed algebra* $(B, \Diamond, \boxminus)$ is a structure (X_B, R_B) where for all $F, G \in X_B$,

$$F\, R_B\, G \overset{\text{def}}{\Leftrightarrow} \Diamond(G) \subseteq F.$$

Thus the relation R_B is defined as the relation $R_\Diamond$ in the canonical frame of the P-Boolean algebra reduct of the mixed algebra under consideration.

Observe that by the axiom (Mix) and Lemma 3.6.4 we have

$$F\, R_B\, G \Leftrightarrow \boxminus(G) \cap F \neq \emptyset.$$

This means that R_B coincides with the relation $R_\boxminus$ in the canonical frame of the S-Boolean algebra reduct of the mixed algebra under consideration.

Hence the axiom (Mix) says that the relation $R_\Diamond$ in the canonical frame of a possibility algebra (see Section 3.2) coincides with the relation $R_\boxminus$ from the canonical frame of a sufficiency algebra (see Section 3.3). Thus we have $R_B = R_\Diamond = R_\boxminus$.

Clearly, the canonical frame of a mixed algebra is a mixed frame. Moreover, the complex algebra of the canonical frame is a mixed algebra.

The representation theorem for mixed algebras will be provided by the Stone mapping h, used in Section 2.3. We show that h preserves the operators $\Diamond$ and $\boxminus$ of mixed algebras.

Lemma 3.6.7 *Let $(B, \Diamond, \boxbar)$ be a mixed algebra. Then the mapping h preserves the operators $\Diamond$ and $\boxbar$.*

Proof:
By way of example we show the preservation of the sufficiency operator $\boxbar$, that is $h(\boxbar a) = \boxbar_{R_B}(h(a))$.
First, observe that for every $F \in X_B$,

$$
\begin{aligned}
F \in \boxbar_{R_B}(h(a)) &\Leftrightarrow F \in [\![R_B]\!]h(a) \\
&\Leftrightarrow \forall G \in X_B,\ G \in h(a) \Rightarrow F\, R_B\, G \\
&\Leftrightarrow \forall G \in X_B,\ a \in G \Rightarrow \boxbar(G) \cap F \neq \emptyset
\end{aligned}
$$

where the last equivalence is due to the axiom (Mix) and Lemma 3.6.4. Then the proof can be easily obtained from the proof of Lemma 3.3.4. The proof of preservation of the possibility operator $\Diamond$ follows from Lemma 3.2.5. □

For the representation theorem for mixed frames (X, R) we follow the developments of Section 3.2 and Section 3.3. We show that the mapping $k : X \to X_{2^X}$ preserves the relation R, that is, $xRy \Leftrightarrow k(x) R_{2^X} k(y)$. Since R_{2^X} is defined as in Section 3.2 we have, for every $F, G \in X_{2^X}$,

$$F\, R_{2^X}\, G \Leftrightarrow \langle R_B \rangle G \subseteq F \Leftrightarrow \{\langle R \rangle A : A \in G\} \subseteq F \Leftrightarrow (A \in G \Rightarrow \langle R \rangle A \in F).$$

Thus we can follow the proof of Theorem 3.2.6. Concluding, we have the following discrete representations for mixed algebras and mixed frames.

Theorem 3.6.8

(a) *Every mixed algebra is embeddable into the complex algebra of its canonical frame;*

(b) *Every mixed frame is embeddable into the canonical frame of its complex algebra.*

Lemma 3.6.9 *Let $(B, \Diamond, \boxbar)$ be a mixed algebra. Then, for every $a, b \in B$, if $a \wedge b \neq 0$, then $\boxbar a \leq \Diamond b$.*

Proof:
Let $(B, \Diamond, \boxbar)$ be a mixed algebra. Due to Lemma 3.6.1, Lemma 3.6.2, and Lemma 3.6.4, $(2^{X_B}, \Diamond_{R_\Diamond}, \boxbar_{R_\boxbar})$ is a complete mixed algebra. Every universal property that holds in 2^{X_B} holds in each of its subalgebras. By the representation theorem, B is isomorphic with a subalgebra of 2^{X_B}. Therefore it is sufficient to prove that the required condition holds in 2^{X_B}.

Let $A, B \in 2^{X_B}$. Since 2^{X_B} is atomic, there is an atom $\{F\}$, $F \in X_B$, such that $\{F\} \subseteq A \cap B$. Since $\Diamond_{R_\Diamond}$ is monotone, $\boxbar_{R_\boxbar}$ is antitone, by Lemma 3.6.3 we have $\boxbar_{R_\boxbar} A \subseteq \boxbar_{R_\boxbar}(A \cap B) \subseteq \boxbar_{R_\boxbar}(\{F\}) = \Diamond_{R_\Diamond}(\{F\}) \subseteq \Diamond_{R_\Diamond}(A \cap B) \subseteq \Diamond_{R_\Diamond} B$. □

As observed earlier in this section, the complex algebra theorem does not hold in a usual logic with possibility and sufficiency operators whose semantics is determined by mixed frames. Therefore in that case Theorem 3.6.8 cannot be extended to duality via truth.

3.7 An axiomatic extension of mixed algebras

Now we present an axiomatic extension of mixed algebras, referred to as *a right ideal mixed algebras*, such that the complex algebras of their frames are in the class under consideration. In their frames (X, R) the relation R is a *right ideal relation on* X, that is a relation satisfying

$$R\,;(X \times X) = R.$$

The right ideal relations were introduced in [233] and they are extensively used in relational proof systems for non-classical logics, see [174].

A *right ideal mixed algebra* is a mixed algebra $(B, \Diamond, \boxdot)$ with the following additional axiom

(Ideal) $f(1) \leq g(1)$.

A *right ideal mixed frame* is a structure (X, R) with a right ideal relation R. We show that discrete duality holds for right ideal mixed algebras and their frames.

The complex algebra of a right ideal frame (X, R) and the canonical frame of the right ideal mixed algebra $(B, \Diamond, \boxdot)$ are defined as for mixed frames and mixed algebras, respectively.

Proposition 3.7.1 *The complex algebra of a right ideal mixed frame is a right ideal mixed algebra.*

Proof:
It is sufficient to show that the axioms (Ideal) and (Mix) hold.
(Ideal) It is easy to show that if R is a right ideal relation, then for every $A \subseteq X$, $\langle R\rangle A \subseteq [\![R]\!]A$. In particular, $\langle R\rangle X \subseteq [\![R]\!]X$.

(Mix) We will prove the equivalent form of this axiom presented in Lemma 3.6.4. In view of Lemmas 3.6.3 and 3.6.5 it suffices to show that for all ultrafilters F and G of 2^X,

$$(\forall A \subseteq X,\ A \in G \Rightarrow \langle R\rangle A \in F) \ \Rightarrow\ (\exists B \subseteq X,\ B \in G\ \&\ [\![R]\!]B \in F).$$

Since G is an ultrafilter, $X \in G$, so $\langle R\rangle X \in G$. Since R is a right ideal relation, $\langle R\rangle X \subset [\![R]\!]X$. Thus taking X for B completes the proof. □

Proposition 3.7.2 *The canonical frame of a right ideal mixed algebra is a right ideal frame.*

Proof:
We show that $R_{\boxdot}\,;(X_B \times X_B) = R_{\Diamond}$ where the relations $R_{\Diamond}$ and $R_{\boxdot}$ are defined, for $F, G \in X_B$, by $FR_{\Diamond}G \Leftrightarrow \Diamond(G) \subseteq F$ and $FR_{\boxdot}G \Leftrightarrow \boxdot(G) \cap F \neq \emptyset$, defined in Section 3.2 and Section 3.3, respectively. Since $R_{\Diamond} = R_{\boxdot}$ by (Mix) and since Lemma 3.6.4 holds, proving the above equality will show that R_B is a right ideal relation.

Assume there is some $H \in X_B$ such that $FR_{\Box}H$ and $H(X_B \times X_B)G$. Then there is some $a \in B$ such that $a \in H$ and $\Box a \in F$. By (Mix) and Lemma 3.6.4, for every $b \in B$, $b \in H$ implies $\Diamond b \in F$. Since $1 \in H$, $\Diamond 1 \in F$. By (Ideal), $\Box 1 \in F$. We show that $FR_{\Diamond}G$ holds. Take $b \in G$. Since $b \leq 1$, $\Box 1 \leq \Box b$. Since $\Box 1 \in F$, $\Box b \in F$. By Lemma 3.6.4, $\Box b \leq \Diamond b$, thus $\Diamond b \in F$. Hence $R_{\Box} ; (X_B \times X_B) \subseteq R_{\Diamond}$. The reverse inclusion follows directly from the property $R \subseteq R; (X \times X)$ of every binary relation R on X. □

The following discrete representations for right ideal mixed algebras and right ideal mixed frames are analogous to those in Theorem 3.3.6.

Theorem 3.7.3

(a) *Every right ideal mixed algebra is embeddable into the complex algebra of its canonical frame;*

(b) *Every right ideal mixed frame is embeddable into the canonical frame of its complex algebra.*

3.8 Preference algebras

A *preference algebra, Pref-algebra,* is a structure $(B, \Diamond_1, \Box_1, \Diamond_2, \Box_2)$ such that $(B, \Diamond_1, \Box_1)$ and $(B, \Diamond_2, \Box_2)$ are mixed algebras satisfying, for every $a \in B$,

(Pref1) $a \wedge \Diamond_1(\Box_1(a)) = 0$

(Pref2) $a \wedge \Diamond_1(\Box_2(a)) = 0$

(Pref3) $a \leq \Diamond_2(a)$

(Pref4) $a \leq \Box_2(\Box_2(a))$.

A *preference frame, Pref-frame,* is a structure (X, P, I) where (X, P) and (X, I) are mixed frames corresponding to mixed algebras $(B, \Diamond_1, \Box_1)$ and $(B, \Diamond_2, \Box_2)$, respectively, and the following conditions are satisfied

(FPref1) $P \cap P^{\smile} = \emptyset$

(FPref2) $P \cap I = \emptyset$

(FPref3) I is reflexive

(FPref4) I is symmetric.

The *complex algebra of a Pref-frame* X is a structure $(2^X, \Diamond_P, \Box_P, \Diamond_I, \Box_I)$ such that $(2^X, \Diamond_P, \Box_P)$ and $(2^X, \Diamond_I, \Box_I)$ are the complex algebras of mixed frames (X, P) and (X, I), respectively, and for every $A \subseteq X$,

$$\Diamond_P(A) \stackrel{\text{def}}{=} \langle P \rangle A \qquad \Box_P(A) \stackrel{\text{def}}{=} [\![P]\!]A$$
$$\Diamond_I(A) \stackrel{\text{def}}{=} \langle I \rangle A \qquad \Box_I(A) \stackrel{\text{def}}{=} [\![I]\!]A.$$

Since the complex algebra of a mixed frame is not necessarily a mixed algebra, we are only able to show the following two lemmas.

Lemma 3.8.1 *The complex algebra of a Pref-frame satisfies (Pref1) – (Pref4).*

Proof:
(Pref1) Let $A \subseteq X$, let $x \in A$, and suppose that $x \in \langle P \rangle [\![P]\!] A$. Then there is some $y \in X$ such that xPy and $y \in [\![P]\!] A$. From the latter we obtain that for every $z \in X$, $z \in A$ implies yPz. Since $x \in A$, we obtain yPx which together with xPy contradicts asymmetry of P expressed with (FPref1).

(Pref2) Let $A \subseteq X$, let $x \in A$, and suppose $x \in \langle P \rangle [\![I]\!] A$. Reasoning in a similar way as in the proof of (Pref1) there is some $y \in X$ such that xPy and yIx. By (FPref4) I is symmetric, so we also have xIy contradicting (FPref2).

(Pref3) Follows from (FPref3).

(Pref4) Follows from (FPref4). □

Proposition 3.8.2 *The complex algebra of the canonical frame of a Pref-algebra is a Pref-algebra.*

Proof:
Axiom (Mix) and Lemma 3.6.4 imply that the complex algebra of the canonical frame of a mixed algebra is a mixed algebra. Then, in view of Lemma 3.8.1, the required result follows. □

The canonical frame of a Pref-algebra B is a structure (X_B, P_B, I_B) where (X_B, P_B) and (X_B, I_B) are the canonical frames of mixed algebras $(B, \Diamond_1, ⊡_1)$ and $(B, \Diamond_2, ⊡_2)$, respectively, defined in Section 3.6.

Proposition 3.8.3 *The canonical frame of a preference algebra is a preference frame.*

Proof:
Since the canonical frame of a mixed algebra is a mixed frame, it is sufficient to show that the frame axioms (FPref1) – (FPref4) are satisfied.

It is well known that (Pref3) implies reflexivity of I and (Pref4) implies symmetry of I which yield (FPref3) and (FPref4).

(FPref1) Suppose that P_B is not asymmetric. Then there are ultrafilters $F, G \in X_B$ such that $\Diamond_1(G) \subseteq F$ and $\Diamond_1(F) \subseteq G$. By the latter condition and Lemma 3.6.4 there is some $a \in B$ such that $⊡_1(a) \in G$. From $\Diamond_1(G) \subseteq F$ we obtain $\Diamond_1(⊡_1(a)) \in F$. Since $a \in F$ and F is a proper filter, we have $a \wedge \Diamond_1 ⊡_1 a \neq 0$ which contradicts (Pref1).

(FPref2) Suppose that there are $F, G \in X_B$ such that $F \, P_B \, G$ and $F \, I_B \, G$. Since I_B is symmetric, we also have $G \, I_B \, F$, that is $\Diamond_1(G) \subseteq F$ and $\Diamond_2(F) \subseteq G$. Then, by Lemma 3.6.4, $G \cap ⊡_2(F) \neq \emptyset$. Thus there is some $a \in F$ such that $⊡_2 a \in G$. Since $\Diamond_1(G) \subseteq F$, we get $\Diamond_1 ⊡_2 a \in F$ and hence $a \wedge \Diamond_1 ⊡_2 a \in F$. Since F is a proper filter, this contradicts (Pref2). □

As usual, the representation theorems for Pref-algebras and for Pref-frames are provided by the mappings h and k, respectively, discussed in Section 2.3. The preservation of the operations of Pref-algebras by the mapping h and the preservation of relations in Pref-frames by the mapping k can be proved as in Section 3.2 and Section 3.3, respectively. Therefore we have the following discrete representations for Pref-algebras and Pref-frames.

Theorem 3.8.4

(a) *Every Pref-algebra is embeddable into the complex algebra of its canonical frame;*

(b) *Every Pref-frame is embeddable into the canonical frame of its complex algebra.*

3.9 Event algebras

In this section we propose a class of algebras such that their associated frames are event structures. Event structures belong to the class of noninterleaving models of concurrency, which take into account the fact that a modeled system may consist of a family of independent subsystems. In an event structure the behavior of a system is modeled in terms of a set of events, which represent occurrences of actions, a causal ordering among events, and a conflict relation modeling exclusions between events. We prove discrete duality between event algebras and event structures inspired by a temporal logic of event structures presented in [196].

An *event algebra, E-algebra,* is a structure $(B, f^{\leqslant}, f^{\geqslant}, g^{\#}, r)$ where B is a Boolean algebra, $r \in B$, and, for every $a \in B$,

(EA1) $f^{\leqslant}$ and $f^{\geqslant}$ are possibility operators

(EA2) $g^{\#}$ is a sufficiency operator

(EA3) $a \leqslant f^{\leqslant}(a)$ and $a \leqslant f^{\geqslant}(a)$

(EA4) $f^{\leqslant}(f^{\leqslant}(a)) \leqslant f^{\leqslant}(a)$ and $f^{\geqslant}(f^{\geqslant}(a)) \geqslant f^{\geqslant}(a)$

(EA5) $a \leqslant -f^{\leqslant}(-f^{\geqslant}(a))$ and $a \leqslant -f^{\geqslant}(-f^{\leqslant}(a))$

(EA6) $r \leqslant g^{\#}(-r)$

(EA7) $g^{\#}(a) \leqslant -a$

(EA8) $a \leqslant g^{\#}(g^{\#}(a))$

(EA9) $g^{\#}(a) \leqslant -f^{\#}(-g^{\#}(a))$.

An *event structure, E-frame,* is a structure $(X, \leqslant, \#, R)$ such that X is a non-empty set of events, $\leqslant$ is a binary relation on X called a *causality relation,* $\#$ is a binary relation on X called a *conflict relation,* $R \subseteq X$ is called a *run,* and the following conditions hold

(E1) $\leqslant$ is reflexive and transitive

(E2) $\#$ is irreflexive and symmetric

(E3) R is a maximal subset of X such that $(R \times R) \cap \# = \emptyset$

(E4) $\forall x, y, z \in X,\ x \# y \ \&\ y \leqslant z \to x \# z$.

Note that the axiom (E3) says that R is a maximal conflict-free subset of X. It can be equivalently replaced by

(E3') $\forall x \in X,\ x \in R \Leftrightarrow \forall y \in X, y \notin R \to x \# y$.

The axiom (E4), known as the *conflict inheritance* property, states that if an event x is in conflict with some event y then it is in conflict with causal successors of y.

The *complex algebra of an E-frame* $(X, \leqslant, \#, R)$ is a structure

$$
\begin{aligned}
f_X^{\leqslant}(A) &\stackrel{\text{def}}{=} \langle\leqslant\rangle A \\
f_X^{\geqslant}(A) &\stackrel{\text{def}}{=} \langle\geqslant\rangle A \\
g_X^{\#}(A) &\stackrel{\text{def}}{=} [\![\#]\!] A \\
r_X &\stackrel{\text{def}}{=} R.
\end{aligned}
$$

Proposition 3.9.1 *The complex algebra of an E-frame is an E-algebra.*

Proof:
(EA1) and (EA2) follow from the definitions of $f^{\leqslant}, f^{\geqslant}$, and $g^{\#}$. Also, from the reflexivity of $\leqslant$ and Lemma 1.10.1(b), we obtain (EA3) and from the transitivity of $\leq$ and Lemma 1.10.1(g), (EA4) follows. Next, by Lemma 1.8.4 we obtain (EA5). (EA6) follows from (E3'). Furthermore, by irreflexivity of $\#$ and Lemma 1.10.1(c), we obtain (EA7). Symmetry of $\#$ and Lemma 1.10.1(f) imply (EA8). Finally, for (EA9), we have to show that $[\![\#]\!]A \subseteq -\langle\leqslant\rangle(-[\![\#]\!]A)$. Let $x \in X$ be such that $x \leqslant y$, $z \in A$, and not $y \# z$. From $z \in A$ and $x \in [\![\#]\!]A$ it follows that $x \# z$ so, by symmetry of $\#$, $z \# x$. Hence, since $x \leqslant y$, by (E4) $z \# y$ which implies $y \# z$, a contradiction. □

The *canonical frame of an E-algebra* $(B, f^{\leqslant}, f^{\geqslant}, g^{\#}, r)$ is a structure

$$(X_B, \leqslant_B, \#_B, R_B)$$

where X_B is the set of ultrafilters of B, $\leqslant$ and $\#_B$ are binary relations on X_B and $R_B \subseteq X_B$ such that, for all $F, G \in X_B$,

$$
\begin{aligned}
F \leq_B G &\Leftrightarrow G \subseteq (F^{\leqslant})^{-1}(F) \\
F \#_B G &\Leftrightarrow F \cap g^{\#}(G) \neq \emptyset \\
R_B &= \{F \in X_B : r \in F\}.
\end{aligned}
$$

Proposition 3.9.2 *The canonical frame of an E-algebra is an E-frame.*

Proof:
(E1) For reflexivity of $\leqslant_B$, take $F \in X_B$ and $a \in B$ such that $a \notin (f^{\leqslant})^{-1}(F)$. Thus $f^{\leqslant}(a) \notin F$. By (EA3), $a \notin F$. For transitivity of $\leqslant_B$, take $F, G, H \in X_B$

such that $G \subseteq (f^{\leqslant})^{-1}(F)$ and $H \subseteq (f^{\leqslant})^{-1}(G)$. Let $a \in H$. Then $f^{\leqslant}(a) \in G$, and furthermore, $f^{\leqslant}(f^{\leqslant}(a)) \in F$. Hence, by (EA4), $f^{\leqslant}(a) \in F$, thus $a \in (f^{\leqslant})^{-1}(F)$. Therefore, $F \leqslant_B H$.

(E2) Let $F \in X_B$ and take $a \in F$. Then $-a \notin F$, so by (EA7), $g^{\#}(a) \notin F$. Hence, $F \cap g^{\#}(F) = \emptyset$, so $\#_B$ is irreflexive. For symmetry of $\#_B$, assume that $F \cap g^{\#}(G) \neq \emptyset$. Then there is some $a \in G$ such that $g^{\#}(a) \in F$. Hence, by (EA8), $g^{\#}(g^{\#}(a)) \in G$. Since $g^{\#}(g^{\#}(a)) \in g^{\#}(F)$ we obtain $G \cap g^{\#}(F) \neq \emptyset$.

(E3) We have to show that, for every $F \in X_B$,

$$r \in F \;\Leftrightarrow\; \forall G \in X_B, F \cap g^{\#}(G) \neq \emptyset \Rightarrow r \in G.$$

Assume that $r \in F$ and take any $G \in X_B$ such that $F \cap g^{\#}(G) = \emptyset$. Then, for $a \in B$, $g^{\#}(a) \in F$ implies $a \notin G$. Since $r \in F$, by (EA6), $g^{\#}(-r) \in F$, so $-r \notin G$, hence $r \in G$. On the other hand, assume that $F \in [\![\#_B]\!](-R_B)$, and take any $G \in X_B$ such that $F \cap g^{\#}(G) = \emptyset$. Then $r \in G$, so by (EA6), $g^{\#}(-r) \in G$. Also, $g^{\#}(g^{\#}(-r)) \notin F$ so, by (EA8), $r \in F$.

(E4) Let $F, G, H \in X_B$. We have to show that $F \cap g^{\#}(G) \neq \emptyset$ and $H \subseteq (f^{\#})^{-1}(G)$ imply $F \cap g^{\#}(H) = \emptyset$. By assumption, there is some $a \in G$ such that $g^{\#}(a) \in F$. Since $a \in G$, by (EA8), $g^{\#}(g^{\#}(a)) \in G$. Thus, by (EA9), $-f^{\leqslant}(-g^{\#}(g^{\#}(a))) \in G$. Then $f^{\leqslant}(-g^{\#}(g^{\#}(a))) \notin G$. Since $H \subseteq (f^{\#})^{-1}(G)$, we have $-g^{\#}(g^{\#}(a)) \notin H$, so $g^{\#}(g^{\#}(a)) \in H$. Moreover, by (EA8), $g^{\#}(a) \in F$ implies $g^{\#}(g^{\#}(g^{\#}(a))) \in F$. Hence $F \cap g^{\#}(H) \neq \emptyset$. □

As usual, the representation theorems for E-algebras and for E-frames are provided by the mappings h and k, respectively, discussed in Section 2.3. The preservation of the operations $f^{\leqslant}, f^{\geqslant}, g^{\#}$ and the distinguished element r of E-algebras by the mapping h is proved in Lemma 3.9.3; while the preservation of relations $\leqslant$ and $\#$ and the distinguished set R of E-frames by the mapping k is proved in Lemma 3.9.4.

Lemma 3.9.3 *Let B be an E-algebra. For every $a \in B$,*

(a) $h(f^{\leqslant}(a)) = f^{\leqslant}_{X_B}(h(a))$;

(b) $h(f^{\geqslant}(a)) = f^{\geqslant}_{X_B}(h(a))$;

(c) $h(g^{\#}(a)) = g^{\#}_{X_B}(h(a))$;

(d) $h(r) = R_B$.

Proof:
(a), (b) and (c) follow from Lemma 3.2.5. For (d), note that, for every $F \in X_B$, $F \in h(r)$ iff $r \in F$ iff $F \in R_B$. □

Lemma 3.9.4 *Let X be an E-frame. For all $x, y \in X$,*

(a) $x \leqslant y \;\Leftrightarrow\; k(x) \leqslant_{2^X} k(y)$;

(b) $x \# y \;\Leftrightarrow\; k(x) \#_{2^X} k(y)$;

(c) $x \in R \;\Leftrightarrow\; k(x) \in R_{2X}$.

Proof:
(a) and (b) follow from Lemma 3.2.6 . For (c), take any $x \in X$ and observe that $x \in R$ iff $R \in k(x)$ iff $r_X \in k(x)$ iff $k(x) \in R_{2X}$. □

We conclude with the discrete representations for E-algebras and E-frames.

Theorem 3.9.5

(a) *Every E-algebra is embeddable into the complex algebra of its canonical frame;*

(b) *Every E-frame is embeddable into the canonical frame of its complex algebra.*

Chapter 4

Boolean algebras with information operators

4.1 Introduction

In this chapter a discrete duality for some structures inspired by the rough set theory originated by Pawlak [195] are considered. The fundamental concepts of the theory are a formal information system, an approximation space derived from an information system, and approximation operations determined by an approximation space. An information system is a collection of objects together with their properties expressed in terms of attributes and their values. Given an information system, an approximation space derived from the system is a pair consisting of the set of objects of the system and an equivalence relation on it, referred to as an indiscernibility relation, such that two objects are indiscernible whenever for every attribute of the information system they assume the same value of the attribute. Equivalence classes of the indiscernibility relation determine granularity of knowledge about the objects derived from information about them provided in the information system. The objects in an equivalence class cannot be distinguished as individual entities on the basis of that information. As a consequence, subsets of objects can be characterized only approximately. A subset of objects is definable in an information system whenever it is the union of some equivalence classes of the indiscernibility relation derived from this system. Usually, not all of the subsets of the set of objects are definable. The approximation operations enable us to define for every subset of objects a pair of definable sets which, intuitively, are as close as possible to the original set. The lower approximation of a subset, say A, of objects is the greatest definable set included in A and the upper approximation of A is the smallest definable set containing A.

In Section 4.2 we consider a general notion of an information system which allows for the objects assuming subsets of values of the attributes, not necessarily single values. In such systems, apart from the indiscernibility relations, many other classes of meaningful relations are derivable, see [40]. The relations are determined by subsets of attributes of the system and therefore they are referred to as relative relations. In Section 4.3 approximation operations determined by the indiscernibility relations and various types of definability of sets of objects are presented. In Section 4.4 a family of knowledge operators derived from an information system is presented and their properties are discussed. The operators enable us to estimate granularity of knowledge. They depend on the subsets of attributes which are taken into account by an agent in the process of derivation of knowledge from an information system. In Section 4.5 Boolean algebras with a knowledge operator introduced in [178] and the corresponding frames are presented and a discrete duality between these two classes is proved. The operator is an abstract counterpart to the knowledge operators discussed in Section 4.4. It is shown that the operator may be seen as the Hintikka operator 'knowing whether', see [129]. In the subsequent three sections the Boolean algebras with operators which enable us a characterization of relations of diversity, weak similarity, and strong right orthogonality are presented. The corresponding classes of frames are defined and discrete dualities between the algebras and frames are proved. Diversity algebras presented in Section 4.6 are axiomatic extensions of Boolean algebras with a sufficiency operator considered in Section 3.3. The relation in the corresponding frames is irreflexive, symmetric, and co-transitive. Algebras of weak similarity considered in Section 4.7 are signature and axiomatic extensions of possibility Boolean algebras discussed in Section 3.2. The corresponding frames are endowed with a family of weak relative relations which are weakly reflexive and symmetric. In the last Section 4.8 the algebras of strong right orthogonality are considered. They are signature and axiomatic extensions of sufficiency Boolean algebras presented in Section 3.3. The relations in the corresponding frames are strong relative relations, co-weakly reflexive, and symmetric.

4.2 Information systems

In many applications the available data are represented as a collection of objects together with their properties. Formally, such an information is a triple $\Sigma = (Ob, \mathcal{A}t, \{ Val_a : a \in \mathcal{A}t\})$ where Ob is a set of objects, $\mathcal{A}t$ is a set of attributes and Val_a, $a \in \mathcal{A}t$, is a set of values of an attribute a. More specifically, each attribute $a \in \mathcal{A}t$ is a mapping $a : Ob \to Val_a$ and the pair $(a, a(x))$ is interpreted as a property of an object x. In some cases an attribute may assign a subset of values to an object. For example, the attribute 'foreign language' may assign to a person several languages he/she speaks. Similarly, if the age of a person is known approximately as belonging to an interval of natural numbers, then the attribute 'age' assigns to a person the set of the elements of that interval. In the theory of rough sets originated by Pawlak ([194, 195]) structures of the form Σ are referred to as *information systems.* In the following we will use the more concise notation $(Ob, \mathcal{A}t)$ instead of $(Ob, \mathcal{A}t, \{ Val_a : a \in \mathcal{A}t\})$.

The queries to an information system often have the form of a request for finding a set of objects whose sets of attribute values satisfy some conditions. This leads to the notion of information relation derived from an information system and determined by a set of attributes of the system. A comprehensive survey of information relations can be found in [40].

Let $a(x)$ and $a(y)$ be sets of values of an attribute a of the objects x and y, respectively. We may want to know a set of those objects from an information system whose sets of values of all (or some) of the attributes from a subset A of attributes are equal (or disjoint, or overlap etc.). To represent such queries we define, first, information relations on the set of objects and, second, information operators determined by those relations. The first natural class of information relations is the class of *indiscernibility relations* which reflect a kind of sameness of objects. An indiscernibility relation determined by $a \in \mathcal{A}t$ is defined as

$$(x, y) \in ind(a) \overset{\text{def}}{\Leftrightarrow} a(x) = a(y).$$

Next, we can extend this relation to any subset A of attributes so that a quantification over A is added:

$$(x, y) \in ind(A) \overset{\text{def}}{\Leftrightarrow} a(x) = a(y) \quad \text{for all (some) } a \in A.$$

Relations defined with the universal (resp. existential) quantifier are referred to as strong (resp. weak) relations and if needed they will be prefixed with 's' or 'w', respectively.

In an abstract setting we take a set of sets 2^{Par} as an index set, where each set $P \subseteq Par$ is intuitively viewed as a set of attributes of objects in an information system. Then strong or weak relations are defined axiomatically. Collectively they are referred to as relative relations.

It is clear that every strong indiscernibility relation is an equivalence relation and hence it determines a partition of the set Ob.

The variety of other relations derived from information systems were studied, see, for example, [40]. In this chapter we consider three of these classes, namely diversity relations, similarity relations, and right orthogonality relations.

The complements of indiscernibility relations are referred to as *diversity relations.* The algebras characterizing diversity relations are presented in Section 4.6.

Similarity relations are defined as

$$(x, y) \in sim(a) \overset{\text{def}}{\Leftrightarrow} a(x) \cap a(y) \neq \emptyset.$$

We extend this relation to any subset A of attributes.

$$(x, y) \in sim(A) \overset{\text{def}}{\Leftrightarrow} a(x) \cap a(y) \neq \emptyset \text{ for all (some) } a \in A.$$

The algebras characterizing the weak similarity relations are considered in Section 4.7.

The other class of relations derived from an information system are relations of *right orthogonality* defined as follows. For objects x and y of an information system and an attribute a,

$$(x, y) \in rort(a) \overset{\text{def}}{\Leftrightarrow} a(x) \subseteq -a(y).$$

It follows that $(x,y)\in rort(a)$ if and only if $a(x)\cap a(y)=\emptyset$. Therefore the right orthogonality relation is sometimes referred to as the disjointness relation.

For a subset A of attributes we may define strong (weak) relations by

$$(x,y)\in rort(A) \overset{\text{def}}{\Leftrightarrow} a(x)\subseteq -a(y) \text{ for all (some) } a\in A.$$

The algebras characterizing the strong right orthogonality relations are considered in Section 4.8.

4.3 Approximation operations

Given a set $X\subseteq Ob$, usually it can be characterized only approximately in terms of the attributes from a set $A\subseteq \mathcal{A}t$. In the rough set theory, two approximation operators are defined for every $X\subseteq Ob$ and for every $A\subseteq \mathcal{A}t$,

$$L_A(X) \overset{\text{def}}{=} \{x\in Ob : \forall y\in Ob,\ (x,y)\in ind(A) \Rightarrow y\in X\}$$
$$U_A(X) \overset{\text{def}}{=} \{x\in Ob : \exists y\in Ob,\ (x,y)\in ind(A)\ \&\ y\in X\}.$$

For any $X\subseteq Ob$, $L_A(X)$ is an A–lower approximation of X and $U_A(X)$ is an A–upper approximation of X. A well known interpretation of these operators is that the objects in $L_A(X)$ are those which certainly belong to X, whereas objects in $U_A(X)$ can only be seen as possible members of X in view of properties of the elements of X determined by the attributes from $A\subseteq \mathcal{A}t$. Note that L_A and U_A are modal operations of necessity and possibility (see Section 1.8), respectively, determined by $ind(A)$.

Given an information system $(Ob, \mathcal{A}t)$ and a set $A\subseteq \mathcal{A}t$ of attributes we define, for any set $X\subseteq Ob$, three regions of certainty determined by A, by

$$Pos_A(X) = L_A(X)$$
$$Neg_A(X) = Ob - U_A(X)$$
$$B_A(X) = U_A(X) - L_A(X).$$

Following [171, 172], these are called *positive*, *negative*, and *borderline instances of* X, respectively. Given a set A of attributes, the elements of $Pos_A(X)$ definitely belong to X, the elements of $Neg_A(X)$ definitely do not belong to X, and $B_A(X)$ is the region of uncertainty: we can only say that the elements of $B_A(X)$ possibly belong to X, but we cannot decide it for certain.

In terms of the approximations, we define several types of definability and indefinability of sets. Given an information system $(Ob, \mathcal{A}t)$ and a set $A\subseteq \mathcal{A}t$ of attributes, we say that a set $X\subseteq Ob$ of objects is

A–definable iff $Pos_A(X)=X$
roughly A-definable iff $Pos_A(X)\neq\emptyset$, $Neg_A(X)\neq\emptyset$, and $B_A(X)\neq\emptyset$
bottom A-indefinable iff $Pos_A(X)=\emptyset$
top A-indefinable iff $Neg_A(X)=\emptyset$
totally A-indefinable iff $Pos_A(X)=Neg_A(X)=\emptyset$.

If X is A-definable, then we can precisely determine the properties of all its elements expressed by attributes in A. However, if it is roughly A-definable, we are only able to say which properties its elements certainly possess and which ones they certainly do not possess but we may not be able to determine all their properties. Accordingly, for the bottom (resp. top) A-indefinable set X we cannot indicate properties which it certainly possesses (resp. does not possess). In the last case we are unable to say anything about properties of X.

The following lemmas provide a characterization of degrees of definability of sets.

Lemma 4.3.1 *For every information system* $(Ob, \mathcal{A}t)$*, for every* $A \subseteq \mathcal{A}t$*, and for every* $X \subseteq OB$*, the following statements are equivalent:*

(a) X *is* A*-definable;*

(b) $U_A(X) = X$*;*

(c) $B_A(X) = \emptyset$*;*

(d) $Pos_A(X) = Ob - Neg_A(X)$*;*

(e) $X = \bigcup_{x \in X} [x]_A$*, where* $[x]_A$ *is an equivalence class of* $ind(A)$ *determined by* $x \in Ob$*.*

Lemma 4.3.2 *For every information system* $(Ob, \mathcal{A}t)$*, for every* $A \subseteq \mathcal{A}t$*, and for every* $X \subseteq Ob$*,*

(a) X *is roughly* A*-definable if and only if* $B_A(X) \neq \emptyset$*;*

(b) $X \neq \emptyset$ *is bottom* A*-indefinable if and only if* $B_A(X) = Ob - Neg_A(X)$ *and* $X \neq \emptyset$*;*

(c) $X \neq Ob$ *is top* A*-indefinable if and only if* $B_A(X) = Ob - Pos_A(X)$ *and* $X \neq Ob$*;*

(d) X *is totally* A*-indefinable if and only if* $B_A(X) = Ob$*.*

4.4 Knowledge operator

In the literature there is an extensive discussion of characteristics and formal representations of knowledge (see, for example, [129, 153, 210, 248]). In artificial intelligence and computer science, logics of knowledge (epistemic logics) are usually modal logics where knowledge operator is identified with the modal operator of necessity. The modal system S5 is a familiar logic to model knowledge ([81, 165, 245, 249]). This system has nice mathematical properties which often motivate researchers to adopt it in their exploration of the field. However, it models an idealized notion of knowledge. In particular, it suffers from the problem of *veridicality* – what the agent knows is true – and the problem of *logical omniscience* – the agent knows all logical consequences of his knowledge. In the logical literature veridicality is represented by a formula $K_a\alpha \to \alpha$, where $K_a\alpha$ stands for 'an agent a knows that α'. Omniscience is represented in various ways, we consider it in the form of the formula $K_a\alpha \wedge K_a(\alpha \to \beta) \to K_a\beta$, see [188, 189].

In [173] another definition of knowledge operator is provided relevant to modelling incomplete or uncertain knowledge derived from an information system. The underlying intuition is that knowledge of an agent is manifested in his abilities to find patterns and regularities in a set of data. In the context of information systems it means that a knowledgeable agent is able to classify the objects in terms of their properties determined by the attributes.

Let $(Ob, \mathcal{A}t)$ be an information system and let $A \subseteq \mathcal{A}t$ be a set of attributes. Following [173], see also [40], we define the *knowledge operator* determined by A for every $X \subseteq Ob$,

$$\text{(K)} \qquad K_A(X) \stackrel{\text{def}}{=} Pos_A(X) \cup Neg_A(X).$$

This operator may be seen as a rough-set style formulation of the operator 'knowing whether' discussed in [129]. Hintikka describes its intuitive meaning as 'Clearly one knows whether α is true if and only if one knows that α is true or knows that α is false'. In the context of information systems, knowledge of an agent whose indiscernibility relation is $ind(A)$, for some $A \subseteq \mathcal{A}t$, is reflected by his ability of classifying the objects in the system according to their properties determined by the attributes in A. The agent knows a set $X \subseteq Ob$ whenever for every $x \in Ob$ he can ascertain whether $x \in X$ or $x \notin X$. The knowledge operator defined by (K) has the property that if an object, say x, is such that the equivalence class $[x]_A = \{x\}$, that is the agent has a crisp information about x, then for any set $X \subseteq Ob$ we have $x \in K_A(X)$. The S5 knowledge operator does not have this property. The disjoint representation of rough sets developed and studied in [185], see also [186], is based on this property.

We say that knowledge about a set X determined by attributes from a set A is

$$\begin{aligned}
\textit{complete}, &\text{ if } K_A(X) = Ob\\
\textit{rough}, &\text{ if } Pos_A(X) \neq \emptyset, Neg_A(X) \neq \emptyset, \text{ and } B_A(X) \neq \emptyset\\
\textit{pos-empty}, &\text{ if } Pos_A(X) = \emptyset\\
\textit{neg-empty}, &\text{ if } Neg_A(X) = \emptyset\\
\textit{empty}, &\text{ if it is pos} - \text{empty and neg} - \text{empty}.
\end{aligned}$$

Lemma 4.4.1 [173] *For every information system $(Ob, \mathcal{A}t)$, for every $A \subseteq \mathcal{A}t$, and for every $X \subseteq OB$,*

(a) *$K_A(X)$ is complete if and only if X is A-definable;*

(b) *$K_A(X)$ is rough if an only if X is roughly A-definable;*

(c) *$K_A(X)$ is pos-empty if and only if X is bottom A-indefinable;*

(d) *$K_A(X)$ is neg–empty if and only if X is top A-indefinable;*

(e) *$K_A(X)$ is empty if and only if X is totally A-indefinable.*

The following lemma gives some facts about interaction of information operators.

Lemma 4.4.2 [40] *For every information system $(Ob, \mathcal{A}t)$, for all $A, B \subseteq \mathcal{A}t$, and for every $X \subseteq OB$,*

(a) $Pos_A(K_A(X)) = K_A(X)$ *and* $Neg_A(K_A(X)) = B_A(X)$

(b) $B_A(K_A(X)) = \emptyset$;

(c) $K_A(Pos_A(X)) = K_A(Neg_A(X)) = Ob$;

(d) $K_A(K_A(X)) = Ob$;

(e) $K_A(X) = K_A(Ob - X)$;

(f) $K_A(\emptyset) = Ob$;

(g) *If* $A \subseteq B$, *then* $K_A(X) \subseteq K_B(X)$;

(h) *If* $ind(A) \subseteq ind(B)$, *then* $K_B(X) \subseteq K_A(X)$.

The knowledge operator K_A defined by (K) does not have the property of veridicality typical for the traditional epistemic logic based on S5 modal system. To see that $K_A(X)$ is not necessarily included in X consider the following example.

Example 4.4.3 *Let* $Ob = \{o_1, \ldots, o_5\}$ *and let* $X = \{o_1, o_2, o_3\}$. *Assume that an indiscernibility relation determined by a set A of attributes generates the following partition:* $\{\{o_1, o_2\}, \{o_3, o_4\}, \{o_5\}\}$. *Then we have*

$$K_A(X) = Pos_A(X) \cup Neg_A(X) = \{o_1, o_2\} \cup \{o_5\}$$

which obviously is not included in X.

Similarly, the knowledge operator K_A does not have the unwanted property of logical omniscience as the following example shows.

Example 4.4.4 *Let* $Ob = \{o_1, o_2, o_3, o_4\}$ *and let an indiscernibility relation determined by a set A of attributes generates the partition* $\{\{o_1, o_2\}, \{o_3, o_4\}\}$. *Take* $X = \{o_1\}$ *and* $Y = \{o_2, o_3\}$. *Then*

$$\begin{aligned} K_A(X) &= Pos_A(X) \cup Neg_A(X) = \emptyset \cup \{o_3, o_4\} = \{o_3, o_4\} \\ K_A(-X \cup Y) &= Pos_A(-X \cup Y) \cup Neg_A(-X \cup Y) = \{o_3, o_4\} \cup \emptyset = \{o_3, o_4\} \\ K_A(Y) &= Pos_A(Y) \cup Neg_A(Y) = \emptyset. \end{aligned}$$

Hence $K_A(X) \cap K_A(-X \cup Y) = \{o_3, o_4\}$. *Clearly, it is not included in* $K_A(Y)$.

However, the agents reasoning according to the knowledge operators K_A are fully introspective: they know what they know (positive introspection) and they know what they do not know (negative introspection). Formally, positive introspection is reflected by $K_A(X) \subseteq K_A(K_A(X))$ and negative introspection by $-K_A(X) \subseteq K_A(-K_A(X))$ for every $X \subseteq Ob$.

4.5 Boolean algebras with a knowledge operator

In this section we present a discrete duality for Boolean algebras with an operator which characterizes in the abstract way any knowledge operator of the form (K) discussed in the previous section. We confine ourselves to the algebras with a single knowledge operator. Thus the algebras enable us to represent and prove properties of every operator K_A for a fixed set A of attributes but they do not provide any means for discussing relationships between the operators which are determined by possibly different sets of attributes. However, following a logic with relative knowledge operators ([39, 40]) an extension to the algebras with multiple knowledge operators determined by subsets of a set of attributes can be easily defined, possibly with some specific axioms reflecting properties of indiscernibility relations determined by the underlying sets of attributes. Such algebras can be further expanded to the algebras with the operators of common knowledge and joint knowledge. For example, the operator $K_{A\cup B}$ determined by the union of two sets of attributes may be seen as the operator of joint knowledge. For every $X \subseteq Ob$ we have $K_A(X), K_B(X) \subseteq K_{A\cup B}(X)$.

A *knowledge algebra*, *K-algebra* for short, is a structure (B, K) where B is a Boolean algebra and K is an unary operator on B satisfying, for any $a, b \in B$,

(K1) $a \leq b \Rightarrow a \wedge Ka \leq Kb$

(K2) $K(a \vee -Ka) = 1$

(K3) $Ka = K(-a)$.

Operator K is not monotone. For example, for all $a, b \in B$ we have $a \wedge b \leq a$ but not necessarily $K(a \wedge b) \leq K(a)$. Furthermore, since $K(1) = 1$ by (K2), we have $K(0) = 1$ by (K3). But it is not true that for every $a \in B$, $1 = K(a \wedge 0) \leq K(a)$. Similarly, K is not antitone. For example, for every $a \in B$ we have $a \leq 1$ but not necessarily $1 = K(1) \leq K(a)$.

A *knowledge frame* (*K-frame*) is a structure (X, R) where X is a Boolean frame and R is an equivalence relation on X.

The complex algebra of a K-frame X is a structure $(2^X, K_R)$ where 2^X is the complex algebra of a Boolean frame X and K_R is a unary operator satisfying, for all $A \subseteq X$,

$$K_R(A) \stackrel{\text{def}}{=} [R]A \cup [R](-A).$$

Proposition 4.5.1 *The complex algebra of a K-frame is a K-algebra.*

Proof:
(K1) We show, for any $A, B \subseteq X$, that if $A \subseteq B$ then $A \cap K_R(A) \subseteq K_R(B)$. Assume $A \subseteq B$ and take $x \in X$ such that $x \in A$ and $x \in [R]A$ or $x \in [R](-A)$. If $x \in [R]A$, then since $[R]$ is monotone we get $x \in [R]B$ and hence $x \in K_R(B)$. Observe that if $x \in A$, then it cannot be $x \in [R](-A)$. For suppose otherwise, then by reflexivity of R we have $x \notin A$, a contradiction.

(K2) We have to show that $K_R(A \cup -K_R(A)) = X$. Suppose that there is some $x \in X$ such that $x \notin K_R(A \cup -K_R(A))$. This means that

$$x \notin [R](A \cup -K_R(A)) \cup [R](-(A \cup -K_R(A))),$$

so (i) $x \notin [R](A \cup -K_R(A))$ and (ii) $x \notin [R](-(A \cup -K_R(A)))$. From (i), there exists some $y \in X$ such that xRy and $y \notin A \cup -K_R(A)$, so $y \notin A$ and $y \in K_R(A)$, i.e., $y \in [R]A \cup [R](-A)$. By reflexivity of R, $[R]A \subseteq A$, so since $y \notin A$, we get $y \notin [R]A$. Hence, since $y \in [R]A \cup [R](-A)$, we get $y \in [R](-A)$. Similarly, (ii) means that there is some $z \in X$ such that xRz and $z \in A \cup -K_R(A)$. By symmetry and transitivity of R, from xRy and xRz it follows that yRz which together with $y \in [R](-A)$ yields $z \notin A$. Since $z \in A \cup -K_R(A)$, we get $z \notin K_R(A)$, that is $z \notin [R]A \cup [R](-A)$ which implies $z \notin [R](-A)$. Hence there is some $u \in X$ such that zRu and $u \in A$. Again by transitivity of R, from yRz and zRu we get yRu which together with $y \in [R](-A)$ gives $u \notin A$, a contradiction.

(K3) Obvious. □

The canonical frame of a K-algebra B is a structure (X_B, R_K) where X_B is the set of prime filters of B and R_K is a binary relation on X_B such that, for all $F, G \in X_B$,

$$F\, R_K\, G \overset{\text{def}}{\Leftrightarrow} K(F) \subseteq G$$

where $K(F) \overset{\text{def}}{=} \{a \in B : a, Ka \in F\}$.

Proposition 4.5.2 *The canonical frame of a K-algebra is a K-frame.*

Proof:
We have to show that R_K is an equivalence relation.
Reflexivity of R_K is obvious. For symmetry, take $F, G \in X_B$ and assume that $K(F) \subseteq G$. We show that for every $a \in B$, $a \notin F$ implies $a \notin G$ or $Ka \notin G$. Take $a \in B$ such that $a \notin F$. Then $-a \in F$. Since $-a \leq -a \vee -Ka$, we get $-a \vee -Ka \in F$. By axioms (K2) and (K3) we also have

$$1 = K(-a \vee -K(-a)) = K(-a \vee -Ka) \in F.$$

Hence $-a \vee -Ka \in K(F)$. By the assumption, $-a \vee -Ka \in G$. Since G is prime, $-a \in G$ or $-Ka \in G$, and hence $a \notin G$ or $Ka \notin G$.
Now we show that R_K is transitive. Let $F, G, H \in X_B$ be such that $K(F) \subseteq G$ and $K(G) \subseteq H$. Take $a \in K(F)$, that is $a \in F$ and $Ka \in F$. Then $a \wedge Ka \in F$ and $a \in G$. Using (K2) and (K3) we get

$$1 = K(-a \vee -K(-a)) = K(-(a \wedge K(-a))) = K(a \wedge Ka) \in F.$$

Since $a \wedge Ka \in F$, we have $a \wedge Ka \in K(F)$. By the assumption, $a \wedge Ka \in G$. Since $a \wedge Ka \leq Ka$, $Ka \in G$. We also have $a \in G$, so $a \in K(G)$. By the assumption $a \in H$, as required. □

Let B be a K-algebra. We now show that the embedding $h : B \to 2^{X_B}$, defined in Section 2.3, preserves the operator K.

Lemma 4.5.3 *For every* $a \in B$, $h(Ka) = K_{R_K}(h(a))$.

Proof:
Observe that, for every $a \in B$ and $F \in X_B$,

$$\begin{aligned} F \in K_{R_K}(h(a)) &\Leftrightarrow F \in [R_K]h(a) \cup [R_K](-h(a)) \\ &\Leftrightarrow (\forall G \in X_B,\ FR_K G \Rightarrow a \in G) \\ &\quad\ \text{or } (\forall G \in X_B,\ FR_K G \Rightarrow a \notin G). \end{aligned}$$

($\Rightarrow$) Assume $Ka \in F$ and take G such that FR_KG, that is for every $b \in B$, $b, Kb \in F$ implies $b \in G$. If $a \notin F$, then $-a \in F$. Since $Ka = K(-a)$ by (K3), $K(-a) \in F$. Hence $-a \in K(F)$. So by the assumption $-a \in G$, thus $a \notin G$. If $a \in F$, then since $Ka \in F$, $a \in K(F)$, whence $a \in G$.

($\Leftarrow$) Assume $Ka \notin F$. We show that there exists some $G_1 \in X_B$ such that $K(F) \subseteq G_1$ and $a \notin G_1$ and there exists some $G_2 \in X_B$ such that $K(F) \subseteq G_2$ and $a \in G_2$. Consider a filter G_1' generated by $K(F) \cup \{a\}$. It is a proper filter, for suppose otherwise, then there is some $b \in K(F)$ such that $b \wedge a = 0$. Hence $b \leq -a$, and by axioms (K1) and (K3), $b \wedge Kb \leq Ka$. Since $Ka \notin F$, $b \wedge Kb \notin F$ which is a contradiction with $b \in K(F)$. Let G_1 be a prime filter containing G_1'. Thus $K(F) \subseteq G_1$ and $a \in G_1$.
Similarly, let G_2' be a filter generated by $K(F) \cup \{-a\}$. It is easy to see that it is a proper filter, so it can be extended to a prime filter including it, say G_2. Then we have $K(F) \subseteq G_2$ and $a \notin G_2$. □

Let X be a K-frame. We now show that the embedding $k : X \to X_{2^X}$, defined in Section 2.3, preserves the relation R.

Lemma 4.5.4 *For all* $x, y \in X$, xRy *if and only if* $k(x) R_{K_R} k(y)$.

Proof:
Observe that, for all $x, y \in X$,

$$\begin{aligned} k(x)R_{K_R}k(y) &\Leftrightarrow K_R(k(x) \subseteq k(y)) \\ &\Leftrightarrow \forall A \subseteq X,\ A \in k(x)\ \&\ K_R(A) \in k(x) \Rightarrow A \in k(y) \\ &\Leftrightarrow \forall A \subseteq X,\ x \in A\ \&\ x \in [R]A \cup [R](-A) \Rightarrow y \in A. \end{aligned}$$

Assume xRy, $x \in A$ and $\forall z$, $xRz \Rightarrow z \in A$. Suppose $y \notin A$. Then taking y for z we obtain a contradiction. Similarly, assume xRy, $x \in A$ and $\forall z$, $xRz \Rightarrow z \notin A$. Suppose $y \notin A$. Then taking x for z we obtain a contradiction. On the other hand, assume that for every $A \subseteq X$, $A \in k(x)$ and $K_R(A) \in k(x)$ imply $A \in k(y)$. Consider $A = \langle R \rangle \{x\}$. Then we have $x \in \{x\} \subseteq \langle R \rangle \{x\}$. Since R is symmetric, $\{x\} \subseteq [R]\langle R \rangle \{x\}$, so $\langle R \rangle \{x\} \in k(x)$ and $K_R(\langle R \rangle \{x\}) \subseteq k(x)$. Then $\langle R \rangle \{x\} \in k(y)$ which gives yRx and by symmetry of R we finally obtain xRy. □

With these results we obtain the following discrete representations for K-algebras and K-frames.

Theorem 4.5.5

(a) *Every K-algebra is embeddable into the complex algebra of its canonical frame;*

(b) *Every K-frame is embeddable into the canonical frame of is complex algebra.*

A logic with the operator K considered in the present section is developed in [38], see also[39, 40].

4.6 Diversity algebras

A *diversity algebra, Div-algebra,* is a Boolean algebra B endowed with a sufficiency operator $\boxdot$, $(B, \boxdot)$, such that the operator $\boxdot$ satisfies the axioms (S1) and (S2) of Section 3.3 and the following axioms, for any $a \in B$,

(Div1) $\boxdot a \leq -a$

(Div2) $a \leq \boxdot\boxdot a$

(Div3) $\boxdot a \leq \boxdot(-\boxdot a)$.

A *diversity frame, Div-frame,* is a sufficiency Boolean frame (X, R) such that the relation R satisfies

(FDiv1) R is irreflexive

(FDiv2) R is symmetric

(FDiv3) R is co-transitive.

The complex algebra of a Div-frame X is a structure $(B_X, \boxdot_R)$ where B_X is a Boolean algebra of all the subsets of X, that is, B_X is the complex algebra of X understood as a Boolean frame, and, for every $A \subseteq X$,

$$\boxdot_R A \stackrel{\text{def}}{=} [\![R]\!]A.$$

In Proposition 3.3.2 it is shown that $(B_X, \boxdot_R)$ is an S-Boolean algebra.

Proposition 4.6.1 *The complex algebra of a Div-frame is a Div-algebra.*

Proof:
In view of Proposition 3.3.2 it is sufficient to show that the specific axioms of diversity algebras are satisfied in $(B_X, \boxdot_R)$. All of them follow easily from the corresponding frame axioms. By way of example we show (Div3). Let $x \in [\![R]\!]A$ which means that for every $y \in X$, $y \in A$ implies xRy. Suppose $x \notin [\![R]\!](-[\![R]\!]A)$. It follows that for some $z \in X$, $z \in -[\![R]\!]A$ and $x(-R)z$. Furthermore, $z \in -[\![R]\!]A$ yields there is some t such that $t \in A$ and $z(-R)t$. Since $x(-R)z$ and $z(-R)t$, by co-transitivity of R we get $x(-R)t$. Since $t \in A$, we have xRt, which provides a required contradiction. □

The canonical frame of a Div-algebra B is a structure $(X_B, R_\boxdot)$ where X_B is the family of all ultrafilters of the Boolean reduct of the Div-algebra, that is, X_B is the canonical frame of that Boolean algebra and, for every $F, G \in X_B$,

$$FR_\boxdot G \stackrel{\text{def}}{\Leftrightarrow} \boxdot(G) \cap F \neq \emptyset.$$

Proposition 4.6.2 *The canonical frame of a Div-algebra is a Div-frame.*

Proof:
Since the canonical frame of an S-algebra is an S-frame, as shown in Proposition 3.3.3, it is sufficient to show that the specific axioms of Div-frames are satisfied in $(X_B, R_{\boxbar})$. (FDiv1) and (FDiv2) follow easily from the axioms of Div-algebras.

For (FDiv3), take any $F, G, H \in X_B$ such that $F(-R_{\boxbar})G$ and $G(-R_{\boxbar})H$. Suppose $FR_{\boxbar}H$. Then we have $\boxbar(G) \cap F = \emptyset$, $\boxbar(H) \cap G = \emptyset$, and $\boxbar(H) \cap F \neq \emptyset$. The latter means that there is some $a \in H$ such that $\boxbar a \in F$. Since $\boxbar(H) \cap G = \emptyset$, $\boxbar a \notin G$ and since G is an ultrafilter of B, $-\boxbar a \in G$. Thus $\boxbar(-\boxbar a) \in \boxbar(G)$. Since $\boxbar(G) \cap F = \emptyset$, $\boxbar(-\boxbar a) \notin F$. Using axiom (Div3) we get $\boxbar a \notin F$, a contradiction. $\Box$

In view of Lemma 3.3.4, Lemma 3.3.5 and Theorem 3.3.6 we obtain the following discrete representations for Div-algebras and Div-frames.

Theorem 4.6.3

(a) *Every Div-algebra is embeddable into the complex algebra of its canonical frame;*

(b) *Every Div-frame is embeddable into the canonical frame of its complex algebra.*

4.7 Algebras of weak similarity

In this section we apply the results of Section 3.2 to a class of information algebras which are signature and axiomatic extensions of possibility algebras such that an indexed family of possibility operators is introduced and the axioms enabling us a characterization of relative similarity relations are postulated.

An *information algebra of weak similarity, wSim-algebra*, is a Boolean algebra B with a family $\{\Diamond_P : P \subseteq Par\}$ of possibility operators satisfying axioms (P1) and (P2) from Section 3.2 and the following axioms, for any $a \in B$,

(wSim1) $\Diamond_{P \cup Q} a = \Diamond_P a \vee \Diamond_Q a$

(wSim2) $\Diamond_{\emptyset} a = 0$

(wSim3) $a \wedge \Diamond_P 1 \leq \Diamond_P a$

(wSim4) $a \leq \Box_P \Diamond_P a$

where $\Box_P = -\Diamond_P -$ for every $P \subseteq Par$.

An *information frame of weak similarity, wSim-frame*, for short, is a relational structure $(X, \{R_P : P \subseteq Par\})$ where the binary relations $R_P \subseteq X \times X$, for each $P \subseteq Par$, satisfy

(FwSim1) $R_{P \cup Q} = R_P \cup R_Q$

(FwSim2) $R_{\emptyset} = \emptyset$

(FwSim3) R_P is weakly reflexive (i.e., $\forall x, y \in X$, $xR_Py \Rightarrow xR_Px$)

(FwSim4) R_P is symmetric.

Axioms (FwSim1) and (FwSim2) reflect the intuition of weak relations, while axioms (FwSim3) and (FwSim4) are the abstract characterization of similarity relations derived from an information system.

The complex algebra of a wSim-frame X is a structure

$$(B_X, \{\Diamond_{R_P} : P \subseteq Par\})$$

where B_X is the complex algebra of the Boolean frame X and, for every $A \subseteq X$,

$$\Diamond_{R_P}(A) \stackrel{\text{def}}{=} \langle R_P \rangle A.$$

Proposition 4.7.1 *Let* $(X, \{R_P : P \subseteq Par\})$ *be an information frame. For every* $P \subseteq Par$*, the operator* $\Diamond_{R_P}$ *on* 2^X *satisfies axioms (wSim1)–(wSim4).*

Proof:
We prove (wSim1), (wSim2), and (wSim3). Condition (wSim4) is well known in modal correspondence theory [243], see also Section 1.8.

(wSim1) For $A \subseteq X$ we show $\langle R_{P \cup Q} \rangle A = \langle R_P \rangle A \cup \langle R_Q \rangle A$. We have

$$\begin{aligned} x \in \langle R_{P\cup Q} \rangle A &\Leftrightarrow R_{P \cup Q}(x) \cap A \neq \emptyset \Leftrightarrow R_P(x) \cap A \cup R_Q(x) \cap A \neq \emptyset. \\ &\Leftrightarrow R_P(x) \cap A \neq \emptyset \text{ or } R_Q(x) \cap A \neq \emptyset. \end{aligned}$$

(wSim2) It is clear that due to (FwSim2) we have $\langle R_\emptyset \rangle A = \emptyset$ for every $A \subseteq X$.

(wSim3) We show $A \cap \langle R_P \rangle X \subseteq \langle R_P \rangle A$. Take any $x \in X$ such that $x \in A \cap \langle R_P \rangle (X)$ and $x \notin \langle R_P \rangle (X)$. Then $x \in A$, $R_P(x) \neq \emptyset$, and by (FwSim3), xR_Px. Suppose $x \notin \langle R_P \rangle A$. Then for every y, xR_Py implies $y \notin A$. Taking x for y we get a contradiction. □

The canonical frame of a wSim-algebra B is a structure

$$(X_B, \{R_{\Diamond_P} : P \subseteq Par\})$$

where X_B is the set of ultrafilters of B and, for all $F, G \in X_B$,

$$FR_{\Diamond_P}G \stackrel{\text{def}}{\Leftrightarrow} G \subseteq \Diamond_P^{-1}(F).$$

Proposition 4.7.2 *Let* $(B, \{\Diamond_P : P \subseteq Par\})$ *be a wSim-algebra. For every* $P \subseteq Par$*, the relation* $R_{\Diamond_P}$ *on* X_B *satisfies axioms (FwSim1)–(FwSim4).*

Proof:
We prove (FwSim1), (FwSim2), and (FwSim3). (FwSim4) is well known from modal correspondence theory [243], see Section 1.8.

(FwSim1) For any $F, G \in X_B$,

$$
\begin{aligned}
FR_{\Diamond_{P\cup Q}}G &\Leftrightarrow G\subseteq \Diamond^{-1}_{P\cup Q}(F) \\
&\Leftrightarrow \forall a,\ a\in G \Rightarrow \Diamond_{P\cup Q}a\in F \\
&\Leftrightarrow \forall a,\ a\in G \Rightarrow \Diamond_P a\vee \Diamond_Q a\in F && \text{(wSim1)}\\
&\Leftrightarrow \forall a,\ a\in G \Rightarrow (\Diamond_P a\in F \text{ or } \Diamond_Q a\in F) && F \text{ is prime}\\
&\Leftrightarrow \forall a,\ a\in G \Rightarrow \Diamond_P a\in F \text{ or } \forall a,\ a\in G \Rightarrow \Diamond_Q a\in F\\
&\Leftrightarrow FR_PG \text{ or } FR_QG.
\end{aligned}
$$

(FwSim2) Note that we have

$$FR_{\Diamond_\emptyset}G \Leftrightarrow G\subseteq \Diamond_\emptyset^{-1}(F) \Leftrightarrow G\subseteq\emptyset.$$

The latter is always false since G is a filter. Thus $R_{\Diamond_\emptyset}=\emptyset$.

(FwSim3) Take any $F, G\in X_B$ with $FR_{\Diamond_P}G$. Then for every a, $a\in G$ implies $\Diamond_P a\in F$. In particular, $\Diamond_P 1\in F$. Suppose not $FR_{\Diamond_P}F$. Then there is some $b\in F$ such that $\Diamond_P b\notin F$. By (wSim3) $b\wedge\Diamond_P 1\leq \Diamond_P b$. Hence $b\wedge\Diamond_P 1\notin F$, a contradiction. □

The discrete representations, analogous to those in Theorem 3.2.7, hold for wSim-algebras and wSim-frames.

Now we briefly mention duality via truth. The language $\mathcal{L}an_{wSim}$ relevant for wSim-algebras and wSim-frames is an extension of the modal language $\mathcal{L}an_P$ with a family $\{\langle R_P\rangle : P\subseteq Par\}$ of modal possibility operators. Algebraic semantics of the language is provided by the class of wSim-algebras, $\mathcal{A}lg_{wSim}$, and the frame semantics by the class of wSim-frames, $\mathcal{F}rm_{wSim}$. The notion of a model based on a frame of $\mathcal{F}rm_{wSim}$, satisfaction relation, and the notions of truth in a model, in a frame and in a class of frames are analogous to the respective notions in Section 2.3. In view of Proposition 4.7.1 the complex algebra theorem mentioned in Step 6 of Section 2.2 holds for $\mathcal{F}rm_{wSim}$. From the representation theorem and the complex algebra theorem we obtain a duality via truth theorem.

4.8 Algebras of strong right orthogonality

The representation results of Theorem 3.3.6 can be extended to some information algebras based on sufficiency algebras. An abstract characterization of strong relations of right orthogonality derived from an information system may be defined as follows.

An *information algebra of strong right orthogonality, sRort-algebra* for short, is a Boolean algebra B with a family $\{\boxdot g_P : P\subseteq Par\}$ of sufficiency operators satisfying axioms (S1) and (S2) from Section 3.3 and the following axioms, for any $a\in B$,

(sRort1) $\boxdot_{P\cup Q}(a) = \boxdot_P(a)\wedge\boxdot_Q(a)$

(sRort2) $\boxdot_\emptyset(a)=1$

(sRort3) $a\wedge\boxdot_P(a)\leq\boxdot_P(1)$

(sRort4) $a \leq \boxminus_P \boxminus_P(a)$.

An *information frame of strong right orthogonality*, *sRort-frame* for short, is a relational structure $(X, \{R_P : P \subseteq Par\})$ where, for all $P, Q \subseteq Par$,

(FsRort1) $R_{P \cup Q} = R_P \cap R_Q$
(FsRort2) $R_\emptyset = X \times X$
(FsRort3) R_P is co-weakly reflexive (i.e., $\forall x, y \in X,\ x(-R_P)y \Rightarrow x(-R_P)x$)
(FsRort4) R_P is symmetric.

The complex algebra of an sRort-frame X is a structure

$$(B_X, \{\boxminus_{R_P} : P \subseteq Par\})$$

where B_X is the complex algebra of the Boolean frame X and, for every $A \subseteq X$,

$$\boxminus_{R_P}(A) \stackrel{\text{def}}{=} [\![R_P]\!]A.$$

Proposition 4.8.1 *Let $(X, \{R_P : P \subseteq Par\})$ be an information frame of strong right orthogonality. For every $P \subseteq Par$, the operator $\boxminus_{R_P}$ on 2^X satisfies axioms (sRort1)–(sRort4).*

Proof:
(sRort1) Note that

$$\begin{aligned} x \in \boxminus_{R_{P\cup Q}}(A) &\Leftrightarrow A \subseteq R_{P \cup Q}(x) \\ &\Leftrightarrow A \subseteq R_P(x) \cap R_Q(x) \qquad \text{(FsRort1)} \\ &\Leftrightarrow A \subseteq R_P(x)\ \&\ A \subseteq R_Q(x) \\ &\Leftrightarrow x \in \boxminus_{R_P}(A) \cap \boxminus_{R_Q}(A). \end{aligned}$$

(sRort2) Observe that

$$x \in \boxminus_{R_\emptyset}(A) \Leftrightarrow A \subseteq R_\emptyset(x) \Leftrightarrow A \subseteq X.$$

The latter is always true so $\boxminus_{R_\emptyset}(A) = X$.

(sRort3) Take any $x \in X$ such that $x \in A \cap \boxminus_{R_P}(A)$ and $x \notin \boxminus_{R_P}(X)$. Then $x \in A$ and $A \subseteq R_P(x)$ and $X \not\subseteq R_P(x)$. So xR_Px and for some $y \in X$, $x(-R_P)y$. Thus by axiom (FsRort3), $x(-R)x$, which provides the required contradiction.

(sRort4) Assume $x \notin \boxminus_{R_P}\boxminus_{R_P}A$. Then for some y, $y \in \boxminus_{R_P}A$ and $x(-R)y$. It follows that for every $z \in X$, $z \in A$ implies yRz. Suppose $x \in A$. Then yRx and $x(-R)y$. Since R is symmetric, we get the required contradiction. □

The canonical frame of a sRort-algebra B is a structure

$$(X_B, \{R_{\boxminus_P} : P \subseteq Par\})$$

where X_B is the set of ultrafilters of B and for all $F, G \in X_B$,

$$FR_{\boxminus_P}G \stackrel{\text{def}}{\Leftrightarrow} F \cap \boxminus_P(G) \neq \emptyset.$$

Proposition 4.8.2 *Let $(B, \{\Box_P : P \subseteq Par\})$ be an information algebra of strong right orthogonality. For every $P \subseteq Par$, the relation $R_{\Box_P}$ on X_B satisfies axioms (FsRort1)–(FsRort4).*

Proof:
Take any $F, G \in X_B$.
(FsRort1) Note that

$$\begin{array}{rcll} FR_{\Box_{P\cup Q}}G & \Leftrightarrow & \Box_{P\cup Q}(G) \cap F \neq \emptyset & \\ & \Leftrightarrow & \Box_P(G) \cap \Box_Q(G) \cap F \neq \emptyset & \text{(FsRort1)} \\ & \Leftrightarrow & \Box_P(G) \cap F \neq \emptyset \ \& \ \Box_Q(G) \cap F \neq \emptyset & \\ & \Leftrightarrow & FR_{\Box_P}G \ \& \ FR_{\Box_Q}G. & \end{array}$$

(FsRort2) We have

$$FR_{\Box_\emptyset}G \Leftrightarrow \Box_\emptyset(G) \cap F \neq \emptyset \Leftrightarrow 2^X \cap F \neq \emptyset \Leftrightarrow F \neq \emptyset,$$

where the last but one equivalence is due to (FsRort2). The latter is always true since F is a filter. Thus $R_{\Box_\emptyset} = X_B \times X_B$.

(FsRort3) Assume $FR_{\Box_P}F$, that is $\Box_P(F) \cap F \neq \emptyset$. Thus there is some $a \in B$ such that $a \in F$ and $\Box a \in F$. By (sRort3) $\Box 1 \in F$. Since $1 \in G$, $\Box 1 \in \Box(G)$ and hence $\Box 1 \in \Box(G) \cap F \neq \emptyset$ which yields $FR_{\Box_P}G$.

(FsRort4) Assume $\Box_{R_P}(G) \cap F \neq \emptyset$ and suppose $\Box_{R_P}(F) \cap G = \emptyset$. Then for some $a \in G$, $\Box_P a \in F$ and for every $b \in B$, $b \in F$ implies $\Box_P b \notin G$. Taking $\Box_P a$ for b we have $\Box_P \Box_P a \notin G$. By (sRort4) $a \notin F$, a contradiction. □

Discrete representations, analogous to the representation Theorem 3.3.6, hold for algebras and frames of strong right orthogonality.

The language $\mathcal{L}an_{sRort}$ relevant for a logic of strong right orthogonality is an extension of the modal language $\mathcal{L}an_S$ with a family of $\{[\![R_P]\!] : P \subseteq Par\}$ of sufficiency operators. Algebraic semantics of the language is provided by the class of sRort-algebras, $\mathcal{A}lg_{sRort}$, and the frame semantics by the class of sRort-frames, $\mathcal{F}rm_{sRort}$. The notion of a model based on a frame of $\mathcal{F}rm_{sRort}$, satisfaction relation, and the notions of truth in a model, in a frame and in a class of frames are analogous to the respective notions in Section 3.2. In view of Proposition 4.8.1 the complex algebra theorem mentioned in Step 6 of Chapter 1 holds for $\mathcal{F}rm_{sRort}$. From the representation theorem and the complex algebra theorem we can obtain a duality via truth theorem.

Chapter 5

Boolean algebras with relational operators

5.1 Introduction

In this chapter we develop discrete duality for relation algebras and cylindric algebras of finite dimension. The basis of our work are the developments in [127, 139, 160, 161, 162]. The notions of relation frame, cylindric frame, and complex algebras of such frames presented in this chapter are derived from the developments there. In the representation theorems for relation algebras and cylindric algebras presented in [127] the representation algebras are the complex algebras of the atom structures of the given algebras. In [162] a representation theorem for relation algebras and cylindric algebras is announced without a proof showing that the representation algebra is the complex algebra of the canonical frame of the given algebra, as it is usual in the discrete duality framework.

Many properties of relations discussed already in [35] and [197] are captured in terms of an abstract algebraic structure defined as a Boolean algebra with operators, see, for example, [233, 163]. Accordingly, in Section 5.2, we define a relation algebra to be a Boolean algebra with certain operators. Following [126, 139, 160, 164] the notion of relation frame is defined as an abstract relational structure whose universe is not necessarily a set of atoms of a relation algebra. We then prove a discrete duality for relation algebras and relation frames, as presented in [181]. The representation for relation algebras is guaranteed by the Sahlqvist theorem [144] adapted to Boolean algebras with operators in the sense of [139] in [213].

The notion of cylindric algebra, invented by Alfred Tarski and presented in [126] and [127], plays a role comparable for first-order logic as Boolean algebras play for propositional logic. In Section 5.3 we consider cylindric algebras of dimension n where $3 \leq n < \omega$. They enable us to study relations of arbitrary finite rank $n > 2$ in an analogous way as relation algebras for binary relations. These are Boolean algebras with additional *cylindrification* operators that model quantification. Following [126, 127, 160] the notion of cylindric frame is defined analogously to that of relation frame. We then prove a discrete duality for cylindric algebras and cylindric frames, as presented in [181].

5.2 Relation algebras

A *relation algebra* is a structure $(B, ;, {}^{\smile}, 1')$ where B is a Boolean algebra, $(B, ;, 1')$ is a monoid, and ; is a binary operator on B and ${}^{\smile}$ is a unary operator on B satisfying, for any $a, b, c \in B$,

(R1) $(a \vee b) ; c = (a ; c) \vee (b ; c)$

(R2) $a^{\smile\smile} = a$

(R3) $(a \vee b)^{\smile} = a^{\smile} \vee b^{\smile}$

(R4) $(a ; b)^{\smile} = b^{\smile} ; a^{\smile}$

(R5) $a^{\smile} ; (-(a ; b)) \leq -b$.

It follows that relation algebras are Boolean lattice-ordered monoids with involution, see Section 1.6. The proofs for the following properties of relations can be found in [25]; the property in (f) is De Morgan Theorem K in [35].

Lemma 5.2.1 *Let $(B, ;, {}^{\smile}, 1')$ be a relation algebra. For any $a, b, c, d \in B$,*

(a) $c ; (a \vee b) = (c ; a) \vee (c ; b)$

(b) $1' ; a = a = a ; 1'$

(c) $(a \wedge b)^{\smile} = a^{\smile} \wedge b^{\smile}$

(d) $a \leq b \Leftrightarrow a^{\smile} \leq b^{\smile}$

(e) *if* $a \leq b$ *and* $c \leq d$*, then* $(a ; c) \leq (b ; d)$

(f) $(a ; b) \wedge c = 0 \Leftrightarrow (a^{\smile} ; c) \wedge b = 0 \Leftrightarrow (c ; b^{\smile}) \wedge a = 0$.

A *relation frame* is a structure (X, R, f, I) where X is a non-empty set, and R is a ternary relation on X, f is a mapping $f : X \to X$ and $I \subseteq X$ is a designated set satisfying, for any $v, w, x, y, z \in X$,

(FR1) $f(f(x)) = x$

(FR2) $R(x, y, z) \Rightarrow R(f(x), z, y)$

(FR3) $R(x, y, z) \Rightarrow R(z, f(y), x)$

(FR4) $x = y \Leftrightarrow \exists z \in X,\ I(x)\ \&\ R(x, z, y)$

(FR5) $R(x, y, z)\ \&\ R(z, v, w) \Rightarrow \exists u \in X,\ R(x, u, w)\ \&\ R(y, v, u)$

(FR6) $R(x, y, z)\ \&\ R(v, z, w) \Rightarrow \exists u \in X,\ R(u, y, w)\ \&\ R(v, x, u)$.

Lemma 5.2.2 *Let (X, R, f, I) be a relation frame. For any $x, y, z \in X$,*

(a) $R(x, y, z) \Leftrightarrow R(f(x), z, y)$;

(b) $R(x, y, z) \Leftrightarrow R(z, f(y), x)$).

Proof:
(a) Note that, by (FR1) and (FR2), for any $x, y, z \in X$,

$$R(x,y,z) \Rightarrow R(f(x),z,y) \Rightarrow R(f(f(x)),y,z) \Leftrightarrow R(x,y,z).$$

Similarly, (b) follows using (FR1) and (FR3). □

The *complex algebra of a relation frame* X is a structure $(2^X, ;_R, {}^{\smile_f}, 1'_I)$ where 2^X is the complex algebra of its Boolean frame reduct X, and the binary operator $;_R$ on 2^X, the unary operator ${}^{\smile_f}$ on 2^X and the designated set $1'_I \subseteq X$ are defined, for any $A, B \subseteq X$, by

$$\begin{aligned} A ;_R B &\stackrel{\text{def}}{=} \{z \in X : \exists x \in A,\ \exists y \in B,\ R(x,y,z)\} \\ A^{\smile_f} &\stackrel{\text{def}}{=} \{f(x) : x \in A\} \\ 1'_I &\stackrel{\text{def}}{=} I. \end{aligned}$$

Proposition 5.2.3 *The complex algebra of a relation frame is a relation algebra.*

Proof:
Clearly, the powerset algebra of the non-empty set X is a Boolean algebra. For associativity of $;_R$, let $B_1, B_2 \subseteq X$. We now use (FR5) to show

$$(A ;_R B_1) ;_R B_2 \subseteq A ;_R (B_1 ;_R B_2).$$

Assume $z \in (A ;_R B_1) ;_R B_2$. Then there are $w \in A ;_R B_1$ and $v \in B_2$ such that $R(w,v,z)$. So, expanding further, for some $x \in A$, for some $y \in B_1$, and for some $v \in B_2$, $R(x,y,w)$ and $R(w,v,z)$. Thus, by (FR5), for some $u \in X$, $R(x,u,z)$ and $R(y,v,u)$. Hence, for some $x \in A$ and for some $u \in B$, $R(x,u,z)$ and $u \in B_1 ;_R B_2$. Therefore $z \in A ;_R (B_1 ;_R B_2)$, as required.
Similarly, using (FR6), $A ;_R (B_1 ;_R B_2) \subseteq (A ;_R B_1) ;_R B_2$.
To show that I is the unit element of $;_R$ note that, using (FR4), we have

$$\begin{aligned} z \in A ;_R I &\Leftrightarrow \exists x \in A,\ \exists y \in I,\ R(x,y,z) \\ &\Leftrightarrow \exists x \in A,\ x = z \\ &\Leftrightarrow z \in A. \end{aligned}$$

Similarly, using (FR1) to (FR4), we have

$$\begin{aligned} z \in I ;_R A &\Leftrightarrow \exists x \in I,\ \exists y \in A,\ R(x,y,z) \\ &\Leftrightarrow \exists x \in I,\ \exists y \in A,\ R(z,f(y),x) \\ &\Leftrightarrow \exists x \in I,\ \exists y \in A,\ R(f(z),x,f(y)) \\ &\Leftrightarrow \exists y \in A,\ f(y) = f(z) \\ &\Leftrightarrow \exists y \in A,\ f(f(y)) = f(f(z)) \\ &\Leftrightarrow \exists y \in A,\ y = z \\ &\Leftrightarrow z \in A. \end{aligned}$$

(R1) By definition of union, it follows that $;_R$ distributes over $\cup$.

(R2) Using (FR1) we have

$$\begin{aligned} z \in A^{\smile_f \smile_f} &\Leftrightarrow \exists y \in A^{\smile_f},\ z = f(y) \\ &\Leftrightarrow \exists x \in A,\ z = f(f(x)) \\ &\Leftrightarrow \exists x \in A,\ z = x \\ &\Leftrightarrow z \in A. \end{aligned}$$

(R3) holds since, for any $z \in X$,

$$\exists x \in A \cup B,\ z = f(x) \Leftrightarrow (\exists x \in A,\ z = f(x)) \text{ or } (\exists x \in B,\ z = f(x)).$$

(R4) We have that for any $z \in X$,

$$\begin{aligned} z \in (A \,;_R B)^{\smile_f} &\Leftrightarrow \exists u \in A \,;_R B,\ z = f(u) \\ &\Leftrightarrow \exists u,\ \exists x \in A,\ \exists y \in B,\ R(x,y,u)\ \&\ z = f(u) \\ &\Leftrightarrow \exists x \in A,\ \exists y \in B,\ R(x,y,f(z)) \end{aligned}$$

since $z = f(u)$ if an only if $u = f(z)$. Also,

$$\begin{aligned} z \in B^{\smile_f} \,;_R A^{\smile_f} &\Leftrightarrow \exists w \in A^{\smile_f},\ \exists v \in B^{\smile_f},\ R(v,w,z) \\ &\Leftrightarrow \exists x \in A,\ \exists y \in B,\ R(f(y),f(x),z). \end{aligned}$$

Now, by Lemma 5.2.2(a) and (b), and (FR1),

$$R(x,y,f(z)) \Leftrightarrow R(f(x),f(z),y) \Leftrightarrow R(y,z,f(x)) \Leftrightarrow R(f(y),f(x),z).$$

Hence, the required result follows.

(R5) Assume $z \in (A^{\smile_f}) \,;_R (-(A \,;_R B))$. Then there is some $x \in A$ and there is some $v \in X$ such that

$$\forall u \forall y,\ u \in A\ \&\ R(u,y,v) \Rightarrow (y \notin B)\ \&\ R(f(x),v,z).$$

Suppose $z \in B$. Take u to be x and y to be z. Then, since $x \in A$, $R(x,z,v)$ does not hold. Thus, by Lemma 5.2.2(a), not $R(f(x),v,z)$ which gives the required contradiction. □

The *canonical frame of a relation algebra* B is a structure $(X_B, R_;, f_\smile, I_{1'})$ such that X_B is the canonical frame of its Boolean reduct, and the ternary relation $R_;$ on X_B, the unary operator $f_\smile$ on X_B, and the designate set $I_{1'} \subseteq X_B$ are defined, for any $F, G, H \in X_B$, by

$$\begin{aligned} R_;(F,G,H) &\overset{\text{def}}{\Leftrightarrow} F \,; G \subseteq H \\ f_\smile(F) &\overset{\text{def}}{=} \{a^\smile : a \in F\} \\ I_{1'} &\overset{\text{def}}{=} \{F \in X_B : 1' \in F\}. \end{aligned}$$

We recall that $F \,; G = \{c \in B : \exists a \in F, \exists b \in G,\ a \,; b \leq c\}$ and, as shown in Section 1.6, it is a filter of B. Since B is a Boolean algebra, $f_\smile$ is well-defined, that is, $f_\smile(F) \in X_B$ for every $F \in X_B$.

Proposition 5.2.4 *The canonical frame of a relation algebra is a relation frame.*

Proof:
(FR1) Using (R2) we have

$$\begin{aligned} f_{\smile}(f_{\smile}(F)) &= \{a^{\smile} : a \in f_{\smile}(F)\} \\ &= \{a^{\smile} : \exists b \in F\ a = b^{\smile}\} \\ &= \{b^{\smile\smile} : b \in F\}, \end{aligned}$$

hence $f_{\smile}(f_{\smile}(F)) = F$.

(FR2) Assume $R_;(F,G,H)$, that is, $F\,;G \subseteq H$. Take any $z \in B$ such that $z \in f_{\smile}(F)\,;H$ and $z \notin G$. Then, for some $a \in F$ and for some $b \in H$, $a^{\smile}\,;b \leq z$ and $z \notin G$, hence $a^{\smile}\,;b \notin G$. Now from $a \in F$ and $-(a^{\smile}\,;b) \in G$ it follows that $a;(-(a^{\smile}\,;b)) \in F\,;G \subseteq H$. Thus, by (R5), $-b \in H$, which gives the required contradiction.

(FR3) Assume $R_;(F,G,H)$, that is, $F\,;G \subseteq H$. Take any $z \in B$ such that $z \in H\,;f_{\smile}(G)$ and $z \notin F$. Then, for some $a \in G$ and for some $b \in H$, $b\,;a^{\smile} \leq z$ and $z \notin F$, and hence $b\,;a^{\smile} \notin F$. Now $-(b\,;a^{\smile})\,;a \in F\,;G \subseteq H$. Furthermore, $b\,;a^{\smile} = (b\,;a^{\smile})^{\smile\smile} = (a\,;b^{\smile})^{\smile}$. By (R5), $a^{\smile}\,;(-(a\,;b^{\smile})) \leq -b^{\smile}$ and by Lemma 5.2.1(d), $(a^{\smile}\,;(-(a\,;b^{\smile})))^{\smile} \leq (-b^{\smile})^{\smile}$, so $-(a\,;b^{\smile})^{\smile}\,;a \leq -b$. Since $-(a\,;b^{\smile})^{\smile}\,;a \in H$, $-b \in H$ which gives the contradiction.

(FR4) Assume $F = G$. Let $\uparrow_{\leq} 1'$ be the principal filter generated by $1'$. Then $F;\ \uparrow_{\leq} 1' \subseteq G$. Hence, by Theorem 1.6.11, there is some $H \in X_B$ such that $1' \in\ \uparrow_{\leq} 1' \subseteq H$ and $F;H \subseteq G$. On the other hand, assume that for some $H \in X_B$, $1' \in H$ and $F;H \subseteq G$. Take any $a \in F$. Then $a = a;1' \in F;H \subseteq G$. So $F \subseteq G$. Now take any $a \notin F$. Then, since $(B,\vee,\wedge,-,0,1)$ is a Boolean algebra, $-a \in F$ and hence $-a = -a;1' \in F;H \subseteq G$. Thus $a \notin G$ and hence $G \subseteq F$.

(FR5) Assume $R_;(F,G,H)$ and $R_;(H,K,M)$, that is, $F;G \subseteq H$ and $H;K \subseteq M$. Then $F;(G;K) \subseteq M$ and $G;K$ is a filter. Therefore by Theorem 1.6.11, there is a prime filter $U \in X_B$ such that $G;K \subseteq U$ and $F;U \subseteq M$. Hence $R_;(G,K,U)$ and $R_;(F,U,M)$, as required.

The proof of (FR6) is similar to that for (FR5). □

Let B be a relation algebra. We now show that the Stone embedding $h : B \to 2^{X_B}$, defined in Section 2.3, preserves the binary operator ;, the unary operator $^{\smile}$ and preserves the designated element $1'$.

Lemma 5.2.5 *For any $a,b \in B$,*

(a) $h(a;b) = h(a)\,;_{R_;}\,h(b)$;
(b) $h(a^{\smile}) = h(a)^{\smile_{f_{\smile}}}$;
(c) $h(1') = 1'_{h(1')}$.

Proof:
(a) We need to show that, for any $H \in X_B$ and any $a, b \in B$,

$$a\,;b \in H \Leftrightarrow \exists F, G \in X_B,\ a \in F\ \&\ b \in G\ \&\ F;G \subseteq H.$$

Assume for some $F, G \in X_B$, $a \in F$, $b \in G$ and $F;G \subseteq H$. Take $c = a;b$. Then $c \in F;G$ and hence $c = a;b \in H$. On the other hand, assume $a;b \in H$. Then $\uparrow_{\leq} a; \uparrow_{\leq} b \subseteq H$ where $\uparrow_{\leq} a$ and $\uparrow_{\leq} b$ are principal filters generated by a and b, respectively. So, by Theorem 1.6.11, there are prime filters of B, say F and G, such that $\uparrow_{\leq} a \subseteq F$, $\uparrow_{\leq} b \subseteq G$, and $F;G \subseteq H$. Clearly, $a \in \uparrow_{\leq} a$, so $a \in F$. Similarly, $b \in G$. Hence $H \in h(a)\ ;_R\ h(b)$.

(b) For any $H \in X_B$ and for any $a \in B$,

$$\begin{aligned} H \in h(a^{\smile}) &\Leftrightarrow a^{\smile} \in H \Leftrightarrow a^{\smile\smile} \in f_{\smile}(H) \Leftrightarrow a \in f_{\smile}(H) \\ &\Leftrightarrow f_{\smile}(f_{\smile}(H)) \in h(a)^{\smile_{f_{\smile}}} \Leftrightarrow H \in h(a)^{\smile_{f_{\smile}}}. \end{aligned}$$

(c) Note that $1'_{h(1')} = 1'_{I_{1'}} = I_{1'} = h(1')$. □

Let X be a relation frame. We now show that the embedding $k : X \to X_{2^X}$, defined in Section 2.3, preserves the relation R, the function f, and the set I.

Lemma 5.2.6 *For any $x, y, z \in X$,*

(a) $R(x, y, z) \Leftrightarrow R_{;_R}(k(x), k(y), k(z))$;
(b) $k(f(x)) = f_{\smile_f}(k(x))$;
(c) $x \in I \Leftrightarrow k(x) \in I_{1'_I}$.

Proof:
(a) For any $x, y, z \in X$,

$$\begin{aligned} & R_{;_R}(k(x), k(y), k(z)) \\ \Leftrightarrow\ & k(x)\ ;_R\ k(y) \subseteq k(z) \\ \Leftrightarrow\ & \forall A \subseteq X,\ [\exists B_1 \in k(x),\ \exists B_2 \in k(y),\ B_1\ ;_R\ B_2 \subseteq A] \Rightarrow A \in k(z) \\ \Leftrightarrow\ & \forall A \subseteq X,\ [\exists B_1, B_2 \subseteq X,\ (x \in B_1)\ \&\ (y \in B_2) \\ & \qquad \&\ \{z \in X : \exists u \in B_1,\ \exists t \in B_2,\ R(u, t, z)\} \subseteq A] \Rightarrow z \in A. \end{aligned}$$

Let $x, y, z \in X$ be such that $R(x, y, z)$ and take any $A \in 2^X$ such that for some $B_1, B_2 \in 2^X$, $x \in B_1$, $y \in B_2$, and $B_1\ ;_R\ B_2 \subseteq A$. Now, since $x \in B_1$, $y \in B_2$, and $R(x, y, z)$, we get $z \in B_1\ ;_R\ B_2$, so $z \in A$. On the other hand, let $x, y, z \in X$ be such that $R_{;_R}(k(x), k(y), k(z))$. Consider $B_1 = \{x\}$ and $B_2 = \{y\}$. Then $B_1\ ;_R\ B_2 = \{u \in X : R(x, y, u)\}$. Put $A = \{u \in X : R(x, y, u)\}$. Hence $z \in A$, thus $R(x, y, z)$.

(b) Note that for every $x \in X$, $f_{\smile_f}(k(x)) = \{A^{\smile_f} : A \in k(x)\} = \{A^{\smile_f} : x \in A\}$. Then for every $A \subseteq X$, we have:

$$\begin{aligned} A \in k(f(x)) &\Leftrightarrow f(x) \in A \Leftrightarrow f(f(x)) \in A^{\smile_f} \Leftrightarrow x \in A^{\smile_f} \\ &\Leftrightarrow A^{\smile_f \smile_f} \in f_{\smile_f}(k(x)) \Leftrightarrow A \in f_{\smile_f}(k(x)). \end{aligned}$$

(c) For every $x \in X$ we have

$$k(x) \in I_{1'_I} \Leftrightarrow 1'_I \in k(x) \Leftrightarrow x \in 1'_I = I$$

which completes the proof. □

With these results we obtain the following discrete representations for relation algebras and relation frames.

Theorem 5.2.7

(a) *Every relation algebra is embeddable into the complex algebra of its canonical frame;*

(b) *Every relation frame is embeddable into the canonical frame of its complex algebra.*

5.3 Cylindric algebras of finite dimension

A *cylindric algebra of dimension* n, $n \in \mathbb{N}$ and $n \geq 3$, is a structure

$$(B, \{d_{ij} : i, j \leq n\}, \{c_i : i \leq n\})$$

such that B is a Boolean algebra and, for all $i, j \leq n$, $d_{ij} \in B$ and $c_i : B \to B$ satisfy, for any $a, b \in B$,

(C1) $c_i(0) = 0$

(C2) $a \leq c_i(a)$

(C3) $c_i(a \wedge c_i(b)) = c_i(a) \wedge c_i(b)$

(C4) $c_i(c_j(a)) = c_j(c_i(a))$

(C5) $d_{ii} = 1$

(C6) $k \neq i, j \Rightarrow d_{ij} = c_k(d_{ik} \wedge d_{kj})$

(C7) $i \neq j \Rightarrow c_i(d_{ij} \wedge a) \wedge c_i(d_{ij} \wedge -a) = 0.$

The proofs of the following properties of the cylindrification can be found in [126] and [127].

Lemma 5.3.1 *Let* $(B, \{d_{ij} : i, j \leq n\}, \{c_i : i \leq n\})$ *be a cylindric algebra. For any* $a, b \in B$ *and for any* $i \leq n$,

(a) $c_i(c_i(a)) = c_i(a)$;

(b) $a \leq b \Rightarrow c_i(a) \leq c_i(b)$;

(c) $c_i(a \vee b) = c_i(a) \vee c_i(b)$;

(d) $c_i(1) = 1$;

(e) $c_i(a) \wedge b = 0 \Leftrightarrow a \wedge c_i(b) = 0$;

(f) $c_i(-c_i(a)) = -c_i(a)$.

A *cylindric frame* is a structure $(X, \{D_{ij} : i, j \leq n\}, \{E_i : i \leq n\})$ where X is a non-empty set and, for all $i, j \leq n$, $D_{ij} \subseteq X$ and $E_i \subseteq X^2$ satisfy

(FC1) E_i is an equivalence relation on X

(FC2) $E_i \,;\, E_j = E_j \,;\, E_i$

(FC3) $D_{ii} = X$

(FC4) $D_{ij} = \{y \in X : \forall k \neq i, j, \exists x \in D_{ik} \cap D_{kj} \,, xE_k y\}$

(FC5) $i \neq j$, $x, y \in D_{ij}$ and $xE_i y$ imply $x = y$.

The *complex algebra of a cylindric frame* X is a structure

$$(2^X, \{d_{D_{ij}} : i, j \leq n\}, \{c_{E_i} : i \leq n\})$$

where 2^X is the complex algebra of its Boolean frame reduct X and, for all $i, j \leq n$, $d_{D_{ij}} \in 2^X$ and $c_{E_i} : 2^X \to 2^X$ are defined, for any $A \subseteq X$, by

$$d_{D_{ij}} \stackrel{\text{def}}{=} D_{ij}$$
$$c_{E_i}(A) \stackrel{\text{def}}{=} \{y \in X : \exists x \in A, \; xE_i y\}.$$

Proposition 5.3.2 *The complex algebra of a cylindric frame is a cylindric algebra.*

Proof:
Clearly, the powerset algebra of the non-empty set X is a Boolean algebra.
(C1) From the definition of c_{E_i} it immediately follows that $c_{E_i}(\emptyset) = \emptyset$.

(C2) Since E_i is reflexive, $yE_i y$ for any $y \in X$, so in particular for any $y \in A$. Hence $A \subseteq c_{E_i}(A)$.

(C3) We have to show that $c_{E_i}(A \cap c_{E_i}(B)) = c_{E_i}(A) \cap c_{E_i}(B)$. Note that for all $A, B \subseteq X$, for every $y \in X$, and for every $i \leq n$,

$$\begin{aligned} y \in c_{E_i}(A \cap c_{E_i}(B)) &\Leftrightarrow \exists x \in X, \; x \in A \cap c_{E_i}(B) \;\&\; xE_i y \\ &\Leftrightarrow \exists x \in X, \; x \in A \;\&\; (\exists z \in X, \; z \in B \;\&\; zE_i x) \;\&\; xE_i y \\ &\Rightarrow (\exists x \in A, \; xE_i y) \;\&\; (\exists z \in B, \; zE_i y) \\ &\Leftrightarrow y \in c_{E_i}(A) \cap c_{E_i}(B). \end{aligned}$$

Hence $c_{E_i}(A \cap c_{E_i}(B) \subseteq c_{E_i}(A) \cap c_{E_i}(B)$. On the other hand, assume that $y \in c_{E_i}(A) \cap c_{E_i}(B)$. Then for some $x \in A$ and for some $z \in B$, $xE_i y$ and $zE_i y$. Since E_i is symmetric and transitive, we get $zE_i x$. Hence $x \in A \cap c_{E_i}(B)$. Since also $xE_i y$, we get $y \in c_{E_i}(A \cap c_{E_i}(B))$, as required.

(C4) Using (FC2) we have for any $A \subseteq X$ and for any $y \in X$,

$$\begin{aligned} y \in c_{E_i}(c_{E_j}(A)) &\Leftrightarrow \exists x \in X, \; x \in c_{E_j}(A) \;\&\; xE_i y \\ &\Leftrightarrow \exists x, z \in X, \; z \in A \;\&\; zE_j x \;\&\; xE_i y \\ &\Leftrightarrow \exists z \in X, \; z \in A \;\&\; z(E_j \,;\, E_i)y \\ &\Leftrightarrow \exists z \in X, \; z \in A \;\&\; z(E_i \,;\, E_j)y \\ &\Leftrightarrow \exists z, x \in X, \; z \in A \;\&\; zE_i x \;\&\; xE_j y \\ &\Leftrightarrow \exists x \in X, \; x \in c_{E_i}(A) \;\&\; xE_j y \\ &\Leftrightarrow y \in c_{E_j}(c_{E_i}(A)). \end{aligned}$$

(C5) By (FC3), $d_{D_{ii}} = D_{ii} = X$.

(C6) Assume $k \neq i, j$. Then, by (FC4), for any $y \in X$,

$$y \in d_{D_{ij}} = D_{ij} \Leftrightarrow \exists x \in D_{ik} \cap D_{kj},\ xE_k y \Leftrightarrow y \in c_{E_k}(D_{ik} \cap D_{kj}).$$

(C7) Assume $i \neq j$. Take any $y \in c_{E_i}(d_{D_{ij}} \cap A)$. Then, for some $x \in d_{D_{ij}} \cap A$, $xE_i y$. Suppose $y \in c_{E_i}(d_{D_{ij}} \cap -A)$. Then, for some $z \in d_{D_{ij}} \cap -A$, $zE_i y$. Thus $x, z \in d_{D_{ij}}$ and by symmetry and transitivity of E_i, $xE_i z$. So, by (FC5), $z = x$. Then $z \in A$ and $x \notin A$ which give the required contradiction. □

The *canonical frame of a cylindric algebra B* is a structure

$$(X_B, \{D_{d_{ij}} : i, j \leq n\}, \{E_{c_i} : i \leq n\})$$

where X_B is the set of all prime filters of the Boolean algebra B and, for all $i, j \leq n$, $D_{d_{ij}} \subseteq X_B$ and $E_{c_i} \subseteq X_B \times X_B$ are defined, for any $F, G \in X_B$, by

$$D_{d_{ij}} \stackrel{\text{def}}{=} \{F \in X_B : d_{ij} \in F\}$$
$$FE_{c_i}G \stackrel{\text{def}}{\Leftrightarrow} c_i(F) \subseteq G,$$

where $c_i(F) = \{c_i(a) : a \in F\}$.

Proposition 5.3.3 *The canonical frame of a cylindric algebra is a cylindric frame.*

Proof:
(FC1) We have to show that E_{c_i} are equivalence relations. For reflexivity of E_{c_i} take $F \in X_B$ and $a \in B$ such that $a \in c_i(F)$. Then there is some $b \in F$ such that $a = c_i(b)$. Since $b \in F$ and F is a filter of B, by (C2) we get $c_i(b) \in F$, thus $a \in F$. Hence $c_i(F) \subseteq F$, as expected.
Suppose that E_{c_i} is not symmetric. Then, for some $F, G \in X_B$, $FE_{c_i}G$ and not $GE_{c_i}F$, that is, $c_i(F) \subseteq G$ and $c_i(G) \cap -F \neq \emptyset$. Hence there is some $a \in B$ such that $a \in -F$ and $a \in c_i(G)$, so $-a \in F$ and, for some $b \in G$, $a = c_i(b)$. Now, by Lemma 5.3.1(f), $-c_i(b) = c_i(-c_i(b)) = c_i(-a)$. Also $c_i(-a) \in c_i(F)$. Then $-c_i(b) \in c_i(F) \subseteq G$, and hence $c_i(b) \notin G$. Since $b \in G$ we have by (C2) that $c_i(b) \in G$ which gives the required contradiction. Therefore, E_{c_i} is symmetric.
For transitivity of E_{c_i}, assume that $c_i(F) \subseteq G$ and $c_i(G) \subseteq H$. Take any $b \in B$ such that $b \in F$. Then $c_i(b) \in G$ and hence $c_i(c_i(b)) \in H$. Now, by Lemma 5.3.1(a), $c_i(b) \in H$, as required.

(FC2) We have to show that $E_{c_i}\,;E_{c_j} = E_{c_j}\,;E_{c_i}$. By (C4) we have for all $F, G \in X_B$,

$$\begin{aligned} F(E_i\,;E_j)G &\Leftrightarrow \exists H \in X_B,\ FE_iH\ \&\ HE_jG \\ &\Leftrightarrow \exists H \in X_B,\ c_i(F) \subseteq H\ \&\ c_j(H) \subseteq G \\ &\Leftrightarrow c_j(c_i(F)) \subseteq G \\ &\Leftrightarrow c_i(c_j(F)) \subseteq G \\ &\Leftrightarrow \exists H \in X_B,\ c_j(F) \subseteq H\ \&\ c_i(H) \subseteq G \\ &\Leftrightarrow \exists H \in X_B,\ FE_jH\ \&\ HE_iG \\ &\Leftrightarrow F(E_j\,;E_i)G. \end{aligned}$$

(FC3) Using (C5) it follows that $D_{d_{ii}} = \{F \in X_B : 1 = d_{ii} \in F\} = X_B$.

(FC4) We need to show that for any $G \in X_B$,

$$G \in D_{d_{ij}} \Leftrightarrow \forall k \neq i, j,\ \exists F \in X_B,\ d_{ik} \wedge d_{kj} \in F\ \&\ FE_{c_k}G.$$

Assume that $G \in D_{d_{ij}}$ and $k \neq i, j$. Then, by the definition of $D_{d_{ij}}$ and by (C6), $c_k(d_{ik} \wedge d_{kj}) \in G$. Then $d_{ik} \wedge d_{kj} \in c_k^{-1}(G)$ and, since $c_k^{-1}(G)$ is up-closed, $\uparrow_\leq (d_{ik} \wedge d_{kj}) \subseteq c_k^{-1}(G)$. Also $-c_k^{-1}(G)$ is an ideal disjoint from the principal filter $\uparrow_\leq (d_{ik} \wedge d_{kj})$. Therefore, by the prime filter theorem, there is a prime filter $F \in X_B$ such that $\uparrow_\leq (d_{ik} \wedge d_{kj}) \subseteq F$ and $F \subseteq c_k^{-1}(G)$. Hence, $d_{ik} \wedge d_{kj} \in F$ and $c_k(F) \subseteq c_k(c_k^{-1}(G)) \subseteq G$. On the other hand, take any $G \in X_B$ such that, for $k \neq i, j$, there is some $F \in X_B$ such that $d_{ik} \wedge d_{kj} \in F$ and $c_k(F) \subseteq G$. Hence, $c_k(d_{ik} \wedge d_{kj}) \in c_k(F)$ and $c_k(F) \subseteq G$. Thus, by (C6), $d_{ij} \in G$, that is, $G \in D_{d_{ij}}$.

(FC5) Assume that $i \neq j$, $F, G \in D_{d_{ij}}$ and $c_i(F) \subseteq G$. Take any $a \in G$. Then $a \wedge d_{ij} \in G$, so by (C2), $c_i(a \wedge d_{ij}) \in G$. Hence, by (C7), $-c_i(-a \wedge d_{ij}) \in G$ and so $c_i(-a \wedge d_{ij}) \notin G$. Thus $-a \wedge d_{ij} \notin F$, that is, $-a \notin F$ since $d_{ij} \in F$. Therefore, $a \in F$. Thus $G \subseteq F$. On the other hand, take any $a \in F$. Then $c_i(a \wedge d_{ij}) \in c_i(F)$ and since $c_i(F) \subseteq G$, we get $c_i(a \wedge d_{ij}) \in G$. Hence, by (C7), $c_i(-a \wedge d_{ij}) \notin G$ and hence, by (C2), $-a \wedge d_{ij} \notin G$, that is, $-a \notin G$. Therefore $a \in G$, thus $F \subseteq G$. □

Let B be a cylindric algebra. We now show that the Stone embedding $h : B \to 2^{X_B}$, defined in Section 2.3, preserves the operators $c_i : B \to B$, for $i \leq n$.

Lemma 5.3.4 *For any $i, j \leq n$, and for any $a \in B$,*

(a) $h(d_{ij}) = d_{D_{d_{ij}}}$;

(b) $h(c_i(a)) = c_{E_{c_i}}(h(a))$.

Proof:
(a) By definition, for any $F \in X_B$,

$$F \in h(d_{ij}) \Leftrightarrow d_{ij} \in F \Leftrightarrow F \in D_{d_{ij}} = d_{D_{d_{ij}}}.$$

(b) We need to show, for any $G \in X_B$ and any $a \in B$,

$$c_i(a) \in G \Leftrightarrow \exists F \in h(a),\ FE_{c_i}G \Leftrightarrow \exists F \in X_B,\ a \in F\ \&\ c_i(F) \subseteq G.$$

The right-to-left direction is obvious. For the left-to-right direction, assume that $c_i(a) \in G$. We need to show that there is some $F \in X_B$ such that $a \in F$ and $c_i(F) \subseteq G$. Since $c_i(a) \in G$, we have $a \in c_i^{-1}(G)$. Now $c_i^{-1}(G)$ is an up-closed set containing the principal filter $\uparrow_\leq a$ and $-c_i^{-1}(G)$ is an ideal. By Theorem 1.6.8 there is a prime filter, say F such that $\uparrow_\leq a \subseteq F$ and $F \subseteq c_i^{-1}(G)$. Hence, for some $F \in X_B$, $a \in F$ and $c_i(F) \subseteq c_i(c_i^{-1}(G)) \subseteq G$. □

Let X be a cylindric frame. We now show that the embedding $k : X \to X_{2^X}$, defined in Section 2.3, preserves the designated sets $D_{ij} \subseteq X$, for $i, j \leq n$, and the binary relations $E_i \subseteq X^2$, for $i \leq n$.

Lemma 5.3.5 *For any $x, y \in X$,*

(a) $x \in D_{ij} \Leftrightarrow k(x) \in D_{d_{D_{ij}}}$;

(b) $xE_iy \Leftrightarrow k(x)E_{c_{E_i}}k(y)$.

Proof:
(a) Note that $k(x) \in D_{d_{D_{ij}}} \Leftrightarrow d_{D_{ij}} \in k(x) \Leftrightarrow D_{ij} \in k(x) \Leftrightarrow x \in D_{ij}$.

(b) Note that, for any $x, y \in X$,

$$\begin{aligned} k(x)E_{c_{E_i}}k(y) &\Leftrightarrow c_{E_i}(k(x)) \subseteq k(y) \\ &\Leftrightarrow \{c_{E_i}(A) : x \in A\} \subseteq k(y) \\ &\Leftrightarrow \forall A \subseteq X,\ x \in A \Rightarrow c_{E_i}(A) \in k(y) \\ &\Leftrightarrow \forall A \subseteq X,\ x \in A \Rightarrow y \in c_{E_i}(A) \\ &\Leftrightarrow \forall A \subseteq X,\ x \in A \Rightarrow \exists u \in A,\ uE_iy. \end{aligned}$$

Assume xE_iy. Take any $A \subseteq X$ such that $x \in A$. Then, for some $u \in A$, uE_iy. In particular, taking u to be x, we get uE_iy. On the other hand, assume xE_iy does not hold. Let $A = \{x\} \subseteq X$. Then $x \in A$ and, for every $u \in A$, uE_iy does not hold. Hence $k(x)E_{c_{E_i}}k(y)$ does not hold. □

With these results we obtain the following discrete representations for cylindric algebras and cylindric frames.

Theorem 5.3.6

(a) *Every cylindric algebra is embeddable into the complex algebra of its canonical frame;*

(b) *Every cylindric frame is embeddable into the canonical frame of its complex algebra.*

Chapter 6

Multirelational structures

6.1 Introduction

A calculus of binary multirelations was introduced in [211] and further developed in [217] as a tool for defining semantics of commands with two types of nondeterminism that cannot be captured by the ordinary relational semantics. A binary multirelation over a non-empty set is a subset of the Cartesian product of that set and the powerset of the set. In this chapter we consider two major classes of multirelational structures of the form (X, R), where R is an up-closed (respectively down-closed) multirelation over a non-empty set X.

In Section 6.2 an arithmetic of multirelations, some of their properties, and their relationship with ordinary binary relations are presented. In Section 6.3 and Section 6.4 discrete representation theorems for the class of up-closed (respectively down-closed) multirelational structures and their algebras are presented based on the results in [66]. The algebras associated with the up-closed multirelational structures are Boolean algebras with a monotone operator and those associated with down-closed multirelational structures are Boolean algebras with an antitone operator.

An algebraic structure of some subclasses of multirelational structures is a subject of investigations in [89, 87] where relationships between Kleene algebras and algebras of multirelations are presented. In [88] a representation theorem for complete idempotent semirings is proved such that the representation algebra is an algebra of multirelations. A duality of Boolean algebras with a monotone operator and monotone general neighbourhood frames is given in [120]. In [42] a duality between Boolean algebras with arbitrary additional unary operators and general neighbourhood frames is established. In [21] an extension of duality presented in [120] to the class of Boolean algebras endowed with n-ary normal monotone operator is developed. It generalizes the Jónsson-Tarski duality presented in [140].

6.2 Multirelations

Let X and Y be non-empty sets. A *multirelation* is a subset of the Cartesian product $X \times 2^Y$, that is, a set of ordered pairs (x, A) where $x \in X$ and $A \subseteq Y$. The image under a binary multirelation R of any $x \in X$ is denoted $R(x)$ and defined to be the set $\{A \subseteq Y : xRA\}$. Mostly we will deal with the case of $X = Y$, and we will refer to binary multirelations simply as multirelations. We distinguish two extreme multirelations, namely the *universal multirelation* $\top = X \times 2^X$ and the *empty multirelation* $\bot = \emptyset$.

For any multirelations $R, T \subseteq X \times 2^X$, their composition may be defined by

$$R\,;T = \{(s, Q) : \exists Q' \subseteq X,\ sRQ' \ \&\ Q' \subseteq \{y \in X : yTQ\}\}.$$

Multirelations have many interesting properties, some of which are given below. Let X be a non-empty set and $R \subseteq X \times 2^X$ be a multirelation. Then

R is total if, for every $x \in X$, $R(x) \neq \emptyset$

R is proper if, for every $x \in X$, $\emptyset \notin R(x)$

R is up-closed if, for every $x \in X$ and for every $A \subseteq X$,

$$xRA \ \&\ A \subseteq A' \Rightarrow xRA'$$

Dually, R is down-closed if, for any $x \in X$ and any $A \subseteq X$,

$$xRA \ \&\ A' \subseteq A \Rightarrow xRA'$$

R is multiplicative if, for every $x \in X$ and for every non-empty set $\mathsf{A} \subseteq 2^X$,

$$xR(\bigcap \mathsf{A}) \Leftrightarrow \forall B \in \mathsf{A},\ xRB$$

R is additive if, for every $x \in X$ and for any non-empty set $\mathsf{A} \subseteq 2^X$

$$xR(\bigcup \mathsf{A}) \Leftrightarrow \exists B \in \mathsf{A},\ xRB$$

R is co-multiplicative if, for every $x \in X$ and for every non-empty set $\mathsf{A} \subseteq 2^X$,

$$xR(\bigcap \mathsf{A}) \Leftrightarrow \exists B \in \mathsf{A},\ xRB$$

R is co-additive if, for every $x \in X$ and for every non-empty set $\mathsf{A} \subseteq 2^X$,

$$xR(\bigcup \mathsf{A}) \Leftrightarrow \forall B \in \mathsf{A},\ xRB.$$

It is easy to show that every multiplicative (or additive) multirelation is up-closed, and that every co-multiplicative (or co-additive) multirelation is down-closed. Let $\mathcal{R}_u$ (respectively, $\mathcal{R}_d$) denote the family of up-closed (respectively, down-closed) multirelations.

If $M \subseteq X$, we let $F_M = \{Y \subseteq X : M \subseteq Y\}$ be the principal filter in 2^X generated by M, and $I_M = \{Y \subseteq X : Y \subseteq M\}$ be the principal ideal in 2^M generated by M.

Lemma 6.2.1 *Let $R \in \mathcal{R}_u$. Then, R is multiplicative if and only if*

$$R = \bigcup\{\{x\} \times F_{\bigcap R(x)} : x \in \mathrm{dom}(R)\}.$$

Proof:
Suppose that R is multiplicative, and let $x \in \operatorname{dom}(R)$. Since R is multiplicative, we have $xR\bigcap R(x)$, and the fact that R is up-closed implies $\{x\} \times F_{\bigcap R(x)} \subseteq R$. Conversely, if xRA, then $\bigcap R(x) \subseteq A$, hence, $(x, A) \in \{x\} \times F_{\bigcap R(x)}$. On the other hand, suppose $R = \bigcup\{\{x\} \times F_{\bigcap R(x)} : x \in \operatorname{dom}(R)\}$. If $\{A_i : i \in I\} \subseteq 2^X$, and $xR\bigcap\{A_i : i \in I\}$, then xRA_i for every $i \in I$, since R is up-closed. Conversely, if xRA_i for every $i \in I$, then $\bigcap R(x) \subseteq A_i$ for every $i \in I$, and hence, $\bigcap R(x) \subseteq \bigcap\{A_i : i \in I\}$. Since $F_{\bigcap R(x)}$ is up-closed and $\{x\} \times F_{\bigcap R(x)} \subseteq R$, we obtain $xR\bigcap\{A_i : i \in I\}$. □

Lemma 6.2.2 *The family of up-closed (respectively down-closed) multirelations is a complete ring of sets in which*

$$\bigwedge\{R_i : i \in I\} = \bigcap\{R_i : i \in I\}$$
$$\bigvee\{R_i : i \in I\} = \bigcup\{R_i : i \in I\}.$$

Proof:
By way of example we give the poof for the family of down-closed multirelations. Consider $\{R_i : i \in I\} \subseteq \mathcal{R}_d$. Let $x(\bigcap\{R_i : i \in I\})A$, that is, xR_iA for every $i \in I$, and $A' \subseteq A$. Since each R_i is down-closed, we have xR_iA' for every $i \in I$, that is, $x(\bigcap\{R_i : i \in I\})A'$. It follows that $\bigcap\{R_i : i \in I\}$ is down-closed. Next, let $x(\bigcup\{R_i : i \in I\})A$. Then there is some $i \in I$ such that xR_iA. If $A' \subseteq A$, then xRA', since $R_i \in \mathcal{R}_d$. Hence, $x(\bigcup\{R_i : i \in I\})A'$, and thus $\bigcup\{R_i : i \in I\})$ is in $\mathcal{R}_d$. □

Note that if $R \subseteq X \times 2^X$ is an up-closed (respectively, down-closed) multirelation whose domain contains more than one element, there is a partition $\{D_0, D_1\}$ of $\operatorname{dom}(R)$. If

$$R_0 = \bigcup\{\{x\} \times R(x) : x \in D_0\} \text{ and } R_1 = \bigcup\{\{x\} \times R(x) : x \in D_1\},$$

then $R_0, R_1 \neq \emptyset$, $R_0 \cap R_1 = \emptyset$, both R_0 and R_1 are up-closed (respectively, down-closed), and $R_0 \cup R_1 = R$. Thus, when proving (a) and one part of (b) below we assume $\operatorname{dom}(R) = \{x\}$.

Lemma 6.2.3 *Let X be a non-empty set.*

(a) *An up-closed non-empty relation R is completely join irreducible in $\mathcal{R}_u$ if and only if, for some $x \in X, M \subseteq X$, $R = \{x\} \times F_M$;*

(b) *A down-closed non-empty relation R is completely join irreducible in $\mathcal{R}_d$ if and only if R is co-additive and* $\operatorname{dom}(R)$ *has exactly one element.*

Proof:
For (a), suppose that R is non-empty, up-closed and completely join irreducible. Since R is up-closed, we have $R = \bigcup\{\{x\} \times F_Q : x \in \operatorname{dom}(R),\ Q \in R(x)\}$, and join irreducibility of R implies that $R = \{x\} \times F_Q$ for some $x \in X$ and $Q \subseteq X$. Conversely, let $R = \{x\} \times F_Q$ for some $x \in X$ and $Q \subseteq X$. If $R = \bigcup\{R_i : i \in I\}$

where each R_i is up–closed, then xR_iQ for some $i \in I$. Since R_i is up-closed, $\{x\} \times F_Q \subseteq R_i \subseteq R = \{x\} \times F_Q$, and thus, $R = R_i$.

For (b), suppose that R is completely join irreducible in $\mathcal{R}_d$, and $R \neq \emptyset$. For each $Q \in R(x)$ let $R_Q = \{\{x\} \times 2^Q\}$. Clearly, each R_Q is down-closed and co-additive. Since R is down-closed $R_Q \subseteq R$, hence $R = \bigcup\{R_Q : Q \in R(x)\}$. Since R is completely join irreducible, there is some $Q \in R(X)$ for which $R = R_Q$, and it follows that R is co-additive. Conversely, let $\text{dom}(R) = \{x\}$, and R be co-additive. Suppose that $R = \bigcup\{R_i : i \in I\}$ and each R_i is down-closed. Then, $xR_i\bigcup R_i(x)$ since R is co-additive, and therefore, since $R_i \subseteq R$ and R is co-additive, $xR\bigcup\{\bigcup R_i(x) : i \in I\}$. Since $R = \bigcup\{R_i : i \in I\}$, there is some $j \in I$ such that $xR_j\bigcup\{\bigcup R_i(x) : i \in I\}$. If xRA, there is some $i \in I$ such that xR_iA, that is, $A \in R_i(x)$. Since $A \subseteq \bigcup\{\bigcup R_i(x) : i \in I\}$ and R_j is down-closed, it follows that xR_jA. Hence, $R \subseteq R_j$. □

Lemma 6.2.4 *Let X be a non-empty set.*

(a) *If R is up-closed and $R \neq X \times 2^X$, then R is completely meet irreducible in $\mathcal{R}_u$ if and only if there are some $x \in X$ and some $M \subseteq X$, $M \neq X$ such that $R = (X \times 2^X) - (\{x\} \times I_M)$;*

(b) *If R is down-closed and $R \neq X \times 2^X$, then R is completely meet irreducible in $\mathcal{R}_d$ if and only if there are some $x \in X$ and some $M \subseteq X$, $M \neq X$, such that $R = (X \times 2^X) - (\{x\} \times F_M)$.*

Proof:
For (a), since $R \neq X \times 2^X$, there is some $(x, A) \notin R$. Let $M = \{y : (x, \{y\}) \notin R\}$. Then, $M \neq \emptyset$, since $R \neq X \times 2^X$ and R is up-closed. Also, $R \cap (\{x\} \times I_M) = \emptyset$. Assume that there is some $A \subseteq M$ such that $(x, A) \in R$. Since R is up-closed, it follows that $(x, M) \in R$. For each $y \in M$ let $R_y = R \cup (\{x\} \times F_{\{y\}})$. By definition of M, $(x, \{y\}) \notin R$ and thus, $R_y \neq R$ for every $y \in M$. Now, $\bigcap\{R_y : y \in M\} = R \cup \bigcap\{\{x\} \times F_{\{y\}} : y \in M\} = R \cup (\{x\} \times F_M) = R$, the latter by $(x, M) \in R$ and the fact that R is up-closed. This contradicts the hypothesis that R is completely meet irreducible. It follows that $R \subseteq (X \times 2^X) - (\{x\} \times I_M)$. Conversely, let $(y, \{z\}) \notin R$. Assume $x \neq y$. Choose some $u \in M$ and set $R_0 = R \cup (\{x\} \times F_{\{u\}}), R_1 = R \cup (\{y\} \times F_{\{z\}})$. Then, both R_0 and R_1 are up-closed, $R_0, R_1 \neq R$ and $R_0 \cap R_1 = R$, contradicting our hypothesis. By definition of M we now have $z \in M$, and thus, $(X \times 2^X) - (\{x\} \times I_M) \subseteq R$. If $M = X$, then $x \notin \text{dom}(R)$ and clearly, R cannot be meet irreducible. On the other hand, suppose that $R = (X \times 2^X) - (\{x\} \times I_M)$ for some $x \in X$, $M \subseteq X$, $M \neq X$. Let $R = \bigcap\{R_i : i \in I\}$ where each R_i is up-closed, and assume that $R_i \neq R$ for every $i \in I$. If $(y, A) \in R_i - R$, then $y = x$ and $A \subseteq M$ by $R \subseteq R_i$ and the definition of R. Since each R_i is up-closed, $(x, M) \in R_i$ for every $i \in I$ and thus $(x, M) \in \bigcap\{R_i : i \in I\} = R$ which contradicts the definition of R.

For (b), since $R \neq X \times 2^X$, there is some $(x, A) \notin R$. Let $M = X - Q$, where $Q = \{y : (x, X - \{y\}) \notin R\}$. If $Q = \emptyset$, then $R = (X \times 2^X) - \{(x, X)\}$, and R has the desired form. Thus, suppose that $Q \neq \emptyset$. Assume that $R \cap (\{x\} \times F_M) \neq \emptyset$. Since R is down-closed $(x, M) \in R$. For every $y \in Q$, let

$$R_y = R \cup (\{x\} \times I_{X - \{y\}}).$$

Then, each R_y is down-closed, $R \neq R_y$ by definition of Q, and $\bigcap\{R_y : y \in Q\} = R \cup \bigcap\{\{x\} \times I_{X-\{y\}} : y \in Q\} = R \cup (\{x\} \times I_M) = R$, the latter since $(x, M) \in R$ by the assumption. This contradicts the fact that R is completely meet irreducible. Conversely, let $(y, B) \notin R$. We need to prove that $(y, B) \in \{x\} \times F_M$. As in the previous lemma it can be shown that $x = y$. Assume that $M \not\subseteq B$ and choose some $z \in M - B$. Since $z \in M$, we have $(x, X - \{z\}) \in R$ by the definition of M. Since $z \notin B$ and R is down-closed, we obtain $(x, B) \in R$, a contradiction. Hence, $(X \times 2^X) - (\{x\} \times F_M) \subseteq R$. On the other hand, suppose that $R = (X \times 2^X) - (\{x\} \times F_M)$ for some $x \in X$, $M \subseteq X$. Let $R = \bigcap\{R_i : i \in I\}$ where each R_i is down-closed, and assume that $R_i \neq R$ for every $i \in I$. If $(y, B) \in R_i$, then $y = x$ and $M \subseteq B$ by the definition of R. Thus, $(x, M) \in R_i$ for every $i \in I$, since R_i is down-closed, and thus, $(x, M) \in R$, a contradiction. □

We conclude this description of multirelations with a correspondence between binary relations and certain multirelations. Let X be a non-empty set. From a multirelation $R \subseteq X \times 2^X$ we may define a binary relation $r_R \subseteq X \times X$ given, for every $x \in X$, by

$$x r_R y \overset{\text{def}}{\Leftrightarrow} xR\{y\}.$$

We will call such a relation r_R the flattening of the multirelation R. Given a binary relation $r \subseteq X \times X$, interesting examples of multirelations R_r defined from r include, for any $x \in X$ and for any $A \subseteq X$,

the $[\,]$-multirelation $R_r \subseteq X \times 2^X$ defined by

$$xR_rA \Leftrightarrow \forall y \in X,\ xry \Rightarrow y \in A \Leftrightarrow x \in [r]A$$

the $\langle\rangle$-multirelation $R_r \subseteq X \times 2^X$ defined by

$$xR_rA \Leftrightarrow \exists y \in X,\ xry \;\&\; y \in A \Leftrightarrow x \in \langle r \rangle A$$

the $[\![\,]\!]$-multirelation $R_r \subseteq X \times 2^X$ defined by

$$xR_rA \Leftrightarrow \forall y \in X,\ y \in A \Rightarrow xry \Leftrightarrow x \in [\![r]\!]A$$

the $\langle\!\langle\rangle\!\rangle$-multirelation $R_r \subseteq X \times 2^X$ defined by

$$xR_rA \Leftrightarrow \exists y \in X,\ \text{not } xry \;\&\; y \notin A \Leftrightarrow x \in \langle\!\langle r \rangle\!\rangle A.$$

Observe that $[\,]$ and $\langle\rangle$ are the usual modal operators of necessity and possibility, respectively, and $[\![\,]\!]$ and $\langle\!\langle\rangle\!\rangle$ are the operators of sufficiency and dual sufficiency, respectively, presented in Section 1.8. It is easy to check that for any relation r, if R_r is a $[\,]$-multirelation, then it is a total, multiplicative and up-closed multirelation, and if R_r is a $\langle\rangle$-multirelation, then it is a proper, additive and up-closed multirelation. Similarly, a $[\![\,]\!]$-multirelation is total, co-additive and down-closed, and a $\langle\!\langle\rangle\!\rangle$-multirelation is proper, co-multiplicative and down-closed.

Lemma 6.2.5 *Let X be a non-empty set.*

(a) *If $r \subseteq X \times X$ is a binary relation, then r is a flattening of the $\langle\rangle$-multirelation and the $[\![\,]\!]$-multirelation defined from r;*

(b) *If R is a proper, additive and up-closed multirelation, then R is a $\langle\rangle$-multirelation defined from its flattening r_R;*

(c) *If R is a co-additive multirelation, then R is a $[\![\,]\!]$-multirelation defined from its flattening r_R.*

Proof:
For (a), given a binary relation $r \subseteq X \times X$, define the $\langle\rangle$-multirelation R_r by: $xR_rA \Leftrightarrow x \in \langle r\rangle A$. Taking its flattening we have $xp_{R_r}y \Leftrightarrow xR_r\{y\}$, for every $y \in X$. It follows that $r = p_{R_r}$ since, for any $x, y \in X$,

$$xry \Leftrightarrow x \in \langle r\rangle\{y\} \Leftrightarrow xR_r\{y\} \Leftrightarrow xp_{R_r}y.$$

Similarly, define the $[\![\,]\!]$-multirelation R_r by $xR_rA \Leftrightarrow x \in [\![r]\!]A$ and consider its flattening p_{R_r}. Then, for any $x, y \in X$,

$$xry \Leftrightarrow x \in [\![r]\!]\{y\} \Leftrightarrow xR_r\{y\} \Leftrightarrow xp_{R_r}y.$$

For (b), given a proper, additive multirelation R, define its flattening r_R and then form the $\langle\rangle$-multirelation P_{r_R} from r_R by setting $xP_{r_R}A \Leftrightarrow x \in \langle r_R\rangle A$. We now show that $R = P_{r_R}$. Take any $x \in X$ and $A \subseteq X$ such that xRA. Since R is proper, $A \neq \emptyset$. Since $A = \bigcup\{\{y\} : y \in A\}$ and R is additive, there is some $y \in A$ such that $xR\{y\}$. Hence, there is some $y \in X$ such that xr_Ry and $y \in A$, which means $x \in \langle r_R\rangle A$. So $R \subseteq P_{r_R}$. On the other hand, take any $x \in X$ and $A \subseteq X$ such that $xP_{r_R}A$. Then, there is some $y \in X$ such that $xR\{y\}$ and $y \in A$. Since R is up-closed, xRA. So $P_{r_R} \subseteq R$.

For (c), given a co-additive multirelation, define its flattening r_R and then form a $[\![\,]\!]$-multirelation P_{r_R} from r_R. Then, for any $x \in X$ and $A \subseteq X$,

$$\begin{aligned} xRA &\Leftrightarrow xR\left(\bigcup\{\{y\} : y \in A\}\right) \\ &\Leftrightarrow \forall y \in A,\ xR\{y\} \text{ since } R \text{ is co-additive} \\ &\Leftrightarrow \forall y \in A,\ xr_Ry \\ &\Leftrightarrow \forall y \in A,\ x \in [\![r_R]\!]\{y\} \\ &\Leftrightarrow x \in \bigcap\{[\![r_R]\!]\{y\} : y \in A\} \\ &\Leftrightarrow x \in [\![r_R]\!]\left(\bigcup\{\{y\} : y \in A\}\right) \\ &\Leftrightarrow xP_{r_R}A. \end{aligned}$$

Thus R is a $[\![\,]\!]$-multirelation defined from the flattening r_R of R. □

6.3 Boolean algebras with a monotone operator

In this section we present a correspondence between the class of up-closed multirelational structures (X, R), that is, a non-empty set X endowed with an up-closed multirelation $R \subseteq X \times 2^X$, and the class of Boolean algebras with a monotone unary operator (B, f).

The *complex algebra of an up-closed multirelational structure* X is $(2^X, f_R)$ where 2^X is the complex algebra of its Boolean frame reduct and the mapping f_R over 2^X is defined, for any $A \subseteq X$, by

$$f_R(A) \stackrel{\text{def}}{=} \{x \in X : xRA\}.$$

Since R is up-closed, f_R is monotone. Hence, the complex algebra of an up-closed multirelational structure is a Boolean algebra with a monotone unary operator.

Let B be a Boolean algebra with a monotone unary operator f. The *multirelational canonical frame* of B is a structure (X_B, R_f) where X_B is the canonical frame of its Boolean reduct and the multirelation $R_f \subseteq X_B \times 2^{X_B}$ is defined, for any $F \in X_B$ and for any $\mathsf{G} \subseteq X_B$, by

$$FR_f\mathsf{G} \stackrel{\text{def}}{\Leftrightarrow} \exists G \in \mathcal{F}_o(B), A_G \subseteq \mathsf{G} \ \& \ G \subseteq f^{-1}(F),$$

where $A_G = \{H \in X_B : G \subseteq H\}$ and $\mathcal{F}_0(B)$ is the family of all filters of B.

It is easy to show that the multirelation R_f is up-closed, and hence the multirelational canonical frame of a Boolean algebra with a monotone unary operator is an up-closed multirelational structure. Thus every up-closed multirelational structure gives rise to a Boolean algebra with a monotone unary operator, and conversely. We now give the discrete representations which together give a discrete duality for up-closed multirelational structures and Boolean algebras with a monotone unary operator.

Theorem 6.3.1 *Any Boolean algebra with a monotone unary operator is embeddable into the complex algebra is of its multirelational canonical frame.*

Proof:
Consider any Boolean algebra with a monotone unary operator (B, f). It suffices to show that the Stone embedding $h : B \to 2^{X_B}$, defined as in Section 2.3, preserves the operator f in the sense that, for any $a \in B$, $f_{R_f}(h(a)) = h(f(a))$. Assume $F \in h(f(a))$. Then $f(a) \in F$ so $a \in f^{-1}(F)$ and hence, since $f^{-1}(F)$ is up-closed, $\uparrow_{\leq} a \subseteq f^{-1}(F)$. Let $G = \uparrow_{\leq} a \in \mathcal{F}_o(B)$. Then, for every $H \in A_G$, $a \in G \subseteq H$ so $H \in h(a)$. Hence, $A_G \subseteq h(a)$ and $G \subseteq f^{-1}(F)$. On the other hand, assume that for some $G \in \mathcal{F}_o(B)$, $A_G \subseteq h(a)$ and $G \subseteq f^{-1}(F)$, where $A_G = \{H \in X_B : G \subseteq H\}$. Then $a \in \bigcap\{H \in X_B : G \subseteq H\} = G$. Thus $a \in f^{-1}(F)$ and hence $f(a) \in F$. □

Theorem 6.3.2 *Any up-closed multirelational structure is embeddable into the multirelational canonical frame of its complex algebra.*

Proof:
Consider any up-closed multirelational structure (X, R). It suffices to show

that the embedding $k : X \to X_{2^X}$, defined as in Section 2.3, preserves the multirelation R in the sense that, for any $x \in X$ and for any $Y \subseteq X$,

$$xRY \Leftrightarrow k(x)R_{f_R}k(Y),$$

where $k(Y) = \{k(y) : y \in Y\} = \{k(y) : Y \in k(y)\} \subseteq X_{2^X}$.

Assume xRY. The set $\mathsf{G} = \{A \subseteq X : Y \subseteq A\}$ is a filter of 2^X. Since R is up-closed, for every $A \in \mathsf{G}$, xRA. Therefore $x \in f_R(A)$, and hence $f_R(A) \in k(x)$. Thus $\mathsf{G} \subseteq f_R^{-1}(k(x))$. Now we show $A_{\mathsf{G}} = \{\mathsf{H} \in X_{2^X} : \mathsf{G} \subseteq \mathsf{H}\} \subseteq k(Y)$. Take any $\mathsf{H} \in A_{\mathsf{G}}$. Since $Y \in \mathsf{G}$, $Y \in \mathsf{H}$. So $\mathsf{H} \in k(Y)$, as required. On the other hand, assume $k(x)R_{f_R}k(Y)$ which means that for some filter G of 2^X, $A_{\mathsf{G}} \subseteq k(Y)$ and $\mathsf{G} \subseteq f_R^{-1}(k(x))$. Then $Y \in \bigcap\{\mathsf{H} \in X_{2^X} : \mathsf{G} \subseteq \mathsf{H}\} = \mathsf{G}$. We also have

$$\begin{aligned}
&\forall A \subseteq X,\ A \in \mathsf{G} \Rightarrow A \in f_R^{-1}(k(x))\\
&\quad \Leftrightarrow \forall A \subseteq X,\ A \in \mathsf{G} \Rightarrow f_R(A) \in k(x)\\
&\quad \Leftrightarrow \forall A \subseteq X,\ A \in \mathsf{G} \Rightarrow x \in f_R(A)\\
&\quad \Leftrightarrow \forall A \subseteq X,\ A \in \mathsf{G} \Rightarrow xRA.
\end{aligned}$$

Since $Y \in \mathsf{G}$, we have xRY, as required. □

6.4 Boolean algebras with an antitone operator

In this section we establish a correspondence between the class of down-closed multirelational frames (X, R), that is, a non-empty set X endowed with a down-closed multirelation $R \subseteq X \times 2^X$, and the class of Boolean algebras with an antitone unary operator (B, g).

The *complex algebra of a down-closed multirelational frame* X is $(2^X, g_R)$ where 2^X is the complex algebra of its Boolean frame reduct and the mapping g_R over 2^X is defined, for any $A \subseteq X$, by

$$g_R(A) \stackrel{\text{def}}{=} \{x \in X : xRA\}.$$

Since R is down-closed, g_R is antitone. Hence, the complex algebra of a down-closed multirelational frame is a Boolean algebra with an antitone unary operator.

On the other hand, the *multirelational canonical frame of a Boolean algebra with an antitone unary operator* B is a structure (X_B, R_g) where X_B is the multirelational canonical frame of its Boolean reduct and, for each antitone operator $g : B \to B$, the multirelation $R_g \subseteq X_B \times 2^{X_B}$ is defined, for any $F \in X_B$ and for any $\mathsf{G} \subseteq X_B$, by

$$FR_g\mathsf{G} \stackrel{\text{def}}{\Leftrightarrow} \exists G \in \mathcal{F}(B),\ \mathsf{G} \subseteq A_G \ \&\ g(G) \cap F \neq \emptyset,$$

where $A_G = \{H \in X_B : G \subseteq H\}$ and $\mathcal{F}(B)$ is the set of all filters of B.

It is easy to show that the multirelation R_g is down-closed, and hence the multirelational canonical frame of a Boolean algebra with an antitone unary operator is a down-closed multirelational frame. Thus every down-closed multirelational frame gives rise to a Boolean algebra with an antitone unary operator, and conversely. We now give the discrete representations for down-closed multirelational structures and Boolean algebras with an antitone unary operator.

Theorem 6.4.1 *Any Boolean algebra with an antitone unary operator is embeddable into the complex algebra of its multirelational canonical frame.*

Proof:
Let (B, g) be a Boolean algebra with an antitone unary operator. It suffices to show that the Stone embedding $h : B \to 2^{X_B}$, defined as in Section 2.3, preserves the operator g, in the sense that, for any $a \in B$, $h(g(a)) = g_{R_g}(h(a))$. This follows if we show that, for any $a \in B$ and any $F \in X_B$,

$$g(a) \in F \Leftrightarrow \exists G \in \mathcal{F}(B),\ h(a) \subseteq A_G \ \&\ g(G) \cap F \neq \emptyset.$$

Take any $F \in X_B$ such that $g(a) \in F$. Let $G = \uparrow_{\leq} a \in \mathcal{F}(B)$. Then $g(a) \in g(G)$, so $g(G) \cap F \neq \emptyset$. Also, take any $F \in h(a)$. Then $G \subseteq F$ since for any $b \in G$, $a \leq b$ so, since $a \in F$ and F is a filter, $b \in F$. Therefore $h(a) \subseteq A_G$. On the other hand, assume that for some $G \in \mathcal{F}(B)$, $h(a) \subseteq A_G$ and $g(G) \cap F \neq \emptyset$. Then, by definition of greatest lower bound, $G \subseteq \bigcap\{H \in X_B : a \in H\}$. Now

$$G \subseteq \bigcap\{H \in X_B : a \in H\} = \bigcap\{H \in X_B : \uparrow_{\leq} a \subseteq H\} = \uparrow_{\leq} a.$$

Hence, $g(G) \subseteq g(\uparrow_{\leq} a)$. Thus $g(\uparrow_{\leq} a) \cap F \neq \emptyset$. So there is some $c \in \uparrow_{\leq} a$ such that $g(c) \in F$. Since g is antitone, $g(c) \leq g(a)$. Thus, since F is up-closed, $g(a) \in F$. □

Theorem 6.4.2 *Any down-closed multirelational frame is embeddable into the multirelational canonical frame of its complex algebra.*

Proof:
Consider any down-closed multirelational frame (X, R). It suffices to show that the embedding $k : X \to X_{2^X}$, defined as in Section 2.3, preserves the multirelation R, that is, for any $x \in X$ and for any $Y \subseteq X$, $xRY \Leftrightarrow k(x)R_{g_R}k(Y)$. Note that

$$k(x)R_{g_R}k(Y) \Leftrightarrow \exists \mathsf{G} \in \mathcal{F}(2^X),\ k(Y) \subseteq A_{\mathsf{G}} \ \&\ g_R(\mathsf{G}) \cap k(x) \neq \emptyset.$$

Assume xRY and define $\mathsf{G} = \{A \subseteq X : Y \subseteq A\}$. Then G is a filter of 2^X. Since $x \in g_R(Y)$, $g_R(Y) \in k(x)$. Since $Y \in \mathsf{G}$, $g_R(Y) \in g_R(\mathsf{G})$. Hence $g_R(\mathsf{G}) \cap k(x) \neq \emptyset$. Now take any $\mathsf{F} \in k(Y)$. Then $\mathsf{F} = k(y)$ for some $y \in Y$. We need to show $\mathsf{F} \in A_{\mathsf{G}} = \{\mathsf{H} \in X_{2^X} : \mathsf{G} \subseteq \mathsf{H}\}$. Take any $A \in \mathsf{G}$. Then $Y \subseteq A$, so $y \in A$, and hence $A \in k(y) = \mathsf{F}$, as required.
On the other hand, assume that $k(x)R_{g_R}k(Y)$, that is, there is some $\mathsf{G} \in \mathcal{F}(2^X)$ such that $k(Y) \subseteq A_{\mathsf{G}}$ and $g_R(\mathsf{G}) \cap k(x) \neq \emptyset$. Then there is some $A \subseteq X$ such that $A \in \mathsf{G}$ and $x \in g_R(A)$. Thus xRA. Furthermore, since for every $\mathsf{H} \in X_{2^X}$, $\mathsf{H} \in k(Y)$ implies $\mathsf{G} \subseteq \mathsf{H}$, we get $\mathsf{G} \subseteq k(y)$ for every $y \in Y$. Since $A \in \mathsf{G}$, $A \in k(y)$ for every $y \in Y$. Hence $Y \subseteq A$. Since R is down-closed, xRY, as required. □

Chapter 7

Boolean algebras with relations

7.1 Introduction

In this chapter we present discrete dualities between some classes of Boolean algebras or their reducts, additionally endowed with a binary relation, and the appropriate classes of frames.

Theories of qualitative spatial reasoning aim at expressing non-numerical relationships among spatial objects. Many interesting relationships can be formulated in terms of relations of proximity and connection, also referred to as a contact relation. Proximity relations between regions model the concept of nearness in metric spaces. The Efremovič paper [74] is an early contribution to formalization of proximity. Since then various kinds of proximity relations have been studied, see, for example, [168, 226, 240, 41, 68]. Whitehead in his work on foundations of mathematics [254] introduced a relation of connection between regions as a basis for his system, based on an earlier work of de Laguna [34]. The system includes the mereology of Leśniewski [154]. A first attempt to formalization of the Whitehead relation of connection is presented in [29]. The further developments and studies of the concept of connection can be found in [205, 52, 69, 41], among others.

In the first part of this chapter Boolean algebras endowed with various kinds of proximity relations or contact relations are considered. The corresponding classes of frames are introduced and discrete representation theorems both for algebras and for frames are presented. In Section 7.2 a weak proximity relation on a Boolean algebra is considered. The axioms of the relation reflect the intuition that the relation may hold only between non-empty regions and it distributes over the union of regions in both arguments. Next, in Section 7.3 various other axioms on the proximity relation are postulated, corresponding to Čech, Efremovič and Perwin proximities, among others, and the representation theorems for the underlying structures are proved. In Section 7.4 Boolean algebras with a contact relation are presented as defined in [52]. These algebras are axiomatic extensions of proximity algebras and their frames are axiomatic extensions of proximity frames. A discrete representation theorems for algebras and frames are proved such that representation structures are defined with ultrafilters as in the case of proximity structures. In Section 7.5 some other representation results for contact algebras and contact frames are proved where the representation structures are defined with clans following the idea of Düntsch [53]. This is in the spirit of representation theorem for contact algebras presented in [69] where, however, the representation algebra is defined with clusters (maximal clans) and not with arbitrary clans and there is no any representation theorem for frames. The notion of cluster seems to be too strong, to get a representation theorem for frames, the canonical frames in Section 7.5 are defined on the family of clans of the Boolean contact algebras. A first order logic for reasoning about contact relation and the duality via truth theorem are presented. The logic is inspired by the logic developed in [239].

An overview of theories of qualitative spatial reasoning can be found in [30, 3]. Topological representation theorems for various classes of Boolean contact algebras are presented in [69, 41]. Dual tableaux deduction systems for several spatial theories can be found in [174].

In the second part of the chapter we consider some of the algebras related to the Aristotelian syllogistic. The typical examples of syllogistic patterns of reasoning are 'all a are b' and 'some a are b'. In modern logic the first axiom systems for various universal first order theories with binary predicates representing syllogistic patterns are presented in [252] and [155]. The predicates in those theories are interpreted as binary relations on non-empty families of sets. The two patterns mentioned above may be understood as the universal and the existential quantifier, respectively, of restricted scope. In more recent studies also generalized quantifiers are related to syllogistic, see [253]. In [219] some of the axiomatic theories of [252] and [155] are presented in an algebraic framework. We consider two of those classes of algebras and develop a discrete duality for them. In Section 7.7 we present discrete representation theorems for $\forall$-algebras and frames. $\forall$-algebras are reducts of Boolean algebras with a negation as the only operator and with a binary relation $\forall$. In case of set $\forall$-algebras the negation is the complement of sets and relation $\forall$ is the set inclusion and hence it may be interpreted as the universal quantifier of restricted scope. In Section 7.8 the $\forall\exists$-algebras are considered with the two binary relations $\forall$ and $\exists$ and the appropriate class of frames. In case of set $\forall\exists$-algebras the relations are set inclusion and non-empty overlap of sets, respectively, which allows for their interpretation as the quantifiers of restricted scope. The discrete representation theorems for the algebras and frames are proved. The content of these sections is from [62]. For further recent developments on syllogistic see [135] and the bibliography therein.

7.2 Proximity algebras

A *proximity algebra*, *Prox-algebra*, is a structure (B, δ) where B is a Boolean algebra and $\delta \subseteq B \times B$ satisfies, for any $a, b, c \in B$,

(Prox1) If $a\delta b$, then $a \neq 0$ and $b \neq 0$

(Prox2) $a\delta(b \vee c) \Leftrightarrow a\delta b$ or $a\delta c$

(Prox3) $(a \vee b)\delta c \Leftrightarrow a\delta c$ or $b\delta c$.

A *proximity frame*, *Prox-frame*, is a structure (X, R) where X is a non-empty set and R is a non-empty binary relation on X.

The *complex algebra of a Prox-frame* X is a structure $(2^X, \delta_R)$ where, for any $A, B \subseteq X$,

$$A\delta_R B \overset{\text{def}}{\Leftrightarrow} \exists x \in A\, \exists y \in B,\ xRy.$$

Proposition 7.2.1 *The complex algebra of a Prox-frame is a Prox-algebra.*

Proof:
We show that the specific axioms of proximity algebras are satisfied in the Boolean algebra 2^X.

(Prox1) Let $A, B \subseteq X$. Then $A\delta_R B$ if and only if there is some $x \in A$ and there is some $y \in B$ such that xRy. Since $R \neq \emptyset$, $A \neq \emptyset$ and $B \neq \emptyset$.

(Prox2) Note that:

$$\begin{array}{rcl} A\delta_R(B \cup C) & \Leftrightarrow & \exists x \in A,\ \exists y \in B \cup C,\ xRy \\ & \Leftrightarrow & \exists x \in A,\ \exists y \in B,\ xRy \ \text{ or } \ \exists x \in A,\ \exists y \in C,\ xRy \\ & \Leftrightarrow & A\delta_R B \ \text{ or } \ A\delta_R C. \end{array}$$

The proof of (Prox3) is similar. □

The *canonical frame of a Prox-algebra* B is the relational structure (X_B, R_δ) where X_B is the set of all ultrafilters of B and R_δ is a binary relation on X_B satisfying, for any $F, G \in X_B$,

$$FR_\delta G \overset{\text{def}}{\Leftrightarrow} F \times G \subseteq \delta.$$

Clearly, we have the following proposition.

Proposition 7.2.2 *The canonical frame of a Prox-algebra is a Prox-frame.*

Let B be a prox-algebra. We now show that the Stone embedding $h : B \to 2^{X_B}$, defined in Section 2.3, preserves relation δ.

Lemma 7.2.3 *For any $a, b \in B$, $a\delta b \Leftrightarrow h(a)\delta_{R_\delta} h(b)$.*

Proof:
Assume $a\delta b$. We need to show that for some $F \in h(a)$ and for some $G \in h(b)$, $F \times G \subseteq \delta$. Consider the filters $F' = \uparrow_\leq a$ and $G' = \uparrow_\leq b$. They are proper filters, so there are prime filters F and G such that $F' \subseteq F$ and $G' \subseteq G$. Let $a', b' \in B$ satisfy $a \leq a'$ and $b \leq b'$. Then $a' \in F$ and $b' \in G$. From axiom (Prox2), $a\delta b$ implies $a\delta(b \vee c)$ for any $c \in B$. In particular, taking b' for c we obtain $a\delta(b \vee b')$. Since $b \vee b' = b'$, we have $a\delta b'$. From axiom (Prox3), $a\delta b'$ implies $(a \vee c)\delta b'$ for any c. In particular, $(a \vee a')\delta b'$. Hence $a'\delta b'$.

On the other hand, assume $h(a)\delta_{R_\delta}h(b)$. Then, for some $F\in h(a)$ and some $G\in h(b)$, $F\times G\subseteq\delta$. It follow that for every $a'\in F$ and for every $b'\in G$, $a'\delta b'$. Now since $a\in F$ and $b\in G$, $a\delta b$. □

Let X be a prox-frame. We now show that the embedding $k: X\to X_{2^X}$, defined in Section 2.3, preserves the relation R.

Lemma 7.2.4 *For any $x,y\in X$, $xRy\Leftrightarrow k(x)R_{\delta_R}k(y)$,*

Proof:
Observe that, for any $x,y\in X$,

$$\begin{aligned} k(x)R_{\delta_R}k(y) &\Leftrightarrow k(x)\times k(y)\subseteq\delta_R \\ &\Leftrightarrow \forall A,B\subseteq X,\ A\in k(x)\ \&\ B\in k(y)\Rightarrow A\delta_R B \\ &\Leftrightarrow \forall A,B\subseteq X\ x\in A\ \&\ y\in B\Rightarrow\exists x'\in A\ \exists y'\in B,\ x'Ry'. \end{aligned}$$

Assume xRy and take $A,B\subseteq X$ such that $x\in A$ and $y\in B$. Then clearly $k(x)R_{\delta_R}k(y)$. Conversely, assume $k(x)R_{\delta_R}k(y)$. Taking A to be $\{x\}$ and B to be $\{y\}$ we have xRy, as required. □

With these results we obtain the following discrete representations for Prox-algebras and Prox-frames.

Theorem 7.2.5

(a) *Every Prox-algebra is embeddable into the complex algebra of its canonical frame;*

(b) *Every Prox-frame is embeddable into the canonical frame of its complex algebra.*

7.3 Axiomatic extensions of proximity algebras

In this section we present discrete duality between Čech (resp. Efremovič, Perwin) proximity algebras and their corresponding frames.

The part of this duality which provides representation theorems for algebras is implicitly included in [68].

Consider proximity algebras defined in Section 7.2 with the additional axiom, for any $a,b\in B$,

(Prox4) $a\wedge b\neq 0\Rightarrow a\delta b$

and the proximity frames of the form (X,R) with the additional axiom

(FProx4) R is reflexive.

Although in proximity frames there are no specific assumption on the relation, we denote the above frame axiom by (FProx4) to stress that it is a counterpart of the axiom (Prox4).

It is clear that if (X,R) is such that R is reflexive, then the complex algebra of such a proximity frame satisfies (Prox4).

Proposition 7.3.1 *Let B be a proximity algebra satisfying (Prox4) and let (X_B, R_δ) be its canonical frame. Then R_δ is reflexive.*

Proof:
If $F \in X_B$, then $0 \notin F$ and for every $a \in F$, $(a,a) \in F \times F$. By (Prox4) for every $a \in F$, $a\delta a$ holds and hence $F \times F \subseteq \delta$, as required. □

Now, let us consider proximity algebras with the axiom

(Prox5) δ is symmetric

and the proximity frames with the axiom

(FProx5) R is symmetric.

Clearly, if R is symmetric, then δ_R is symmetric and hence the complex algebra of a proximity frame satisfying (FProx5) satisfies (Prox5).

Proposition 7.3.2 *Let B be a proximity algebra satisfying (Prox5) and let (X_B, R_δ) be its canonical frame. Then R_δ is symmetric.*

Proof:
Let $F, G \in X_B$ be such that $FR_\delta G$. By the definition of R_δ, $F \times G \subseteq \delta$. Since δ is symmetric, $G \times F \subseteq \delta$, as required. □

The proximity relation satisfying (Prox4) and (Prox5) is known as *Čech proximity.*

Consider a class of proximity algebras satisfying

(Prox6) $a(-\delta)b \Rightarrow \exists c,\ a(-\delta)c \ \&\ -c(-\delta)b$

and its corresponding class of proximity frames with the axiom

(FProx6) R is transitive.

Proposition 7.3.3 *The complex algebra of a proximity frame X with a transitive relation R satisfies (Prox6).*

Proof:
Let (X, R) be a proximity frame satisfying (FProx6). We show that for all $A, B \subseteq X$,

$$A(-\delta_R)B \Rightarrow \exists C \subseteq X,\ A(-\delta_R)C \ \&\ (-C)(-\delta_R)B.$$

Let $A, B \subseteq X$ be such that for all $x, y \in X$, $x \in A$ and $y \in B$ imply $x(-R)y$. Consider the set $C = \langle R \rangle B$. Suppose that $A\delta_R C$. Then there is some $x \in A$ and some $y \in C$ such that xRy. Since $y \in C$, there exists $z \in X$ such that yRz and $z \in B$. By transitivity of R, xRz holds. Then by the assumption we get $x(-R)z$, a contradiction.
Now we show that $(-C)(-\delta_R)B$ holds. Let $x \notin C$ and let $y \in B$. Then in view of the definition of C, for every $z \in X$, xRz implies $z \notin B$. Taking y for z we get $z(-R)y$, as required. □

Proposition 7.3.4 *Let B be a proximity algebra satisfying (Prox6) and let (X_B, R_δ) be its canonical frame. Then R_δ is transitive.*

Proof:
Let $F, G, H \in X_B$, let $F \times G \subseteq \delta$, and let $G \times H \subseteq \delta$. Suppose $F(-R_\delta)H$. Then $F \times H \nsubseteq \delta$ and therefore for some $a \in F$ and $b \in H$, $a(-\delta)b$. By (Prox6) there is some $c \in B$ such that $a(-\delta)c$ and $-c(-\delta)b$. Since $b \in H$ and $G \times H \subseteq \delta$, $-c \notin G$. But G is an ultrafilter of B, so $c \in G$. Since $a \in F$ and $F \times G \subseteq \delta$, we have $a\delta c$, a contradiction. □

The proximity satisfying (Prox4) and (Prox6) is known in the literature as a *Perwin proximity.*

Now, let us consider a class of proximity algebras satisfying

(Prox7) If $a \neq 0$ and $-a \neq 0$, then $a\delta(-a)$

and let a class of proximity frames (X, R) satisfies

(FProx7) R is connected.

Proposition 7.3.5 *Let X be a proximity frame satisfying (FProx7). Then its complex algebra satisfies (Prox7).*

Proof:
We show that for every $A \subseteq X$, if $A \neq \emptyset$ and $-A \neq \emptyset$, then $A\delta_R(-A)$. Let $x \in A$ and let $y \in -A$. Then $x \neq y$. Since R is connected, there are $z_0, \ldots, z_n \in X$, $n \geq 1$, such that $z_0 R z_1, \ldots, z_{n-1} R z_n$ where $z_0 = x$ and $z_n = y$. Let i be the smallest element of the set $\{0, 1, \ldots, n\}$ such that $z_i \in A$ and $z_{i+1} \notin A$. Such an element i exists because $z_0 \in A$ and $z_n \notin A$. Then, since $z_i R z_{i+1}$, we have $A\delta_R(-A)$, as required. □

In [68] an example is presented of a proximity algebra satisfying (Prox7) such that its canonical frame does not satisfy (FProx7). Therefore discrete duality between these algebras and frames does not hold.

For the remaining classes of structures discussed in this section we obtain discrete representations proceeding as in Section 7.2.

Theorem 7.3.6

(a) *Every Čech (resp. Efremovič, Perwin) proximity algebra is embeddable into the complex algebra of its canonical frame;*

(b) *Every proximity frame corresponding to Čech (resp. Efremovič, Perwin) proximity is embeddable into the canonical frame of its complex algebra.*

7.4 Boolean contact algebras

A *Boolean contact algebra, BC-algebra,* is a structure (B, C) where B is a Boolean algebra and $C \subseteq B \times B$ satisfies, for all $a, b \in B$,

(BC1) $aCb \Rightarrow a, b \neq 0$
(BC2) $a \neq 0 \Rightarrow aCa$
(BC3) C is symmetric
(BC4) aCb and $b \leq c \Rightarrow aCc$
(BC5) $aC(b \vee c) \Rightarrow aCb$ or aCc

where $\leq$ is the natural ordering on B and for every $a \in B$, $C(a) = \{b \in B : aCb\}$.

Note that (BC1) (resp. (BC3), (BC5)) coincide with (Prox1) (resp. (Prox5), (Prox2)) and (BC1) is a particular case of (Prox4). Since relation C is symmetric, (BC5) implies (Prox3). It follows that every BC-algebra is a Prox-algebra. Therefore it is natural to expect that a discrete duality for BC-algebras can be obtained along the lines of Section 7.2 and Section 7.3. Indeed, we show that if BC-frames are appropriate axiomatic extensions of Prox-frames, and if the canonical frames of BC-algebras are defined in terms of ultrafilters of the underlying Boolean algebras, then a representation of BC-algebras is provided by the usual Stone mapping as presented in Section 7.2. Similarly, if the complex algebras of BC-frames are defined as the complex algebras of Prox-frames, then a representation of BC-frames is provided by the usual mapping k as shown in Section 7.2.

In this section we define a *Boolean contact frame*, *BC-frame*, as a Prox-frame (X, R) such that R is a reflexive and symmetric relation on X. The complex algebra of a BC-frame X is defined in the same way as the complex algebra of a Prox-frame, that is as a structure $(2^X, C_R)$ where for all $A, B \subseteq X$,

$$AC_R B \Leftrightarrow \exists x \in A\ \exists y \in B,\ xRy.$$

Similarly, the canonical frame of a BC-algebra B is defined as the canonical frame of a Prox-algebra, that is as a structure (X_B, R_C), where X_B is the set of all ultrafilters of B and for all $F, G \in X_B$,

$$FR_C G \Leftrightarrow F \times G \subseteq C.$$

Proposition 7.4.1 *The complex algebra of a BC-frame is a BC-algebra.*

Proof:
In view of Proposition 7.2.1 it is sufficient to show that axioms (BC2), (BC3), and (BC4) are satisfied in 2^X. The axioms (BC2) and (BC3) follow directly from reflexivity and symmetry of the relation R in BC-frames, respectively. It is also easy to see that (BC4) is satisfied. □

Similarly, in view of Proposition 7.2.2, Proposition 7.3.1, and symmetry of the contact relation C in BC-algebras, it is easy to see that the following proposition holds.

Proposition 7.4.2 *The canonical frame of a BC-algebra is a BC-frame.*

Proceeding as in Section 7.2, we obtain discrete representations for BC-algebras and BC-frames.

Theorem 7.4.3

(a) *Every BC-algebra is embeddable into the complex algebra of its canonical frame;*

(b) *Every BC-frame is embeddable into the canonical frame of its complex algebra.*

7.5 Boolean contact algebras represented with clans

In this section we present another representation results for Boolean contact structures. Here BC-algebras are defined as in Section 7.4, BC-frames are posets, canonical frames of BC-algebras are defined with clans, and the representation of BC-algebras is provided by a mapping defined in terms of clans instead of ultrafilters. The idea of representing BC-algebras with clans is due to Düntch [53] and the properties of clans listed in Lemma 7.5.1, Lemma 7.5.4, and Lemma 7.5.5 are from [69].

Let B be a BC-algebra. A *clan of* B is a non–empty set $\Gamma \subseteq B$ such that, for all $a, b \in B$,

(Γ1) $a, b \in \Gamma \Rightarrow aCb$

(Γ2) $a \vee b \in \Gamma \Rightarrow a \in \Gamma$ or $b \in \Gamma$

(Γ3) $a \in \Gamma$ & $a \leq b \Rightarrow b \in \Gamma$.

Observe that by (Γ1) and (BC1), for every clan Γ, $0 \notin \Gamma$. Similarly, by (Γ3), $1 \in \Gamma$ for every clan Γ.

Let $Clan(B)$ be the family of all clans of B.

Lemma 7.5.1

(a) *Every ultrafilter of* B *is a clan of* B*;*

(b) *The complement of a clan of* B *is an ideal of* B*;*

(c) *For every clan* Γ *of* B *and for every* $a \in \Gamma$ *there is an ultrafilter* F *of* B *such that* $a \in F$ *and* $F \subseteq \Gamma$*;*

(d) *Every clan* Γ *of* B *is the union of all ultrafilters contained in* Γ.

Proof:
By way of example we prove (c).
Let F' be the principal filter generated by a. By (Γ3), $F' \subseteq \Gamma$. Hence $F' \cap (-\Gamma) = \emptyset$. Since $-\Gamma$ is an ideal of B, by Theorem 1.6.13 there is an ultrafilter F of B such that $F' \subseteq F$ and $F \cap (-\Gamma) = \emptyset$. Thus $F \subseteq \Gamma$. □

A *BC-frame* is a structure $(X, \leqslant)$ where $X \neq \emptyset$ and $\leqslant$ is a reflexive, transitive, and antisymmetric relation on X.

The *complex algebra of a BC-frame X* is a structure

$$(B_X, \cup, \wedge_X, -_X, \emptyset, X, C_X)$$

where $B_X \overset{\text{def}}{=} \{A \subseteq X : A = \langle\geqslant\rangle[\geqslant]A\}$ and, for all $A, B \subseteq X$,

$$A \wedge_X B \overset{\text{def}}{=} \langle\geqslant\rangle[\geqslant](A \cap B)$$
$$-_X A \overset{\text{def}}{=} \langle\geqslant\rangle(-A)$$
$$AC_X B \overset{\text{def}}{\Leftrightarrow} A \cap B \neq \emptyset.$$

Proposition 7.5.2 *The complex algebra of any BC-frame is a BC-algebra.*

Proof:
We show that B_X is closed under operations $\cup$, $\wedge_X$, and $-_X$. Let $A, B \subseteq X$ be such that $A = \langle\geqslant\rangle[\geqslant]A$ and $B = \langle\geqslant\rangle[\geqslant]B$.
($\cup$) We show that $A \cup B = \langle\geqslant\rangle[\geqslant](A \cup B)$. Note that by properties of operators $\langle\,\rangle$ and $[\,]$ we have

$$A \cup B = \langle\geqslant\rangle[\geqslant]A \cup \langle\geqslant\rangle[\geqslant]B = \langle\geqslant\rangle([\geqslant]A \cup [\geqslant]B) \subseteq \langle\geqslant\rangle[\geqslant](A \cup B).$$

For the reverse inclusion, by reflexivity of $\geqslant$, $[\geqslant](A \cup B) \subseteq A \cup B$. Hence, since $\langle\,\rangle$ is monotone, we get $\langle\geqslant\rangle[\geqslant](A \cup B) \subseteq \langle\geqslant\rangle(A \cup B)$. Since $\langle\,\rangle$ is additive, $\langle\geqslant\rangle(A \cup B) = \langle\geqslant\rangle A \cup \langle\geqslant\rangle B$. By the assumption, $A, B \in B_X$, so $\langle\geqslant\rangle(A \cup B) = \langle\geqslant\rangle\langle\geqslant\rangle[\geqslant]A \cup \langle\geqslant\rangle\langle\geqslant\rangle[\geqslant]B$. Finally, by transitivity and reflexivity of $\geqslant$, $\langle\geqslant\rangle\langle\geqslant\rangle Z = \langle\geqslant\rangle Z$ for every $Z \subseteq X$, so we have $\langle\geqslant\rangle(A \cup B) = \langle\geqslant\rangle[\geqslant]A \cup \langle\geqslant\rangle[\geqslant]B = A \cup B$. Thus $\langle\geqslant\rangle[\geqslant](A \cup B) \subseteq A \cup B$.

($\wedge_X$) We show that $A \wedge_X B = \langle\geqslant\rangle[\geqslant](A \wedge_X B)$. Indeed, by the properties of modal operators listed in Section 1.8 we have

$$\begin{aligned} A \wedge_X B &= \langle\geqslant\rangle[\geqslant](A \cap B) \\ &= \langle\geqslant\rangle[\geqslant]\langle\geqslant\rangle[\geqslant](A \cap B) \\ &= \langle\geqslant\rangle[\geqslant](A \wedge_X B). \end{aligned}$$

($-_X$) We need to show $-_X A = \langle\geqslant\rangle[\geqslant](-_X A)$. Since $A \in B_X$,

$$[\geqslant]A = [\geqslant]\langle\geqslant\rangle[\geqslant]A.$$

Thus $-[\geqslant]A = -[\geqslant]\langle\geqslant\rangle[\geqslant]A$. Equivalently, $\langle\geqslant\rangle(-A) = \langle\geqslant\rangle[\geqslant]\langle\geqslant\rangle(-A)$, as required.

It is easy to verify that $(B_X, \cup, \wedge_X, -_X, \emptyset, X)$ is a Boolean algebra and the axioms (BC1)–(BC5) are satisfied in (B_X, C_X). By way of example we show that $A \wedge_X (-_X A) = \emptyset$ for every $A \in B_X$.

$$\begin{aligned} A \wedge_X (-_X A) &= \langle\geqslant\rangle[\geqslant](A \cap \langle\geqslant\rangle(-A)) \\ &= \langle\geqslant\rangle([\geqslant]A \cap [\geqslant]\langle\geqslant\rangle(-A)) = \langle\geqslant\rangle([\geqslant]A \cap -\langle\geqslant\rangle[\geqslant]A) \\ &= \langle\geqslant\rangle([\geqslant]A \cap (-A)) \subseteq \langle\geqslant\rangle(A \cap -A) = \langle\geqslant\rangle\emptyset = \emptyset. \end{aligned}$$

□

The *canonical frame of a BC-algebra* B is a relational structure $(X_B, \subseteq)$ where $X_B = Clan(B)$. It is clear that the following proposition holds.

Proposition 7.5.3 *The canonical frame of a BC-algebra is a BC-frame.*

Lemma 7.5.4 *Let B be a BC-algebra. Let $F, G \subseteq B$ be filters of B such that $F \times G \subseteq C$. Then there is a clan containing $F \cup G$.*

Proof:
For any $a \in B$ let $Z_a = \{b \in B : a(-C)b\}$ and for any $M \subseteq B$ let I_M be the ideal generated by $\bigcup_{a \in M} Z_a$. It follows that there is no element $b \in I_M$ such that aCb for some $a \in M$. Since $F \times G \subseteq C$, $G \cap I_F = \emptyset$. Thus there is some ultrafilter G' of B such that $G \subseteq G'$ and $G' \cap I_F = \emptyset$.
Since $G' \cap I_F = \emptyset$, $b \in G'$ implies that there is some $a \in F$ such that aCb. It follows that $F \cap I_{G'} = \emptyset$. For suppose otherwise, then there is some $a \in F$ such that a is in the relation C with none of the elements of G', a contradiction. Therefore there is an ultrafilter F' of B such that $F \subseteq F'$ and $F' \cap I_{G'} = \emptyset$. Thus $F' \times G' \subseteq C$ and clearly $F' \cup G'$ is a clan including $F \cup G$. □

As a corollary we obtain the following lemma.

Lemma 7.5.5 *For all $a, b \in B$, if aCb, then there is a clan Γ such that $a, b \in \Gamma$.*

Proof:
Since aCb, by (BC1) $a, b \neq 0$. Consider two principal filters F and G generated by a and b, respectively. Then by (BC4), $F \times G \subseteq C$. By Lemma 7.5.4 there is a clan Γ of B including $F \cup G$ and hence $a, b \in \Gamma$. □

Let B be a BC-algebra. Let the mapping $h : B \to B_{X_B}$ be defined, for every $a \in B$, by

$$h(a) \stackrel{\text{def}}{=} \{\Gamma \in Clan(B) : a \in \Gamma\}.$$

Lemma 7.5.6

(a) *The mapping h is well-defined;*
(b) *The mapping h is injective;*
(c) *The mapping h preserves the operations and the contact relation.*

Proof:
(a) We show $h(a) = \langle\supseteq\rangle[\supseteq]h(a)$ for every $a \in B$. Observe that

$$\begin{aligned} \Gamma \in \langle\supseteq\rangle[\supseteq]h(a) &\Leftrightarrow \exists\Gamma' \in Clan(B),\ \Gamma' \subseteq \Gamma \ \&\ \Gamma' \in [\supseteq]h(a) \\ &\Leftrightarrow \exists\Gamma' \in Clan(B),\ \Gamma' \subseteq \Gamma \ \& \\ &\qquad \forall\Delta \in Clan(B),\ \Delta \subseteq \Gamma' \Rightarrow \Delta \in h(a). \end{aligned}$$

Assume $a \in \Gamma$ and let Γ' be an ultrafilter included in Γ such that $a \in \Gamma'$. By Lemma 7.5.1 such a Γ' exists. Take any clan Δ such that $\Delta \subseteq \Gamma'$ and suppose $a \notin \Delta$. Since Γ' is an ultrafilter and $a \in \Gamma'$, $-a \notin \Gamma'$. Then $-a \notin \Delta$. But $1 =$

$a \vee -a \in \Delta$, so necessarily $a \in \Delta$, a contradiction. On the other hand, assume $\Gamma \in \langle\supseteq\rangle[\supseteq]h(a)$ and let Γ' be as in the equivalences above. Suppose $a \notin \Gamma$. Then $a \notin \Gamma'$. Take Γ' for Δ. Then $a \in \Gamma'$. Since $\Gamma' \subseteq \Gamma$, $a \in \Gamma$, a contradiction.

(b) Assume $a \neq b$. Then there is an ultrafilter F such that $a \in F$ and $b \notin F$. Since F is a clan, by Lemma 7.5.1, $F \in h(a)$ and $F \notin h(b)$. Hence $h(a) \neq h(b)$, as required.

For (c) first observe that since 0 is not contained in any clan by (Γ1) and (BC1), we have $h(0) = \emptyset$. Similarly, since 1 is contained in every clan by (Γ3), we have $h(1) = X_B$.
Now we consider the following four cases.
($\vee$) We show that $h(a \vee b) = h(a) \cup h(b)$. We have $\Gamma \in h(a \vee b)$ if and only if $a \vee b \in \Gamma$. Thus by (Γ2), $a \in \Gamma$ or $b \in \Gamma$. Hence $\Gamma \in h(a)$ or $\Gamma \in h(b)$. Conversely, let $\Gamma \in h(a) \cup h(b)$. Then $a \in \Gamma$ or $b \in \Gamma$. If $a \in \Gamma$, then by Lemma 7.5.1(c) there is an ultrafilter F of B such that $F \subseteq \Gamma$ and $a \in F$. Since $a \leq a \vee b$, we have $a \vee b \in F$. It follows that $a \vee b \in \Gamma$, hence $\Gamma \in h(a \vee a)$. Reasoning in the similar way we can show that if $b \in \Gamma$, then $\Gamma \in h(a \vee b)$.

($-$) We show that $h(-a) = -_{X_B} h(a)$ where $-_{X_B} h(a) = \langle\supseteq\rangle(X_B - h(a))$. Assume $-a \in \Gamma$. Then by Lemma 7.5.1(c) there is an ultrafilter $F \subseteq \Gamma$ and $-a \in F$. Since F is an ultrafilter, $a \notin F$. But F is a clan of L by Lemma 7.5.1, so $\Gamma \in \langle\supseteq\rangle(-h(a))$. Conversely, assume $\Gamma \in \langle\supseteq\rangle(-h(a))$. Then there is a clan $\Delta \in Clan(B)$ such that $\Delta \subseteq \Gamma$ and $a \notin \Delta$. Since $1 \in \Delta$, $-a \in \Delta$. Thus $-a \in \Gamma$ and hence $\Gamma \in h(-a)$.

($\wedge$) Since B_{X_B} is a Boolean algebra, $\wedge_{X_B}$ is definable in terms of $\cup$ and $-_{X_B}$. Thus we have

$$\begin{aligned} h(a) \wedge_{X_B} h(b) &= -_{X_B}(-_{X_B} h(a) \cup -_{X_B} h(b)) = -_{X_B}(h(-a) \cup h(-b)) \\ &= -_{X_B} h(-a \vee -b) = h(-(-a \vee -b)) = h(a \wedge b). \end{aligned}$$

(C) We show that aCb if and only if $h(a) \cap h(b) \neq \emptyset$. Assume aCb. By Lemma 7.5.5, there is a clan Γ such that $a, b \in \Gamma$, and hence $\Gamma \in h(a) \cap h(b)$. Conversely, if $a \in \Gamma$ and $b \in \Gamma$ for a clan Γ, then aCb by (Γ1). □

Let $k : X \to Clan(B_X)$ be a mapping defined by:

$$k(x) \stackrel{\text{def}}{=} \{A \in B_X : x \in A\}.$$

Lemma 7.5.7

(a) *For every $x \in X$, $k(x)$ is a clan of B_X;*

(b) *The mapping k is injective;*

(c) *The mapping k preserves the relation $\leqslant$.*

Proof:
(a) Straightforward verification.

(b) Let $x, y \in X$ be such that $x \neq y$. Since $\leqslant$ is antisymmetric, without loss of generality we may assume that $x \nleqslant y$. Consider the set $A = \langle\geqslant\rangle[\geqslant]\{x\}$.

Note that $A = \langle\geqslant\rangle[\geqslant]\langle\geqslant\rangle[\geqslant]\{x\}$, so $A \in B_X$. Since $\langle\geqslant\rangle\{x\} = [\geqslant]\{x\}$, we have $A = \langle\geqslant\rangle\langle\geqslant\rangle\{x\}$. Since $\geqslant$ is reflexive and transitive, $A = \langle\geqslant\rangle\{x\}$. Clearly, $x \in A$ and $y \notin A$. Thus $A \in k(x)$ and $A \notin k(y)$, as required.

(c) We need to show that for all $x, y \in X$, $x \leqslant y$ implies $k(x) \subseteq k(y)$. Assume $x \leqslant y$ and take any $A \in B_X$ such that $A \in k(x)$. Then $x \in \langle\geqslant\rangle[\geqslant]A$ which means that there is some $y' \in X$ such that $y' \leqslant x$ and for every $z \in X$, $z \leqslant y'$ implies $z \in A$. We show

$$\exists y'' \in X,\ y'' \leqslant y\ \&\ \forall t \in X,\ t \leqslant y'' \Rightarrow t \in A.$$

Take y' for y''. Since $y' \leqslant x$ and $x \leqslant y$, we have $y' \leqslant y$. Now, take t such that $t \leqslant y'$. Taking t for z we get $t \in A$, as required. □

With these results we obtain the following discrete representations for BC-algebras and BC-frames.

Theorem 7.5.8

(a) *Every BC-algebra is embeddable into the complex algebra of its canonical frame;*

(b) *Every BC-frame is embeddable into the canonical frame of its complex algebra.*

The language for BC-logic is a first order language with function symbols $\vee$, $\wedge$, and $-$ denoting the operations of BC-algebras and with two binary predicate symbols C and $\leq$ denoting the contact relation and the ordering relation of BC-algebras, respectively. The set of terms of the language is the smallest set which includes an infinite denumerable set *Var* of individual variables, individual constants 0 and 1, and is closed with respect to the operations $\vee$, $\wedge$, and $-$. Terms represent elements of BC-algebras. Atomic formulas are of the form aCb and $a \leq b$, where a and b are terms. Compound formulas include atomic formulas and if α and β are formulas, then so are $\neg\alpha$, $\alpha \vee \beta$, $\alpha \wedge \beta$, $\forall z\, \alpha$, $\exists z\, \alpha$, where $z \in Var$. We slightly abuse the language here using the symbols $\vee$ and $\wedge$ both for join and meet, respectively, in BC-algebras and for disjunction and conjunction, respectively, in the logic.

The algebraic semantics of the BC-logic is determined by the class $\mathcal{A}lg_{BC}$ of complete BC-algebras. Let $(B, C) \in \mathcal{A}lg_{BC}$. A valuation in B is a mapping $v : Var \to B$ which extends to all the terms as follows:

$$\begin{aligned} v(a \vee b) &= v(a) \vee v(b) \\ v(a \wedge b) &= v(a) \wedge v(b) \\ v(-a) &= -v(a) \\ v(0) &= 0 \\ v(1) &= 1. \end{aligned}$$

Next, we define inductively a satisfaction relation for formulas. We say that a formula α is satisfied by a valuation v in an algebra B whenever the following hold:

$$\begin{array}{lcl}
B,v\models aCb & \stackrel{\text{def}}{\Leftrightarrow} & (v(a),v(b))\in C \\
B,v\models a\leq b & \stackrel{\text{def}}{\Leftrightarrow} & v(a)\leq v(b) \\
B,v\models \neg\alpha & \stackrel{\text{def}}{\Leftrightarrow} & B,v\not\models\alpha \\
B,v\models \alpha\vee\beta & \stackrel{\text{def}}{\Leftrightarrow} & B,v\models\alpha \text{ or } B,v\models\beta \\
B,v\models \alpha\wedge\beta & \stackrel{\text{def}}{\Leftrightarrow} & B,v\models\alpha \ \&\ B,v\models\beta \\
B,v\models \forall z\,\alpha & \stackrel{\text{def}}{\Leftrightarrow} & \text{for every valuation } v' \text{ in } B \text{ such that } v(t)=v'(t) \\
 & & \text{for every } t\neq z,\ B,v'\models\alpha \\
B,v\models \exists z\,\alpha & \stackrel{\text{def}}{\Leftrightarrow} & \text{there is a valuation } v' \text{ in } B \text{ such that } v(t)=v'(t) \\
 & & \text{for every } t\neq z, \text{ and } B,v'\models\alpha.
\end{array}$$

A formula α is true in an algebra $B\in\mathcal{A}lg_{BC}$ if $B,v\models\alpha$ for every valuation v in B and α is $\mathcal{A}lg_{BC}$-valid if it is true in every algebra $B\in\mathcal{A}lg_{BC}$.

The frame semantics is provided by the class $\mathcal{F}rm_{BC}$ of BC-frames. Given a frame $(X,\leqslant)$, a model based on X is a system $M=(X,\leqslant,m)$ where m is a meaning function $m : Var\rightarrow\{A\subseteq X : A=\langle\geqslant\rangle[\geqslant]A\}$ which extends to all terms

$$\begin{aligned}
m(a\vee b) &\stackrel{\text{def}}{=} m(a)\cup m(b)\\
m(a\wedge b) &\stackrel{\text{def}}{=} \langle\geqslant\rangle[\geqslant](m(a)\cap m(b))\\
m(-a) &\stackrel{\text{def}}{=} \langle\geqslant\rangle(X-m(a)).
\end{aligned}$$

It can be shown that for every term a, $m(a)=\langle\leqslant\rangle[\geqslant]m(a)$. The proof is similar to the proof of Proposition 7.5.2. Truth of a formula α in a model $M=(X,\leqslant,m)$, written $M\models\alpha$, is inductively defined as

$$\begin{array}{lcl}
M\models aCb & \stackrel{\text{def}}{\Leftrightarrow} & m(a)\cap m(b)\neq\emptyset \\
M\models a\leq b & \stackrel{\text{def}}{\Leftrightarrow} & m(a)\subseteq m(b) \\
M\models \neg\alpha & \stackrel{\text{def}}{\Leftrightarrow} & M\not\models\alpha \\
M\models \alpha\vee\beta & \stackrel{\text{def}}{\Leftrightarrow} & M\models\alpha \text{ or } M\models\beta \\
M\models \alpha\wedge\beta & \stackrel{\text{def}}{\Leftrightarrow} & M\models\alpha \ \&\ M\models\beta \\
M\models \forall z\,\alpha & \stackrel{\text{def}}{\Leftrightarrow} & \text{for every model } M'=(X,\leqslant,m') \text{ such that } m'(t)=m(t) \\
 & & \text{for every } t\neq z, M'\models\alpha \\
M\models \exists z\,\alpha & \stackrel{\text{def}}{\Leftrightarrow} & \text{there is a model } M'=(X,\leqslant,m') \text{ such that } m'(t)=m(t) \\
 & & \text{for every } t\neq z \text{ and } M'\models\alpha.
\end{array}$$

A formula α is true in a frame $X\in\mathcal{F}rm_{BC}$ if α is true in all models based on X and α is $\mathcal{F}rm_{BC}$-valid if it is true in all frames $X\in\mathcal{F}rm_{BC}$. It is easy to see that the meaning functions are the valuations in the complex algebras of BC-frames. Therefore we have the complex algebra theorem.

Theorem 7.5.9 (Complex algebra theorem)
For every frame $X\in\mathcal{F}rm_{BC}$ and for every formula α, α is true in X if and only if α is true in B_X.

Theorem 7.5.10 (Duality via truth)
A formula α is true in every BC-algebra $B \in Alg_{BC}$ if and only if it is true in every frame $X \in \mathcal{F}rm_{BC}$.

7.6 Syllogistic ∀-algebras

In [219] an algebraic approach to the Aristotelian syllogistic is presented. The algebras proposed there may be seen as reducts of Boolean algebras endowed with some binary relations relevant for syllogistic. In the present section we consider one of those classes of algebras, namely, ∀-algebras, of the form $\langle L, \neg, \forall \rangle$ with a binary relation which we denote with the symbol of the universal quantifier for the following reason: in the abstract setting, the ∀-algebras are defined with the set of axioms which say that the relation ∀ is reflexive and transitive, and determine its action on the negated elements. The axioms are chosen so that in case L is a set algebra and $\neg$ is the operator of set complementation, that is, $\neg = -$, the relation ∀ may be interpreted as set inclusion. From a logical perspective, sets correspond to unary predicates, and set inclusion $A \subseteq B$ is defined with the well known formula $(\forall z)[A(z) \Rightarrow B(z)]$. In this formula, the quantifier ∀ is a universal quantifier with a restricted scope.

Following [219], with some abuse of language we call a structure $\langle L, \neg, \forall \rangle$ an *∀-algebra*, if L is a non-empty set, $\neg$ is a unary operation on L and ∀ a binary relation on L such that, for any $a, b, c \in L$,

(∀1) $a \forall \neg\neg a$
(∀2) $\neg\neg a \forall a$
(∀3) $a \forall b$ and $b \forall c$ imply $a \forall c$
(∀4) $a \forall b$ implies $\neg b \forall \neg a$
(∀5) $a \forall \neg a$ implies $a \forall b$.

Note that (∀1) – (∀3) imply that ∀ is reflexive and transitive, and therefore, $[\forall]$ is an interior operator. An ∀-frame is a Boolean frame X. Its complex algebra is a structure $\langle 2^X, -, \subseteq \rangle$.

If L is an ∀-algebra, $A \subseteq L$ is called *∀-closed*, if $A = [\forall]A$. A non-empty subset A of L is called an *∀-set*, if it is ∀-closed and satisfies, for all $a, b \in A$,

$$a(-\forall)\neg b.$$

Lemma 7.6.1 *The intersection of two ∀-sets is an ∀-set.*

Let X'_L be the collection of all ∀-sets. Observe that L is not an ∀-set, since L is non-empty and (∀1) holds. Furthermore, X'_L may be empty, e.g. if ∀ is the universal relation on L, in particular, if $|L| = 1$.

Example 7.6.2 Suppose that $\langle L, \neg, \forall \rangle$ is an ∀-algebra, and $a, a' \notin L, a \neq a'$. Let $L' = L \cup \{a, a'\}$, and extend ∀ and $\neg$ by

$$\forall' = \forall \cup (\{a\} \times L') \cup (L' \times \{a'\}), \tag{7.1}$$
$$\neg' = \neg \cup \{\langle a, a' \rangle, \langle a', a \rangle\}. \tag{7.2}$$

Then, $\langle L', \forall', \neg' \rangle$ is an $\forall$-algebra since all the axioms are satisfied in L':
($\forall$1) If $b \in L$, then $\neg' b = \neg b$ and $b \forall \neg\neg b$. Otherwise, $a \forall' \neg' \neg' a$, since $a = \neg' \neg' a$ and $a \forall' a$. Similarly, $a' \forall' \neg' \neg' a'$.
($\forall$2) Similarly.
($\forall$3) Let $b \forall' c \forall' d$. If $b = a$, then $b \forall' d$ since $\{a\} \times L' \subseteq \forall'$. If $b = a'$, then $c = d = a'$. Thus, let $b \notin \{a, a'\}$. Then, $c \neq a$. If $c = a'$, then $a' \forall' d$ implies $d = a'$, and therefore, $b \forall' d$ since $L' \times \{a'\} \subseteq \forall'$. Thus, let $\{b, c\} \cap \{a, a'\} = \emptyset$; then, $d \neq a$. If $d = a'$, then, as before, $b \forall' d$. Otherwise, $b \forall c$, $c \forall d$, and $b \forall d$ by ($\forall$3).
($\forall$4) and ($\forall$5) follow immediately from the definition of $\forall'$ and $\neg'$. Observe that $\{a'\}$ is an $\forall$-set, and that no $\forall$-set contains a. □

The following lemma characterizes when an element of an $\forall$-algebra is contained in an $\forall$-set.

Lemma 7.6.3 *Let $a \in L$. Then, $\forall(a)$ is an $\forall$-set if and only if $\{a\} \times L \not\subseteq \forall$. Furthermore, $\forall(a)$ is a subset of each $\forall$-set containing a.*

Proof:
($\Rightarrow$) Suppose that $\forall(a)$ is an $\forall$-set. Then, $a(-\forall)\neg a$ and hence $\{a\} \times L \not\subseteq \forall$.
($\Leftarrow$) Suppose that $\forall(a)$ is not an $\forall$-set. Since $\forall(a)$ is $\forall$-closed, there are $b, c \in L$ such that $a \forall b, a \forall c$, and $b \forall \neg c$. Then, $a \forall \neg c$ by transitivity of $\forall$, and therefore, $\neg\neg c \forall \neg a$ by ($\forall$4). From ($\forall$1) we obtain $c \forall \neg\neg c$, and, again by transitivity, we have $c \forall \neg a$. Using $a \forall c$ and again transitivity we obtain $a \forall \neg a$, and therefore, $a \forall d$ for all $d \in L$ by ($\forall$5).

If A is an $\forall$-set containing a, then the fact that A is $\forall$-closed implies that $\forall(a) \subseteq A$. □

Lemma 7.6.4 *$\forall$ is the universal relation if and only if there is some $a \in L$ such that $(\{a\} \times L) \cup (\{\neg a\} \times L) \subseteq \forall$.*

Proof:
The left-to-right direction is obvious. Conversely, let $(\{a\} \times L) \cup (\{\neg a\} \times L) \subseteq \forall$, and $b, c \in L$. Then, $a \forall b$, $a \forall \neg b$, and $\neg a \forall c$ by the hypothesis. From ($\forall$4) we obtain $\neg\neg b \forall \neg a$, and by ($\forall$1) we have $b \forall \neg\neg b$. Thus, $b \forall \neg\neg b$, $\neg\neg b \forall \neg a$ and $\neg a \forall c$, and the claim follows from transitivity of $\forall$. □

If X'_L is non-empty, then, clearly, X'_L is closed under union of chains, and thus, each element of X'_L is contained in a maximal element. Let X_L be the set of all maximal elements of X'_L which we call the *canonical frame of L*. Our aim is to show under which conditions an $\forall$-algebra can be embedded into the complex algebra of its canonical frame. This will follow from a sequence of lemmas.

Lemma 7.6.5

(a) *If $A \in X'_L$, then there is no $a \in L$ such that $a \in A$ and $\neg a \in A$.*
(b) *If $A \in X_L$, then $a \in A$ or $\neg a \in A$ for every $a \in L$.*

Proof:
For (a), suppose that $A \in X'_L$ and $a, \neg a \in A$. Since A is an $\forall$-set, $a(-\forall)\neg\neg a$, contradicting ($\forall$1).

For (b), let A be a maximal $\forall$-set, and assume there is some $a \in L$ such that $\neg a \notin A$. Let $A' = A \cup \forall(a)$. If we can show that A' is an $\forall$ − set, then $A = A'$ because of the maximality of A, and it follows that $a \in A$. Since $[\forall]$ is an interior operator, we have $[\forall](A') \subseteq A'$. Conversely, let $b \in A'$, and $b\forall c$; we need to show that $c \in A'$. If $b \in A$, then $c \in A \subseteq A'$ since $[\forall](A) = A$. Otherwise, $a\forall b$ since $A' = A \cup \forall(a)$ and $b \in A'$, and now the transitivity of $\forall$ implies $a\forall c$; hence, $c \in A'$.

Next, let $c, d \in A'$ and assume that $c\forall\neg d$. Then, not both c and d can be in A. Suppose that $c \notin A$; then, $a\forall c$. By transitivity of $\forall$ we get $a\forall\neg d$ which implies $\neg d \in A'$, a contradiction with (a). Similarly, if $d \notin A$ then $a\forall\neg d$, a contradiction. Now since A' is an $\forall$-set and A is maximal, necessarily $A = A'$. It follows that $a \in A$ as required. □

Let L be a $\forall$-algebra and consider the Stone mapping $h : L \to 2^{X_L}$, used in Section 2.3.

Lemma 7.6.6 *For any $a, b \in L$,*

(a) $[\subseteq](h(a)) = h(a)$;

(b) $a\forall b$ *if and only if* $h(a) \subseteq h(b)$;

(c) $h(\neg a) = X_L - h(a)$.

Proof:
(a) It is sufficient to consider $\supseteq$, since $\subseteq$ is reflexive. If $A \in h(a)$ and $A \subseteq B$, then clearly $a \in B$.

(b) Let $a\forall b$, and $A \in h(a)$, that is, $a \in A$. Since $A \in X_L$, A is $\forall$-closed, and therefore, $b \in A$ which implies $A \in h(b)$. Conversely, assume that $a(-\forall)b$. Then, $\{a\} \times L \not\subseteq \forall$, and therefore, $\forall(a)$ is an $\forall$-set by Lemma 7.6.3. Suppose that $\neg b\forall\neg a$. Then, by ($\forall$4), $\neg\neg a\forall\neg\neg b$. Furthermore, $a\forall\neg\neg a$ by ($\forall$1), and $\neg\neg b\forall b$ by ($\forall$2), so that $a\forall\neg\neg a, \neg\neg a\forall\neg\neg b$, and $\neg\neg b\forall b$. Transitivity of $\forall$ now implies $a\forall b$, contradicting the assumption. Therefore, $\neg b(-\forall)\neg a$, and thus, $\forall(\neg b)$ is an $\forall$-set as well. Now, by Lemma 7.6.1, $A = \forall(a) \cap \forall(\neg b)$ is an $\forall$-set, and so there is a maximal $\forall$-set A' containing A. By reflexivity of $\forall$ we have $a, \neg b \in A'$. Hence, $A' \in h(a)$, and $b \notin A'$ by Lemma 7.6.5. It follows that $h(a) \not\subseteq h(b)$.
(c) Let $A \in X_L$. If $\neg a \in A$ then, by Lemma 7.6.5, we have $a \notin A$, i.e. $A \notin h(a)$. If $a \notin A$ then, by Lemma 7.6.5, we have $\neg a \in A$. □

In general, h is not injective. We have, however,

Theorem 7.6.7 *h is injective if and only if $\forall$ is a partial order.*

Proof:
($\Rightarrow$) Suppose that $h(a) = h(b)$ implies $a = b$. All we need to show is that $\forall$ is antisymmetric. Suppose that there are $a, b \in L$ such that $a\forall b$ and $b\forall a$. If

$A \in X_L$ and $a \in A$, then $a\forall b$ and the fact that A is $\forall$-closed imply that $b \in A$. It follows that $h(a) \subseteq h(b)$, and, similarly, $b\forall a$ implies $h(b) \subseteq h(a)$. Injectivity of h now implies $a = b$.
($\Leftarrow$) Suppose that $\forall$ is antisymmetric and $a \neq b$. Then, $a(-\forall)b$ or $b(-\forall)a$, suppose without loss of generality the former. As shown in the proof of Lemma 7.6.6, $\neg b\forall\neg a$ implies $a\forall b$, and therefore, by contraposition, $a(-\forall)b$ implies $\neg b(-\forall)\neg a$. Thus, $\forall(a)$ and $\forall(\neg b)$ are $\forall$-sets by Lemma 7.6.3, and therefore, so is $A = \forall(a) \cap \forall(\neg b)$ by Lemma 7.6.1. It follows that any maximal $\forall$-set A' with $A \subseteq A'$ contains a and $\neg b$. By Lemma 7.6.5(a), $A' \in h(a)$ and $A' \notin h(b)$. □

Thus we have a necessary and sufficient condition for a discrete representation for $\forall$-algebras.

Theorem 7.6.8 *An $\forall$-algebra $\langle L, \neg, \forall\rangle$ can be embedded into the complex algebra of its canonical frame if and only if $\forall$ is a partial order.*

We now have the following discrete representation for $\forall$-frames.

Theorem 7.6.9 *Every $\forall$-frame is embeddable into the canonical frame of its complex algebra.*

Proof:
The complex algebra of an $\forall$-frame X is the structure $\langle 2^X, -, \subseteq\rangle$, and the elements of its canonical frame X_{2^X} are the maximal $\forall_X$-sets, that is, those subsets A of 2^X maximal with the properties

1. If $Y \in \mathsf{A}$ and $Y \subseteq Z$, then $Z \in \mathsf{A}$.

2. If $Y, Z \in \mathsf{A}$, then $Y \cap Z \neq \emptyset$.

Let $k : X \to 2^{2^X}$ be defined by $k(x) = \{A \subseteq X : x \in A\}$. Then, k is injective, and $k(x)$ satisfies 1. and 2. since it is a maximal filter of the set algebra 2^X. □

7.7 $\forall\exists$-algebras

In this section we consider the structures with two relations $\forall$ and $\exists$ interpreted as universal and existential quantifier, respectively, of restricted scope.

Following [219] by an *$\forall\exists$-algebra* we mean a structure $\langle L, \forall, \exists\rangle$ such that $L \neq \emptyset$, and $\forall$, $\exists$ are binary relations on L satisfying, for any a, $b \in L$,

($\forall\exists$1) $\forall$ is reflexive

($\forall\exists$2) $\forall$ is transitive

($\forall\exists$3) $a \,\forall\, b$ and $a \,\exists\, c$ imply $c \,\exists\, b$

($\forall\exists$4) $a \,\exists\, b$ implies $a \,\exists\, a$

($\forall\exists$5) $a \,\exists\, a$ or $a \,\forall\, b$.

Note that by (∀∃1) and (∀∃3), the relation ∃ is symmetric. Although it is not natural to qualify a structure without operations as an algebra, we follow [219] in this respect because it enables us to formulate relationships between ∀∃-algebras and ∀∃-structures.

An ∀∃-*frame* is just a Boolean frame $X \neq \emptyset$.

The *complex algebra* of an ∀∃-frame X is the structure $(2^X, \forall_X, \exists_X)$ such that, for any A, $B \subseteq X$,

$$A \ \forall_X \ B \text{ iff } A \subseteq B$$
$$A \ \exists_X \ B \text{ iff } A \cap B \neq \emptyset.$$

The *canonical frame of an* ∀∃-*algebra* L is the set

$$X_L = \{A \subseteq L : A = [\forall]A \text{ and } A \times A \subseteq \exists\}.$$

The elements of X_L are referred to as ∀∃-sets.

∀∃-algebras and ∀-algebras considered in Section 7.6 are related as presented in the following lemmas based on [219].

Lemma 7.7.1 *Given an* ∀-*algebra* $\langle L, \neg, \forall\rangle$, *define a binary relation* ∃ *on* L *by* $a \exists b$ *if and only if* $a(-\forall)\neg b$. *Then* $\langle L, \forall, \exists\rangle$ *is an* ∀∃-*algebra.*

Proof:
It suffices to show that the axioms (∀∃1) – (∀∃5) are satisfied. Note that with the assumed definition of the relation ∃, the axiom (∀∃3) has the form

$$a\forall b \text{ and } c\forall\neg b \Rightarrow a\forall\neg c.$$

To show that this form of (∀∃3) holds, assume $a\forall b$ and $c\forall\neg b$. By (∀4) we get $\neg\neg b\forall\neg c$ which together with (∀1) gives $b\forall\neg c$ by (∀3). Applying (∀3) again we obtain $a\forall\neg c$ as required. (∀∃4) and ($\forall\exists_5$) follow from (∀5). □

Conversely, suppose that $\langle L, \forall, \exists\rangle$ is an ∀∃-algebra. Let $f : L \to L'$ be a $1-1$ map of L onto a set L' with $L \cap L' = \emptyset$, and let $L \cup L'$ be the (disjoint) union of L and its "copy" L'. Define operations ¬ and ∀′, for any $a \in L \cup L'$, by

$$\neg a = \begin{cases} f(a) & \text{if } a \in L \\ f^{-1}(a) & \text{if } a \in L' \end{cases}$$

$$a \ \forall' \ b \quad \text{if and only if} \quad \begin{cases} a \ \forall \ b & \text{if } a, b \in L \\ f^{-1}(b) \ \forall \ f^{-1}(a) & \text{if } a, b \in L' \\ a(-\exists) \ f^{-1}(b) & \text{if } a \in L, \ b \in L' \\ false & \text{if } a \in L', \ b \in L \end{cases}$$

Lemma 7.7.2 *Let* $\langle L, \forall, \exists\rangle$ *be an* ∀∃-*algebra. For any* a, $b \in L$ *the following conditions are equivalent*

(a) *$a \exists b$ holds in L;*

(b) *$a \forall' \neg b$ does not hold in $L \cup L'$.*

Proof:
Let a, b, $\in L$. Then $\neg b = f(b) \in L'$. By definition of $\forall'$, a $\forall' f(b)$ if and only if a $(-\exists) f^{-1}(f(b))$ and the latter holds if and only if a $(-\exists)\, b$. Hence the required equivalence holds. □

The following two lemmas provide some "extreme" examples of $\forall\exists$-algebras.

Lemma 7.7.3 *Let $\langle L, \forall, \exists\rangle$ be an $\forall\exists$-algebra.*

(a) *If $\forall$ is the universal relation, then $\exists$ is universal or empty;*

(b) *If $\exists \cap Id_L = \emptyset$, then $\forall$ is universal, where Id_L is the identity on L.*

Proof:
(a) Assume that $\forall$ is universal and suppose that $\exists \neq \emptyset$, say $a\exists b$ holds. Since $\exists$ is symmetric, $b\exists a$. Let $c \in L$. Then $b\forall c$ and by $(\forall\exists 3)$ we get $a\exists c$. It follows that $\{a\} \times L \subseteq \exists$. Furthermore, symmetry of $\exists$ and universality of $\forall$ imply that $\exists$ is universal.

(b) If $\exists \cap Id_L = \emptyset$, then $a(-\exists)a$ for all $a \in L$. Hence by $(\forall\exists 5)$, for all $a, b \in L$ we have $a\forall b$. This completes the proof. □

Lemma 7.7.4 *Let $\langle L, \forall, \exists\rangle$ be an $\forall\exists$-algebra. Then $\exists$ is reflexive if and only if $\forall \subseteq \exists$.*

Proof:
If $Id_L \subseteq \exists$, then since $\exists$ is symmetric we get $\forall = Id_L; \forall \subseteq \exists; \forall = \exists^{\smile}; \forall \subseteq \exists$ where the latter inclusion follows from $(\forall\exists 3)$. Conversely, if $\forall \subseteq \exists$, then $Id_L \subseteq \exists$ since $\forall$ is reflexive. □

Similarly as in Section 7.6 we show that a discrete representation theorem can be proved for the class of $\forall\exists$-algebras such that the relation $\forall$ is a partial order.

Lemma 7.7.5 *For any $a, b \in L$, $a \exists b$ implies $\forall(a) \times \forall(b) \subseteq \exists$.*

Proof:
First, observe that $\forall^{\smile} ; \exists ; \forall \subseteq \exists$ because the symmetry of $\exists$ and axiom $(\forall\exists 3)$ imply that $\forall^{\smile} ; \exists \subseteq \exists$, and therefore using $(\forall\exists 3)$ again we obtain $\forall^{\smile} ; \exists ; \forall \subseteq \exists ; \forall \subseteq \exists$. Now assume $a \exists b$ and let $c \in \forall(a)$ and $d \in \forall(b)$. Thus $a \forall c$ and $b \forall d$. Then $(c, d) \in \forall^{\smile} ; \exists ; \forall \subseteq \exists$. □

Lemma 7.7.6 *If $a \in L$ and $a \exists a$ then $\forall(a)$ is the smallest $\forall\exists$-set containing a.*

Proof:
Since $\forall$ is reflexive, we have $[\forall]\,\forall(a) \subseteq \forall(a)$. Conversely, assume that $b \in \forall(a)$ and $c \in \forall(b)$, that is $a\,\forall\,b$ and $b\,\forall\,c$. By transitivity of $\forall$ we have $a\,\forall\,c$ and hence $\forall(b) \subseteq \forall(a)$. It follows that $\forall(a) \subseteq [\forall]\,\forall(a)$. Next, by $a\,\exists\,a$ and Lemma 7.7.5, $\forall(a) \times \forall(a) \subseteq \exists$. Thus $\forall(a)$ is an $\forall\exists$-set. Clearly, any $\forall\exists$-set containing a must contain $\forall(a)$ as a subset. □

Let L be a $\forall\exists$-algebra and consider the Stone mapping $h : L \rightarrow 2^{X_L}$, used in Section 2.3.

Lemma 7.7.7 *For all $a, b \in L$,*

(a) $a\,\forall\,b$ *if and only if* $h(a) \subseteq h(b)$

(b) $a\,\exists\,b$ *if and only if* $h(a) \cap h(b) \neq \emptyset$.

Proof:
(a) Assume $a\,\forall\,b$ and take $A \in X_L$ such that $a \in A$. Since $A \times A \subseteq \exists$, $a\,\exists\,a$. By Lemma 7.7.6 $\forall(a) \subseteq A$ and hence $b \in A$ as required. Conversely, consider the following two cases. If $a\,(-\exists)\,a$, then by ($\forall\exists$5) we have $a\,\forall\,b$. If $a\,\exists\,a$, then by Lemma 7.7.6 we have $\forall(a) \in h(a)$ and hence $\forall(a) \in h(b)$ which implies $a\,\forall\,b$.

(b) Assume $a\,\exists\,b$. By symmetry of $\exists$ we have $b\,\exists\,a$ and by ($\forall\exists$4) we get $a\,\exists\,a$ and $b\,\exists\,b$. By Lemma 7.7.6 $\forall(a), \forall(b) \in X_L$. Consider $A = \forall(a) \cup \forall(b)$. We show that $A \in X_L$, that is $[\forall]A = A$ and $A \times A \subseteq \exists$. $[\forall]A \subseteq A$ by reflexivity of $\forall$. Conversely, $\forall(a) \cup \forall(b) = [\forall]\forall(a) \cup [\forall]\forall(b) \subseteq [\forall](\forall(a) \cup \forall(b))$. Now, from $a\,\exists\,a$, $b\,\exists\,b$, and Lemma 7.7.5 we have $\forall(a) \times \forall(a) \subseteq \exists$ and $\forall(b) \times \forall(b) \subseteq \exists$. From $a\,\exists\,b$ and Lemma 7.7.5 $\forall(a) \times \forall(b) \subseteq \exists$. By symmetry of $\exists$, $\forall(b) \times \forall(a) \subseteq \exists$. □

Lemma 7.7.8 *The mapping h is injective if and only if $\forall$ is a partial order.*

Proof:
Assume that h is injective and let $a\,\forall\,b$ and $b\,\forall\,a$. Then $\forall(a) = \forall(b)$. If $A \in X_L$ and $a \in A$, then $\forall(a) \subseteq A$ and hence $\forall(b) \subseteq A$. Since $\forall$ is reflexive, $b \in A$. Thus $h(a) \subseteq h(b)$. Similarly, we obtain $h(b) \subseteq h(a)$. By injectivity of h, $a = b$.

Conversely, assume that $\forall$ is antisymmetric and let $a, b \in L$ with $a \neq b$. We may assume without loss of generality that $a\,(-\forall)\,b$. Then $b \notin \forall(a)$, so $\forall(a) \notin h(b)$. Clearly, $\forall(a) \in h(a)$, and hence $h(a) \neq h(b)$. □

Lemmas 7.7.7 and 7.7.8 lead to the necessary and sufficient condition for a representation for $\forall\exists$-algebras.

Theorem 7.7.9 *The $\forall\exists$-algebra $\langle L, \forall, \exists\rangle$ is embeddable into the complex algebra of its canonical frame if and only if the relation $\forall$ is a partial order.*

We have the following discrete representation for $\forall\exists$-frames.

Theorem 7.7.10 *Every $\forall\exists$-frame is embeddable into the canonical frame of its complex algebra.*

Proof:
Consider $k : X \to 2^{2^X}$ defined as $k(x) = \{A \subseteq X : x \in A\}$. We show that k is well defined, that is $k(x) = [\forall_X]\, k(x)$ and $k(x) \times k(x) \subseteq \exists_X$. Indeed, if $x \in A$, then $x \in B$ for every $B \supseteq A$. Similarly, if $x \in A$ and $x \in B$ then clearly, $A \cap B \neq \emptyset$. Next, note that k is injective because if for every $A \subseteq X$, $x \in A$ if and only if $y \in A$, then considering $A = \{x\}$ we get $x = y$. □

Part II

Distributive lattices with operators

Chapter 8

Bounded distributive lattices with modal operators

8.1 Introduction

For bounded distributive lattices with operators, the basis for our discrete duality is that any distributive lattice is isomorphic to a ring of sets, meaning a set of sets closed under unions and intersections. We consider this in the style of Stone [228] and Priestley [202], working with prime filters partially ordered by set inclusion. We disregard the topology which is not needed for our purposes of obtaining representation results which we refer to as discrete dualities. On the other hand from a poset, the up-closed sets form a bounded distributive lattice.

Following [184], in Sections 8.2 to 8.5 we present discrete dualities for four classes of bounded distributive lattices with a single modal operator – namely, necessity, possibility, sufficiency and dual sufficiency. Contrary to Boolean algebras with modal operators (see Section 8.2 and Section 8.4) neither the operators of possibility and necessity nor the operators of sufficiency and dual sufficiency are mutually definable in lattices. In Section 8.6 and Section 8.7 we discuss discrete duality for positive modal algebras and their axiomatic extensions. Positive modal algebras are bounded distributive lattices endowed both with a possibility operator and a necessity operator. The operators are linked with the two axioms which allow for a single binary relation in the corresponding frames. The relation suffices for a construction of these two operators in the complex algebra. A positive modal logic originated in [48] and its Kripke-style semantics was developed in [22]. Next, in Section 8.8 a discrete duality for negative possibility algebras is presented, obtained from the developments in [20]. Negative modal algebras are bounded distributive lattices endowed with a sufficiency operator and a dual sufficiency operator. As in the case of positive modal algebras, two axioms are postulated enabling us to introduce a single relation in the corresponding frames.

A logic defined as a 'join' of positive and negative modal logics, referred to as a distributive modal logic is considered in [106], among others.

In Section 8.9, Section 8.10, and Section 8.11 we consider an abstraction of the Galois-style connections established in Theorem 1.8.3, namely, we define classes of bounded distributive lattices with Galois-style pairs of operators, along the lines of [182]. In Section 1.4 four Galois-style connections are defined, namely, a Galois pair, a dual Galois pair, a residuated pair, and a dual residuated pair. Here for a Galois pair we consider a pair of sufficiency operators satisfying the abstract counterparts to conditions (c) and (c') from Lemma 1.8.2; for a dual Galois pair, a pair of dual sufficiency operators satisfying the abstract counterparts to conditions (d) and (d') from Lemma 1.8.2; for a residuated pair, a necessity and possibility operators satisfying the abstract counterparts to conditions (a) and (a') from Lemma 1.8.2. As a consequence of the observation on a relationship between a residuated pair and a dual residuated pair, the duality in case of a dual residuated pair of operators easily follows from the residuated case.

Modal operators which form the Galois-style connections have applications in methods of data analysis and reasoning with incomplete information. In formal concept analysis which originated in Wille [256], see also Ganter and Wille [96], the operators of extent and intent of a concept determined by a context are sufficiency operators forming a Galois connection. Context algebras and their discrete duality are discussed in Section 3.4. Dual residuated pairs of modal operators are used in formal models of relationships between problems and skills needed to solve them, see, for example, [99]. Residuated and dual residuated modal operators are extensively used in the field of temporal logic and temporal knowledge representation.

8.2 Possibility distributive lattices

A *possibility lattice*, *P-lattice*, is an algebra $(L, \Diamond)$ where L is a bounded distributive lattice and $\Diamond$ is a unary operator on L satisfying, for any $a, b \in L$,

(P1) $\Diamond(a \vee b) = \Diamond a \vee \Diamond b$ additive

(P2) $\Diamond 0 = 0$ normal.

These lattices are denoted as P-lattices indicating that they are endowed with a possibiity operator.

Lemma 8.2.1 *Let $(L, \Diamond)$ be a P-lattice. Then, for all $a, b \in L$ and for all $A, B \subseteq L$,*

(a) *If $a \leq b$, then $\Diamond a \leq \Diamond b$;*

(b) *If A is an ideal of L, then so is $\Diamond^{-1}(A)$;*

(c) *If A is a prime filter of L, then $-\Diamond^{-1}(A)$ is an ideal of L.*

A *possibility frame*, *P-frame*, is a structure (X, R) where X is a bounded distributive lattice frame and R is a binary relation on X satisfying

(FP) $(\geqslant; R; \geqslant) \subseteq R$.

The *complex algebra of a P-frame* X is a structure $(L_X, \Diamond_R)$ where L_X is the complex algebra of its lattice frame reduct, defined in Section 2.5, and, for every $A \in L_X$,

$$\Diamond_R(A) \stackrel{\text{def}}{=} \langle R \rangle A.$$

Proposition 8.2.2 *The complex algebra of a P-frame is a P-lattice.*

Proof:
As shown in Lemma 2.5.1, L_X is closed under the lattice operations. We need to show that, for any $A \in L_X$, $\Diamond_R(A) = [\leqslant]\Diamond_R(A)$. The right-to-left inclusion follows from the reflexivity of $\leqslant$. For the reverse inclusion, suppose $x \in \langle R \rangle A$ and $x \notin [\leqslant]\langle R \rangle A$. Then, for some y, xRy and $y \in A$, and for some z, $x \leqslant z$ and $z \notin \langle R \rangle A$. Hence, $z(\geqslant; R)y$. The frame axiom (FP) implies $(\geqslant; R) \subseteq R$, so zRy. We also have $y \in A$, thus $z \in \langle R \rangle A$, a contradiction. Axioms (P1) and (P2) are satisfied, since $\Diamond_R = \langle R \rangle$. □

The *canonical frame of a P-lattice* L is a structure $(X_L, R_\Diamond)$ where X_L is the canonical frame of its lattice reduct and $R_\Diamond$ is a binary relation on X_L defined, for all $F, G \in X_L$, by

$$F R_\Diamond G \stackrel{\text{def}}{\Leftrightarrow} G \subseteq \Diamond^{-1}(F).$$

Proposition 8.2.3 *The canonical frame of a P-lattice is a P-frame.*

Proof:
In view of Theorem 2.5.2 we only need to show $(\supseteq; R_\Diamond; \supseteq) \subseteq R_\Diamond$. Take any $F, G \in X_L$ such that $F(\supseteq; R_\Diamond; \supseteq)G$. Then, for some $F', G' \in X_L$, $F \supseteq F'$, $G \subseteq \Diamond^{-1}(F') \subseteq G'$, and $G' \supseteq G$. In order to show $F R_\Diamond G$, take any $a \in L$ such that $a \in G$. Then $a \in G'$, so $\Diamond a \in F'$, hence $\Diamond a \in F$, as required. □

Let $(L, \Diamond)$ be a P-lattice. We now show that the embedding $h : L \to L_{X_L}$, used in Section 2.5, preserves the operator $\Diamond$.

Lemma 8.2.4 *For any $a \in L$, $h(\Diamond a) = \Diamond_{R_\Diamond}(h(a))$.*

Proof:
For this we show that for every $F \in X_L$ and for every $a \in L$,

$$\Diamond a \in F \Leftrightarrow \exists G \in X_L,\ G \subseteq \Diamond^{-1}(F)\ \&\ a \in G.$$

Assume $G \subseteq \Diamond^{-1}(F)$ and $a \in G$. Then $G \subseteq \{a : \Diamond a \in F\}$ and hence $\Diamond a \in F$. On the other hand, suppose $\Diamond a \in F$. Then $a \in \Diamond^{-1}(F)$. Since $\Diamond$ is monotone, $\Diamond^{-1}(F)$ is up-closed. Since F is a prime filter, $-\Diamond^{-1}(F)$ is an ideal. Now $-\Diamond^{-1}(F) \cap \uparrow_{\leq} a = \emptyset$. Then by the prime filter theorem for bounded distributive lattices, there is some prime filter $H \in X_L$ such that $-\Diamond^{-1}(F) \cap H = \emptyset$ and $\uparrow_{\leq} a \subseteq H$. Thus $H \subseteq \Diamond^{-1}(F)$ and $a \in H$. □

Let (X, R) be a P-frame. We now show that the embedding $k : X \to X_{L_X}$, used in Section 2.5, preserves the relation R.

Lemma 8.2.5 *For any $x, y \in X$, $xRy \Leftrightarrow k(x) R_{\Diamond_R} k(y)$.*

Proof:
We show, for any $x, y \in X$, that

$$xRy \Leftrightarrow \forall A \subseteq X,\ y \in A \Rightarrow x \in \langle R \rangle A.$$

Assume that xRy. Take any $A \in L_X$ such that $y \in A$. Hence $x \in \langle R \rangle A$. On the other hand, assume that for every $A \subseteq X$, $y \in A \Rightarrow x \in \langle R \rangle A$. Take $A = \uparrow_{\leqslant} y$. Clearly, $[\leqslant]\uparrow_{\leqslant} y = \uparrow_{\leqslant} y$, so $A \in L_X$. Moreover, $y \in A$. Then, by the assumption, $x \in \langle R \rangle \uparrow_{\leqslant} y$, so there is some $z \in X$ such that xRz and $y \leqslant z$. Hence, $x(R\,;\,\geqslant)y$. By reflexivity of $\leqslant$, we have $(R\,;\,\geqslant) \subseteq (\geqslant\,;\,R\,;\,\geqslant)$ and, by (FP), $(\geqslant\,;\,R\,;\,\geqslant) \subseteq R$, Thus $(R\,;\,\geqslant) \subseteq R$ and hence xRy, as required. □

Invoking Theorem 2.5.5, we obtain the discrete representations for P-lattices and P-frames.

Theorem 8.2.6

(a) *Every P-lattice is embeddable into the complex algebra of its canonical frame;*

(b) *Every P-frame is embeddable into the canonical frame of its complex algebra.*

In order to extend the discrete duality in Theorem 8.2.6 to a duality via truth, we follow a similar approach to that in Section 3.2 building a modal language from $\mathcal{L}an_L$ and the possibility operator ($\Diamond$). Analogously, we may obtain a duality via truth for necessity distributive lattices, sufficiency distributive lattices, dual sufficiency lattices in Section 8.3, Section 8.4, and Section 8.5, respectively, by extending the language $\mathcal{L}an_L$ with the necessity operator ($\Box$), the sufficiency operator ($\boxminus$), and the dual sufficiency operator ($\diamondminus$), respectively. In each case, the complex algebra theorem and the duality via truth theorem will hold.

8.3 Necessity distributive lattices

A *necessity lattice*, *N-lattice*, is an algebra $(L, \Box)$ where L is a bounded distributive lattice and $\Box$ is a unary operation on L satisfying for all $a, b, c \in L$,

(N1) $\Box(a \wedge b) = \Box a \wedge \Box b$ multiplicative

(N2) $\Box 1 = 1$ dually normal.

Using similar reasoning to that in Lemma 8.2.1, we obtain

Lemma 8.3.1 *For all $a, b \in L$ and for all $A, B \subseteq L$,*

(a) *If $a \leq b$, then $\Box a \leq \Box b$;*

(b) *If A is a filter of L, then so are $\Box(A)$ and $\Box^{-1}(A)$;*

(c) *If A is a prime ideal of L, then $-\Box^{-1}(A)$ is a filter of L;*

A *necessity frame*, *N-frame*, is a structure (X, R) where X is a bounded distributive lattice frame and R is a binary relation on X satisfying

(FN) $(\leqslant; R; \leqslant) \subseteq R$,

The *complex algebra of an N-frame* X is a structure $(L_X, \Box_R)$ where L_X is the complex algebra of its lattice frame reduct and, for every $A \in L_X$,

$$\Box_R A \stackrel{\text{def}}{=} [R]A.$$

Proposition 8.3.2 *The complex algebra of an N-frame is an N-lattice.*

Proof:
As shown in Section 2.5, L_X is closed under the lattice operations. We show that for any $A \in L_X$, $\Box_R(A) = [\leqslant]\,\Box_R A$. The right-to-left inclusion follows from the reflexivity of $\leqslant$. For the other inclusion note that from the frame axiom (FN1) we get $(\leqslant; R) \subseteq R$ which implies

$$\Box_R A = [R]A \subseteq [\leqslant; R]A = [\leqslant][R]A = [\leqslant]\,\Box_R A,$$

for any $A \in L_X$. Axioms (N1) and (N2) are satisfied, since $\Box_R = [R]$. □

The *canonical frame of an N-lattice* L is a structure $(X_L, R_\Box)$ where X_L is the canonical frame of its lattice reduct and $R_\Box$ is a binary relation on X_L defined for all $F, G \in X_L$ by

$$F R_\Box G \stackrel{\text{def}}{\Leftrightarrow} \Box^{-1}(F) \subseteq G.$$

Proposition 8.3.3 *The canonical frame of an N-lattice is an N-frame.*

Proof:
By Proposition 2.5.2 it suffices to show $(\subseteq; R_\Box; \subseteq) \subseteq R_\Box$. Take any $F, G \in X_L$ such that $F(\subseteq; R_\Box; \subseteq)G$. Then, for some $F', G' \in X_L$, $F \subseteq F'$, $\Box^{-1}(F') \subseteq G'$, and $G' \subseteq G$. In order to show $F R_\Box G$, take any $a \in L$ such that $\Box a \in F$. Then $\Box a \in F'$, so $a \in G'$, hence $a \in G$, as required. □

Let L be an N-lattice. We now show that the embedding $h : L \to L_{X_L}$, used in Section 2.5, preserves the operator $\Box$.

Lemma 8.3.4 *For any $a \in L$, $h(\Box a) = \Box_{R_\Box}(h(a))$.*

Proof:
We show that, for any $F \in X_L$ and for any $a \in L$,

$$\Box a \notin F \Leftrightarrow \exists G \in X_L,\ \Box^{-1}(F) \subseteq G \ \&\ a \notin G.$$

Suppose $\Box^{-1}(F) \subseteq G$ and $a \notin G$. Then $\{a \in L : \Box a \in F\} \subseteq G$ and hence $\Box a \notin F$. On the other hand, suppose $\Box a \notin F$. Then $a \notin \Box^{-1}(F)$. Since $\Box$ is monotone, $\Box^{-1}(F)$ is up-closed and since $\Box$ is multiplicative $\Box^{-1}(F)$ is a filter. Thus $\Box^{-1}(F) \cap \downarrow_{\leq} a = \emptyset$. So, by the prime filter theorem for bounded distributive lattices, there is a prime filter $G \in X_L$ such that $\Box^{-1}(F) \subseteq G$ and $\downarrow_{\leq} a \notin G$. Thus $\Box^{-1}(F) \subseteq G$ and $a \notin G$, as required. □

Let X be an N-frame. We now show that the embedding $k : X \to X_{L_X}$, used in Section 2.5, preserves the relation R.

Lemma 8.3.5 *For any $x, y \in X$, $xRy \Leftrightarrow k(x) R_{\Box_R} k(y)$.*

Proof:
We show, for any $x, y \in X$, that

$$xRy \Leftrightarrow \forall A \in L_X,\ x \in [R](A) \Rightarrow y \in A.$$

Assume xRy. Take any $A \in L_X$ such that $x \in \Box_R A$. Then $y \in A$, as required. Assume that, for every $A \in L_X$, $x \in [R]A$ implies $y \in A$. Take $A = [\leqslant](-\{y\})$. By transitivity of $\leqslant$, $A \subseteq [\leqslant]A$ and, by reflexivity of $\leqslant$, $[\leqslant]A \subseteq A$. Thus $A \in L_X$. Moreover, $y \notin [\leqslant](-\{y\})$. Then, by the assumption, $x \notin [R][\leqslant](-\{y\})$. Thus $x \in \langle R\rangle\langle\leqslant\rangle\{y\} = \langle R;\leqslant\rangle\{y\}$, and hence $x(R;\leqslant)y$ holds. Now $(R;\leqslant) \subseteq (\leqslant;R;\leqslant)$ and, by (FN), $(\leqslant;R;\leqslant) \subseteq R$. Hence xRy, as required. □

Invoking Theorem 2.5.5, we obtain the discrete representations for N-lattices and N-frames.

Theorem 8.3.6

(a) *Every N-lattice is embeddable into the complex algebra of its canonical frame;*

(b) *Every N-frame is embeddable into the canonical frame of its complex algebra.*

8.4 Sufficiency distributive lattices

A *sufficiency lattice*, *S-lattice*, is an algebra $(L, \boxminus)$ where L is a bounded distributive lattice and $\boxminus$ is a unary operator on L satisfying, for all $a, b \in L$,

(S1) $\boxminus(a \vee b) = \boxminus a \wedge \boxminus b$ co-additive

(S2) $\boxminus 0 = 1$ co-normal.

Lemma 8.4.1 *Let* $(L, \boxminus)$ *be an S-lattice. Then, for all* $a, b \in L$ *and for all* $A, B \subseteq L$,

(a) *If* $a \leq b$, *then* $\boxminus b \leq \boxminus a$;

(b) *If* A *is a prime filter of* L, *then* $\boxminus^{-1}(A)$ *is an ideal of* L.

Proof:
The proof of (a) is straightforward. For (b), take a prime filter A of L and $a, b \in L$ such that $a \in \boxminus^{-1}(A)$ and $b \leq a$. Thus $\boxminus a \in A$. By (a), $\boxminus$ is antitone, so $\boxminus a \leq \boxminus b$, and hence $\boxminus b \in A$, that is $b \in \boxminus^{-1}(A)$. Next, take $a, b \in \boxminus^{-1}(A)$, so $\boxminus a \in A$ and $\boxminus b \in A$. Then $\boxminus a \wedge \boxminus b \in A$, since A is a filter of L. Hence, by (S1), $\boxminus(a \vee b) \in A$, thus $a \vee b \in \boxminus^{-1}(A)$. □

A *sufficiency frame*, *S-frame*, is a structure (X, R) where X is a bounded distributive lattice frame and R is a binary relation on X satisfying

(FS) $(\leqslant; R; \geqslant) \subseteq R$.

The *complex algebra of an S-frame* X is a structure $(L_X, \boxminus_R)$ where L_X is the complex algebra of its lattice frame reduct and, for every $A \in L_X$,

$$\boxminus_R(A) \stackrel{\text{def}}{=} [R](-A).$$

The *canonical frame of an S-lattice* L is a structure $(X_L, R_{\boxminus})$ where X_L is the canonical frame of its lattice reduct and $R_{\boxminus}$ is a binary relation on X_L defined, for all $F, G \in X_L$, by

$$FR_{\boxminus}G \stackrel{\text{def}}{\Leftrightarrow} \boxminus^{-1}(F) \subseteq -G.$$

Using analogous reasoning to that for the proofs of Theorem 8.3.2 and Theorem 8.3.3 it follows that the complex algebra of any S-frame is an S-lattice, and that the canonical frame of any S-lattice is an S-frame.

Let L be an S-lattice. We now show that the embedding $h : L \to L_{X_L}$, used in Section 2.5, preserves the operator $\boxminus$.

Lemma 8.4.2 *For every* $a \in L$, $h(\boxminus a) = \boxminus_{R_{\boxminus}}(h(a))$.

Proof:
We show, for any $F \in X_L$ and for any $a \in L$, that

$$\boxminus a \in F \Leftrightarrow \forall G \in X_L,\ \boxminus^{-1}(F) \subseteq -G \Rightarrow a \notin G.$$

Assume $\boxminus a \in F$. Then $a \in \boxminus^{-1}(F)$, so if $\boxminus^{-1}(F) \subseteq -G$, then $a \notin G$, as required. On the other hand, assume $\boxminus a \notin F$. Since $\boxminus$ is antitone, $\boxminus^{-1}(F)$ is down-closed. By Lemma 8.4.1(b), $\boxminus^{-1}(F)$ is an ideal. Since $0 \in \boxminus^{-1}(F)$, we have $\boxminus^{-1}(F) \neq \emptyset$. Now $\boxminus^{-1}(F) \cap \uparrow_{\leq} a = \emptyset$. So, by the prime filter theorem for bounded distributive lattices, there is a prime filter $G \in X_L$ such that $\uparrow_{\leq} a \subseteq G$ and $\boxminus^{-1}(F) \cap G = \emptyset$. Hence $a \in G$ and $\boxminus^{-1}(F) \subseteq -G$. □

Let X be an S-frame. We now show that the embedding $k : X \to X_{L_X}$, used in Section 2.5, preserves the relation R.

Lemma 8.4.3 *For any* $x, y \in X$, $xRy \Leftrightarrow k(x) R_{\boxminus_R} k(y)$.

Proof:
We show, for any $x, y \in X$, that

$$xRy \Leftrightarrow \forall A \in L_X,\ x \in [R](-A) \Rightarrow y \in -A.$$

This follow using analogous reasoning to that in the proof of Theorem 8.3.5, □

Invoking Theorem 2.5.5, we obtain the discrete representations for S-lattices and S-frames.

Theorem 8.4.4

(a) *Every S-lattice is embeddable into the complex algebra of its canonical frame;*

(b) *Every S-frame is embeddable into the canonical frame of its complex algebra.*

8.5 Dual sufficiency distributive lattices

A *dual sufficiency lattice*, *dS-lattice*, is an algebra $(L, \diamondminus)$ where L is a bounded distributive lattice and $\diamondminus$ is a unary operator on L satisfying, for all $a, b \in L$,

(dS1) $\diamondminus(a \wedge b) = \diamondminus a \vee \diamondminus b$ co-multiplicative

(dS2) $\diamondminus 1 = 0$ dually co-normal.

Lemma 8.5.1 *For all* $a, b \in L$ *and for all* $A, B \subseteq L$,

(a) *If* $a \leq b$, *then* $\diamondminus b \leq \diamondminus a$;

(b) *If* A *is a prime filter of* L, *then* $-\diamondminus^{-1}(A)$ *is a filter of* L.

Proof:
We prove (b). Let A be a prime filter of L. Take $a \in -\diamondminus^{-1}(A)$ and $a \leq b$. Then $\diamondminus a \notin A$ and, by (a), $\diamondminus b \leq \diamondminus a$. Hence $\diamondminus b \notin A$, that is $b \in -\diamondminus^{-1}(A)$. Also, let $a \in -\diamondminus^{-1}(A)$ and $b \in -\diamondminus^{-1}(A)$. Then $\diamondminus a \notin A$ and $\diamondminus b \notin A$, so since A is prime, $\diamondminus \vee \diamondminus b \notin A$. Thus by (dS1), $\diamondminus(a \wedge b) \notin A$, that is $a \wedge b \in -\diamondminus^{-1}(A)$. □

A *dual sufficiency frame*, *dS-frame*, is a structure (X, R) where X is a bounded distributive lattice frame and P is a binary relation on X satisfying

(FdS) $(\geqslant; R; \leqslant) \subseteq R$,

The *complex algebra of a dS-frame* X is a structure $(L_X, \diamondminus_R)$ where L_X is the complex algebra of its lattice frame reduct and, for every $A \in L_X$,

$$\diamondminus_R(A) \stackrel{\text{def}}{=} \langle R \rangle(-A) = \{x \in X : \exists y \in X,\ xRy \ \&\ y \notin A\}.$$

The *canonical frame of dS-lattice* L is a structure $(X_L, R_\diamondsuit)$ where X_L is the canonical frame of its lattice reduct and $R_\diamondsuit$ is a binary relation on X_L defined, for all $F, G \in X_L$, by

$$FR_\diamondsuit G \overset{\text{def}}{\Leftrightarrow} -G \subseteq \diamondsuit^{-1}(F).$$

Using analogous reasoning to that for the proofs of Theorems 8.2.2 and 8.2.3 it follows that the complex algebra of any dS-frame is a dS-lattice, and that the canonical frame of any dS-lattice is a dS-frame.

Let L be a dS-lattice. We now show that the embedding $h : L \to L_{X_L}$, used in Section 2.5, preserves the operator $\diamondsuit$.

Lemma 8.5.2 *For every* $a \in L$, $h(\diamondsuit a) = \diamondsuit_{R_\diamondsuit}(h(a))$.

Proof:
We show, for any $F \in X_L$ and for any $a \in L$,

$$\diamondsuit a \in F \Leftrightarrow \exists G \in X_L,\ -G \subseteq \diamondsuit^{-1}(F) \ \&\ a \notin G.$$

Assume $\diamondsuit a \notin F$. Take any $G \in X_L$ such that $-G \subseteq \diamondsuit^{-1}(F)$. Then we have $-G \subseteq \{a : \diamondsuit a \in F\}$, so $a \notin -G$ and hence $a \in G$, as required. On the other hand, assume $\diamondsuit a \in F$. Then $a \in \diamondsuit^{-1}(F)$, so $a \notin -\diamondsuit^{-1}(F)$. Since $1 \notin \diamondsuit^{-1}(F)$, $-\diamondsuit^{-1}(F) \neq \emptyset$. Since $\diamondsuit$ is antitone, $-\diamondsuit^{-1}(F)$ is up-closed. Since F is prime, $-\diamondsuit^{-1}(F)$ is closed under meets. Hence $-\diamondsuit^{-1}(F)$ is a filter. Also $\downarrow_\leq a \cap -\diamondsuit^{-1}(F) = \emptyset$. Thus, by the prime filter theorem for bounded distributive lattices, there is some $G \in X_L$ such that $-\diamondsuit^{-1}(F) \subseteq G$ and $\downarrow_\leq a \cap G = \emptyset$. Therefore $-G \subseteq \diamondsuit^{-1}(F)$ and $a \notin G$. □

Let X be a dS-frame. We now show that the embedding $k : X \to X_{L_X}$, used in Section 2.5, preserves the relation R.

Lemma 8.5.3 *For any* $x, y \in X$, $xRy \Leftrightarrow k(x) R_{\diamondsuit_R} k(y)$.

Proof:
We show, for any $x, y \in X$,

$$xRy \Leftrightarrow \forall A \in L_X,\ y \in -A \Rightarrow x \in \langle R \rangle(-A).$$

This follow using analogous reasoning to that in the proof of Theorem 8.2.5. □

Invoking Theorem 2.5.5, we obtain the discrete representations for dS-algebras and dS-frames.

Theorem 8.5.4

(a) *Every dS-lattice is embeddable into the complex algebra of its canonical frame;*

(b) *Every dS-frame is embeddable into the canonical frame of its complex algebra.*

8.6 Positive modal algebras

A *positive modal algebra, PN-algebra,* is a structure $(L, \Diamond, \Box)$ where L is a bounded distributive lattice and $\Diamond$ and $\Box$ are modal operators of possibility and necessity satisfying the axioms (P1), (P2) from Section 8.2 and (N1), (N2) from Section 8.3, respectively, and, for all $a, b \in L$,

(PN1) $\Box a \wedge \Diamond b \leq \Diamond(a \wedge b)$

(PN2) $\Box(a \vee b) \leq \Box a \vee \Diamond b$.

These axioms were proposed in [48].

Following [22], a *PN-frame* is a relational structure (X, R) where X is a bounded distributive lattice frame and R is a binary relation on X satisfying

(FPN1) $(\leqslant; R) \subseteq (R; \leqslant)$

(FPN2) $(\geqslant; R) \subseteq (R; \geqslant)$

(FPN3) $(R; \leqslant) \cap (R; \geqslant) \subseteq R$

where as usual $\geqslant = \leqslant^{-1}$. The axiom (FPN3) added to the two axioms of [22] is needed for proving a representation theorem for PN-frames.

The *complex algebra of a PN-frame* X is a structure $(L_X, \Diamond_R, \Box_R)$ where L_X is the complex algebra of its lattice frame reduct and $\Diamond_R$ and $\Box_R$ are defined, for every $A \subseteq X$, by

$$\Diamond_R(A) \stackrel{\text{def}}{=} \langle R; \geqslant \rangle A$$
$$\Box_R(A) \stackrel{\text{def}}{=} [R; \leqslant] A.$$

We show that these operators coincide with the operators in the complex algebras of a P-frame and an N-frame, as defined in Sections 8.2 and 8.3, respectively.

Lemma 8.6.1 *Let (X, R) be a PN-frame and let $(L_X, \Diamond_R, \Box_R)$ be its complex algebra. Then, for every $A \in L_X$,*

(a) $\Diamond_R(A) = \langle R \rangle A$;

(b) $\Box_R(A) = [R] A$.

Proof:
(a) Let $x \in X$ be such that $x \in \langle R; \geqslant \rangle A$. Then there are $y, z \in X$ such that xRy, $y \geqslant z$, and $z \in A$. Since $A = [\leqslant] A$, we get $y \in A$. Hence $x \in \langle R \rangle A$. On the other hand, by reflexivity of $\leqslant$, $R \subseteq (R; \geqslant)$, so by Lemma 1.8.4(i), $\langle R \rangle A \subseteq \langle R; \geqslant \rangle A$. The proof of (b) is similar. □

Proposition 8.6.2 *The complex algebra of a PN-frame is a PN-algebra.*

Proof:
(PN1) In view of Lemma 8.6.1 it suffices to show that $[R]A \cap \langle R\rangle B \subseteq \langle R\rangle(A\cap B)$ for any $A, B \in L_X$. Let $x \in X$ be such that $x \in [R]A \cap \langle R\rangle B$. Since $x \in \langle R\rangle B$, there is some $y \in X$ such that xRy and $y \in B$. Now, xRy and $x \in [R]A$ yield $y \in A$. Then $y \in A \cap B$, which together with xRy gives $x \in \langle R\rangle(A \cap B)$.

(PN2) In view of Lemma 8.6.1 it suffices to show that $[R](A\cup B) \subseteq [R]A \cup \langle R\rangle B$ for all $A, B \in L_X$. Take $x \in X$ such that $x \notin [R]A \cup \langle R\rangle B$. Since $x \notin [R]A$, there is some $y \in X$ such that xRy and $y \notin A$. Also, xRy and $x \notin \langle R\rangle B$ imply $y \notin B$. Hence $y \notin A \cup B$, which together with xRy yields $x \notin [R](A \cup B)$. □

The *canonical frame* of a PN-algebra L is a structure (X_L, R_L) such that X_L is the canonical frame of its lattice reduct and R_L is a binary relation on X_L defined, for all $F, G \in X_L$,

$$F\, R_L G \overset{\text{def}}{\Leftrightarrow} \Box^{-1}(F) \subseteq G \subseteq \Diamond^{-1}(F).$$

Lemma 8.6.3 *Let $F, G \in X_L$. Then*

(a) $\Box^{-1}(F) \subseteq G \Leftrightarrow F(R_L; \subseteq)G$;
(b) $F \subseteq \Diamond^{-1}(G) \Leftrightarrow F(R_L; \supseteq)G$.

Proof:
(a) Assume that $F(R_L, \subseteq)G$. This means that there is some $H \in X_L$ such that $\Box^{-1}(F) \subseteq H \subseteq \Diamond^{-1}(F)$ and $H \subseteq G$. Hence $\Box^{-1}(F) \subseteq G$. For the reverse inclusion, assume that $\Box^{-1}(F) \subseteq G$ and consider an ideal I of L generated by $-G \cup -\Diamond^{-1}(F)$. We show that $\Box^{-1}(F) \cap I = \emptyset$. For suppose otherwise, then there is some $a \in L$ such that $\Box a \in F$ and $a \in I$. Since $a \in I$, there are $b, c \in L$ such that $b \notin G$, $\Diamond c \notin F$, and $a \leq b \vee c$. By Lemma 8.3.1(a), $\Box$ is monotone, so we get $\Box a \leq \Box(b \vee c)$. Next, by axiom (PN2), $\Box(b \vee c) \leq \Box b \vee \Diamond c$. Then $\Box a \leq \Box b \vee \Diamond c$. Since $\Box a \in F$, we get $\Box b \vee \Diamond c \in F$. However, $b \notin G$, so by assumption, $\Box b \notin F$. Hence, since $\Diamond c \notin F$, by primeness of F we get $\Box b \vee \Diamond c \notin F$, a contradiction. Now, since $\Box^{-1}(F)$ is a filter of L, by the prime filter theorem for distributive lattices there exists a prime filter of L, say H, such that $\Box^{-1}(F) \subseteq H$ and $H \cap I = \emptyset$, that is $H \subseteq -I$. Since $-\Diamond^{-1}(F) \subseteq I$ and $-G \subseteq I$, we finally get $\Box^{-1}(F) \subseteq H \subseteq \Diamond^{-1}(F)$ and $H \subseteq G$, as required.

(b) Assume $F \subseteq \Diamond^{-1}(G)$. Consider a filter K of L generated by $F \cap \Box^{-1}(G)$. We show that $K \subseteq \Diamond^{-1}(G)$. For suppose otherwise, then there is $a \in K$ such that $\Diamond a \notin G$. Since $a \in K$, there is some $b \in \Box^{-1}(G)$ and there is some $c \in F$ such that $b \wedge c \leq a$. Since $\Diamond$ is monotone, $\Diamond(b \wedge c) \leq \Diamond a$. Hence, by axiom (PN1), $\Box b \wedge \Diamond c \leq \Diamond a$. Since $c \in F$, by assumption $\Diamond c \in G$. Also, $\Box b \in G$. Hence $\Box b \wedge \Diamond c \in G$, thus $\Diamond a \in G$, a contradiction. Therefore $K \cap -\Diamond^{-1}(G) = \emptyset$. Since $-\Diamond^{-1}(G)$ is an ideal of L, applying the prime filter theorem for distributive lattices we get a prime filter of L, say H, such that $H \cap -\Diamond^{-1}(G) = \emptyset$ and $K \subseteq H$. Then $H \subseteq \Diamond^{-1}(G)$ and, since $\Box^{-1}(G) \subseteq K$, we have $\Box^{-1}(G) \subseteq H$. Moreover, $F \subseteq K$, thus $F \subseteq H$, as required. □

This lemma says that in any PN-lattice L, the relation $R_L; \subseteq$ coincides with the relation $R_\Box$ in the canonical frames of the N-lattice reduct of L, see

Section 8.3. Similarly, the relation $R_L;\supseteq$ coincides with the relation $R_\Diamond$ in the canonical frame of the P-lattice reduct of L, see Section 8.2.

Proposition 8.6.4 *The canonical frame of a PN-algebra is a PN-frame.*

Proof:
(FPN1) Let $F, G \in X_L$ be such that $F(\subseteq; R_L)G$. Then there is some $H \in X_L$ such that $F \subseteq H$ and $\Box^{-1}(H) \subseteq G \subseteq \Diamond^{-1}(H)$. By Lemma 8.3.1(b), $\Box^{-1}$ is monotone, so $\Box^{-1}(F) \subseteq \Box^{-1}(H)$, thus $\Box^{-1}(F) \subseteq G$. Hence, by Lemma 8.6.3(a), there is some $H \in X_L$ such that $\Box^{-1}(F) \subseteq H \subseteq \Diamond^{-1}(F)$ and $H \subseteq G$ which means that $F(R_L; \subseteq)G$ holds.

(FPN2) Assume that $F(\supseteq; R_L)G$. Then there is some $H \in X_L$ such that $H \subseteq F$ and $\Box^{-1}(H) \subseteq G \subseteq \Diamond^{-1}(H)$. Hence, using Lemma 8.2.1(b), $\Diamond^{-1}(H) \subseteq \Diamond^{-1}(F)$, so $G \subseteq \Diamond^{-1}(F)$. By Lemma 8.6.3(b) there exists $H \in X_L$ such that $\Box^{-1}(F) \subseteq H \subseteq \Diamond^{-1}(F)$ and $G \subseteq H$ which means that $F(R_L; \supseteq)G$ holds. □

Let L be a PN-algebra. We now show that the embedding $h : L \to L_{X_L}$, used in Section 2.5, preserves the operators $\Diamond$ and $\Box$.

Lemma 8.6.5 *For every $a \in L$,*

(a) $h(\Diamond a) = \Diamond_{R_L} h(a)$;
(b) $h(\Box a) = \Box_{R_L} h(a)$.

Proof:
By Proposition 8.6.4, the canonical frame of a PN-algebra is a PN-frame, so applying Lemma 8.6.1, $\Diamond_{R_L}(h(a)) = \langle R_L \rangle h(a)$ and $\Box_{R_L}(h(a)) = [R_L]h(a)$ for any $a \in L$.

(a) We have to show that for every $F \in X_L$,

$$\Diamond a \in F \Leftrightarrow \exists G \in X_L,\ \Box^{-1}(F) \subseteq G \subseteq \Diamond^{-1}(F)\ \&\ a \in G.$$

Take any $F \in X_L$. Assume that for some $G \in X_L$, $\Box^{-1}(F) \subseteq G \subseteq \Diamond^{-1}(F)$ and $a \in G$. Then, by Lemma 8.2.4 $\Diamond a \in F$. On the other hand, assume that $\Diamond a \in F$. Consider the filter H of L generated by $\Box^{-1}(F) \cup \{a\}$. Suppose $H \cap -\Diamond^{-1}(F) \neq \emptyset$. Then there is some $b \in H$ such that $\Diamond b \notin F$. Hence there is some $c \in \Box^{-1}(F)$ such that $c \wedge a \leq b$. Since $\Diamond$ is monotone, $\Diamond(c \wedge a) \leq \Diamond b$. Hence, by axiom (PN1) we get $\Box c \wedge \Diamond a \leq \Diamond b$. Since $\Box c \in F$ and $\Diamond a \in F$, $\Box c \wedge \Diamond a \in F$, whence $\Diamond b \in F$, a contradiction. Therefore $H \cap -\Diamond^{-1}(F) = \emptyset$. Since F is a filter of L, $-\Diamond^{-1}(F)$ is an ideal of L. By applying the prime filter theorem for distributive lattices we get a prime filter of L, say G, such that $H \subseteq G$ and $G \cap -\Diamond^{-1}(F) = \emptyset$. Hence $G \subseteq \Diamond^{-1}(F)$ and, moreover, since $\Box^{-1}(F) \subseteq H$ and $a \in H$, $\Box^{-1}F \subseteq G$ and $a \in G$.

(b) We show, for every $F \in X_L$, that

$$\Box a \notin F \Leftrightarrow \exists G \in X_L,\ \Box^{-1}(F) \subseteq G \subseteq \Diamond^{-1}(F)\ \&\ a \notin G.$$

Take any $F \in X_L$. Assume that $\exists G \in X_L$, $\Box^{-1}(F) \subseteq G \subseteq \Diamond^{-1}(F)$ & $a \notin G$. Then, by Lemma 8.3.4, $\Box a \notin F$. On the other hand, assume that $\Box a \notin F$.

Consider an ideal I generated by $-\Diamond^{-1}(F)\cup\{a\}$. Suppose that $\Box^{-1}(F)\cap I\neq\emptyset$. Then there are $b,c\in L$ such that $\Box b\in F$, $c\notin\Diamond^{-1}(F)$, and $b\leq a\vee c$. Since $\Box$ is monotone, $\Box b\leq\Box(a\vee c)$. Hence, by axiom (PN2), $\Box b\leq\Box a\vee\Diamond c$. Since F is a prime filter, the conditions $\Box a\notin F$ and $\Diamond c\notin F$ imply $\Box a\vee\Diamond c\notin F$. Thus $\Box b\notin F$, a contradiction. Then $\Box^{-1}(F)\cap I=\emptyset$. By the prime filter theorem for distributive lattices, there exists a prime filter, say G, such that $\Box^{-1}(F)\subseteq G$ and $G\cap I=\emptyset$. But $-\Diamond^{-1}(F)\subseteq I$, so $G\subseteq\Diamond^{-1}(F)$. Moreover, $a\in I$, whence $a\notin G$. □

Let X be a PN-frame. We now show that the embedding $k: X\to X_{L_X}$, used in Section 2.5, preserves the relation R.

Lemma 8.6.6 *For all $x,y\in X$, $xRy \Leftrightarrow k(x)R_{L_X}k(y)$.*

Proof:
We show, for all $x,y\in X$, that

$$xRy \Leftrightarrow (\forall A\in L_X,\ x\in[R]A\Rightarrow y\in A)\ \&\ (\forall B\in L_X,\ y\in B\Rightarrow x\in\langle R\rangle B).$$

Assume that xRy and take any $A,B\in L_X$ such that $x\in[R]A$ and $y\in B$. Then $y\in A$ and $x\in\langle R\rangle B$.
On the other hand, assume that for every $A\in L_X$, $x\in[R]A\Rightarrow y\in A$ and for every $B\in L_X$, $y\in B\Rightarrow x\in\langle R\rangle B)$. Then, by Lemma 8.3.5, $x(R;\leqslant)y$ and, by Lemma 8.2.5, $x(R;\geqslant)y$. Hence, by (FPN3), xRy, as required. □

Lemma 8.6.5 and Lemma 8.6.6 lead to the following discrete representations for PN-algebras and PN-frames.

Theorem 8.6.7

(a) *Every PN-algebra is embeddable into the complex algebra of its canonical frame;*

(b) *Every PN-frame is embeddable into the canonical frame of its complex algebra.*

8.7 Axiomatic extensions of positive modal algebras

Following [22] in this section we consider PN-algebras endowed with additional axioms corresponding to the well-known axioms of classical modal algebras based on Boolean algebras. More specifically, we invoke the following conditions

(D)	$\Box a \leq \Diamond a$
($T_\Diamond$)	$a \leq \Diamond a$
($T_\Box$)	$\Box a \leq a$
($4_\Diamond$)	$\Diamond\Diamond a \leq \Diamond a$
($4_\Box$)	$\Box a \leq \Box\Box a$
($B_\Diamond$)	$\Diamond\Box a \leq a$
($B_\Box$)	$a \leq \Box\Diamond a$
($E_\Diamond$)	$\Diamond a \leq \Box\Diamond a$
($E_\Box$)	$\Diamond\Box a \leq \Box a$.

Let α stand for any one of the above axioms. A PN-algebra $(L, \Diamond, \Box)$ is called a *PNα-algebra* if it satisfies the axiom (α).

Let (X, R) be a PN-frame. We consider the following additional axioms

(FD)	R is serial
(F$T_\Diamond$)	$R;\geqslant$ is reflexive
(F$T_\Box$)	$R;\leqslant$ is reflexive
(F$4_\Diamond$)	$R;\geqslant$ is transitive
(F$4_\Box$)	$R;\leqslant$ is transitive
(F$B_\Diamond$)	$(R;\geqslant)^{-1} \subseteq (R;\leqslant)$
(F$B_\Box$)	$(R;\leqslant) \subseteq (R;\geqslant)^{-1}$
(F$E_\Diamond$)	$\forall x, y, z \in X,\ x(R;\leqslant)y\ \&\ x(R;\geqslant)z \Rightarrow y(R;\geqslant)z$
(F$E_\Box$)	$\forall x, y, z \in X,\ x(R;\geqslant)y\ \&\ x(R;\leqslant)z \Rightarrow y(R;\leqslant)z$.

Let (Fα) denote any one of the above frame axioms. A PN-frame (X, R) is called *PNα-frame* if the axiom (Fα) holds in X.

For $\alpha \in \{D, T_\Diamond, T_\Box, 4_\Diamond, 4_\Box, B_\Diamond, B_\Box, E_\Diamond, E_\Box\}$, the *complex algebra of a PNα-frame*, is defined in the same way as the complex algebra of a PN-algebra.

Proposition 8.7.1 *Let $\alpha \in \{D, T_\Diamond, T_\Box, 4_\Diamond, 4_\Box, B_\Diamond, B_\Box, E_\Diamond, E_\Box\}$. The complex algebra of a PNα-frame is a PNα-algebra.*

Proof:
(D) Let $A \in L_X$ and take $x \in [R;\leqslant]A$. By Lemma 8.6.1, for every $A \in L_X$, $[R;\leqslant]A = [R]A$. Hence $x \in [R]A$. Since R is serial, by the correspondence result from Lemma 1.10.1(a) we get $x \in \langle R \rangle A$, so $x \in \langle R;\geqslant \rangle A$.

For ($T_\Diamond$), ($T_\Box$), ($4_\Diamond$), and ($4_\Box$) the results follow from Lemma 1.10.1.

($B_\Diamond$) Take $x \in \langle R;\geqslant \rangle [R;\leqslant]A$. Then there is some $y \in X$ such that $x(R;\geqslant)y$ and $y \in [R;\leqslant]A$. By (F$B_\Diamond$) we get $y(R;\leqslant)x$, thus since $y \in [R;\leqslant]A$, $x \in A$. Then $\Diamond_R \Box_R(A) \subseteq A$, as required. The proof of ($B_\Box$) is similar.

($E_\Diamond$) Assume $x \notin [R;\leqslant]\langle R;\geqslant \rangle A$. Then there exists $y \in X$ such that $x(R;\leqslant)y$ and $y \notin \langle R;\geqslant \rangle A$. Suppose $x \in \langle R;\geqslant \rangle A$. Thus there is some $z \in X$ such that

$x(R;\geqslant)z$ and $z\in A$. Now by (FB$_\Diamond$ we get $y(R;\geqslant)z$. Hence, since $y\notin\langle R;\geqslant\rangle A$, we get $z\notin$, a contradiction. The proof of (FB$_\Box$) is similar. □

The *canonical frame of a PNα-algebra,* $\alpha\in\{\mathrm{D},\mathrm{T}_\Diamond,\mathrm{T}_\Box,4_\Diamond,4_\Box,\mathrm{B}_\Diamond,\mathrm{B}_\Box,\mathrm{E}_\Diamond,\mathrm{E}_\Box\}$, is just the canonical frame of a PN-algebra.

Proposition 8.7.2 *Let* $\alpha\in\{\mathrm{D},\mathrm{T}_\Diamond,\mathrm{T}_\Box,4_\Diamond,4_\Box,\mathrm{B}_\Diamond,\mathrm{B}_\Box,\mathrm{E}_\Diamond,\mathrm{E}_\Box\}$. *The canonical frame of a PNα-algebra is a PNα-frame.*

Proof:
(FD) Take $F\in X_L$. Note that $\Box^{-1}(F)\subseteq\Diamond^{-1}(F)$. Indeed, if $a\in\Box^{-1}(F)$, then $\Box a\in F$, so by axiom (D), $\Diamond a\in F$, whence $a\in\Diamond^{-1}(F)$. Then $\Box^{-1}(F)\cap-\Diamond^{-1}(F)=\emptyset$. Also, $\Box^{-1}(F)$ is a filter of L and $-\Diamond^{-1}(F)$ is an ideal of L. By the prime filter theorem for distributive lattices we get a prime filter of L, say G, such that $G\cap-\Diamond^{-1}(F)=\emptyset$ and $\Box^{-1}(F)\subseteq G$. Hence $\Box^{-1}(F)\subseteq G\subseteq\Diamond^{-1}(F)$ which means that FR_LG, as required.

(FT$_\Diamond$) Let $F\in X_L$. Observe that $F\subseteq\Diamond^{-1}(F)$. Indeed, if $a\in F$, then by (T$_\Diamond$), $\Diamond a\in F$, thus $a\in\Diamond^{-1}(F)$. By Lemma 8.6.3(b), $F(R_L;\supseteq)F$.
Analogously, using Lemma 8.6.3(a), (FT$_\Box$) can be proved.

(F4$_\Diamond$) Take $F,G,H\in X_L$ such that $F(R_L;\supseteq)G$ and $G(R_L;\supseteq)H$. By the assumption, there is $F',G'\in X_L$ such that $\Box^{-1}(F)\subseteq F'\subseteq\Diamond^{-1}(F)$, $G\subseteq F'$, and $\Box^{-1}(G)\subseteq G'\subseteq\Diamond^{-1}(G)$, $H\subseteq G'$.
Analogously, using Lemma 8.6.3(a), (F4$_\Box$) can be proved.

(FB$_\Box$) We show that $(R_L;\subseteq)\subseteq(R_L;\supseteq)^{-1}$. Let $F,G\in X_L$ and assume that $F(R_L;\subseteq)G$. Then there exists $F'\in X_L$ such that $\Box^{-1}(F)\subseteq F'\subseteq\Diamond^{-1}(F)$ and $F'\subseteq G$. Let $a\in F$. By (B$_\Box$) we get $\Box\Diamond a\in F$, so $\Diamond a\in\Box^{-1}(F)$, thus $\Diamond a\in G$, that is $a\in\Diamond^{-1}(G)$. Hence $F\subseteq\Diamond^{-1}(G)$. Applying Lemma 8.6.3(a) we get $G(R_L;\supseteq)F$, so $F(R_L;\supseteq)^{-1}G$, as required.
Analogously, using Lemma 8.6.3(b), (TB$_\Diamond$) can be proved

(FE$_\Diamond$) We show that $F(R_L;\subseteq)G$ and $F(R_L;\supseteq)H$ imply $G(R_L;\supseteq)H$ for all $F,G,\ H\in X_L$. Assume that $F(R_L;\subseteq)G$ and $F(R_L;\supseteq)H$. Then there exist $F',G'\in X_L$ such that $\Box^{-1}(F)\subseteq F;\subseteq\Diamond^{-1}(F)$, $F'\subseteq G$, $\Box^{-1}(F)\subseteq G'\subseteq\Diamond^{-1}(F)$, and $H\subseteq G'$. Take $a\in L$ such that $a\in H$. Then $a\in G'$ and since $G'\subseteq\Diamond^{-1}(F)$, we get $\Diamond a\in G$. Hence, by (E$_\Diamond$), $\Box\Diamond a\in G$, so since $\Box^{-1}(F)\subseteq G$, we have $\Diamond a\in G$, that is, $a\in\Diamond^{-1}(G)$. Therefore $H\subseteq\Diamond^{-1}(G)$. Now using Lemma 8.6.3(b), $G(R_L;\supseteq)H$.
Analogously, using Lemma 8.6.3(a), (FE$_\Box$) can be proved. □

Invoking Theorem 8.6.7 we obtain the discrete representations for the axiomatic extensions of PNα-algebras and PNα-frames

Theorem 8.7.3 *Let* $\alpha\in\{\mathrm{D},\mathrm{T}_\Diamond,\mathrm{T}_\Box,4_\Diamond,4_\Box,\mathrm{B}_\Diamond,\mathrm{B}_\Box,\mathrm{E}_\Diamond,\mathrm{E}_\Box\}$.

(a) *Every PNα-algebra is embeddable into the complex algebra of its canonical frame;*

(b) *Every PNα-frame is embeddable into the complex algebra of its canonical frame.*

8.8 Negative modal algebras

In [20] *negative modal algebras* are defined as an extension of distributive lattices obtained by endowing them with the operations of sufficiency $\boxbar$ and dual sufficiency $\diamondbar$, and some representation results for these algebras are presented. In this section we adapt these results to our framework.

An *SdS-algebra* is a structure $(L, \boxbar, \diamondbar)$ such that $(L, \boxbar)$ is an S-algebra, $(L, \diamondbar)$ is a dS-algebra and, for all $a, b \in L$,

(SdS1) $\boxbar(a \wedge b) \leq \boxbar a \vee \diamondbar b$

(SdS2) $\boxbar a \wedge \diamondbar b \leq \diamondbar(a \vee b)$.

An *SdS-frame* is a relational structure (X, R) where X is a bounded distributive lattice frame and R is a binary relation on X satisfying

(FSdS1) $(\leqslant; R) \subseteq (R; \geqslant)$

(FSdS2) $(\geqslant; R) \subseteq (R; \leqslant)$

(FSdS3) $(R; \geqslant) \cap (R; \leqslant) \subseteq R$.

Note that (FSdS3) coincides with (FPN3) in Section 8.6.

The *complex algebra of an SdS-frame* X is a structure $(L_X, \boxbar_R, \diamondbar_R)$ where L_X is the complex algebra of its lattice frame reduct and $\boxbar_R$ and $\diamondbar_R$ are defined, for every $A \subseteq X$, by

$$\boxbar_R A \stackrel{\text{def}}{=} [R; \geqslant](-A)$$

$$\diamondbar_R A \stackrel{\text{def}}{=} \langle R; \leqslant\rangle(-A).$$

Analogous to Lemma 8.6.1, we have

Lemma 8.8.1 *For every* $A \in L_X$,

(a) $\boxbar_R A = [R](-A)$;

(b) $\diamondbar_R A = \langle R\rangle(-A)$.

Proposition 8.8.2 *The complex algebra of an SdS-frame is an SdS-algebra.*

Proof:
First, we show that $\boxbar_R A \in L_X$ and $\diamondbar_R A \in L_X$. In view of Lemma 8.8.1 it suffices to show that $[R](-A) = [\leqslant][R](-A)$ and $\langle R\rangle(-A) = [\leqslant]\langle R\rangle(-A)$. In both cases the right-to-left inclusion follows from reflexivity of $\leqslant$, see Lemma 1.10.1(b). We now show, for any $A \in L_X$, that $[R](-A) \subseteq [\leqslant][R](-A)$. Assume that $x \in [R](-A)$. Suppose $x \notin [\leqslant][R](-A)$. Then there are $y, z \in X$ such that $x \leqslant y$, yRz, and $z \in A$. Thus $x(\leqslant; R)z$, so by (FSdS1), $x(R; \geqslant)z$. Then, for some u, xRu and $z \leqslant u$. Since $A = [\leqslant]A$, from $z \leqslant u$ and $z \in A$ we get $u \in A$. However, xRu and $x \in [R](-A)$ imply $u \notin A$, a contradiction. Now, assume that $x \in \langle R\rangle(-A)$. This means that, for some y, xRy and $y \notin A$. Suppose $x \notin [\leqslant]\langle R\rangle(-A)$. Then there is some $z \in X$ such that $x \leqslant z$ and $z \in [R]A$. Since $z \geqslant x$ and xRy, that is, $z(\geqslant; R)y$, by (FSdS2) we get $z(R; \leqslant)y$, so there is some

$u \in X$ such that zRu and $u \leqslant y$. Now, zRu and $z \in [R]A$ imply $u \in A$, which together with $u \leqslant y$ gives $y \in A$, a contradiction.

The proofs that the axioms (SdS1) and (SdS2) hold in L_X are straightforward. □

The *canonical frame of an SdS-algebra L* is a structure (X_L, R_L) where X_L is the canonical frame of its lattice reduct and R_L is a binary relation on X_L defined, for all $F, G \in X_L$, by

$$FR_L G \overset{\text{def}}{\Leftrightarrow} \Box^{-1}(F) \subseteq -G \subseteq \Diamond^{-1}(F).$$

Lemma 8.8.3 *For any $F, G \in X_L$,*

(a) $\Box^{-1}(F) \subseteq -G \Leftrightarrow F(R_L; \supseteq)G$;

(b) $-G \subseteq \Diamond^{-1}(F) \Leftrightarrow F(R_L; \subseteq)G$.

Proof:
(a) Let $F, G \in X_L$ be such that $F(R_L; \supseteq)G$. Then there is some $G' \in X_L$ such that $G' \supseteq G$ and $\Box^{-1}(F) \subseteq -G' \subseteq \Diamond^{-1}(F)$. Since $-G' \subseteq -G$, $\Box^{-1}(F) \subseteq -G$. On the other hand, let $\Box^{-1}(F) \subseteq -G$ and consider a filter K of L generated by $-\Diamond^{-1}(F) \cup G$. We show that $K \cap \Box^{-1}(F) = \emptyset$. Suppose otherwise, that is, there exist $a, b, c \in L$ such that $a \in G$, $b \in -\Diamond^{-1}(F)$, $c \in \Box^{-1}(F)$, and $a \wedge b \leq c$. Hence, $\Diamond b \notin F$ and by the assumption $\Box a \notin F$. Next, since $\Diamond$ is antitone, we have $\Diamond c \leq \Diamond(a \wedge b)$, thus by (SdS1), $\Diamond c \leq \Box a \vee \Diamond b$. Hence, since $\Box c \in F$, $\Diamond a \vee \Box b \in F$. But F is a prime filter of L, so $\Box a \in F$ or $\Diamond b \in F$, a contradiction. Therefore by the prime filter theorem for distributive lattices we get a prime filter of L, say H, such that $H \cap \Box^{-1}(F) = \emptyset$ and $K \subseteq H$. Hence, $\Box^{-1}(F) \subseteq -H$. Since $G \subseteq K$ and $-\Diamond^{-1}(F) \subseteq K$, we obtain $-\Diamond^{-1}(F) \subseteq H$ and $G \subseteq H$. Hence, $\Box^{-1}(F) \subseteq -H \subseteq \Diamond^{-1}(F)$ and $G \subseteq H$, as required.

(b) Let $F, G \in X_L$ be such that $F(R_L; \subseteq)G$. Then there is some $G' \in X_L$ such that $G' \subseteq G$ and $\Box^{-1}(F) \subseteq -G' \subseteq \Diamond^{-1}(F)$. Since $-G \subseteq -G'$, $-G \subseteq \Diamond^{-1}(F)$. On the other hand, let $-G \subseteq \Diamond^{-1}(F)$ and consider the ideal I of L generated by $\Box^{-1}(F) \cup -G$. We show that $-\Diamond^{-1}(F) \cap I = \emptyset$. Suppose otherwise, that is, there exist $a \in \Box^{-1}(F)$, $b \notin G$, $c \notin \Diamond^{-1}(F)$ such that $c \leq a \vee b$. Hence $\Box a \in F$ and by the assumption, $\Diamond b \in F$, so $\Box a \wedge \Diamond b \in F$. Moreover, since $\Diamond$ is antitone, the condition $c \leq a \vee b$ implies $\Diamond(a \vee b) \leq \Diamond c$, which by (SdS2) yields $\Box a \wedge \Diamond b \leq \Diamond c$. But $\Diamond c \notin F$, whence $\Box a \wedge \Diamond b \notin F$, a contradiction. Therefore $-\Diamond^{-1}(F) \cap I = \emptyset$. By applying the prime filter theorem for distributive lattices we get a prime filter H of L such that $-\Diamond^{-1}(F) \subseteq H$ and $H \cap I = \emptyset$. Hence $-H \subseteq \Diamond^{-1}(F)$ and $I \subseteq -H$. Clearly, $\Box^{-1}(F) \subseteq I$ and $-G \subseteq I$, so $\Box^{-1}(F) \subseteq -H \subseteq \Diamond^{-1}(F)$ and $H \subseteq G$, as required. □

This lemma says that in any SdS-algebra L, the relation $R_L; \supseteq$ coincides with the relation $R_{\Box}$ in the canonical frames of the S-lattice reduct of L, see Section 8.4. Similarly, the relation $R_L; \subseteq$ coincides with the relation $R_{\Diamond}$ in the canonical frame of the dS-lattice reduct of L, see Section 8.5.

Proposition 8.8.4 *The canonical frame of an SdS-algebra is an SdS-frame.*

Proof:
Invoking Lemma 8.8.3 and reasoning analogously to that in Proposition 8.6.4, we can prove (FSdS1) and (FSdS2). For (FSdS3), let $F, G \in X_L$ be such that $F(R_L;\supseteq)G$ and $F(R_L;\subseteq)G$. Then there are $H, K \in X_L$ such that $\boxbar^{-1}(F) \subseteq -H \subseteq \diamondbar^{-1}(F)$, $\boxbar^{-1}(F) \subseteq -K \subseteq \diamondbar^{-1}(F)$, $G \subseteq H$, and $K \subseteq G$. Hence $-G \subseteq \diamondbar^{-1}(F)$ and $\boxbar^{-1}(F) \subseteq -G$, thus FR_LG holds. □

Let L be a SdS-algebra. We now show that the embedding $h : L \to L_{X_L}$, used in Section 2.5, preserves the operators $\boxbar$ and $\diamondbar$.

Lemma 8.8.5 *For every $a \in L$,*

(a) $h(\boxbar a) = \boxbar_{R_L} h(a)$;
(b) $h(\diamondbar a) = \diamondbar_{R_L} h(a)$.

Proof:

(a) We show, for any $F \in X_L$ and for any $a \in L$, that

$$\boxbar a \notin F \Leftrightarrow \exists G \in X_L,\ \boxbar^{-1}(F) \subseteq -G \subseteq \diamondbar^{-1}(F) \ \&\ a \in G.$$

Take any $a \in L$ and $F \in X_L$. Assume that $\boxbar a \in F$, or equivalently, $a \in \boxbar^{-1}(F)$, and take any $G \in X_L$ such that $-\diamondbar^{-1}F \subseteq G \subseteq -\boxbar^{-1}(F)$. Hence $a \notin G$.
On the other hand, assume that $\boxbar a \notin F$. Consider a filter H of L generated by $-\diamondbar^{-1}(F) \cup \{a\}$. Suppose $H \cap \boxbar^{-1}(F) \neq \emptyset$. Thus there is some $b \in H$ such that $\boxbar b \in F$. Also, there is some $c \in -\diamondbar^{-1}(F)$ such that $a \wedge c \leq b$. Hence, since $\boxbar$ is antitone, $\boxbar b \leq \boxbar(a \wedge c)$, so by (SdS1), $\boxbar b \leq \boxbar a \vee \diamondbar c$. As $\boxbar b \in F$, we get $\boxbar a \vee \diamondbar c \in F$, which by primeness of F yields $\boxbar a \in F$ or $\diamondbar c \in F$, a contradiction. Therefore $H \cap \boxbar^{-1}(F) = \emptyset$. By Lemma 8.4.1(c), $\boxbar^{-1}(F)$ is an ideal of L. Applying the prime filter theorem for distributive lattices we get a prime filter G of L such that $H \subseteq G$ and $G \cap \boxbar^{-1}(F) = \emptyset$, that is, $-G \subseteq -H$. Then $\boxbar^{-1}(F) \subseteq -G$ and, since $-\diamondbar^{-1}(F) \subseteq H$ and $a \in H$, we also have $-G \subseteq \diamondbar^{-1}(F)$ and $a \in G$, as required.

(b) We show, for any $F \in X_L$ and for any $a \in L$, that

$$\diamondbar a \in F \Leftrightarrow \exists G \in X_L,\ \boxbar^{-1}(F) \subseteq -G \subseteq \diamondbar^{-1}(F) \ \&\ a \notin G.$$

Take any $F \in X_L$ and any $a \in L$. Assume that there is some $G \in X_L$ such that $\boxbar^{-1}(F) \subseteq -G \subseteq \boxbar^{-1}(F)$ and $a \notin G$. Then, by Lemma 8.5.2, $\diamondbar a \in F$.
On the other hand, assume $\diamondbar a \in F$. Consider an ideal I of L generated by $\boxbar^{-1}(F) \cup \{a\}$. Suppose $I \cap -\diamondbar^{-1}(F) \neq \emptyset$. Thus there is some $b \in I$ such that $\diamondbar b \notin F$. Also, there is some $c \in \boxbar^{-1}(F)$ such that $b \leq c \vee a$. As $\diamondbar$ is antitone, we get $\diamondbar(c \vee a) \leq \diamondbar b$, and hence by (SdS2), $\boxbar c \wedge \diamondbar a \leq \diamondbar b$. But $\boxbar c \in F$ and $\diamondbar a \in F$, so $\boxbar c \wedge \diamondbar a \in F$, thus $\diamondbar b \in F$, a contradiction. Therefore $I \cap -\diamondbar^{-1}(F) = \emptyset$. By Lemma 8.5.1(c), $-\diamondbar^{-1}(F)$ is a filter of L. Now, by the prime filter theorem for distributive lattice we get a prime filter G of L such

that $-\Diamond^{-1}(F) \subseteq G$ and $I \cap G = \emptyset$, that is, $I \subseteq -G$. Thus $-G \subseteq \Diamond^{-1}(F)$ and, since $\Box^{-1}(F) \subseteq I$ and $a \in I$, we also have $\Box^{-1}(F) \subseteq -G$ and $a \notin G$, as required. □

Let X be a SdS-frame. We now show that the embedding $k : X \to X_{L_X}$, used in Section 2.5, preserves the relation R.

Lemma 8.8.6 *For any $x, y \in X$, $xRy \Leftrightarrow k(x)R_{L_X}k(y)$.*

Proof:
We show, for all $x, y \in X$, that

$$xRy \Leftrightarrow (\forall A \in L_X, x \in [R](-A) \Rightarrow y \in -A) \ \& \\ (\forall B \in L_X, y \in -B \Rightarrow x \notin \langle R\rangle(-B))$$

Assume xRy. Take any $A, B \in L_X$ be such that $x \notin \langle R\rangle(-A)$ and $y \in B$. Then $y \in A$ and $x \notin [R](-B)$.
On the other hand, assume that for every $A \in L_X, x \in [R](-A) \Rightarrow y \in -A$ and for every $B \in L_X, y \in -B \Rightarrow x \notin \langle R\rangle(-B)$. Then, by Theorem 8.4.3, $x(R; \leqslant)y$ and, by Theorem 8.5.3, $x(R; \geqslant)y$. Hence, by (FSdS3), xRy. □

With these results we obtain the discrete representations for the axiomatic extensions of SdS-algebras and SdS-frames.

Theorem 8.8.7

(a) *Every SdS-algebra is embeddable into the complex algebra of its canonical frame;*

(b) *Every SdS-frame is embeddable into the canonical frame of its complex algebra.*

8.9 Galois algebras

A *Galois algebra* is a structure (L, f, g) where L is a bounded distributive lattice and f and g are two sufficiency operators on L satisfying, for every $a \in L$,

(GM1) $a \leq f(g(a))$

(GM2) $a \leq g(f(a))$.

As shown in Section 8.4 each sufficiency operator f corresponds to a certain order-respecting binary relation R_f over a poset. The properties of the sufficiency operators in a Galois pair correspond to the relationships between their respective relations.

Accordingly, a *Galois frame* (X, R, S) is a bounded distributive lattice frame X endowed with binary relations R and S satisfying

(FGM1) $S = R^{-1}$

(FGM2) $(\geqslant ; R ; \leqslant) \subseteq R$

(FGM3) $(\geqslant ; S ; \leqslant) \subseteq S$.

Note that (FGM3) follows from (FGM1) and (FGM2).

The *complex algebra of a Galois frame* X is a structure (L_X, f_R, g_S) where L_X is the complex algebra of its lattice frame reduct and, for every $A \in L_X$,

$$f_R(A) \stackrel{\text{def}}{=} [R](-A) = \{x \in X : \forall y \in X,\ xRy \Rightarrow y \notin A\}$$
$$g_S(A) \stackrel{\text{def}}{=} [S](-A) = \{x \in X : \forall y \in X,\ xSy \Rightarrow y \notin A\}.$$

Proposition 8.9.1 *The complex algebra of a Galois frame is a Galois algebra.*

Proof:
We show (GM1), the proof of (GM2) is similar. For any $A \in L_X$, we need to show that $A \subseteq f_R(g_S(A))$. Take any $x \in A$ and take any $y \in X$ such that xRy. We need to show that for some $z \in X$, ySz and $z \in A$. Since $S = R^{-1}$, take z to be x. □

The *canonical frame of a Galois algebra* L is a structure (X_L, R_f, S_g) where X_L is the canonical frame of its lattice reduct and R_f and S_g are binary relations on X_L defined, for all $F, G \in X_L$, by

$$FR_fG \stackrel{\text{def}}{\Leftrightarrow} f^{-1}(F) \subseteq -G$$
$$FS_gG \stackrel{\text{def}}{\Leftrightarrow} g^{-1}(F) \subseteq -G,$$

where $f^{-1}(F) = \{a \in L : f(a) \in F\}$ and $g^{-1}(F) = \{a \in L : g(a) \in F\}$.

Proposition 8.9.2 *The canonical frame of a Galois algebra is a Galois frame.*

Proof:
By analogous reasoning to Proposition 8.3.3, conditions (FGM2) and (FGM3) hold. We show (FGM1) that is $(R_f)^{-1} = S_g$. Assume $(F, G) \in (R_f)^{-1}$. Then $(G, F) \in R_f$, that is, $f^{-1}(G) \subseteq -F$. We need to show that FS_gG, that is, $g^{-1}(F) \subseteq -G$. Assume $a \in g^{-1}(F)$. Then $g(a) \in F$, hence $g(a) \notin -F$. Thus $g(a) \notin f^{-1}(G)$, and so $fg(a) \notin G$. Hence, by (GM1) and since G is up-closed, $a \notin G$. Thus $a \in -G$, as required. On the other hand, assume FS_gG, that is, $g^{-1}(F) \subseteq -G$. We need to show GR_fF, that is, $f^{-1}(G) \subseteq -F$. Assume $a \in f^{-1}(G)$. Then $f(a) \in G$, hence $f(a) \notin -G$. Thus $f(a) \notin g^{-1}(F)$, and so $gf(a) \notin F$. Hence, by (GM2) and since F is up-closed, $a \notin F$. Thus $a \in -F$, as required. □

Invoking Theorems 8.4.2 and 8.4.3, we have discrete representations for Galois algebras and Galois frames.

Theorem 8.9.3

(a) *Every Galois algebra is embeddable into the complex algebra of its canonical frame;*

(b) *Every Galois frame is embeddable into the canonical frame of its complex algebra.*

8.10 Dual Galois algebras

A *dual Galois algebra* is an algebra (L, f, g) such that L is is a bounded distributive lattice and f and g are dual sufficiency operators satisfying, for every $a \in L$,

(dGM1) $f(g(a)) \leq a$

(dGM2) $g(f(a)) \leq a$.

As shown in Section 8.5 each dual sufficiency operator f corresponds to a certain order-respecting binary relation R_f over a poset. The properties of the dual sufficiency operators in a dual Galois pair correspond to the relationships between their respective relations.

Accordingly, a *dual Galois frame* is a structure (X, R, S) where X is a bounded distributive lattice frame and R and S are binary relations on X satisfying

(FdGM1) $S = R^{-1}$

(FdGM2) $(\geqslant; R; \leqslant) \subseteq R$

(FdGM3) $(\geqslant; S; \leqslant) \subseteq S$.

Note that (FdGM3) follows from (FdGM1) and (FdGM2).

The *complex algebra of a dual Galois frame* X is a structure (L_X, f_R, g_S) where L_X is the complex algebra of its lattice frame reduct and, for every $A \in L_X$,

$$f_R(A) \stackrel{\text{def}}{=} \langle R\rangle(-A) = \{x \in X : \exists y \in X,\ y \in -A \ \&\ xRy\}$$
$$g_S(A) \stackrel{\text{def}}{=} \langle S\rangle(-A) = \{x \in X : \exists y \in X,\ y \in -A \ \&\ xSy\}.$$

Proposition 8.10.1 *The complex algebra of a dual Galois frame is a dual Galois algebra.*

Proof:
For (dGM1) we need to show, for any $A \in L_X$, $\langle R\rangle[S]A \subseteq A$. This follows from (FdGM1) and Lemma 1.8.2(d). Similarly, for (dGM2) we need to show, for every $A \in L_X$, $\langle S\rangle[R]A \subseteq A$. This follows from (FdGM1) and Lemma 1.8.2(a). □

The *canonical frame of a dual Galois algebra* L is a structure (X_L, R_f, S_g) where X_L is the complex algebra of its lattice reduct and R_f and S_g are binary relations on X_L defined, for all $F, G \in X_L$, by

$$FR_fG \stackrel{\text{def}}{\Leftrightarrow} -G \subseteq f^{-1}(F)$$
$$FS_gG \stackrel{\text{def}}{\Leftrightarrow} -G \subseteq g^{-1}(F).$$

Proposition 8.10.2 *The canonical frame of a dual Galois algebra is a dual Galois frame.*

Proof:
By reasoning analogous to that in the proof of Proposition 8.2.3, conditions (FdGM2) and (FdGM3) can be proved. For (FdGM1), we show that $R_f^{-1} = S_g$. Let $(F,G) \in R_f^{-1}$. Then $(G,F) \in R_f$, that is, $-F \subseteq f^{-1}(G)$. We need to show that $(F,G) \in S_g$, that is, $-G \subseteq g^{-1}(F)$. Take any $a \in L$ such that $a \notin G$. Then, since $f(g(a)) \leq a$ and G is up-closed, $f(g(a)) \notin G$. Hence, $g(a) \notin f^{-1}(G)$ so, since $(G,F) \in R_f$, $g(a) \notin -F$. Thus $g(a) \in F$, hence $a \in g^{-1}(F)$. Similarly, assume $(F,G) \in S_g$, that is, $-G \subseteq g^{-1}(F)$. We need to show that $(G,F) \in R_f$, that is, $-F \subseteq f^{-1}(G)$. Take any $a \in L$ such that $a \notin F$. Then, since $g(f(a)) \leq a$ and F is up-closed, $g(f(a)) \notin F$. Hence, $f(a) \notin g^{-1}(F)$ so, since $(F,G) \in S_g$, $f(a) \notin -G$. Thus $f(a) \in G$, hence $a \in f^{-1}(G)$. □

Invoking Theorems 8.5.2 and 8.5.3 we obtain discrete representations for dual Galois algebras and dual Galois frames.

Theorem 8.10.3

(a) *Every dual Galois algebra is embeddable into the complex algebra of its canonical frame;*
(b) *Every dual Galois frame is embeddable into the canonical frame of its complex algebra.*

8.11 Residuated algebras

A *residuated algebra*is bounded distributive lattice L endowed with a necessity operator f and a possibility operator g satisfying, for every $a \in L$,

(RM1) $g(f(a)) \leq a$
(RM2) $a \leq f(g(a))$.

As the corresponding frame, we define a *residuated frame* to be a relational structure (X,R,S) where X is a bounded distributive lattice frame and R and S are binary relations on X satisfying

(FRM1) $S = R^{-1}$
(FRM2) $(\leqslant; R; \leqslant) \subseteq R$
(FRM3) $(\geqslant; S; \geqslant) \subseteq S$.

Note that (FRM3) follows from (FRM1) and (FRM2).

The *complex algebra of a residuated frame* X is a structure (L_X, f_R, g_S) where L_X is the complex algebra of its lattice frame reduct and, for every $A \in L_X$,

$$f_R(A) \stackrel{\text{def}}{=} [R]A$$
$$g_S(A) \stackrel{\text{def}}{=} \langle S\rangle A.$$

Proposition 8.11.1 *The complex algebra of a residuated frame is a residuated algebra.*

Proof:
As shown in Proposition 2.5.1 and Proposition 8.2.2 L_X is closed under the lattice operations and under the operators f_R and g_S.

(RM1) We need to show, for every $A \in L_X$, that $\langle S\rangle[R](A) \subseteq A$. This follows from Lemma 1.8.2(a) since, by (FRM1), $S = R^{-1}$.

(RM2) We need to show, for any $A \in L_X$, that $A \subseteq [R](\langle S\rangle A$. This follows from Lemma 1.8.2(b) since, by (FRM1), $S = R^{-1}$. □

The *canonical frame of a residuated algebra L* is the structure (X_L, R_f, S_g) where X_L the canonical frame of its lattice reduct and R_f and S_g are binary relations on X_L defined, for all $F, G \in X_L$, by

$$FR_fG \overset{\text{def}}{\Leftrightarrow} f^{-1}(F) \subseteq G$$
$$FS_gG \overset{\text{def}}{\Leftrightarrow} G \subseteq g^{-1}(F).$$

Proposition 8.11.2 *The canonical frame of a residuated algebra is a residuated frame.*

Proof:
By Proposition 2.5.2 (FRM2) holds, and by Proposition 8.2.3 (FRM3) holds. For (FRM1) we show that $S_g = (R_f)^{-1}$. Assume FS_gG. Take any $b \in L$ such that $b \in f^{-1}(F)$, that is $f(b) \in G$. Then, since $G \subseteq g^{-1}(F)$, $f(b) \in g^{-1}(F)$, that is $g(f(b)) \in F$. Hence, since $g(f(b)) \leq b$ and F is up-closed, $b \in F$, as required. On the other hand, assume $FR_f^{-1}(G)$, that is, GR_fF. Take any $a \in L$ such that $a \in G$. Then, since $a \leq fg(a)$ and G is up-closed, $f(g(a)) \in G$. So $g(a) \in f^{-1}(G)$. Since $f^{-1}(G) \subseteq F$, $g(a) \in F$, that is, $a \in g^{-1}(F)$, as required. □

Invoking Theorems 2.5.5(a), 8.2.4, 2.5.5(b), and 8.2.5, we obtain discrete representations for residuated algebras and residuated frames.

Theorem 8.11.3

(a) *Every residuated algebra is embeddable into the complex algebra of its canonical frame.*

(b) *Every residuated frame is embeddable into the canonical frame of its complex algebra.*

Chapter 9

Distributive lattices with negations

9.1 Introduction

In this chapter we consider distributive lattices endowed with a unary operator which reflects an intuition of negation: the properties of a negated element of a structure are opposite to the properties of that element. Majority of the negations treated in the chapter are, in a sense, most basic ones, they are defined with as small as possible number of axioms. For each class of the lattices with negations we define an appropriate class of frames and we present representation theorems for all these structures within a discrete framework. The representation theorems for algebras are, in many cases, reconstructed from the topological representations known in the literature.

Not necessarily distributive lattices with negations are presented in Chapter 14.

In Section 9.2 distributive lattices with a relative pseudocomplement introduced in [14] are presented. The relative pseudocomplement is a binary operator which may be viewed both as a negation of an element of a structure relative to some other element and as an implication. The well known pseudocomplement in the intuitionistic logic and in Heyting algebras is definable in terms of the relative pseudocomplement. In Sections 9.3, 9.4, 9.5, and in Section 9.7 relatively pseudocomplemented lattices with negations which are weaker than the pseudocomplement are considered together with the corresponding frames and the discrete representations are developed both for algebras and frames. In Section 9.3 relatively pseudocomplemented lattices with seminegation introduced in [207] are considered. The corresponding class of frames is defined and representation theorems are proved. In Section 9.4 and Section 9.5 two approaches to a minimal negation are considered, namely, relatively pseudocomplemented lattices with contrapositional negation presented in [206] and with minimal negation presented in [138], respectively. The corresponding frames are defined and the representation theorems are proved. The pseudocomplement in a distributive lattice is a negation which is both a semicomplement and a contrapositional negation. In Section 9.6 and Section 9.7 representation theorems for lattices with negations weaker than De Morgan negation and for their frames are presented. The regular negation discussed in Section 9.6 was investigated in [238] and a pre-De Morgan negation from Section 9.7 was investigated on a Heyting algebra in [43]. In Section 9.8 distributive lattices with De Morgan negation and their frames are discussed. In Section 9.9 De Morgan lattices with possibility and sufficiency operators are introduced together with the corresponding frames. They are weakening of Boolean algebras with possibility and sufficiency operators presented in Section 3.5. In Section 9.10 distributive lattices endowed with a pseudocomplement and dual pseudocomplement are considered. These algebras, in the literature referred to as *double p-algebras*, are studied in [246, 146, 147, 236, 214] among others.

Topological representation theorems for distributive lattices with relative pseudocomplement, with seminegation, and with contrapositional negation, respectively, are presented in [206], and for distributive lattices with De Morgan negation in [12]. Observe that the operator of sufficiency discussed in Section 3.3, Section 3.4, Section 8.4, Section 8.9, and Section 13.7 may be also viewed as a negation. Discrete representation theorems for distributive lattices with relative pseudocomplement, with minimal negation, with De Morgan negation, and with sufficiency operator, respectively, are presented in [72]. A variety of approaches to negation are surveyed in [250] and [92]. A Priestley duality for p-algebras is presented in [201].

9.2 Relatively pseudocomplemented lattices

A *relatively pseudocomplemented lattice*, *RP-lattice*, is an algebra $(L, \rightarrow)$ where L is a distributive lattice and $\rightarrow$ is a binary operation on L satisfying, for any $a, b, c \in L$,

(RP) $a \wedge c \leq b \Leftrightarrow c \leq a \rightarrow b$.

For properties of relatively pseudocomplemented lattices, see [208] and [7]. It is known that every relatively pseudo-complemented lattice is distributive and has a greatest element 1 definable by $1 = a \rightarrow a$. Thus we include 1 in the signature and write $(L, \rightarrow, 1)$ for an RP-lattice. If an RP-lattice L is treated as a residuated lattice (see Section 1.6) such that its product is the meet $\wedge$ of L and the unit element of product is 1, then the operation $\rightarrow$ is the residual of $\wedge$.

Lemma 9.2.1 *For all $a, b, c \in L$,*

(a) $a \rightarrow b = 1 \Leftrightarrow a \leq b$;

(b) $1 \rightarrow b = b$ *and* $a \rightarrow 1 = 1$;

(c) $a \rightarrow b = 1$ *and* $a = 1$ *imply* $b = 1$;

(d) $a \rightarrow (b \rightarrow c) = b \rightarrow (a \rightarrow c)$;

(e) $a \wedge (a \rightarrow b) = a \wedge b$;

(f) $b \leq a \rightarrow b$;

(g) $a \leq b \rightarrow (a \wedge b)$;

(h) $a \leq b$ *implies* $c \rightarrow a \leq c \rightarrow b$.

A *relatively pseudocomplemented lattice frame, RP-frame*, is $(X, \leqslant)$ where X is non-empty set and $\leqslant$ is a partial order.

The *complex algebra* of an RP-frame X is a structure $(L_X, \rightarrow_{\leqslant}, X)$ where L_X is the complex algebra of the distributive lattice frame X and

$$A \rightarrow_{\leqslant} B \stackrel{\text{def}}{=} [\leqslant](-A \cup B) \text{ for all } A, B \subseteq X.$$

Proposition 9.2.2 *The complex algebra of an RP-frame is an RP-lattice.*

Proof:
It is clear that L_X is closed under $\cap$ and $\cup$ and that L_X with these operations forms a distributive lattice with greatest element X. We need only show that $\rightarrow_{\leqslant}$ is the residual of $\cap$, that is, for all $A, B, C \in L_X$,

$$A \cap C \subseteq B \Leftrightarrow C \subseteq [\leqslant](-A \cup B).$$

Assume that $A \cap C \subseteq B$ and let $x \in C$. Take any $y \in X$ such that $x \leqslant y$. Then $y \in C$ since $C \in L_X$. If $y \in A$ then $y \in A \cap C$ and hence $y \in B$ so $y \in -A \cup B$. If $y \notin A$ then, clearly, $y \in -A \cup B$. Conversely, assume $C \subseteq [\leqslant](-A \cup B)$ and let $x \in A \cap C$. Then $x \in C$, hence $x \in [\leqslant](-A \cup B)$. Since $x \leqslant x$, we have $x \in -A \cup B$, but $x \in A$, so we must have $x \in B$, as required. □

The *canonical frame* of an RP-lattice L is defined in the same way as the canonical frame of a bounded distributive lattice, that is a structure $(X_L, \subseteq)$ where X_L is the set of all prime filters of L ordered by set inclusion. It follows that due to Proposition 2.5.2 we have the analogous result for RP-frames.

Proposition 9.2.3 *The canonical frame of an RP-lattice is an RP-frame.*

The above propositions imply that the complex algebra of the canonical frame of an RP-lattice is an RP-lattice and the canonical frame of the complex algebra of an RP-frame is an RP-frame.

For the proof of the next theorem we need the following observation.

Lemma 9.2.4 *Let F be a filter of an RP-lattice L. Then for all $a, b \in L$, $a \in F$ and $a \rightarrow b \in F$ imply $b \in F$.*

Let L be a RP-lattice. We now show that the embedding $h : L \rightarrow L_{X_L}$, as presented in Section 2.5, preserves the binary operation $\rightarrow$.

Lemma 9.2.5 *For any $a, b \in L$, $h(a \rightarrow b) = h(a) \rightarrow_{\subseteq} h(b)$.*

Proof:
We show that, for any $a, b \in L$, $h(a \rightarrow b) = [\subseteq](-h(a) \cup h(b))$ and $h(1) = X_L$. Let $F \in h(a \rightarrow b)$, which means $a \rightarrow b \in F$. Then $a \notin F$ or $b \in F$, whence $F \notin h(a)$ or $F \in h(b)$. Thus we have $F \in -h(a) \cup h(b)$, so $h(a \rightarrow b) \subseteq -h(a) \cup h(b)$. Since

for every $a \in L$, $h(a) = [\subseteq]h(a)$, by monotonicity of the operator $[\,]$ we have $h(a \to b) = [\subseteq]h(a \to b) \subseteq [\subseteq](-h(a) \cup h(b))$. For the converse inclusion, suppose $F \in [\subseteq](-h(a) \cup h(b))$. Then, for all $G \in L_X$, $F \subseteq G \Rightarrow a \notin G$ or $b \in G$. In particular, $a \notin F$ or $b \in F$. If $b \in F$ then, since $b \leq a \to b$, we have $a \to b \in F$. If $b \notin F$, then $a \notin F$. We show that also in this case $a \to b \in F$. Suppose otherwise, that is $a \to b \notin F$. Set $H = \{c \in L : a \to c \in F\}$. Since $a \to (c \wedge d) = (a \to c) \wedge (a \to d)$, it follows that H is closed under meets. Since $c \leq d$ implies $a \to c \leq a \to d$, H is up-closed. Thus, H is a filter of L. Moreover, $F \subseteq H$, $a \in H$ and $b \notin H$. Thus, we may extend H to a prime filter H' such that $b \notin H'$, but $F \subseteq H'$ and $a \in H'$, which is a contradiction. □

Let X be a RP-frame.The mapping $k : X \to X_{L_X}$ providing a representation of RP-frames is the same as in Section 2.5. It is shown to be injective and order-preserving.

Concluding, we have the discrete representations for RP-lattices and RP-frames.

Theorem 9.2.6

(a) *Every RP-lattice is embeddable into the complex algebra of its canonical frame;*

(b) *Every RP-frame is embeddable into the canonical frame of its complex algebra.*

9.3 RP-lattices with seminegation

A *relatively pseudocomplemented lattice with seminegation, RPSemi-lattice*, is a structure $(L, \to, \neg, 1)$ where $(L, \to, 1)$ is an RP-lattice and $\neg$ satisfies, for all $a, b \in L$,

(RPSemi) $\neg(a \to a) \to b = 1$.

Lemma 9.3.1 *Let $(L, \to, 1)$ be an RP-lattice and let $\neg$ be a unary operator on L. Then (RPSemi) holds in L if and only if $\neg 1$ is the smallest element of L.*

Proof:
By Lemma 9.2.1(a), $a \to a = 1$. Thus by (RPSemi), $\neg 1 \to b = 1$ and hence $\neg 1 \leq b$ for every $b \in L$. Conversely, let $\neg 1 = \neg(a \to a)$ be the smallest element of L. Then $\neg(a \to a) \leq b$ for every $b \in L$. By Lemma 9.2.1(a) (RPSemi) holds. □

An *RPSemi-frame* is an RP-frame X.

The *complex algebra* of an RPSemi-frame X is a structure $(L_X, \to_\leqslant, \neg_\leqslant, X)$ where $(L_X, \to_\leqslant, X)$ is the complex algebra of an RP-frame and the negation $\neg_\leqslant$ on L_X is defined, for every $A \subseteq X$, by

$$\neg_\leqslant A \stackrel{\text{def}}{=} \begin{cases} \emptyset & \text{if } A = X \\ \text{any } Z \in L_X & \text{otherwise} \end{cases}$$

Proposition 9.3.2 *The complex algebra of an RPSemi-frame is an RPSemi-lattice.*

Proof:
First, observe that L_X is closed on $\neg_{\leqslant}$, that is for every $A \in L_X$, $\neg_{\leqslant} A \in L_X$. If $A = X$, then $\neg_{\leqslant} X = \emptyset$ and $[\leqslant]\emptyset = \emptyset$ since $\leqslant$ is reflexive. If $A \neq X$, then $\neg_{\leqslant} A \in L_X$ by definition. Now we show that axiom (RPSemi) is satisfied. Indeed, we have

$$\begin{aligned}\neg_{\leqslant}(A \to_{\leqslant} A) \to_{\leqslant} B &= [\leqslant](-\neg_{\leqslant}([\leqslant](-A \cup A)) \cup B)\\ &= [\leqslant](-\neg_{\leqslant}[\leqslant]X \cup B)\\ &= [\leqslant](-\neg_{\leqslant}X \cup B)\\ &= [\leqslant](-\emptyset \cup B)\\ &= X.\end{aligned}$$

□

The *canonical frame of an RPSemi-lattice* is the RP-frame $(X_L, \subseteq)$. In view of Proposition 9.2.3 it is clear that it is also an RPSemi-frame.

It follows that the complex algebra of the canonical frame of an RPSemi-lattice is an RPSemi-lattice and the canonical frame of the complex algebra of an RPSemi-frame is an RPSemi-frame.

Note that in $(L_{X_L}, \to_{\subseteq}, \neg_{\subseteq}, X_L)$, which is the complex algebra of the canonical frame of an RPSemi-lattice $(L, \to, \neg, 1)$, the negation $\neg_{\subseteq}$ is defined, for every $A \subseteq X_L$, by

$$\neg_{\subseteq} A = \begin{cases} \emptyset & \text{if } \forall a \in L,\ A \neq h(a)\\ h(\neg a) & \text{if } \exists a \in L,\ A = h(a),\end{cases}$$

where $h : L \to L_{X_L}$ is the embedding, as presented in Section 2.5. Observe that this definition is a special case of the definition of the seminegation in complex algebras. If $A = X_L$, then $A = h(1)$ and according to the definition of $\neg_{\subseteq}$,

$$\neg_{\subseteq} X_L = h(\neg 1) = h(0) = \emptyset.$$

If $A \neq X_L$, then we have two cases: if $A = h(a)$ for some $a \in L$, then

$$\neg_{\subseteq} A = h(\neg a) = [\subseteq]h(\neg a);$$

if $A \neq h(a)$ for every $a \in L$, then $\neg_{\subseteq} A = \emptyset = [\subseteq]\emptyset$.

Let L be an RP-lattice. Now we show that the embedding $h : L \to L_{X_L}$, used in Section 2.5, preserves the unary operation $\neg$.

Lemma 9.3.3 *For every $a \in L$,*

$$h(\neg a) = \neg_{\subseteq} h(a).$$

Similarly, as in the previous section, we have the discrete representations for RPSemi-lattices and RPSemi-frames.

Theorem 9.3.4

(a) *Every RPSemi-lattice is embeddable into the complex algebra of its canonical frame;*

(b) *Every RPSemi-frame is embeddable into the canonical frame of its complex algebra.*

9.4 RP-lattices with contrapositional negation

A *relatively pseudocomplemented lattice with contrapositional negation, RPCon-lattice*, is a structure $(L, \rightarrow, \neg, 1)$ such that $(L, \rightarrow, 1)$ is an RP-lattice and $\neg$ is a unary operator in L satisfying, for all $a, b \in L$,

(RPCon) $a \rightarrow \neg b \leq b \rightarrow \neg a$.

Note that (RPCon) is equivalent to $a \rightarrow \neg b = b \rightarrow \neg a$ and corresponds to a Galois-style connection $a \leq \neg b \Rightarrow b \leq \neg a$.

Lemma 9.4.1 *(RPCon) is equivalent to*

(RPCon') $\neg a = a \rightarrow \neg 1$.

Proof:
By Lemma 9.2.1(b) we have $\neg a = 1 \rightarrow \neg a$. By (RPCon), $\neg a = a \rightarrow \neg 1$. Conversely, by (RPCon') and Lemma 9.2.1(d) we obtain the following equalities

$$a \rightarrow \neg b = a \rightarrow (b \rightarrow \neg 1) = b \rightarrow (a \rightarrow \neg 1) = b \rightarrow \neg a$$

which completes the proof. □

Lemma 9.4.2 *Let L be an RPCon-lattice. For all $a, b \in L$,*

(a) $a \leq \neg\neg a$;

(b) $a \rightarrow b \leq \neg b \rightarrow \neg a$;

(c) $\neg(a \vee b) = \neg a \wedge \neg b$;

(d) (RPCon) $\Leftrightarrow$ (a) *and* (b);

(e) (RPCon) $\Leftrightarrow$ (b) *and* (c).

An *RPCon-frame* is an RP-frame X.

The *complex algebra* of an RPCon-frame X is the structure

$$(L_X, \rightarrow_{\leqslant}, \neg_{\leqslant}, X)$$

such that $(L_X, \rightarrow_{\leqslant}, X)$ is the complex algebra of an RP-frame and the negation $\neg_{\leqslant}$ is defined, for every $A \subseteq X$, by

$$\neg_{\leqslant} A \stackrel{\text{def}}{=} \begin{cases} \text{any } Z \subseteq X \text{ such that } Z \in L_X & \text{if } A = X \\ A \rightarrow_{\leqslant} \neg_{\leqslant} X & \text{otherwise.} \end{cases}$$

Proposition 9.4.3 *The complex algebra of an RPCon-frame is an RPCon-algebra.*

Proof:
First, we show that for every $A \in L_X$, $\neg_{\leqslant} A \in L_X$. If $A = X$, then by definition $\neg_{\leqslant} X = [\leqslant]\neg_{\leqslant} X$. If $A \neq X$, then by transitivity of $\leqslant$, $A \rightarrow_{\leqslant} \neg_{\leqslant} X = [\leqslant](A \rightarrow_{\leqslant} \neg_{\leqslant} X)$. Now we show that axiom (RPCon') is satisfied, that is $\neg_{\leqslant} A = A \rightarrow_{\leqslant} \neg_{\leqslant} X$. If $A \neq X$, then by the definition of $\neg_{\leqslant}$ the required equality holds. If $A = X$, then $\neg_{\leqslant} X = Z \subseteq X$ such that $Z = [\leqslant]Z$. Then $X \rightarrow_{\leqslant} Z = [\leqslant](-X \cup Z) = [\leqslant]Z = Z = \neg_{\leqslant} X$. □

The *canonical frame* of an RPCon-lattice L is the RP-frame $(X_L, \subseteq)$. Such a frame is an RPCon-frame.

Note that in L_{X_L} the negation $\neg_{\subseteq}$ is defined, for every $A \subseteq X_L$, by

$$\neg_{\subseteq} A \stackrel{\text{def}}{=} A \rightarrow_{\subseteq} h(\neg 1),$$

where $h : L \rightarrow L_{X_L}$ is the embedding, as presented in Section 2.5. Observe that this definition is a special case of the definition of the contrapositional negation in complex algebras. If $A = X_L$, then

$$\neg_{\subseteq} X_L = X_L \rightarrow_{\subseteq} h(\neg 1) = [\subseteq](-X_L \cup h(\neg 1)) = [\subseteq]h(\neg 1) = h(\neg 1).$$

Thus the condition $\neg_{\subseteq} X_L = [\subseteq]\neg_{\subseteq} X_L$ is satisfied. If $A \neq X_L$, then

$$\neg_{\subseteq} A = A \rightarrow_{\subseteq} h(\neg 1) = A \rightarrow_{\subseteq} \neg_{\subseteq} X_L.$$

Let $(L, \rightarrow, \neg, 1)$ be a RPCon-lattice. Now we show that the embedding $h : L \rightarrow L_{X_L}$, as used in Section 2.5, preserves the contrapositional negation.

Lemma 9.4.4 *For every $a \in L$,*

$$h(\neg a) = \neg_{\subseteq} h(a).$$

Proof:
Note that for every $a \in L$,

$$\begin{aligned} \neg_{\subseteq} h(a) &= h(a) \rightarrow_{\subseteq} h(\neg 1) \\ &= h(a \rightarrow \neg 1) && \text{since } h \text{ preserves } \rightarrow \\ &= h(\neg a) && \text{(RPCon')}, \end{aligned}$$

which completes the proof. □

We now have the discrete representations for RPCon-lattices and RPCon-frames. For a representation of RPCon-frames we use the embedding k discussed in Section 9.2.

Theorem 9.4.5

(a) *Every RPCon-lattice is embeddable into the complex algebra of its canonical frame;*

(b) *Every RPCon-frame is embeddable into the canonical frame of its complex algebra.*

A negation which is both a seminegation and a contrapositional negation is known as the pseudocomplement, or the Heyting negation. Discrete dualities for Heyting algebras with operators are presented in Chapter 11.

9.5 RP-lattices with minimal negation

In this section we present another formalization of RP-lattices with contrapositional negation. We endow the RP-lattices with a constant which enables us to define the negation in terms of the relative pseudocomplement. As a consequence, it will be possible to avoid the 'underspecified' definition of negation in the complex algebra. In the literature the negation formalized in this way is referred to as a minimal negation of Johansson [138].

By a *relatively pseudocomplemented lattice with minimal negation*, *RPJ-lattice*, we mean an algebra $(L, \rightarrow, \neg, \partial, 1)$ where $(L, \rightarrow, 1)$ is an RP-lattice, $\partial \in L$ is a distinguished element which is not necessarily the smallest element, and $\neg$ is a unary operator satisfying, for all $a, b \in L$,

(RPJ1) $(a \rightarrow \neg b) \leq (b \rightarrow \neg a)$;

(RPJ2) $\neg 1 = \partial$.

Lemma 9.5.1 *Let L be an RPJ-lattice. Then, for every $a \in L$,*

(a) $\neg a = a \rightarrow \partial$;

(b) *if $\neg a = a \rightarrow \partial$, then $\neg$ is a minimal negation.*

Proof:
(a) For all $a \in L$ we have $\neg a = 1 \rightarrow \neg a = a \rightarrow \neg 1 = a \rightarrow \partial$.

(b) For (RPJ1), we have $a \rightarrow \neg b = a \rightarrow (b \rightarrow \partial) = b \rightarrow (a \rightarrow \partial) = b \rightarrow \neg a$ for all $a, b \in L$. For (RPJ2), we have $\neg 1 = 1 \rightarrow \partial = \partial$. □

It is easy to see that every RPJ-lattice is an RPCon-lattice and every RPCon-lattice endowed with a constant satisfying (RPJ2) is an RPJ-lattice.

An *RPJ-frame* is a relational structure (X, D) where X is an RP-frame and $D \subseteq X$ satisfies, for all $x, y \in X$,

(FRPJ) $x \leqslant y \ \&\ x \in D \Rightarrow y \in D$.

The *complex algebra* of an RPJ-frame X is a structure $(L_X, \neg_\leqslant, \partial_D)$ where $(L_X, \rightarrow_\leqslant, X)$ is the complex algebra of its RP-frame reduct and

$$\neg_\leqslant A \stackrel{\text{def}}{=} A \rightarrow_\leqslant \partial_D \text{ for every} A \in L_X$$

$$\partial_D \stackrel{\text{def}}{=} [\leqslant] D.$$

Proposition 9.5.2 *The complex algebra of an RPJ-frame is an RPJ-lattice.*

Proof:
Since $[\leqslant][\leqslant]D = [\leqslant]D$, we have $\partial_D \in L_X$ and hence L_X is also closed under $\neg_\leqslant$. Since L_X is a relatively pseudo-complemented lattice, (RPJ1) follows from properties of $\to$. For (RPJ2) note that

$$\neg_\leqslant X = [\leqslant](-X \cup [\leqslant]D) = [\leqslant][\leqslant]D = [\leqslant]D = \partial_D.$$

□

The *canonical frame* of an RPJ-lattice L is a structure (X_L, D_∂) where X_L is the canonical frame of its RP-lattice reduct and

$$D_\partial \stackrel{\text{def}}{=} \{F \in X_L : \partial \in F\}.$$

Proposition 9.5.3 *The canonical frame of an RPJ-lattice is an RPJ-frame.*

Proof:
It is sufficient to show that the condition (FRPJ) is satisfied. Indeed, it is easy to see that for all $F, G \in X_L$, if $F \subseteq G$ and $F \in D_\partial$, then $G \in D_\partial$. □

We have the following discrete representations for RPJ-lattices and RPJ-frames.

Theorem 9.5.4

(a) *Every RPJ-lattice is embeddable into the complex algebra of its canonical frame;*

(b) *Every RPJ-frame is embeddable into the canonical frame of its complex algebra.*

Proof:
(a) Invoking Theorem 9.2.6(a) it suffices to show that the embedding of RP-lattices $h : L \to L_{X_L}$ defined, for any $a \in L$, by $h(a) = \{F \in X_L : a \in F\}$ preserves the distinguished element ∂ and the unary operator $\neg$. We have $h(\partial) = \{F \in X_L : \partial \in F\} = D_\partial$ and $h(\partial)$ is up-closed subset of X_L, so $h(\partial) = [\subseteq]D_\partial = \partial^C$. Since $\to$ is preserved as shown in Section 9.2, it follows that $\neg$ is too.

(b) Invoking Theorem 9.2.6(a) it suffices to show that the embedding $k : X \to X_{L_X}$ preserves the distinguished set D, that is, for every $x \in X$, $x \in D$ if and only if $k(x) \in D_{\partial_D}$.
Note that for any $x \in D$,

$$k(x) \in D_{\partial_D} \Leftrightarrow \partial_D \in k(x) \Leftrightarrow [\leqslant]D \in k(x) \Leftrightarrow x \in [\leqslant]D.$$

Let $x \in D$ and take y such that $x \leqslant y$. Then by (FRPJ) $y \in D$ and hence we get $x \in [\leqslant]D$. The converse implication follows from reflexivity of $\leqslant$. □

9.6 Lattices with regular negation

A *lattice with regular negation*, *Reg-lattice*, is a structure $(L, \neg, 0)$ where L is a distributive lattice, 0 is the smallest element, and $\neg$ is a unary operator in L, called a *regular negation*, satisfying, for all $a, b \in L$,

(Reg) $\neg(a \vee b) = \neg a \wedge \neg b$.

Lemma 9.6.1 *For every Reg-lattice L and for all $a, b \in L$,*

(a) *If $a \leq b$, then $\neg b \leq \neg a$;*

(b) $\neg a \vee \neg b \leq \neg(a \wedge b)$.

A *regular lattice frame*, *Reg-frame*, is a structure (X, M, R) where X is a RP-frame, $M \subseteq X$ and R is a binary relation on X such that

(FReg1) $M = [\leqslant]M$

(FReg2) $(\leqslant; R; \geqslant) \subseteq R$

(FReg3) $\forall x,\ x \notin M \Rightarrow R(x) = X$.

The *complex algebra* of an Reg-frame X is an algebra $(L_X, \neg_R, \emptyset)$ where L_X is the complex algebra of the distributive lattice frame and $\neg_R$ is defined, for every $A \subseteq X$, by

$$\neg_R A \stackrel{\text{def}}{=} M \cap [R](-A).$$

Observe that $M \in L_X$ and $\neg_R \emptyset = M \cap [R](-\emptyset) = M \cap X = M$.

Proposition 9.6.2 *The complex algebra of an Reg-frame is an Reg-lattice.*

Proof:
First, we show that L_X is closed on $\neg_R$, that is, $\neg_R A = [\leqslant](\neg_R A)$ for every $A \in L_X$. The right-to-left inclusion follows from reflexivity of $\leqslant$. Now, assume $x \in M$ and $x \in [R](-A)$. Since $x \in M$, by (FReg1) we get $x \in [\leqslant]M$. We also have $x \in [\leqslant][R](-A)$. For suppose conversely, then $x \in \langle\leqslant\rangle\langle R\rangle A$, so $x \in \langle\leqslant; R\rangle A$. By (FReg2) and reflexivity of $\leqslant$, we have $(\leqslant; R) \subseteq R$. Thus $x \in \langle R\rangle A$ and hence $x \notin [R](-A)$, a contradiction. Then

$$x \in [\leqslant]M \cap [\leqslant][R](-A) = [\leqslant](M \cap [R](-A)) = [\leqslant]\neg_R A.$$

Furthermore, we show that the axiom (Reg) is safisfied in L_X.

$$\begin{aligned}\neg_R(A \cup B) &= M \cap [R](-(A \cup B)) = M \cap [R](-A \cap -B)\\ &= M \cap [R](-A) \cap [R](-B) = \neg_R A \cap \neg_R B.\end{aligned}$$

□

The *canonical frame* of an Reg-lattice L is a structure $(X_L, M_\neg, R_\neg)$ where X_L is the canonical frame of the lattice reduct of L and

$$M_\neg \stackrel{\text{def}}{=} \{F \in X_L : \exists a \in L,\ \neg a \in F\}$$

$$FR_\neg G \stackrel{\text{def}}{\Leftrightarrow} \neg^{-1}(F) \subseteq -G \quad \text{for all } F, G \in X_L.$$

Observe that if $F \in M_\neg$, then since $0 \leq a$ for every $a \in L$, we have $\neg 0 \in F$ by Lemma 9.6.1(a).

Proposition 9.6.3 *The canonical frame of a Reg-lattice is a Reg-frame.*

Proof:
(FReg1) We show $M_\neg = [\leqslant]M_\neg$. The right-to-left inclusion follows from reflexivity of $\leqslant$. Now, let $F \in M_\neg$ and take $G \in X_L$ such that $F \subseteq G$. Since $F \in M_\neg$, $\neg a \in F$ for some $a \in L$. Thus $\neg a \in G$ and hence $G \in M_\neg$.

(FReg2) We show that if $F \subseteq H$, $HR_\neg H'$, and $H' \supseteq G$, then $FR_\neg G$ for all $F, G, H, H' \in X_L$.
Suppose not $FR_\neg G$. Then there is some $a \in L$ such that $\neg a \in F$ and $a \in G$. Then $\neg a \in H$. Since $HR_\neg H'$, $a \notin H'$. Thus $a \notin G$, a contradiction.

(FReg3) Take $F \in X_L$ such that $F \notin M_\neg$. This means that every $a \in L$ is such that $\neg a \notin F$, that is, $a \notin \neg^{-1}(F)$. Hence $\neg^{-1}(F) = \emptyset$, so $\neg^{-1}(F) \subseteq -G$ for any $G \in X_L$. Thus $FR_\neg G$ holds, as required. □

Let L be a Reg-lattice. We now show that the embedding $h : L \to L_{X_L}$, used in Section 2.5, preserves the regular negation $\neg$.

Lemma 9.6.4 *For every $a \in L$, $h(\neg a) = \neg_{R_\neg} h(a)$.*

Proof:
We have to show that, for every $a \in L$,

$$\neg a \in F \Leftrightarrow F \in M_\neg \ \&\ \forall G \in X_L,\ \forall b,\ (\neg b \in F \Rightarrow b \notin G) \Rightarrow a \notin G.$$

Assume $\neg a \in F$. Then $F \in M_\neg$. Take $G \in X_L$ such that for every $b \in L$, $\neg b \in F$ implies $b \notin G$. Taking b to be a we get $a \notin G$. On the other hand, assume $\neg a \notin F$, $F \in M_\neg$, and for every $G \in X_L$, for every $b \in L$ $(\neg b \in F \Rightarrow b \notin G) \Rightarrow a \notin G$. Then $\neg^{-1}(F) \neq \emptyset$. Since $\neg$ is antitone, $\neg^{-1}(F)$ is down-closed. Furthermore, if $a, b \in \neg^{-1}(F)$, then $\neg a, \neg b \in F$. Since F is a filter, $\neg a \wedge \neg b \in F$. Thus by (Reg) $\neg(a \vee b) \in F$ and hence $a \vee b \in \neg^{-1}(F)$. It follows that $\neg^{-1}(F)$ is an ideal and hence $0 \in \neg^{-1}(F)$. Then $\neg 0 \in F$. Since $\neg a \notin F$, $a \notin \neg^{-1}(F)$. Thus $\neg^{-1}(F) \cap \uparrow_\leq a = \emptyset$. By the prime filter theorem there is some $G \in X_L$ such that $\uparrow_\leq a \subseteq G$ and $\neg^{-1}(F) \cap G = \emptyset$, that is $\neg^{-1}(F) \subseteq -G$. By the assumption, $a \notin G$, a contradiction. □

Let X be a Reg-frame. We now show that the embedding $k : X \to X_{L_X}$, defined in Section 2.5, preserves the set M and the binary relation R.

Lemma 9.6.5 *For all $x, y \in X$,*

(a) $x \in M \Leftrightarrow k(x) \in M_{\neg_R}$;

(b) $xRy \Leftrightarrow k(x) R_{\neg_R} k(y)$.

Proof:
(a) Observe that, for every $x \in X$,

$$k(x) \in M_{\neg_R} \Leftrightarrow \exists A \in L_X,\ x \in M \ \&\ x \in [R](-A).$$

Thus if $k(x) \in M_{\neg_R}$, then $x \in M$. Conversely, let $x \in M$ and take A to be $\emptyset$. Then $[R](-\emptyset) = X$, and hence $k(x) \in M_{\neg_R}$.

(b) We show, for all $x, y \in X$, that

$$xRy \Leftrightarrow \forall A \in L_X,\ x \in M \ \&\ x \in [R](-A) \Rightarrow y \notin A.$$

Assume xRy and take any $A \in L_X$ such that $x \in M$ and $x \in [R](-A)$. It follows that $\forall z \in X,\ xRz \Rightarrow z \notin A$. Taking z to be y we get $y \notin A$. On the other hand, assume that, for every $A \in L_X$, $x \in M$ & $x \in [R](-A) \Rightarrow y \notin A$. Taking A to be $[\leqslant](\uparrow_{\leqslant} \{y\})$ we have $A \in X_L$ and $\forall z,\ z \in A \Leftrightarrow y \leqslant z$. In particular, $y \in A$. Then by the assumption, $x \notin M$ or $x \notin [R](-A)$. If $x \notin M$, then by (FReg3), xRy. If $x \notin [R](-A)$, then $x \in \langle R \rangle A$, that is there exists $z \in X$ such that xRz and $z \geqslant y$. By (FReg2) and reflexivity of $\leqslant$ we get $(R; \geqslant) \subseteq R$. Thus we have xRy, as required. □

We conclude with the discrete representations for Reg-lattices and Reg-frames.

Theorem 9.6.6

(a) *Every Reg-lattice is embeddable into the complex algebra of its canonical frame;*

(b) *Every Reg-frame is embeddable into the canonical frame of its complex algebra.*

9.7 RP-lattices with pre-De Morgan negation

A *relatively pseudocomplemented lattice with pre-De Morgan negation*, *RPpDeM-lattice*, is a structure $(L, \rightarrow, \neg)$ where $(L, \rightarrow)$ is a relatively pseudocomplemented lattice and an external negation $\neg$ satisfying, for all $a, b \in L$,

(RPpDeM1) $a \leq b \Rightarrow \neg b \leq \neg a$

(RPpDeM2) $\neg a \wedge \neg b \leq \neg(a \vee b)$.

Note that $\neg$ is not necessarily the pseudocomplement. Also (RPpDeM1) implies $\neg(a \vee b) \leq \neg a \vee \neg b$. Indeed, since $a \leq a \vee b$, we have $\neg(a \vee b) \leq \neg a$. Similarly, $\neg(a \vee b) \leq \neg b$. Hence $\neg(a \vee b) \leq \neg a \wedge \neg b$. Therefore, in view of (RPpDeM2) we have $\neg(a \vee b) = \neg a \wedge \neg b$. This motivates the name *pre-De Morgan negation*.

Lemma 9.7.1 *Let L be an RPpDeM-lattice. For all $F, G \subseteq L$,*

(a) *If $F \subseteq G$, then $\neg^{-1}(F) \subseteq \neg^{-1}(G)$;*

(b) $-\neg^{-1}(F) = \neg^{-1}(-F)$;

(c) *If F is a filter of L, then $\neg^{-1}(F)$ is an ideal of L.*

Proof:
(a) Take any $F, G \subseteq L$ such that $F \subseteq G$. Assume that $a \in \neg^{-1}(F)$. This means that $\neg a \in F$, so by assumption, $\neg a \in G$, that is, $a \in \neg^{-1}(F)$.

(b) Let $a \in L$. Then we have

$$a \in -\neg^{-1}(F) \Leftrightarrow a \notin \neg^{-1}(F) \Leftrightarrow \neg a \notin F \Leftrightarrow \neg a \in -F \Leftrightarrow a \in \neg^{-1}(-F).$$

(c) Let F be a filter of L. Take $a, b \in L$ such that $a \in \neg^{-1}(F)$ and $b \leq a$. Then $\neg a \in F$ and by (RPpDeM1), $\neg a \leq \neg b$. Hence $\neg b \in F$, that is $b \in \neg^{-1}(F)$. Also, let $a, b \in \neg^{-1}(F)$, that is $\neg a \in F$ and $\neg \in F$. Then $\neg a \wedge \neg b \in F$, so by (RPpDeM2), $\neg(a \vee b) \in F$, thus $a \vee b \in \neg^{-1}(F)$. □

An *RPpDeM-frame* (X, R) is a RP-frame X endowed with a binary relation R on X satisfying,

(FRPpDeM1) $(\leqslant; R; \geqslant) \subseteq R$.

The *complex algebra* of an RPpDeM-frame X is a structure $(L_X, \rightarrow_{\leqslant}, \neg_R)$ where $(L_X, \rightarrow_{\leqslant})$ is the complex algebra of its lattice frame reduct $(X, \leqslant)$ and $\neg_R$ is defined, for every $A \in L_X$, by

$$\neg_R(A) \stackrel{\text{def}}{=} [R](-A).$$

Proposition 9.7.2 *The complex algebra of an RPpDeM-frame is an RPpDeM-lattice.*

Proof:
First, we show that $\neg_R(A) = [\leqslant]\neg_R(A)$. Inclusion $\supseteq$ follows from reflexivity of $\leqslant$. For the reverse inclusion, since by axiom (FRPpDeM1) $(\leqslant; R) \subseteq R$, we obtain $[R](-A) \subseteq [\leqslant; R](-A) = [\leqslant][R](-A)$.

Second, we show that the conditions (RPpDeM1) and (RPpDeM2) hold.

(RPpDeM1) Let $A, B \in L_X$ be such that $A \subseteq B$. Then $-B \subseteq -A$, and hence $[R](-B) \subseteq [R](-A)$, as required.

(RPpDeM2) The condition $\neg_R A \cap \neg_R B \subseteq \neg_R(A \cup B)$ holds for all $A, B \in L_X$, since $[R](-A) \cap [R](-B) = [R](-A \cap -B) = [R](-(A \cup B))$. □

The *canonical frame* of an RPpDeM-lattice L is a structure $(X_L, \subseteq, R_\neg)$ where $(X_L, \subseteq)$ is the canonical frame of its lattice reduct and $R_\neg$ is a binary relation on X_L defined, for all $F, G \in X_L$, by

$$FR_\neg G \stackrel{\text{def}}{\Leftrightarrow} \forall a \in L, \neg a \in F \Rightarrow a \notin G.$$

Proposition 9.7.3 *The canonical frame of an RPpDeM-lattice is an RPpDeM-frame.*

Proof:
Take any $F, F', G \in X_L$ such that $FR_{\neg}, G$, $F' \subseteq F$ and $G' \subseteq G$. Let $\neg a \in F'$. Then $\neg a \in F$, so $a \notin G$, and hence $a \notin G'$. □

Let L be a RPpDeM-lattice. We now show that the embedding $h : L \rightarrow L_{X_L}$, used in Section 2.5, preserves the pre-De Morgan negation $\neg$.

Lemma 9.7.4 *For every* $a \in L$, $h(\neg a) = \neg_{R_\neg} h(a)$.

Proof:
Assume $F \in h(\neg a)$, that is, $\neg a \in F$. Take $G \in X_L$ such that $F R_\neg G$. Since $\neg a \in F$, we have $a \notin G$. Hence $G \notin h(a)$ and $F \in \neg_R h(a)$. On the other hand, take any $F \in X_L$ with $F \in \neg_R h(a)$. Now $\uparrow_\leq a \in X_L$ and $a \in \uparrow_\leq a$. Hence, for some $b \in L$, $\neg b \in F$ and $b \in \uparrow_\leq a$, that is, $a \leq b$. By (RPpDeM1), we have $\neg b \leq \neg a$, hence $\neg a \in F$. So $F \in h(a)$. □

We now show that the lattice frame embedding $k : X \rightarrow X_{L_X}$, used in Section 2.5, preserves the binary relation R.

Lemma 9.7.5 *For all* $x, y \in X$,

$$xRy \Leftrightarrow k(x)\, R_{\neg_R}\, k(y).$$

Proof:
We show, for all $x, y \in X$, that

$$xRy \quad \Leftrightarrow \quad \forall A \in L_X, (\forall z \in X, xRz \Rightarrow z \notin A) \Rightarrow y \notin A.$$

Assume xRy and that, for every $z \in X$, xRz implies $z \notin A$. Then, $y \notin A$. On the other hand, assume that for every $A \in L_X$, $(\forall z \in X, xRz \Rightarrow z \notin A) \Rightarrow y \notin A$. Now, $\uparrow_\leqslant y \in L_X$ and $y \in \uparrow_\leqslant y$. So, for some $z \in X$, xRz and $z \in \uparrow_\leqslant y$, that is, $y \leqslant z$. Hence, by the condition (FRPpDeM1), xRy. □

With these results we have the discrete representations for RPpDeM-lattices and RPpDeM-frames.

Theorem 9.7.6

(a) *Every RPpDeM-lattice is embeddable into the complex algebra of its canonical frame;*

(b) *Every RPpDeM-frame is embeddable into the canonical frame of its complex algebra.*

Now, let us consider an *RPpDeM1-lattice* which is an RPpDeM-lattice satisfying, for every $a \in L$,

(RPpDeM3) $a \leq \neg\neg a$

The RPpDeM-frame appropriate for this lattice, *RPpDeM1-frame*, satisfies the additional axiom

(FRPpDeM3) R is symmetric.

Proposition 9.7.7 *The complex algebra of an RPpDeM1-frame is an RPpDeM1-lattice.*

Proof:
Note that $[R](-[R](-A)) = [R]\langle R\rangle A$. Since R is symmetric, $A \subseteq [R]\langle R\rangle A$, so $A \subseteq \neg_R \neg_R A$. □

Proposition 9.7.8 *The canonical frame of an RPpDeM1-lattice is an RPpDeM1-frame.*

Proof:
Let $F, G \in X_L$ be such that $\neg^{-1}(F) \subseteq -G$. Take $a \in \neg^{-1}(G)$, that is $\neg a \in G$, so by assumption, $\neg a \notin \neg^{-1}(F)$, that is, $\neg\neg a \notin F$. By (RPpDeM3), $a \notin F$. Hence $\neg^{-1}(G) \subseteq -F$, so $GR_\neg F$ holds. □

Now consider an *RPpDeM2-algebra*, that is an RPpDeM-algebra which satisfies, for every $a \in L$,

(RPpDeM4) $\neg\neg a \leq a$.

Lemma 9.7.9 *Let L be an RPpDeM-lattice. For every $F \subseteq L$, if F is a prime filter of L, then so is $-\neg^{-1}(F)$.*

Proof:
Let F be a prime filter of L. Consider $G = -\neg^{-1}(F)$. Using (RPpDeM4) we get $0 \leq \neg\neg 0 \leq 0$, so $\neg\neg 0 = 0$. Since F is prime, $\neg\neg 0 \notin F$, that is $\neg 0 \in -\neg^{-1}(F) = G$, hence $G \neq \emptyset$. Let $a \in G$ and $a \leq b$. Then $\neg a \notin F$ and, by (RPpDeM1), $\neg b \leq \neg a$. Then $\neg b \notin F$, which means $b \in G$. Now, let $a, b \in G$. Then $\neg a \notin F$ and $\neg b \notin F$, so since F is prime, $\neg a \vee \neg b \notin F$. Observe that $\neg(a \wedge b) \leq \neg a \vee \neg b$. Indeed, since $\neg a \leq \neg a \vee \neg b$, by (RPpDeM1) we get $\neg(\neg a \vee \neg b) \leq \neg\neg a \leq a$. Similarly, $\neg(\neg a \vee \neg b) \leq b$. Hence $\neg(\neg a \vee \neg b) \leq a \wedge b$, so $\neg(a \wedge b) \leq \neg\neg(\neg a \vee \neg b) \leq \neg a \vee \neg b$ by (RPpDeM1) and (RPpDeM4). It follows that $\neg(a \wedge b) \notin F$, whence $a \wedge b \in G$. For primeness, let $a \vee b \in G$. Then $\neg(a \vee b) \notin F$, so by (RPpDeM2), $\neg a \wedge \neg b \notin F$. Hence $a \in G$ or $b \in G$. □

The frame appropriate for this algebra, *RPpDeM2-frame*, is an RPpDeM-frame that satisfies the additional condition

(FRPpDeM4) $\forall x \in X\, \exists y \in X,\ xRy\ \&\ (\forall z \in X,\ yRz \Rightarrow z \leqslant x)$.

Proposition 9.7.10 *The complex algebra of an RPpDeM2-frame is an RPpDeM2-algebra.*

Proof:
We show that, for every $A \in L_X$, $[R]\langle R\rangle A \subseteq A$. Let $x \in [R]\langle R\rangle A$. Then, by (FRPpDeM4), there is some $y \in X$ such that $y \in \langle R\rangle A$, so there exists $z \in X$ such that yRz and $z \in A$. Also, by (FRPpDeM4), $z \leqslant x$. Since $A = [\leqslant]A$, we get $x \in A$, as required. □

Proposition 9.7.11 *The canonical frame of an RPpDeM2-algebra is an RPpDeM2-frame.*

Proof:
Take any $F \in X_L$. Then, by Lemma 9.7.9, $G = \neg^{-1}(F) \in X_L$. By definition, $\neg^{-1}(F) \subseteq -G$. We now show that, for every $H \in X_L$, $\neg^{-1}(G) \subseteq -H \Rightarrow H \subseteq F$. Assume $\neg^{-1}(G) \subseteq -H$ and let $a \in H$. Since $\neg^{-1}(G) = \neg^{-1}(-\neg^{-1}(F))$, by the assumption we get $a \notin \neg^{-1}(-\neg^{-1}(F))$, which means $\neg\neg a \in F$. Hence, $a \in F$, by (RPpDeM4), as required. □

By Theorem 9.7.6 and Propositions 9.7.7 and 9.7.8 (resp. Proposition 9.7.10 and 9.7.8) we obtain the discrete representations between RPpDeM1-lattices and RPpDeM1-frames (resp. RPpDeM2-lattices and RPpDeM2-frames).

9.8 Lattices with De Morgan negation

A *De Morgan lattice, DeM-lattice,* is a structure $(L, \neg, 1)$ such that $(L, 1)$ is a distributive lattice with the greatest element 1 and negation $\neg$ satisfies, for all $a, b \in L$,

(DeM1) $\neg(a \vee b) = \neg a \wedge \neg b$
(DeM2) $\neg\neg a = a$.

Equivalently, DeM-lattices may be defined with the axioms

(DeM1') $a \leq \neg b \Rightarrow b \leq \neg a$
(DeM2') $\neg\neg a \leq a$.

Lemma 9.8.1 *For all $a, b \in L$,*

(a) $a \leq b \Rightarrow \neg b \leq \neg a$;
(b) $a \leq \neg b \Leftrightarrow b \leq \neg a$;
(c) *there is the smallest element* 0 *in* L *and* $\neg 0 = 1$, $\neg 1 = 0$;
(d) $\neg(a \wedge b) = \neg a \vee \neg b$.

In view of Lemma 9.8.1(c) we include 0 in the signature of DeM-lattices, and write $(L, \neg, 0, 1)$.

For a DeM-lattice L and $A \subseteq L$, let $\neg A = \{\neg a : a \in A\}$.

Lemma 9.8.2 *Let L be a DeM-lattice. Then for every $A \subseteq L$,*

(a) $\neg A = \neg^{-1}(A)$;
(b) $\neg(-A) = -(\neg A)$;
(c) $\neg\neg A = A$;
(d) *A is a prime filter if and only if $\neg A$ is a prime ideal.*

By a *DeM-frame* we mean a relational structure (X, N) where X is an RP-frame and $N : X \to X$ is a function satisfying, for all $x, y \in X$,

(FDeM1) $x \leqslant y \Rightarrow N(y) \leqslant N(x)$

(FDeM2) $N(N(x)) = x$.

Given a DeM-frame (X, N), let $N(A) = \{N(x) : x \in A\}$ for any $A \subseteq X$.

Lemma 9.8.3 *For all $A, B \subseteq X$,*

(a) $N(A) = \{x : N(x) \in A\}$;
(b) $N(-A) = -N(A)$;
(c) $N(A \cup B) = N(A) \cup N(B)$;
(d) $N(N(A)) = A$.

Proof:
The only non-trivial property is (b), but this follows since:

$$x \in N(-A) \Leftrightarrow N(x) \in -A \Leftrightarrow N(x) \notin A \Leftrightarrow x \notin N(A)$$

which completes the proof. □

The *complex algebra* of a DeM-frame X is a structure $(L_X, \neg_N, \emptyset, X)$ where L_X is the complex algebra of its lattice frame reduct and for every $A \subseteq X$,

$$\neg_N A = -N(A)$$

where $N(A) = \{N(x) : x \in A\}$.

Proposition 9.8.4 *The complex algebra of a DeM-frame is a DeM-lattice.*

Proof:
We show that if $A \in L_X$, then $\neg_N A \in L_X$, that is $\neg_N A = [\leqslant] \neg_N A$. Let $x \in \neg_N A$, so $N(x) \notin A$. Suppose that $x \notin [\leqslant] \neg_N A$. Then there is y such that $x \leqslant y$ and $y \notin \neg_N A$, thus $N(y) \in A = [\leqslant] A$, that is $\forall z,\ N(y) \leqslant z \Rightarrow z \in A$. Since $x \leqslant y$, we have $N(y) \leqslant N(x)$, and taking $z = N(x)$ we get $N(x) \in A$, a contradiction. For the converse, we invoke reflexivity of $\leqslant$

(DeM1) We show $\neg_N(A \cup B) = \neg_N A \cap \neg_N B$. By Lemma 9.8.3(c) we have

$$x \in -N(A \cup B) \Leftrightarrow N(x) \notin A \ \&\ N(x) \notin B \Leftrightarrow x \in \neg_N A \cap \neg_N B,$$

(DeM2) Now we show $\neg_N \neg_N A = A$. Using Lemma 9.8.3(b) and 9.8.3(d) we have $-N(\neg_N A) = -N(-N(A)) = -(-NN(A)) = N(N(A)) = A$. □

The *canonical frame* of a DeM-lattice L is the relational structure $(X_L, N_\neg)$ where X_L is the canonical frame of its lattice reduct and for every $F \in X_L$,

$$N_\neg(F) \stackrel{\text{def}}{=} L - (\neg F).$$

Note that $N_\neg(F) \in X_L$ in view of Lemma 9.8.2(d).

Proposition 9.8.5 *The canonical frame of a DeM-lattice is a DeM-frame.*

Proof:
We first show that $N_\neg$ is a function from X_L to X_L. Let $F \in X_L$. It is routine to check that $N_\neg(F)$ is a filter. For primeness, suppose that $a \vee b \in N_\neg(F) = -(\neg F)$. Then $a \vee b \notin \neg F$, so $\neg(a \vee b) = \neg a \wedge \neg b \notin F$. Thus, either $\neg a \notin F$ or $\neg b \notin F$, so $a \notin \neg F$ or $b \notin \neg F$, hence $a \in -(\neg F)$ or $b \in -(\neg F)$.

(FDeM1) Suppose $F, G \in X_L$ and $F \subseteq G$. By Lemma 9.8.2(a), 9.8.2(b) and the respective definitions we have:

$$a \in N_\neg(G) \Leftrightarrow a \in -(\neg G) \Leftrightarrow a \notin \neg G \Leftrightarrow \neg a \notin G.$$

Hence, by the assumption,

$$\neg a \notin F \Leftrightarrow a \notin \neg F \Leftrightarrow a \in -(\neg F) \Leftrightarrow a \in N_\neg(F).$$

(FDeM2) By Lemma 9.8.3(b) and 9.8.3(c) we have $N_\neg(N_\neg(F)) = -(\neg N_\neg(F))$ $= -(\neg(-(\neg F))) = -(-\neg\neg F) = \neg\neg F = F$. □

The above propositions imply that the complex algebra of the canonical frame of a DeM-lattice is a DeM-lattice and the canonical frame of the complex algebra of a DeM-frame is a DeM-frame.

Let L be a DeMPS-lattice. We now show that the embedding $h : L \to L_{X_L}$, used in Section 2.5, preserves the negation on L.

Lemma 9.8.6 *For any $a \in L$,*

$$h(\neg a) = \neg_{N_\neg} h(a).$$

Proof:
For any $a \in L$,

$$\begin{aligned} F \in \neg_{N_\neg} h(a) &\Leftrightarrow F \in X_L - N_\neg h(a) \\ &\Leftrightarrow F \notin N_\neg h(a) \\ &\Leftrightarrow N_\neg(F) \notin h(a) \\ &\Leftrightarrow a \notin N_\neg((F) \\ &\Leftrightarrow \neg a \in F \\ &\Leftrightarrow F \in h(\neg a). \end{aligned}$$

which completes the proof. □

Let X be a DeMPS-frame. We show that the embedding $k : X \to X_{L_X}$, used in Section 2.5 preserves the function N.

Lemma 9.8.7 *For any $x \in X$,*

$$k(N(x)) = N_{\neg_N}(k(x)).$$

Proof:
For every $x \in X$ and for every $A \in L_X$,

$$\begin{aligned} A \in N_{\neg_N}(k(x)) &\Leftrightarrow A \in L_X - (\neg_N k(x)) \\ &\Leftrightarrow A \notin \neg_N k(x) \\ &\Leftrightarrow A \in N(k(x)) \\ &\Leftrightarrow A \in N^{-1}(k(x)) \qquad \text{Lemma 9.8.2(a)} \\ &\Leftrightarrow N(A) \in k(x) \\ &\Leftrightarrow x \in N^{-1}(A) \\ &\Leftrightarrow N(x) \in A \\ &\Leftrightarrow A \in k(N(x)), \end{aligned}$$

which completes the proof. □

As a consequence we obtain the discrete representations for DeM-lattices and DeM-frames.

Theorem 9.8.8

(a) *Every DeM-lattice is embeddable into the complex algebra of its canonical frame;*

(b) *Every DeM-frame is embeddable into the canonical frame of its complex algebra.*

9.9 De Morgan lattices with possibility and sufficiency operators

In this section we consider DeM-lattices endowed with both possibility and sufficiency operators. Bounded distributive lattices with those operators are considered in Section 8.2 and Section 8.4, respectively.

A *DeMPS-lattice* is a structure $(L, \neg, \Diamond, \boxminus, 0, 1)$ such that $(L, \neg, 0, 1)$ is a DeM-lattice, $(L, \Diamond)$ is a P-lattice, $(L, \boxminus)$ is a S-lattice, and, for every $a \in L$,

(DeMPS) $\Diamond a = \neg \boxminus a$.

In the frames there are two relations, R and S, linked with an axiom which is a De Morgan analogue of condition $R = -S$.

A *DeMPS-frame* is a structure (X, N, R, S) such that (X, N) is a DeM-frame, (X, R) is a P-frame, (X, S) is a S-frame, and for all $x, y \in X$,

(FDeMPS) $xRy \Longleftrightarrow$ *not* $N(x)Sy$.

The *complex algebra* of a DeMPS-frame X is a structure

$$(L_X, \neg_N, \Diamond_R, \boxminus_S, \emptyset, X)$$

where the operators $\neg_N$, $\Diamond_R$, and $\boxminus_S$ are defined, for every $A \in L_X$, by

$$\neg_N A \stackrel{\text{def}}{=} -N(A)$$
$$\Diamond_R A \stackrel{\text{def}}{=} \langle R \rangle A$$
$$\boxminus_S A \stackrel{\text{def}}{=} [S](-A).$$

Proposition 9.9.1 *The complex algebra of a DeMPS-frame is a DeMPS-lattice.*

Proof:
We show that (DeMPS) holds, that is for any $A \in L_X$, $\langle R \rangle A = -N([S](-A))$. For every $x \in X$, we have:

$$\begin{aligned} x \notin N([S](X-A)) &\Leftrightarrow N(x) \notin [S](-A) \\ &\Leftrightarrow \exists y \in X,\ y \in A \ \&\ \text{ not } N(x)Sy \\ &\Leftrightarrow \exists y \in X,\ y \in A \ \&\ xRy \qquad \text{(FDeMPS)} \\ &\Leftrightarrow x \in \langle R \rangle A. \end{aligned}$$

□

The *canonical frame* of a DeMPS-lattice L is a structure $(X_L, N_\neg, R_\Diamond, S_\boxminus)$ where, for all $F, G \in X_L$,

$$N_\neg(F) \stackrel{\text{def}}{=} -(\neg F)$$
$$F R_\Diamond G \stackrel{\text{def}}{\Leftrightarrow} G \subseteq \Diamond^{-1}(F)$$
$$F S_\boxminus G \stackrel{\text{def}}{\Leftrightarrow} \boxminus^{-1}(F) \subseteq G$$

defined as in Section 9.8, Section 8.3, and Section 8.4, respectively.

Proposition 9.9.2 *The canonical frame of a DeMPS-lattice is a DeMP- frame.*

Proof:
It is sufficient to show that the axiom (FDeMPS) holds, that is for all $F, G \in X_L$, $G \subseteq \Diamond^{-1}(F) \Leftrightarrow \boxminus^{-1}(N_\neg(F)) \subseteq -G$. Observe that

$$\begin{aligned} a \in \boxminus^{-1}(N_\neg(F)) &\Leftrightarrow \boxminus a \in N_\neg(F) \\ &\Leftrightarrow \boxminus a \notin \neg F \\ &\Leftrightarrow \boxminus a \notin \neg^{-1}(F) \qquad \text{Lemma 9.8.2(a)} \\ &\Leftrightarrow \neg \boxminus a \notin F \\ &\Leftrightarrow \Diamond a \notin F. \qquad \text{(DeMPS)} \end{aligned}$$

Assume that for every $a \in L$, $a \in G \Rightarrow \Diamond a \in F$ and take any $a \in \boxminus^{-1}(N_\neg(F))$. Then $\Diamond a \notin F$ and hence, by the assumption, $a \notin G$. Conversely, let $a \in G$ and suppose $\Diamond a \notin F$. Then $a \notin \boxminus^{-1}(N_\neg(F))$ and by the assumption, $a \notin G$, a contradiction. □

In view of the above propositions and the discrete representation theorems for DeM-lattices, P-lattices, and S-lattices we get the following discrete representations for DeMPS-lattices and DeMPS-frames.

Theorem 9.9.3

(a) *Every DeMPS-lattice is embeddable into the complex algebra of its canonical frame;*

(b) *Every DeMPS-frame is embeddable into the canonical frame of its complex algebra.*

9.10 Pseudocomplemented algebras

A *pseudocomplemented algebra, p-algebra,* is a structure $(L, \vee, \wedge, 0, 1, \neg)$ where $(L, \vee, \wedge, 0, 1)$ is a distributive lattice and $\neg$ is a pseudocomplement, that is, for all $a, b \in L$,

$(\neg)$ $a \wedge b = 0 \Leftrightarrow b \leq \neg a.$

This axiom can be equivalently replaced by

$(\neg 1)$ $a \wedge \neg(a \wedge b) = a \wedge \neg b$

$(\neg 2)$ $a \wedge \neg 0 = a$

$(\neg 3)$ $\neg\neg 0 = 0.$

A *double p-algebra* is a structure $(L, \vee, \wedge, 0, 1, \neg, \ulcorner)$ where $(L, \vee, \wedge, 0, 1, \neg)$ is a p-algebra and $\ulcorner$ is a dual pseudocomplement satisfying, for all $a, b \in L$,

$(\ulcorner)$ $a \vee b = 1 \Leftrightarrow \ulcorner a \leq b.$

It follows that $\ulcorner$ is the pseudocomplement in the lattice L^{op}.

Lemma 9.10.1 *For all $a, b \in L$,*

(a) $\neg$ *and* $\ulcorner$ *are antitone;*

(b1) $a \leq \neg\neg a;$ (b2) $\ulcorner\ulcorner a \leq a;$

(c1) $\neg\neg\neg a = a;$ (c2) $\ulcorner\ulcorner\ulcorner a = a;$

(d1) $a \wedge \neg a = 0;$ (d2) $a \vee \ulcorner a = 1;$

(e1) $\neg(a \vee b) = \neg a \wedge \neg b;$ (e2) $\ulcorner(a \vee b) \leq \ulcorner a \wedge \ulcorner b;$

(f1) $\neg a \vee \neg b \leq \neg(a \wedge b);$ (f2) $\ulcorner a \vee \ulcorner b = \ulcorner(a \wedge b);$

(g1) $\neg\neg(a \wedge b) = \neg\neg a \wedge \neg\neg b;$ (g2) $\ulcorner\ulcorner(a \vee b) = \ulcorner\ulcorner a \vee \ulcorner\ulcorner b;$

(h1) $\neg\neg(a \vee b) = \neg\neg(\neg\neg a \vee \neg\neg b);$ (h2) $\ulcorner\ulcorner(a \wedge b) = \ulcorner\ulcorner(\ulcorner\ulcorner a \wedge \ulcorner\ulcorner b);$

(i) $\neg a \leq \ulcorner a.$

Proof:
(b2) By $(\neg)$, $\ulcorner a \leq \ulcorner a$ if and only if $a \vee \ulcorner a = 1$ if and only if $\ulcorner\ulcorner a \leq a.$

(g2) By (d2) we have $a \vee b \vee \ulcorner(a \vee b) = 1$. Then applying repeatedly $(\neg)$ and (c2) we have

$$\begin{aligned}
a \vee b \vee \ulcorner(a \vee b) &\Leftrightarrow \ulcorner a \leq b \vee \ulcorner(a \vee b) \\
&\Leftrightarrow \ulcorner\ulcorner\ulcorner a \leq b \vee \ulcorner(a \vee b) \\
&\Leftrightarrow \ulcorner\ulcorner a \vee b \vee \ulcorner(a \vee b) = 1 \\
&\Leftrightarrow \ulcorner b \leq \ulcorner\ulcorner a \vee \ulcorner(a \vee b) \\
&\Leftrightarrow \ulcorner\ulcorner\ulcorner b \leq \ulcorner\ulcorner a \vee \ulcorner(a \vee b) \\
&\Leftrightarrow \ulcorner\ulcorner a \vee \ulcorner\ulcorner b \vee \ulcorner(a \vee b) = 1 \\
&\Leftrightarrow \ulcorner\ulcorner(a \vee b) \leq \ulcorner\ulcorner a \vee \ulcorner\ulcorner b.
\end{aligned}$$

For the other inequality, since $a \leq a \vee b$ and $b \leq a \vee b$, $\ulcorner\ulcorner a \leq \ulcorner\ulcorner(a \vee b)$ and $\ulcorner\ulcorner b \leq \ulcorner\ulcorner(a \vee b)$ which implies $\ulcorner\ulcorner a \vee \ulcorner\ulcorner b \leq \ulcorner\ulcorner(a \vee b)$.

(i) $\neg a = \neg a \wedge 1 = \neg a \wedge (a \vee \ulcorner a) = (\neg a \wedge a) \vee (\neg a \wedge \ulcorner a) = \neg a \wedge \ulcorner a$. Hence $\neg a \leq \ulcorner a$. □

A *double p-frame* is a RP-frame X.

The *complex algebra* of a double p-frame X is the structure $(L_X, \neg_\leqslant, \ulcorner_\leqslant)$ where L_X is the complex algebra of its lattice reduct (see Section 2.5) and, for every $A \subseteq X$,

$$\begin{aligned}
\neg_\leqslant A &\stackrel{\text{def}}{=} [\leqslant](-A) \\
\ulcorner_\leqslant A &\stackrel{\text{def}}{=} \langle\geqslant\rangle(-A).
\end{aligned}$$

Proposition 9.10.2 *The complex algebra of a double p-frame is a double p-algebra.*

Proof:
As shown in Section 2.5, L_X is a bounded distributive lattice. First, we show that L_X is closed on $\neg_\leqslant$ and $\ulcorner_\leqslant$. Condition $[\leqslant](-A) = [\leqslant][\leqslant](-A)$ easily follows from transitivity and reflexivity of $\leqslant$. Condition $[\leqslant]\langle\geqslant\rangle(-A) \subseteq \langle\geqslant\rangle(-A)$ follows from reflexivity of $\leqslant$. For the reverse inclusion, assume that there exists some $y \in X$ such that $x \geqslant y$ and $y \in -A$. Take $z \in X$ such that $x \leqslant z$. Since $y \leqslant x$ and $x \leqslant z$, we get $y \leqslant z$. Thus y satisfies $z \geqslant y$ and $y \in X - A$, as required.

Now, we show that the complex algebra of a double p-frame satisfies axioms $(\neg)$ and $(\ulcorner)$.

$(\neg)$ We show that $A \subseteq [\leqslant](-B)$ if and only of $A \cap B = \emptyset$. Let $A \subseteq [\leqslant](-B)$. Since $[\leqslant](-B) \subseteq -B$ by reflexivity of $\leqslant$, we have $A \subseteq -B$. Conversely, if $A \cap B = \emptyset$, then $A \subseteq -B$. Since $[\,]$ is monotone, $[\leqslant]A \subseteq [\leqslant](-B)$. Since $A = [\leqslant]A$, we get $A \subseteq [\leqslant](-B)$.

$(\ulcorner)$ We show that $\langle\geqslant\rangle(-A) \subseteq B$ if and only if $-A \subseteq B$. If $\langle\geqslant\rangle(-A) \subseteq B$, then $-B \subseteq -\langle\geqslant\rangle(-A) = [\geqslant]A$. Since $\geqslant$ is reflexive, $[\geqslant]A \subseteq A$. Hence $X - B \subseteq A$, as required. Conversely, if $-A \subseteq B$, then since $\langle\,\rangle$ is monotone, $\langle\geqslant\rangle(-A) \subseteq \langle\geqslant\rangle B$. Since $B = [\leqslant]B$, we have $B = \langle\geqslant\rangle B$ by Lemma 1.8.6(c). Thus $\langle\geqslant\rangle(-A) \subseteq B$. □

The *canonical frame* of a double p-algebra L is just the canonical frame X_L of a distributive lattice. Therefore following Proposition 2.5.2 we have the following fact.

Proposition 9.10.3 *The canonical frame of a double p-algebra is a double p-frame.*

Let L be double p-algebra. We now show that the $h : L \to L_{X_L}$, used in Section 2.5, preserves the negations $\neg$ and $\ulcorner$.

Lemma 9.10.4 *For every $a \in L$,*

(a) $h(\neg a) = \neg_{\subseteq} h(a)$;

(b) $h(\ulcorner a) = \ulcorner_{\subseteq} h(a)$.

Proof:
(a) We show $F \in h(\neg a) \Leftrightarrow F \in [\subseteq](-h(a))$ for every $F \in X_L$. Assume $\neg a \in F$. Take $G \in X_L$ such that $F \subseteq G$. Then $\neg a \in G$ and necessarily $a \notin G$, for otherwise $0 = a \wedge \neg a \in G$, a contradiction. For the converse implication assume $\neg a \notin F$. We show that there is some $G \in X_L$ such that $F \subseteq G$ and $a \in G$. Let G' be a filter generated by $F \cup \{a\}$. Note that G' is proper, for suppose otherwise, then there is some $b \in F$ such that $a \wedge b = 0$. Then by $(\neg)$, $b \leq \neg a$. Since $\neg a \notin F$, $b \notin F$, a contradiction. Thus G' can be extended to a prime filter G such that $F \subseteq G$ and $a \in G$, as required.

(b) We show $F \in h(\ulcorner a) \Leftrightarrow F \in \langle \supseteq \rangle(-h(a))$ for every $F \in X_L$. Assume that $\ulcorner a \in F$ and show that there is some $G \in X_L$ such that $G \subseteq F$ and $a \notin G$. Consider an ideal I' generated by $(-F) \cup \{a\}$. Then $-F \subseteq I'$ and $a \in I'$. We have $\ulcorner a \notin I'$, for otherwise $a \vee \ulcorner a = 1 \in I'$, a contradiction. By the prime ideal theorem, there is a prime ideal I such that $I' \subseteq I$ and $\ulcorner a \notin I$. Then $-F \subseteq I$ and $a \in I$. Thus $-I$ is a prime filter such that $-I \subseteq F$ and $a \notin -I$. For the converse implication assume that there is some $G \in X_L$ such that $G \subseteq F$ and $a \notin G$. Suppose $\ulcorner a \notin F$. Then $\ulcorner a \notin G$. But $1 = a \vee \ulcorner a \in G$, a contradiction. □

In Section 2.5 it is shown that the mapping $k : X \to X_{L_X}$ defined, for any $x \in X$, as $k(x) = \{A \in L_X : x \in A\}$ is injective and preserves the ordering.

We conclude with discrete representations for double p-algebras and double p-frames.

Theorem 9.10.5

(a) *Every double p-algebra is embeddable into the complex algebra of its canonical frame;*

(b) *Every double p-frame is embeddable into the canonical frame of its complex algebra.*

Chapter 10

Stone algebras

10.1 Introduction

In this chapter some classes of Stone algebras are studied. Stone algebras are p-algebras (see Section 9.10) such that the pseudocomplement is assumed to satisfy a specific axiom which is a weaker version of the law of excluded middle, namely $\neg a \vee \neg\neg a = 1$. The algebras were first investigated in [115]. Double Stone algebras are Stone algebras endowed with the operator of dual pseudocomplement $\ulcorner$ satisfying the specific axiom $\ulcorner a \wedge \ulcorner\ulcorner a = 0$ which is weaker than $a \wedge \ulcorner a = 0$. Regular double Stone algebras are double Stone algebras such that $\neg a = \neg b$ & $\ulcorner a = \ulcorner b \Rightarrow a = b$. In Section 10.2 double Stone algebras and several concepts needed for their characterization are recalled. Next, the corresponding frames are defined. In Section 10.2 discrete representation theorems for double Stone algebras and frames are given and duality via truth is estabished.

Furthermore, in Section 10.3 discrete representation theorems for regular double Stone algebras and frames are presented. A relationship between regular double Stone algebras and Łukasiewicz 3-valued logic is briefly described based on [167]. A topological representation of Stone algebras can be found in [200] and representations of Stone algebras and double Stone algebras in terms of canonical extensions are presented in [125].

In Section 10.4 rough relation algebras, rough relation frames, and discrete representations for them are studied. Rough relations were introduced in [192] as a particular example of rough sets [193], see also [195]. An approximation space is a structure (W, R), where W is a non-empty set and R is an equivalence relation on W. In the context of reasoning with incomplete information a typical example of an approximation space is a set of objects of an information system with the indiscernibility relation determined by the attributes of the system (see Chapter 4). Given a subset X of W, the lower approximation of X in an approximation space W is defined as $L_R(X) = \{x \in W : R(x) \subseteq X\}$. Similarly, the upper approximation of X in W is $U_R(X) = \{x \in W : R(x) \cap X \neq \emptyset\}$. These notions are obvious generalizations of the approximations of subsets of objects of an information system. A rough subset in an approximation space

(W, R) is any pair $(L_R(X), U_R(X))$ for some X included in W. Then the collection of rough subsets of W is $Rough(W) = \{(L_R(X), U_R(X)) : X \subseteq W\}$. In [199] it is shown that for every approximation space (W, R), $Rough(W)$ forms a Stone algebra. It is observed in [31] that this result carries over to regular double Stone algebras where the operations are defined by

$$\begin{aligned}
(L_R(X), U_R(X)) \vee (L_R(Y), U_R(Y)) &\stackrel{\text{def}}{=} (L_R(X) \cup L_R(Y), U_R(X) \cup U_R(Y)) \\
(L_R(X), U_R(X)) \wedge (L_R(Y), U_R(Y)) &\stackrel{\text{def}}{=} (L_R(X) \cap L_R(Y), U_R(X) \cap U_R(Y)) \\
\neg(L_R(X), U_R(X)) &\stackrel{\text{def}}{=} (W - U_R(X), W - U_R(X)) \\
\ulcorner(L_R(X), U_R(X)) &\stackrel{\text{def}}{=} (W - L_R(X), W - L_R(X)) \\
0 &\stackrel{\text{def}}{=} (\emptyset, \emptyset) \\
1 &\stackrel{\text{def}}{=} (W, W).
\end{aligned}$$

Moreover, in [31] a representation theorem for regular double Stone algebras is proved showing that every such algebra is embeddable into the algebra of all rough subsets of an approximation space. Given an approximation space (W, R) the system (W^2, R^2) such that R^2 is a relation on W^2 defined as $(x, y)R^2(x', y') \Leftrightarrow xRx'$ & yRy' is an approximation space as well. A rough (binary) relation on an approximation space (W, R) is defined as a rough subset of the approximation space (W^2, R^2). Tarski originated an approach to ordinary binary relations in terms of the class of relation algebras [233]. Relation algebras are Boolean algebra-ordered involutive monoids. Following this approach in [31] a class of rough relation algebras is proposed. Rough relation algebras differ from relation algebras in that the underlying Boolean algebras are replaced by regular double Stone algebras. Rough relation algebras are studied in [49, 51, 70]. The contents of Section 10.2 and Section 10.3 are based on [58]. The content of Section 10.4 is based on [60].

10.2 Double Stone algebras

Stone algebras are axiomatic extensions of p-algebras considered in Section 9.10. A *Stone algebra* is a p-algebra such that, for all $a, b \in L$,

(St1) $\neg a \vee \neg\neg a = 1$.

Among the distributive pseudocomplemented lattices, Stone algebras are characterized by the following theorem proved in [115].

Theorem 10.2.1 *A distributive pseudocomplemented lattice is a Stone algebra if and only if the join of any two distinct minimal prime ideals in the lattice of all prime ideals of L equals L.*

We recall that this join, $I \vee J$, is the smallest prime ideal including I and J. In case of a distributive lattice we have that

$$I \vee J = \{z \in L : \exists x \in I, \exists y \in J,\ z \leq x \vee y\}.$$

Lemma 10.2.2 [24] *Let $(L, \neg)$ be a Stone algebra. Then*

(a) *Every prime filter of L is contained in exactly one maximal (with respect to set inclusion) prime filter of L;*

(b) *Every prime ideal of L includes exactly one minimal prime ideal of L.*

A *double Stone algebra* is a double p-algebra satisfying the axiom (St1) and, for every $a, \in L$,

(St2) $\ulcorner a \wedge \ulcorner\ulcorner a = 0$.

Observe that if $(L, \neg, \ulcorner)$ is a double Stone algebra, then the algebra $(L^{op}, \ulcorner)$ is a Stone algebra based on the lattice opposite to L with the ordering $\geq = \leq^{-1}$.

Lemma 10.2.3 *For all $a, b \in L$,*

(a1) $\ulcorner\neg a = \neg\neg a$; (a2) $\neg\ulcorner a = \ulcorner\ulcorner a$;

(b1) $\neg\neg(a \vee b) = \neg\neg a \vee \neg\neg b$; (b2) $\ulcorner\ulcorner(a \wedge b) = \ulcorner\ulcorner a \wedge \ulcorner\ulcorner b$.

Let $(L, \neg, \ulcorner)$ be a double Stone algebra. An element $a \in L$ is called *dense* (resp. *dually dense*) if it is of the form $a = b \vee \neg b$ (resp. $a = b \wedge \ulcorner b$) for some $b \in L$. The set of dense elements of L, denoted by $D^{\neg}$, and the set of dually dense elements of L, denoted by $D^{\ulcorner}$, are defined as

$$D^{\neg} \stackrel{\text{def}}{=} \{a \vee \neg a : a \in L\}$$
$$D^{\ulcorner} \stackrel{\text{def}}{=} \{a \wedge \ulcorner a : a \in L\}.$$

The *centre* of L is a set

$$B(L) \stackrel{\text{def}}{=} \{\neg a : a \in L\}.$$

It is easy to show the following properties.

Lemma 10.2.4 *Let L be a double Stone algebra.*

(a) *$D^{\neg}$ is a filter of L and $D^{\ulcorner}$ is an ideal of L;*

(b) *$(B(L), \vee, \wedge, \neg, 0, 1)$ is a subalgebra of L and a Boolean algebra;*

(c) *$B(L) = \{a \in L : a = \neg\neg a\} = \{a \in L : a = \ulcorner\ulcorner a\}$.*

Lemma 10.2.5 [117] *Let L be a double Stone algebra. For every prime filter F of L there are a unique minimal prime filter, $\underline{F}$, and a unique maximal prime filter, $\overline{F}$, such that $\underline{F} \subseteq F \subseteq \overline{F}$.*

Lemma 10.2.6 *Let F be a prime filter of a double Stone algebra L.*

(a) *F is maximal if and only if $D^{\neg} \subseteq F$ if and only if $a \in F$ or $\neg a \in F$ for every $a \in L$;*

(b) *F is minimal if and only if $D^{\ulcorner} \cap F = \emptyset$.*

Lemma 10.2.7 *If F and F' are prime filters of a double Stone algebra L and $F \subseteq F'$, then $F \cap B(L) = F' \cap B(L)$.*

Proof:
The left-to-right inclusion is immediate. Now, let $a \in F' \cap B(L)$. Then $\neg a \notin F'$ and therefore $\neg a \notin F$. Since $a \in B(L)$, by Lemma 10.2.4(b), $a \vee \neg a = 1$. Since F is a prime filter, $a \in F$, and hence $a \in F \cap B(L)$. □

Lemma 10.2.8 *Let F be a prime filter of a double Stone algebra L.*

(a) *$\underline{F}$ is the filter generated by $F \cap B(L)$;*

(b) *$\overline{F}$ is the filter generated by $F \cup D^{\neg}$.*

Proof:
(a) Let G be a filter generated by $F \cap B(L)$. First, we show that G is a prime filter of L. Let $a \vee b \in G$. Then for some $c \in F \cap B(L)$, $c \leq a \vee b$. Since $c \in B(L)$, we may assume that $c = \ulcorner\ulcorner c$ by Lemma 10.2.4(c). Therefore $\ulcorner\ulcorner c \leq \ulcorner\ulcorner(a \vee b) = \ulcorner\ulcorner a \vee \ulcorner\ulcorner b$ by Lemma 9.10.1(a) and (g2). Since $c \in F$ and F is prime, $\ulcorner\ulcorner a \in F$ or $\ulcorner\ulcorner b \in F$. By Lemma 10.2.3(b2), $\ulcorner\ulcorner a \leq a$ and $\ulcorner\ulcorner b \leq b$. Thus $a \in F$ or $b \in F$, as required. Now, by Lemma 10.2.7, $G \cap B(L) = F \cap B(L) = \underline{F} \cap B(L)$. By definition of G, $G \subseteq \underline{F}$. Since $\underline{F}$ is minimal, $G = \underline{F}$.

(b) follows easily from Lemma 10.2.6(a). □

A *double Stone frame* is a bounded distributive lattice frame $(X, \leqslant)$ satisfying, for all $x, y \in X$,

(FSt1) If x and y have a common lower bound in X, then they have a common upper bound in X

(FSt2) If x and y have a common upper bound in X, then they have a common lower bound in X.

Lemma 10.2.9 *Let $(X, \leqslant)$ be a poset. Then*

(a) *If for every $x \in X$ there exists exactly one $y \in X$ such that $x \leqslant y$ and y is maximal in X, then (FSt1) holds in X;*

(b) *If for every $x \in X$ there exists exactly one $y \in X$ such that $y \leqslant x$ and y is minimal in X, then (FSt2) holds in X.*

Proof:
(a) Let for any $x \in X$, $\overline{x}$ denote a maximal element in X such that $x \leqslant \overline{x}$. Let y and z have a common lower bound, say u. Then $u \leqslant y$ and the uniqueness of $\overline{u}$ implies $\overline{u} = \overline{y}$. Similarly, $\overline{z} = \overline{u}$ and thus $\overline{u}$ is a common upper bound of y and z.
The proof of (b) is similar. □

Observe that the converse implications in (a) and (b) of the above lemma are not true. For if $(X, \leqslant)$ is an infinite chain without endpoints, then X satisfies (FSt1) and (FSt2), but neither of the premises of (a) and (b).

The *complex algebra of a double Stone frame* X is defined in the same way as the complex algebra of a double p-frame, that is a structure $(L_X, \neg_{\leqslant}, \ulcorner_{\leqslant})$

where L_X is the complex algebra of its distributive lattice frame reduct and, for every $A \subseteq X$,

$$\neg_{\leqslant} A \stackrel{\text{def}}{=} [\leqslant](-A)$$
$$\ulcorner_{\leqslant} A \stackrel{\text{def}}{=} \langle\geqslant\rangle(-A).$$

Proposition 10.2.10 *The complex algebra of a double Stone frame is a double Stone algebra.*

Proof:
In Proposition 9.10.2 it is shown that L_X is the complex algebra of a double p-frame. Therefore here it is sufficient to show that the axioms (St1) and (St2) are satisfied in L_X.

(St1) We show that $[\leqslant](-A) \cup [\leqslant]\langle\leqslant\rangle A = X$. Suppose that $x \notin [\leqslant](-A)$ and $x \notin [\leqslant]\langle\leqslant\rangle A$ for some $x \in X$. Then there is some $y \in X$ such that $x \leqslant y$ and $y \in A$ and there is some $z \in X$ such that $x \leqslant z$ and $z \notin \langle\leqslant\rangle A$. Since y and z have a common lower bound x, by (FSt1) they have a common upper bound, say t. Thus $y \leqslant t$ and $z \leqslant t$. Since $z \notin \langle\leqslant\rangle A$, $t \notin A$. It follows that $y \notin A$, a contradiction.

(St2) We show that $\langle\geqslant\rangle(-A) \cap \langle\geqslant\rangle[\geqslant]A = \emptyset$. Suppose there is some $x \in X$ such that $x \in \langle\geqslant\rangle(-A)$ and $x \in \langle\geqslant\rangle[\geqslant]A$. Then there are $y \in X$ and $z \in X$ such that x is their common upper bound. Using (FSt2) we easily get a contradiction. □

The *canonical frame of a double Stone algebra* L is just the canonical frame of a distributive lattice.

Proposition 10.2.11 *The canonical frame of a double Stone algebra is a double Stone frame.*

Proof:
In view of Proposition 2.5.2 and Proposition 9.10.3 it suffices to show that the axioms (FSt1) and (FSt2) are satisfied.

(FSt1) By Lemma 10.2.2(b) every prime ideal J of L includes a unique minimal prime ideal, say $\overline{J}$. Then $-J \subseteq -\overline{J}$, and by Lemma 1.6.6(a) and Lemma 10.2.9, condition (FSt1) is satisfied.

(FSt2) Applying Lemma 10.2.2 to the opposite lattice L^{op} we deduce that for every prime ideal J of L^{op} there is a unique minimal prime ideal $\overline{J}$ of L^{op} such that $\overline{J} \subseteq J$. Then, by Lemma 10.2.9(b), condition (FSt2) follows. □

Let L be a double Stone algebra. We now show that the lattice embedding $h : L \to L_{X_L}$, used in Section 2.5, preserves the operators $\neg$ and $\ulcorner$.

Lemma 10.2.12 *For every $a \in L$*

(a) $h(\neg a) = \neg_{\subseteq} h(a)$;
(b) $h(\ulcorner a) = \ulcorner_{\subseteq} h(a)$.

Let X be a double Stone frame. In Section 2.5 it is shown that the mapping $k : X \to X_{L_X}$ defined, for any $x \in X$, as $k(x) = \{A \in L_X : x \in A\}$ is injective and preserves the ordering. We conclude with the discrete representations for double Stone algebras and double Stone frames.

Theorem 10.2.13

(a) *Every double Stone algebra is embeddable into the complex algebra of its canonical frame;*

(b) *Every double Stone frame is embeddable into the canonical frame of its complex algebra.*

We now propose a logic *DSL* based on double Stone algebras. The formulas of the logic are built from the language $\mathcal{L}an_L$, defined in Section 2.7 for the underlying bounded distributive lattice and two negations $\neg$ and $\ulcorner$. For the sake of simplicity we use the same symbols for the operations in logic and in the algebras.

An algebraic semantics of the logic is determined by the class of double Stone algebras. This is the algebraic semantics of the $\mathcal{L}an_L$ where the valuation is extended homomorphically to all formulas of *DSL* by:

$$v(\neg\alpha) = \neg v(\alpha)$$
$$v(\ulcorner\alpha) = \ulcorner v(\alpha).$$

Now we define a Kripke-style semantics for the logic *DSL* in terms of double Stone frames X. This is as in Section 2.7 with definition of the satisfaction relation extended to include the two types of negated formulas, as follows:

$$\mathcal{M}, x \models \neg\alpha \overset{\text{def}}{\Leftrightarrow} \forall y \in X,\ x \leqslant y \Rightarrow \mathcal{M}, y \not\models \alpha$$
$$\mathcal{M}, x \models \ulcorner\alpha \overset{\text{def}}{\Leftrightarrow} \exists y \in X,\ x \geqslant y\ \&\ \mathcal{M}, y \not\models \alpha.$$

We define $m(\alpha) = \{x \in X : \mathcal{M}, x \models \alpha\}$. Invoking Theorem 10.2.10, the following observation is easily verified.

Lemma 10.2.14 *For every DSL model $\mathcal{M} = (X, \leqslant, m)$ and for every formula α, $m(\alpha)$ is $\leqslant$-increasing.*

The next lemma follows from the fact that meaning functions in Kripke models based on a frame can be viewed as valuations in the complex algebra of that frame.

Lemma 10.2.15 *For every double Stone frame $(X, \leqslant)$ the following conditions are equivalent:*

(a) *The sequent $\alpha \vdash \beta$ is true in all models based on $(X, \leqslant)$;*

(b) *The sequent $\alpha \vdash \beta$ is true in the complex algebra L_X.*

We conclude with the result which states the equivalence of the algebraic and the frame semantics for the logic DSL.

Theorem 10.2.16 *For all DSL formulas α and β the following conditions are equivalent:*

(a) *The sequent $\alpha \vdash \beta$ is true in all double Stone algebras;*

(b) *The sequent $\alpha \vdash \beta$ is true in all DSL models.*

Proof:
(a) $\Rightarrow$ (b) It follows from (a) and Theorem 10.2.10 that $\alpha \vdash \beta$ is true in the complex algebras of the double Stone frames. Then, by Lemma 10.2.15, condition (b) follows.

(b) $\Rightarrow$ (a) Consider the canonical frame X_L of L and let $h : L \to L_{X_L}$ be the embedding provided by Theorem 10.2.13. Assume that the sequent $\alpha \vdash \beta$ is true in all DSL models, and suppose that it is not true in some double Stone algebra L. Then, there is a valuation v in L such that $v(\alpha) \not\leq v(\beta)$, and by Corollary 1.6.9(b) there is a filter $F \in X_L$ such that $v(\alpha) \in F$ and $v(\beta) \notin F$. Consider a DSL model $\mathcal{M}$ based on X_L such that its meaning function is defined by $m(p) = h(v(p))$ for every propositional variable p. By induction on the structure of a formula one can show that for every formula α, $m(\alpha) = h(v(\alpha))$. Thus $\alpha \vdash \beta$ is not true in M, a contradiction. □

10.3 Regular double Stone algebras

A double Stone algebra is *regular* if it satisfies, for all $a, b \in L$,

(St3) $a \wedge \ulcorner a \leq b \vee \neg b$.

It is known (see [246]) that the axiom (St3) of regularity has equivalent formulations as given in the following lemma.

Lemma 10.3.1 *The following conditions are equivalent*

(a) *L is a regular double Stone algebra;*

(b) *$\neg a = \neg b$ and $\ulcorner a = \ulcorner b$ imply $a = b$ for all $a, b \in L$;*

(c) *Every chain of prime filters of L has at most two elements.*

A *regular double Stone frame* is a double Stone frame $(X, \leqslant)$ satisfying

(FSt3) $\forall x, y \in X,\ x \leq y \Rightarrow (x = y$ or $\forall z \in X,\ (z \leq x \Rightarrow z = x))$.

The axiom (FSt3) says that each chain of X has at most length 2.

The complex algebra of a regular double Stone frame is defined in the same way as the complex algebra of a double Stone frame. Similarly, the canonical frame of a regular double Stone algebra is defined as that for a double Stone algebra. Therefore to prove the representation theorems for regular double Stone algebras and frames it is sufficient to show the following two propositions.

Proposition 10.3.2 *The complex algebra of a regular double Stone frame satisfies (St3).*

Proof:
We show $A \cap \langle\geqslant\rangle(-A) \subseteq B \cup [\leqslant](-B)$ for all $A, B \subseteq X$. Assume $x \in A$ and there exists some $y \in X$ such that $x \geqslant y$ and $y \notin A$. Suppose $x \notin B$ and there is some $z \in X$ such that $x \leqslant z$ and $z \in B$. Since $x \leqslant z$, by (FSt3) $x = z$ or for every $t \in X$, $t \leqslant x$ implies $t = x$. If $x = z$, then $x \notin B$ and $x \in B$, a contradiction. If the second disjunct holds, then taking y for t we get $x = y$. Then $x \in A$ and $x \notin A$, a contradiction. □

Proposition 10.3.3 *The canonical frame of a regular double Stone algebra satisfies (FSt3).*

Proof:
The proof follows from Lemma 10.3.1. □

Regular double Stone algebras are related to Łukasiewicz's 3-valued logic. Łukasiewicz proposed his 3–valued logic for the first time during a lecture presented at the University of Warsaw on the 7th of March 1918. He defined his logic semantically in terms of matrices and interpreted intuitively the third logical value different from 'true' and 'false' as 'possible'. The axiomatization of his logic was attempted by many authors and various axiom systems were developed, for a comprehensive survey see [8]. Moisil [166, 167] proposed a class of algebras referred to as *3-valued Łukasiewicz algebras*. These are bounded distributive lattices with a De Morgan negation N and a possibility operator M. The operation M is assumed to be an $S4$ modality which, moreover, distributes over meet. Furthermore, N and M are related by the axiom $NMNM(a) = M(a)$ for every element a of the lattice. Then the following hold:

Theorem 10.3.4 *If L is a regular double Stone algebra, then the Moisil axioms of the 3-valued Łukasiewicz algebras with N and M defined by $N(a) = \neg\neg a$ and $M(a) = \neg a \vee (a \wedge \urcorner a)$ are true in L.*

Theorem 10.3.5 *If L is a Moisil 3-valued Łukasiewicz algebra, then the axioms of regular double Stone algebras with the negations defined by $\neg a = NM(a)$ and $\urcorner a = MN(a)$ are true in L.*

A logic based on regular double Stone algebras and a representation of these algebras developed in [147] are presented in [50].

10.4 Rough relation algebras

A *rough relation algebra* is a structure

$$(L, \vee, \wedge, \neg, \urcorner, ;, {}^{\smile}, 1')$$

where $(L, \vee, \wedge, \neg, \urcorner)$ is a regular double Stone algebra, $(L, ;, 1')$ is a monoid, and the following specific axioms are assumed, for all $a, b, c \in L$,

(R^{2}1) $(a \vee b) ; c = (a ; c) \vee (b ; c)$ and $c ; (a \vee b) = (c ; a) \vee (c ; b)$

($R^2$2) $a^{\smile\smile} = a$

($R^2$3) $(a \vee b)^{\smile} = a^{\smile} \vee b^{\smile}$

($R^2$4) $(a\,;b)^{\smile} = b^{\smile}\,; a^{\smile}$

($R^2$5) $a^{\smile}\,; \neg(a\,;b) \leq \neg b$

($R^2$6) $\neg\neg(\neg a\,;\neg b) = \neg a\,;\neg b$

($R^2$7) $\neg\neg 1' = 1'$.

Axioms ($R^2$1) – ($R^2$4) are the same as their counterparts in relation algebras considered in Section 5.2. Axiom ($R^2$5) is a modification of (R5) such that the Boolean complement is replaced by the pseudocomplement. Axioms ($R^2$6) and ($R^2$7) further characterize action of the pseudocomplement with composition of rough relations and with the unit element of the monoid, respectively.

The original axiom system in [31] includes also the axiom

($R^2$8) $(a\,;b) \wedge c \leq a^{\smile}\,;c$.

It is not assumed here for the following reasons. It is not required for a characterization of representability of rough relation algebras. In [49] it is shown without ($R^2$8) that a rough relation algebra is representable if and only if it satisfies $\ulcorner\ulcorner(a\,;b) = \ulcorner\ulcorner a\,;\ulcorner\ulcorner b$ and $B(L)$ is a representable relation algebra. Furthermore, none of the conditions in Lemma 10.4.1 below required ($R^2$8). Condition (h) of the lemma is a weaker version of ($R^2$8).

Lemma 10.4.1 *For all $a, b, c \in L$,*

(a) $a \leq b \Leftrightarrow a^{\smile} \leq b^{\smile}$

(b) $(a \wedge b)^{\smile} = a^{\smile} \wedge b^{\smile}$

(c) $(\neg a)^{\smile} = \neg(a^{\smile})$ *and* $(\ulcorner a)^{\smile} = \ulcorner(a^{\smile})$

(d) $\neg\neg(a\,;b) = \neg\neg a\,;\neg\neg b$

(e) $a \leq b$ *implies* $a\,;c \leq b\,;c$ *and* $c\,;a \leq c\,;b$

(f) $\neg(a\,;b) \wedge b^{\smile} \leq \neg a$

(g) $(a\,;b) \wedge c = 0 \Leftrightarrow (a^{\smile}\,;c) \wedge b = 0 \Leftrightarrow (c\,;b^{\smile}) \wedge a = 0$

(h) $(a\,;b) \wedge c \leq a\,;\neg\neg(a^{\smile}\,;c)$

(i) $D^{\neg}$ *is closed under* ; *and* $^{\smile}$.

Proof:
Proofs of conditions (a) – (g), and (i) can be found in [49].

(h) Since $\neg(a^{\smile}\,;c) \wedge b \leq b$ and ; is monotone in both arguments by (e), we get

(1) $a\,;(\neg(a^{\smile}\,;c) \wedge b) \leq a\,;b$,

Since $(a^{\smile}\,;c) \wedge (\neg(a^{\smile}\,;c) \wedge b) = 0$, by (g) we have

$$(a\,;(\neg(a^{\smile}\,;c) \wedge b)) \wedge c = 0.$$

By the definition ($\neg$) of the pseudocomplement (see Section 9.10),

(2) $a\,;(\neg(a^{\smile}\,;c) \wedge b) \leq \neg c$.

By (1) and (2) we obtain

(3) $a\,;(\neg(a^{\smile}\,;c)\wedge b)\leq(a\,;b)\wedge(\neg c)$.

Now we join $a\,;(\neg\neg(a^{\smile}\,;c)\wedge b)$ to both sides of (3) and get

(4) $a\,;(\neg(a^{\smile}\,;c)\wedge b)\vee a\,;(\neg\neg(a^{\smile}\,;c)\wedge b)\leq(a\,;b)\wedge\neg c\vee a\,;(\neg\neg(a^{\smile}\,;c)\wedge b)$.

Since $\vee$ distributes over ; by ($\mathrm{R}^2 1$), we have

(5)
$$\begin{aligned} &a\,;(\neg(a^{\smile}\,;c)\wedge b)\vee a\,;(\neg\neg(a^{\smile}\,;c)\wedge b)\\ &=a\,;((\neg(a^{\smile}\,;c)\wedge b)\vee(\neg\neg(a^{\smile}\,;c)\wedge b))\\ &=a\,;((\neg(a^{\smile}\,;c)\vee\neg\neg(a^{\smile}\,;c))\wedge b)\\ &=a\,;b.\end{aligned}$$

By (4) and (5) we get

$$a\,;b\leq(a\,;b)\wedge(\neg c)\vee a\,;(\neg\neg(a^{\smile}\,;c)\wedge b).$$

Thus

$$(a\,;b)\wedge c\leq(a\,;(\neg\neg(a^{\smile}\,;c)\wedge b))\wedge c\leq a\,;(\neg\neg(a^{\smile}\,;c)\wedge b)\leq a\,;\neg\neg(a^{\smile}\,;c).$$

which completes the proof. □

A *rough relation frame* is a structure (X,R,f,I) where X is a regular double Stone frame satisfying conditions (FSt1), (FSt2) from Section 10.2 and (FSt3) from Section 10.3, R is a ternary relation on X, $f:X\to X$ is a function, $I\subseteq X$, and the following axioms hold, for all $x,x',y,y',z,z'\in X$,

($\mathrm{FR}^2 1$) $R(x,y,z)$ & $x'\leqslant x$ & $y'\leqslant y$ & $z\leqslant z'\Rightarrow R(x',y',z')$

($\mathrm{FR}^2 2$) $x\leqslant y\Rightarrow f(x)\leqslant f(y)$

($\mathrm{FR}^2 3$) $x\in I$ & $x\leqslant x'\Rightarrow x'\in I$ and $x'\in I$ & $x\leqslant x'\Rightarrow x\in I$

($\mathrm{FR}^2 4$) $R(x,y,z)$ & $R(z,v,w)\Rightarrow\exists u\,(R(x,u,w)$ & $R(y,v,u))$

($\mathrm{FR}^2 5$) $R(x,y,z)$ & $R(v,z,w)\Rightarrow\exists u\,(R(v,x,u)$ & $R(u,y,w))$

($\mathrm{FR}^2 6$) $f(f(x))=x$

($\mathrm{FR}^2 7$) $f(\overline{x})=\overline{f(x)}$ and $f(\underline{x})=\underline{f(x)}$ where for every $z\in X$, $\overline{z}$ is the maximal element such that $z\leqslant\overline{z}$ and $\underline{z}$ is the minimal element such that $\underline{z}\leqslant z$

($\mathrm{FR}^2 8$) $R(x,y,z)\Rightarrow R(f(x),z,\overline{y})$

($\mathrm{FR}^2 9$) $R(x,y,z)\Rightarrow R(z,f(y),\overline{x})$

($\mathrm{FR}^2 10$) $R(x,y,z)\Rightarrow R(\underline{x},\underline{y},\underline{z})$

($\mathrm{FR}^2 11$) $x\leqslant y\Leftrightarrow\exists z,\ z\in I$ & $R(x,z,y)$

($\mathrm{FR}^2 12$) $x\leqslant y\Leftrightarrow\exists z,\ z\in I$ & $R(z,x,y)$.

Lemma 10.4.2 *Let X be a rough relation frame. For all $x,y,z\in X$,*

$$R(x,y,z)\Rightarrow R(\overline{x},\overline{y},\overline{z}).$$

Proof:
Assume that $R(x,y,z)$ holds. Then by (FR^{2}8) we get $R(f(x),z,\overline{y})$. Applying (FR^{2}8) again and then (FR^{2}6) we obtain $R(x,\overline{y},\overline{z})$. Now by (FR^{2}9) we get $R(\overline{z},f(\overline{y}),\overline{x})$ which yields $R(\overline{x},\overline{y},\overline{z})$ by the application of (FR^{2}9) and (FR^{2}6). □

The *complex algebra of a rough relation frame* X is the structure

$$(L_X, ;_R, ^{\smile_f}, 1'_I)$$

where L_X is the complex algebra of the regular double Stone frame reduct of X (see Section 10.3) and for all $A, B \in L_X$,

$$\begin{aligned}
A ;_R B &\stackrel{\text{def}}{=} \{z \in X : \exists x \in A\ \exists y \in B,\ R(x,y,z)\} \\
A^{\smile_f} &\stackrel{\text{def}}{=} \{f(x) : x \in A\} \\
1'_I &\stackrel{\text{def}}{=} I.
\end{aligned}$$

Lemma 10.4.3 *Let X be a rough relation frame. For every $A \in L_X$,*

$$\neg_{\leqslant}\neg_{\leqslant} A = \langle\leqslant\rangle A.$$

Proof:
Assume $x \in \neg_{\leqslant}\neg_{\leqslant} A = [\leqslant]\langle\leqslant\rangle A$. Since $\leqslant$ is reflexive, $[\leqslant]\langle\leqslant\rangle A \subseteq \langle\leqslant\rangle A$. Now let $x \in \langle\leqslant\rangle A$. Thus there is $y \in X$ such that $x \leqslant y$ and $y \in A$. Take $z \in X$ such that $x \leqslant z$. Since y and z have a common lower bound x, by (FSt1) they have a common upper bound, say u. Thus $y \leqslant u$ and $z \leqslant u$. Since $y \in A$, $u \in A$. From $z \leqslant u$ and $u \in A$ we get $z \in \langle\leqslant\rangle A$. Hence $x \in [\leqslant]\langle\leqslant\rangle A$, as required. □

Lemma 10.4.4 *Let X be a rough relation frame.*

(a) *L_X is closed under the operations $;_R$ and $^{\smile_f}$;*

(b) *$1'_I \in L_X$.*

Proof:
(a) For the operation $;_R$, take z, z' such that $z \in A ;_R B$ and $z \leqslant z'$. Then there are $x \in A$ and $y \in B$ such that $R(x,y,z)$. From (FR^{2}1) we obtain $R(x,y,z')$ and hence $z' \in A ;_R B$.
For $^{\smile_f}$, let $z \in A^{\smile_f}$ which means that there is some $y \in A$ such that $z = f(y)$. Let $z \leqslant z'$. Then $f(z) \leqslant f(z')$ by (FR^{2}2). By (FR^{2}6) $f(z) = y$. Since $A \in L_X$, $f(z') \in A$ and hence $z' \in A^{\smile_f}$.

(b) By (FR^{2}3) it is clear that $1'_I \in L_X$. □

Proposition 10.4.5 *The complex algebra of a rough relation frame is a rough relation algebra.*

Proof:
In Section 10.2 and Section 10.3 it is shown that L_X is a regular double Stone algebra. We show that $(L_X, ;_R, 1'_I)$ is a monoid.

Using (FR^{2}4) and (FR^{2}5) and proceeding as in the proof of Proposition 5.2.3. one can show associativity of $;_R$.

We show that $1'_I$ is the unit element, that is $A ;_R 1'_I = A = 1'_I ;_R A$ for every $A \in L_X$. For the first equality, take $z \in A ;_R I$. Then there are $x \in A$ and $y \in I$ such that $R(x, y, z)$. By the right-to left direction of (FR^{2}11), $x \leqslant z$. Since A is increasing, $z \in A$. Hence $A ;_R 1'_I \subseteq A$. On the other hand, let $z \in A$. Since $z \leqslant z$, by the left-to-right direction of (FR^{2}11) there is some $y \in I$ such that $R(z, y, z)$ which implies $z \in A ;_R 1'_I$.
For the second equality, let $z \in I ;_R A$. Then there are $x \in I$ and $y \in A$ such that $R(x, y, z)$. By the right-to-left direction of (FR^{2}12) we obtain $y \leqslant z$. Since A is increasing, $z \in A$, hence $1'_I ;_R A \subseteq A$. For the reverse inclusion, let $z \in A$. Since $z \leqslant z$, by the left-to-right direction of (FR^{2}12) there is some $x \in I$ such that $R(x, z, z)$. This implies $z \in 1'_I ;_R A$.

Now, it is sufficient to show that the axioms (R^{2}1) – (R^{2}7) are satisfied in L_X.

The proofs of (R^{2}1) – (R^{2}4) are the same as for relation algebras and can be found in Section 5.2.

(R^{2}5) Let $A, B \in L_X$. We show $A^{\smile_f} ;_R \neg_{\leqslant}(A ;_R B) \subseteq \neg_{\leqslant} B$. Assume there are $x \in A$ and $y \in \neg_{\leqslant}(A ;_R B)$ such that $R(f(x), y, z)$. Let $\overline{z}$ be the maximal element such that $z \leqslant \overline{z}$. By (FR^{2}1) we get $R(f(x), y, \overline{z})$. By (FR^{2}8) $R(f(f(x)), \overline{z}, \overline{y})$ where $\overline{y}$ is the maximal element such that $y \leqslant \overline{y}$. By (FR^{2}6) $R(x, \overline{z}, \overline{y})$. Suppose $z \notin \neg_{\leqslant} B = [\leqslant](-B)$. Thus there is some z' such that $z \leqslant z'$ and $z' \in B$. Necessarily, $z' \leqslant \overline{z}$, so $\overline{z} \in B$. Thus we have $x \in A$ and $\overline{z} \in B$ such that $R(x, \overline{z}, \overline{y})$ which yields $\overline{y} \in A ;_R B$. Since $y \in \neg_{\leqslant}(A ;_R B)$ and $\neg_{\leqslant}(A ;_R B)$ is up-closed, $\overline{y} \in \neg_{\leqslant}(A ;_R B)$, a contradiction.

(R^{2}6) Take $A, B \in L_X$ and show that $\neg_{\leqslant}\neg_{\leqslant}(\neg_{\leqslant} A ;_R \neg_{\leqslant} B) = \neg_{\leqslant} A ;_R \neg_{\leqslant} B$. Let $z \in \neg_{\leqslant}\neg_{\leqslant}(\neg_{\leqslant} A ;_R \neg_{\leqslant} B)$. By Lemma 10.4.3 $z \in \langle\leqslant\rangle(\neg_{\leqslant} A ;_R \neg_{\leqslant} B)$. Thus there is some $t \in X$ such that $z \leqslant t$ and there are $x \in [\leqslant](-A)$ and $y \in [\leqslant](-B)$ such that $R(x, y, t)$. It follows that $x \notin A$, $y \notin B$ and by (FR^{2}10), $R(\underline{x}, \underline{y}, \underline{t})$. Since $\underline{t} \leqslant z$, $R(\underline{x}, \underline{y}, z)$ by (FR^{2}1). Suppose $\underline{x} \notin [\leqslant](-A)$. Then there is some $x' \in X$ such that $\underline{x} \leqslant x'$ and $x' \in A$. But by (FSt3) $x' = x$ and hence $x' \notin A$, a contradiction. Similarly, $\underline{y} \in [\leqslant](-B)$. Hence $z \in \neg_{\leqslant} A ;_R \neg_{\leqslant} B$.
The reverse inclusion is immediate from reflexivity of $\leqslant$.

(R^{2}7) We show $\neg_{\leqslant}\neg_{\leqslant} 1'_I = 1'_I$. Let $x \in [\leqslant]\langle\leqslant\rangle I$. By reflexivity of $\leqslant$ and (FR^{2}3) we have $[\leqslant]\langle\leqslant\rangle I \subseteq \langle\leqslant\rangle I \subseteq I$. For the other inclusion, since $I \subseteq \langle\leqslant\rangle I$ by reflexivity of $\leqslant$, $[\leqslant] I \subseteq [\leqslant]\langle\leqslant\rangle I$. Since by (FR^{2}3) we have $I \subseteq [\leqslant] I$, the required inclusion follows. □

The *canonical frame of a rough relation algebra* L is a structure

$$(X_L, R_;, f_\smile, I_{1'})$$

where X_L is the canonical frame of the regular double Stone algebra reduct of L and, for all $F, G, H \in X_L$,

$$R_;(F,G,H) \stackrel{\text{def}}{\Leftrightarrow} F\,;G \subseteq H$$
$$f_\smile(F) \stackrel{\text{def}}{=} F^\smile$$
$$I_{1'} \stackrel{\text{def}}{=} \{F \in X_L : 1' \in F\}$$

where, for all $A, B \subseteq L$,

$$A\,;B = \{c \in L : \exists a \in A\ \exists b \in B,\ a\,;b \leq c\}$$
$$A^\smile = \{a^\smile : a \in A\}.$$

Proposition 10.4.6 *The canonical frame of a rough relation algebra is a rough relation frame.*

Proof:
It is shown in Section 10.2 and Section 10.3 that X_L is a regular double Stone frame. We show that X_L satisfies axioms (FR^{2}1) – (FR^{2}12).

Axioms (FR^{2}1), (FR^{2}2), (FR^{2}3), and (FR^{2}6) are immediate from the definitions and Lemma 10.4.1. Axioms (FR^{2}4) and (FR^{2}5) can be proved as in Section 5.2.

(FR^{2}7) Let $F \in X_L$. We show that $(\overline{F})^\smile = \overline{F^\smile}$. Let $a \in (\overline{F})^\smile$. Then $a^\smile \in \overline{F}$. Since $\overline{F}$ is a filter generated by $F \cup D^\neg$ by Lemma 10.2.8(b), there are $b \in F$ and $c \in D^\neg$ such that $b \wedge c \leq a^\smile$. Since $D^\neg$ is closed under $^\smile$ by Lemma 10.4.1(i), $c^\smile \in D^\neg$. Also, $b^\smile \in F^\smile$. Thus $b^\smile \wedge c^\smile \in F^\smile$, $b^\smile \wedge c^\smile \leq a$, and hence $a \in \overline{F^\smile}$. Conversely, let $a \in \overline{F^\smile}$. There there are $b \in F^\smile$ and $c \in D^\neg$ such that $b \wedge c \leq a$, so that $b^\smile \wedge c^\smile \leq a^\smile$. Since $b^\smile \in F$ and $c^\smile \in D^\neg$, we have $a^\smile \in \overline{F}$ and therefore $a \in (\overline{F})^\smile$.
Now we show that $(\underline{F})^\smile = \underline{F^\smile}$. Let $a \in (\underline{F})^\smile$. Then $a^\smile \in \underline{F}$. Since $\underline{F}$ is a filter generated by $F \cap B(L)$ by Lemma 10.2.8(a), there is some $b \in F \cap B(L)$ such that $b \leq a^\smile$. Thus $b^\smile \in F^\smile \cap B(L)$. Since $b^\smile \leq a$, $a \in \underline{F^\smile}$. Conversely, let $a \in \underline{F^\smile}$. Then there is some $b \in F^\smile \cap B(L)$ such that $b \leq a$, so that $b^\smile \leq a^\smile$. Since $b^\smile \in F \cap B(L)$, $a^\smile \in \underline{F}$, and hence $a \in (\underline{F})^\smile$.

(FR^{2}8) Let $F, G, H \in X_L$. We show $R_;(F,G,H) \Rightarrow R_;(f_\smile(F), H, \overline{G})$. Assume $F\,;G \subseteq H$ and let $c \in F^\smile\,;H$. Then there is some $a \in F$ and there is some $b \in H$ such that $a^\smile\,;b \leq c$. Suppose $c \notin \overline{G}$ which implies $a^\smile\,;b \notin \overline{G}$. Since $\overline{G}$ is maximal, $\neg(a^\smile\,;b) \in \overline{G}$. Since $\overline{G} \cap B(L) = G \cap B(L)$ and $\neg(a^\smile\,;b) \in B(L)$, $\neg(a^\smile\,;b) \in G$. Now, $a\,;\neg(a^\smile\,;b) \in F\,;G \subseteq H$. By (R^{2}6), $\neg b \in H$, a contradiction.

(FR^{2}9) We show $R_;(F,G,H) \Rightarrow R_;(H, f_\smile(G), \overline{F})$. Assume $F\,;G \subseteq H$ and let $c \in H\,;G^\smile$. Then there are $a \in H$ and $b \in G$ such that $a\,;b^\smile \leq c$. Suppose $c \notin \overline{F}$. Then $a\,;b^\smile \notin \overline{F}$. Since $\overline{F}$ is maximal, $\neg(a\,;b^\smile) \in \overline{F}$. Hence, since $\neg(a\,;b^\smile) \in B(L)$, $\neg(a\,;b^\smile) \in F$. Then $\neg(a\,;b^\smile)\,;b \in F\,;G \subseteq H$. By Lemma 10.4.1(f), $\neg a \in H$, a contradiction.

(FR^{2}10) We show $R_;(F,G,H) \Rightarrow R_;(\underline{F}, \underline{G}, \underline{H})$. Assume $F\,;G \subseteq H$ and let $c \in \underline{F}\,;\underline{G}$. Then there are $a \in \underline{F}$ and $b \in \underline{G}$ such that $a\,;b \leq c$. Since $\underline{F}$ and $\underline{G}$

are minimal, by Lemma 10.2.8(a) we may assume that $a, b \in B(L)$. By Lemma 10.4.1(d), $a\,;b \in B(L)$. Since $F\,;G \subseteq H$, $a\,;b \in B(L) \cap H$. Now it follows from Lemma 10.2.8(a) that $a\,;b \in \underline{H}$. Since $a\,;b \leq c$, $c \in \underline{H}$, as required.

(FR^{2}11) We show $F \subseteq G \Rightarrow \exists H,\ H \in I_{1'}\ \&\ R_;(F, H, G)$.
Assume $F \subseteq G$ and let $\uparrow_{\leq} 1'$ be the principal filter of L generated by $1'$. Then $F\,;\uparrow_{\leq} 1' \subseteq G$ and by Theorem 1.6.11 there is a prime filter H such that $1' \in H$ and $F\,;G \subseteq H$.

The proof of (FR^{2}12) is similar. □

Let L be a rough relation algebra. We now show that the lattice embedding $h : L \to L_{X_L}$, used in Theorem 2.5.5, preserves ;, $\smile$ and the constant $1'$.

Lemma 10.4.7 *For any $a, b \in L$,*

(a) $h(a;b) = h(a)\,;_{R_;}\,h(b)$.
(b) $h(a^{\smile}) = (h(a))^{\smile_{f_\smile}}$.
(c) $h(1') = 1'_{h(1')}$.

Proof:
(a) Assume $H \in h(a;b)$, that is $a;b \in H$. Consider the principal filters $\uparrow_{\leq} a$ and $\uparrow_{\leq} b$ generated by a and b, respectively. Then $\uparrow_{\leq} a\,;\uparrow_{\leq} b \subseteq H$. By Theorem 1.6.11 there are prime filters F and G such that $\uparrow_{\leq} a \subseteq F$, $\uparrow_{\leq} b \subseteq G$, and $F\,;G \subseteq H$. Clearly, $a \in F$ and $b \in G$, so $H \in h(a)\,;_{R_;}\,h(b)$.
Now assume $F, G \in X_L$, $a \in F$, $b \in G$, and $F\,;G \subseteq H$. Then $a\,;b \in H$ and hence $H \in h(a\,;b)$.

(b) Note that

$$\begin{aligned} H \in h(a^{\smile}) &\Leftrightarrow a^{\smile} \in H \\ &\Leftrightarrow a^{\smile\smile} \in f_\smile(H) \\ &\Leftrightarrow a \in f_\smile(H) \\ &\Leftrightarrow f_\smile(f_\smile(H)) \in (h(a))^{\smile_{f_\smile}} \\ &\Leftrightarrow H \in (h(a))^{\smile_{f_\smile}}. \end{aligned}$$

(c) Observe that $H \in h(1') \Leftrightarrow 1' \in H \Leftrightarrow H \in I_{1'} \Leftrightarrow H \in 1'_{h(1')}$. □

Let X be a rough relation frame. We now show that the embedding $k : X \to X_{L_X}$, used in Theorem 2.5.5, preserves R, f, and the set I.

Lemma 10.4.8 *For any $x, y, z \in X$,*

(a) $R(x, y, z) \Leftrightarrow R_{;_R}(k(x), k(y), k(z))$;
(b) $h(f(x)) = f_{\smile_f}(h(x))$;
(c) $I(x) = I_{1'_I}(h(x))$

Proof:
(a) Note, for any $x, y, z \in X$, that

$$\begin{aligned}
&R_{;_R}(k(x), k(y), k(z))\\
&\quad \Leftrightarrow k(x) \,;_R\, k(y) \subseteq k(z)\\
&\quad \Leftrightarrow \{C \in L_X : \exists A \in k(x)\, \exists B \in k(y),\ A \,;_R\, B \subseteq C\} \subseteq k(z)\\
&\quad \Leftrightarrow \forall A, B \subseteq X,\ A \in k(x)\ \&\ B \in k(y) \Rightarrow A \,;_R\, B \in k(z)\\
&\qquad \text{since } k(z) \text{ is a filter of } L\\
&\quad \Leftrightarrow \forall A, B \subseteq X,\ x \in A\ \&\ y \in B \Rightarrow z \in \{p \in X : \exists u \in A\, \exists t \in B,\ R(u, t, p)\}.
\end{aligned}$$

Assume $R(x, y, z)$, $x \in A$, and $y \in B$. Then $z \in A \,;_R\, B$. Conversely, assume $R_{;_R}(k(x), k(y), k(z))$ and take $\uparrow_{\leqslant} x$ and $\uparrow_{\leqslant} y$ for A and B, respectively.

(1) If $x = \underline{x}$ and $y = \underline{y}$, then $A = \{\underline{x}, \overline{x}\}$ and $B = \{\underline{y}, \overline{y}\}$. Take $u \in A$ and $t \in B$ such that $R(u, t, z)$. By (FR^{2}1) $R(\underline{u}, \underline{t}, \underline{z})$, and hence $R(x, y, z)$.

(2) If $x = \overline{x}$ and $y = \overline{y}$, then $A = \{x\}$ and $B = \{y\}$, and therefore $R(x, y, z)$.

(3) If $x = \underline{x}$ and $y = \overline{y}$, then $A = \{\underline{x}, \overline{x}\}$ and $B = \{y\}$. Let $u, t \in X$ be such that $u \in A$, $t \in B$, and $R(u, t, z)$. Then $t = y$. By (FR^{2}1), $R(\underline{u}, y, z)$ and hence $R(x, y, z)$.

(4) If $x = \overline{x}$ and $y = \underline{y}$, then the reasoning analogous to that in case (3) leads to $R(x, y, z)$.

(b) We have, for every $A \in L_X$,

$$A \in h(f(x)) \text{ iff } f(x) \in A \text{ iff } x \in A^{\smile_f} \text{ iff } A^{\smile_f} \in h(x) \text{ iff } A \in f_{\smile_f}(h(x))$$

(c) Note that $x \in I$ iff $I \in h(x)$ iff $1'_I \in h(x)$ iff $h(x) \in I_{1'_I}$. □

We conclude with the discrete representations for rough relation algebras and rough relation frames.

Theorem 10.4.9

(a) *Every rough relation algebra is embeddable into the complex algebra of its canonical frame;*

(b) *Every rough relation frame is embeddable into the canonical frame of its complex algebra.*

Chapter 11

Heyting algebras

11.1 Introduction

In this chapter we apply the methodology of discrete duality to Heyting algebras with some additional operators. In Section 11.2 we recall a classical duality for plain Heyting algebras restricting it to a discrete duality. The frames corresponding to Heyting algebras are the usual frames of the intuitionistic logic. In Sections 11.3 and 11.4 we deal with Heyting algebras with modal operators. In the literature there are many papers presenting various kinds of modalities on Heyting algebras, see, for example, [9, 10, 83, 84]. In most cases the authors consider monadic Heyting algebras, where the added modality is a counterpart to the S5 modal operator. We start in Section 11.3 with the weakest structures, namely with Heyting algebras with not necessarily normal operators of possibility and necessity which are independent of each other, as introduced in [225]. Consequently, in the corresponding class of frames there are two independent relations which enable us to define the modal operators in a complex algebra. Then in Section 11.4 we consider some of axiomatic extensions of these algebras and frames. We focus on the axioms which enable us to have the same frame relation for the two modalities of possibility $\Diamond$ and necessity $\Box$. For distributive lattices with these operators this problem was solved in [48]. The intuitionistic logic endowed with the operators $\Diamond$ and $\Box$ was introduced in [82] and further studied in [80, 187, 6, 198, 220]. The semantics of this logic is based on the Heyting frames endowed with a single relation which is adequate for both modalities due to specific axioms of the logic. In Section 11.5 we deal with the necessity-like operator on a Heyting algebra introduced in [77]. In Section 11.6 we consider symmetric Heyting algebras with a family of unary Post-like operators as presented in [133], see also [222, 223, 134]. In Section 11.7, we present a discrete duality for Heyting-Brouwer algebras introduced in [209]. While the Heyting implication is the residuum of meet, in Heyting-Brouwer algebras an operator of difference is added which is the residuum of join. Consequently, in these algebras two negations can be naturally defined – a pseudocomplement defined in terms of the implication and a Brouwer negation defined in terms of the difference. Finally, in Section 11.8 we present apartness algebras and apartness frames introduced in [56] as the abstract counterparts to apartness spaces. Apartness spaces were introduced in [18] as a companion to proximity spaces, see Section 7.2. Apartness algebras are Heyting algebras with a sufficiency operator satisfying some specific axioms such that the operator captures the properties of the apartness complement discussed in [212, 247]. Since apartness spaces can be axiomatized purely from the apartness complement, the two notions of apartness algebra and apartness frame capture in an axiomatic way all the relevant properties of the apartness spaces and, moreover, they provide a clear division between the properties whose nature and origin are relational or algebraic. Then the discrete duality between the apartness algebras and the apartness frames clarifies how these two kinds of properties are related.

11.2 Heyting algebras

A *Heyting algebra*, *H-algebra*, is a relatively pseudocomplemented lattice

$$(L, \to, \neg, 1)$$

(see Section 9.2) with the negation referred to as a *pseudocomplement* which is defined as being a contrapositive negation and a seminegation, that is, for any $a, b \in L$,

(H1) $a \to \neg b = b \to \neg a$

(H2) $\neg(a \to a) \to b = 1$.

It follows that every Heyting algebra has the smallest element $0 = \neg 1$. Therefore we include 0 in the signature and write $(L \to, \neg, 0, 1)$ for a Heyting algebra.

This definition of the pseudocomplement [206] is equivalent to its traditional definitions

(H3) $\neg a = a \to 0$

(H4) $a \wedge b = 0 \Leftrightarrow a \leq \neg b$.

Lemma 11.2.1 *Let* $(L, \to, 1)$ *be an RP-lattice and let* $\neg$ *be a unary operator on* L. *Then the following are equivalent*

(a) *(H1) and (H2) hold;*

(b) *L has the smallest element* 0 *and (H3) holds in L;*

(c) *L has the smallest element* 0 *and (H4) holds in L.*

Proof:
By Lemma 9.3.1 and Lemma 9.4.1 (a) is equivalent to (b). Assume that (b) holds. Since $\to$ is the residuum of $\wedge$, if $a \wedge b = 0$, then $a \leq b \to 0 = \neg b$. Conversely, if $a \leq b \to 0$, then $a \wedge b \leq b \wedge (b \to 0)$. By Lemma 9.2.1(e) we get $b \wedge (b \to 0) = 0$ and hence (H4) holds.
Assume that (c) holds. Since $\wedge$ is commutative, we have $a \wedge b = 0 \Leftrightarrow b \leq \neg a$. Therefore $a \leq \neg b \Leftrightarrow b \leq \neg a$ which is equivalent to (H1). Now, for $a = \neg 1$ and $b = 1$ we get $\neg 1 \wedge 1 = 0$, that is $\neg 1 = 0$ which is equivalent to (H2) by Lemma 9.3.1. □

A *Heyting frame*, *H-frame*, is the same as a frame of the relatively pseudocomplemented lattice, that is $(X, \leqslant)$. Therefore the *canonical frame of an H-algebra* is the same as in Section 9.2.

The *complex algebra of an H-frame* is defined as for the frames of relatively pseudocomplemented lattices with the additional condition $\neg_{\leqslant} A = [\leqslant](-A)$ which implies $\neg_{\leqslant}\emptyset = X$.

Due to Proposition 9.2.2 and Proposition 9.2.3 we have the following facts.

Proposition 11.2.2 *The complex algebra of any H-frame is an H-algebra.*

Proposition 11.2.3 *The canonical frame of any H-algebra is an H-frame.*

The lattice embedding $h : L \to L_{X_L}$, used in Section 2.5, is an embedding of relatively pseudocomplemented lattices as shown in Section 9.2. Since the pseudocomplement in a Heyting algebra is definable in terms of relative pseudocomplement, it is clear that h preserves $\neg$.

The mapping $k : X \to X_{L_X}$ providing a representation of H-frames is the same as in Section 2.5. It is shown to be injective and order-preserving.

The developments of this section yield the discrete representations for H-algebras and H-frames.

Theorem 11.2.4

(a) *Every H-algebra is embeddable into the complex algebra of its canonical frame;*

(b) *Every H-frame is embeddable into the canonical frame of its complex algebra.*

Furthermore, we have a duality via truth for H-algebras. The formulas of the language $\mathcal{L}an_H$ of Heyting intuitionistic logic are built from propositional variables of a denumerable set *Var* and propositional constants 0 and 1 with the operations $\vee$, $\wedge$, $\to$, and $\neg$.

Let $\mathcal{A}lg_H$ be the class of Heyting algebras $(L, \vee, \wedge, \to, \neg, 0, 1)$. A valuation $v : Var \to L$ of propositional variables in an algebra L extends in the usual homomorphic way to all the formulas:

$$v(1) = 1, \quad v(0) = 0$$
$$v(\alpha \vee \beta) = v(\alpha) \vee v(\beta)$$
$$v(\alpha \wedge \beta) = v(\alpha) \wedge v(\beta)$$
$$v(\alpha \to \beta) = v(\alpha) \to v(\beta)$$
$$v(\neg\alpha) = \neg v(\alpha).$$

A formula α is true in a Heyting algebra L if and only if $v(\alpha) = 1$ for every v in L and α is $\mathcal{A}lg_H$-valid if and only if it is true in every algebra of $\mathcal{A}lg_H$.

Let $\mathcal{F}rm_H$ be the class of H-frames. A model based on a frame $(X, \leqslant)$ is a structure $\mathcal{M} = (X, \leqslant, m)$ where $m : Var \to \{A \subseteq X : A = [\leqslant]A\}$ is a meaning function. It is easy to see that m satisfies what is called a heredity condition in the intuitionistic logic, that is, $m(p) = [\leqslant]m(p)$. The satisfaction relation is defined for every $x \in X$:

$$\begin{aligned}
\mathcal{M}, x \models p &\overset{\text{def}}{\Leftrightarrow} x \in m(p) \text{ for every } p \in \mathit{Var} \\
\mathcal{M}, x \models \alpha \vee \beta &\overset{\text{def}}{\Leftrightarrow} M, x \models \alpha \text{ or } M, x \models \beta \\
\mathcal{M}, x \models \alpha \wedge \beta &\overset{\text{def}}{\Leftrightarrow} M, x \models \alpha \ \& \ M, x \models \beta \\
\mathcal{M}, x \models \alpha \rightarrow \beta &\overset{\text{def}}{\Leftrightarrow} \forall y \in X,\ x \leqslant y \ \& \ M, x \models \alpha \Rightarrow M, y \models \beta \\
\mathcal{M}, x \models \neg \alpha &\overset{\text{def}}{\Leftrightarrow} \forall y \in X,\ x \leqslant y \Rightarrow M, y \not\models \alpha.
\end{aligned}$$

A formula α is true in a model $\mathcal{M}$ if and only if $\mathcal{M}, x \models \alpha$ for every $x \in X$ and α is true in a frame X if and only if it is true in every model based on X. A formula α is $\mathcal{F}rm_H$-valid if and only if it is true in every frame $X \in \mathcal{F}rm_H$.

The meaning function m can be extended to all formulas by setting

$$m(\alpha) = \{x \in X : \mathcal{M}, x \models \alpha\}.$$

Then

$$\begin{aligned}
m(\alpha \vee \beta) &= m(\alpha) \cup m(\beta) \\
m(\alpha \wedge \beta) &= m(\alpha) \cap m(\beta) \\
m(\alpha \rightarrow \beta) &= [\leqslant](-m(\alpha) \cup m(\beta)) \\
m(\neg \alpha) &= [\leqslant](-m(\alpha)).
\end{aligned}$$

It is easy to verify that $m(\alpha) = [\leqslant]m(\alpha)$ for every formula α which results in the complex algebra theorem.

Theorem 11.2.5 *For every formula $\alpha \in \mathcal{L}an_H$ and for every frame $X \in \mathcal{F}rm_H$, α is true in X if and only if α is true in the complex algebra L_X.*

Theorem 11.2.6 Duality via truth *A formula α is true in every algebra of $\mathcal{A}lg_H$ if and only if it is true in every frame of $\mathcal{F}rm_H$.*

11.3 Heyting algebras with modal operators

A *Heyting algebra with modal operators, HM-algebra*, is a structure $(L, \Box, \Diamond)$ where L is a Heyting algebra and the unary operators $\Box$ and $\Diamond$ satisfy, for any $a, b \in L$,

(HM1) $\Box a \wedge \Box b \leq \Box(a \wedge b)$

(HM2) $\Diamond(a \vee b) \leq \Diamond a \vee \Diamond b$

(HM3) if $a \leq b$, then $\Box a \leq \Box b$

(HM4) if $a \leq b$, then $\Diamond a \leq \Diamond b$.

Note that (HM3) and (HM4) may be replaced, respectively, by the equivalent conditions

(HM3') $\Box(a\wedge b)\leq\Box a\wedge\Box b$

(HM4') $\Diamond a\vee\Diamond b\leq\Diamond(a\vee b)$.

A *co-filter* of an HM-algebra L is a non–empty subset $F\subseteq L$ such that

- $F\neq L$
- if $a\in F$ and $a\leq b$, then $b\in F$
- if $a\vee b\in F$, then $a\in F$ or $b\in F$.

Lemma 11.3.1 *Let L be a HM-algebra. Then*

(a) *If F is a filter of L and $\Box 1\in F$, then $\Box^{-1}(F)$ is a filter of L.*

(b) *If F is a co-filter of L and $\Diamond 0\notin F$, then $\Diamond^{-1}(F)$ is a co-filter of L*

(c) *F is a prime filter if and only if F is a filter and a co-filter.*

Proof:
By way of example we show (a) and (b).

(a) Since $\Box 1\in F$, $\Box^{-1}(F)\neq\emptyset$. Let $a,b\in L$ be such that $a\in\Box^{-1}(F)$ and $a\leq b$. Then $\Box a\in F$. Also, by (HM3), $\Box a\leq\Box b$. Since F is a filter, we have $\Box b\in F$, that is $b\in\Box^{-1}(F)$. Next, let $a,b\in\Box^{-1}(F)$. Then $\Box a\in F$ and $\Box b\in F$, so $\Box a\wedge\Box b\in F$, and hence, by (HM1), $\Box(a\wedge b)\in F$, that is, $a\wedge b\in\Box^{-1}(F)$.

(b) Let F be a co-filter of L. Since $\Diamond 0\notin F$, $0\notin\Diamond^{-1}(F)$ and hence $\Diamond^{-1}(F)\neq L$. First, take $a,b\in L$ such that $a\in\Diamond^{-1}(F)$ and $a\leq b$. Then $\Diamond a\in F$ and by (HM4), $\Diamond a\leq\Diamond b$, so $\Diamond b\in F$, that is $b\in\Diamond^{-1}(F)$. Now, let $a,b\in L$ be such that $a\vee b\in\Diamond^{-1}(F)$, that is $\Diamond(a\vee b)\in F$. By (HM2), $\Diamond(a\vee b)\leq\Diamond a\vee\Diamond b$. Since F is a co-filter of L, $\Diamond a\vee\Diamond b\in F$, and furthermore, $\Diamond a\in F$ or $\Diamond b\in F$. Whence $a\in\Diamond^{-1}(F)$ or $b\in\Diamond^{-1}(F)$. □

The following lemma is adapted from [225].

Lemma 11.3.2 *If F is a filter of a Heyting algebra, G is a co-filter of a Heyting algebra, and $F\subseteq G$, then there is a prime filter H such that $F\subseteq H\subseteq G$.*

Proof:
Consider the set $\{Z : Z \text{ is a filter}, F\subseteq Z\subseteq G\}$. By Zorn's lemma there is a maximal element in this set, say H. Clearly, $H\subseteq G$. Let $a\vee b\in H$ and suppose that $a\notin H$ and $b\notin H$. Then consider the filters H_a and H_b generated by $H\cup\{a\}$ and $H\cup\{b\}$, respectively. They are proper filters. At least one of H_a and H_b is included in G. For suppose otherwise, then there are $c\in H_a$ and $d\in H_b$ such that $c\notin G$ and $d\notin G$. Then $c\vee d\notin G$ since G is a co-filter. Since $a\leq c$ and $b\leq d$, we have $a\vee b\leq c\vee d$. Hence $a\vee b\notin H$, a contradiction. □

An *HM-frame* is a structure (X,M,N,R,S) where X is an H-frame together with designated subsets $M,N\subseteq X$ and binary relations $R,S\subseteq X\times X$ satisfying the following conditions

(FHM1) $M=[\leqslant]M$

(FHM2) $N = [\geqslant]N$
(FHM3) $(\leqslant; R; \leqslant) \subseteq R$
(FHM4) $(\geqslant; S; \geqslant) \subseteq S$
(FHM5) $x \notin M \Rightarrow R(x) = X$
(FHM6) $x \notin N \Rightarrow S(x) = X$.

The elements of sets M and N are intuitively interpreted as the normal states.

The *complex algebra of an HM-frame* X is a structure $(L_X, \Box_R, \Diamond_S)$ where L_X is the complex algebra of its H-frame reduct and for every $A \in L_X$,

$$\Box_R(A) \stackrel{\text{def}}{=} M \cap [R]A$$

$$\Diamond_S(A) \stackrel{\text{def}}{=} (X–N) \cup \langle S\rangle A.$$

Proposition 11.3.3 *The complex algebra of an HM-frame is an HM-algebra.*

Proof:
We need to to show that L_X is closed under the operations $\Box_R$ and $\Diamond_S$, that is $\Box_R(A) = [\leqslant]\Box_R(A)$ and $\Diamond_S(A) = [\leqslant]\Diamond_S(A)$. For both cases the $\supseteq$ inclusion follows from reflexivity of $\leqslant$.
Assume $x \in \Box_R(A)$, that is $x \in M \cap [R]A$. Take any $y \in X$ such that $x \leqslant y$ and take any $z \in X$ such that yRz. By (FHM1) $y \in M$ and by reflexivity of $\leqslant$ and (FHM3) we get xRz. Then by the assumption $z \in A$. Thus $x \in [\leqslant]\Box_R(A)$.
Now assume $x \in \Diamond_S(A)$, that is if $x \in N$, then $x \in \langle S\rangle A$. Take any $y \in X$ such that $x \leqslant y$ and $y \in N$. Then by (FHM2), $x \in [\geqslant]N = N$. By the assumption $x \in \langle S\rangle A$, that is for some $z \in X$, xSz and $z \in A$. Since $y \geqslant x$ and xSz, by (FHM4) we have ySz which together with $z \in A$ yields $y \in \langle S\rangle A$. It follows that $x \in [\leqslant]\Diamond_S(A)$.

(HM1) Note that

$$\Box_R(A) \cap \Box_R(B) = M \cap [R]A \cap M \cap [R]B = M \cap [R](A \cap B) = \Box_R(A \cap B).$$

(HM2) Observe that

$$\Diamond_S(A \cup B) = -N \cup \langle S\rangle(A \cup B) = -N \cup \langle S\rangle A \cup \langle S\rangle B = \Diamond_S(A) \cup \Diamond_S(B).$$

(HM3) If $A \subseteq B$, then $[R]A \subseteq [R]B$ and hence $M \cap [R]A \subseteq M \cap [R]B$, that is $\Box_R(A) \subseteq \Box_R(B)$.

(HM4) Similarly, if $A \subseteq B$, then $\langle S\rangle \subseteq \langle S\rangle B$ for all $S \subseteq X \times X$ and for all $A, B \subseteq X$. Thus $-N \cup \langle S\rangle A \subseteq -N \cup \langle S\rangle B$, that is $\Diamond_S(A) \subseteq \Diamond_S(B)$. $\Box$

The *canonical frame of an HM-algebra* L is a structure $(X_L, M_\Box, N_\Diamond, R_\Box, S_\Diamond)$ such that X_L is the canonical frame of its H-algebra reduct and

$$M_\Box \stackrel{\text{def}}{=} \{F \in X_L : \exists a \in L,\ \Box a \in F\}$$

$$N_\Diamond \stackrel{\text{def}}{=} \{F \in X_L : \exists a \in L,\ \Diamond a \notin F\}$$

$$F\, R_\Box\, G \stackrel{\text{def}}{\Leftrightarrow} \Box^{-1}(F) \subseteq G \text{ for any } F, G \in X_L$$

$$F\, S_\Diamond\, G \stackrel{\text{def}}{\Leftrightarrow} G \subseteq \Diamond^{-1}(F) \text{ for any } F, G \in X_L$$

where for any $A \subseteq L$, $\Box^{-1}(A) = \{a \in L : \Box a \in A\}$ and $\Diamond^{-1}(A) = \{a \in L : \Diamond a \in A\}$.

Proposition 11.3.4 *The canonical frame of an HM-algebra is an HM-frame.*

Proof:
First, we show (FHM1), that is $M_\Box = [\subseteq]M_\Box$. The right-to-left inclusion follows from reflexivity of $\supseteq$. Let us take any $F, G \in X_L$ such that $F \in M_\Box$ and $F \subseteq G$. Then there exists $a \in L$ such that $\Box a \in F$, thus $\Box a \in G$, whence $G \in M_\Box$, as required.

(FHM2) We need to show $N_\Diamond = [\supseteq]N_\Diamond$. As in the previous case, the right-to-left inclusion follows from reflexivity of $\supseteq$. Take any $F \in X_L$ satisfying $F \notin [\supseteq]N_\Diamond$. Then there is some $G \in X_L$ such that $G \subseteq F$ and $G \notin N_\Diamond$. Then $\Diamond^{-1}(G) = L$. Since $G \subseteq F$, we have $\Diamond^{-1}(G) \subseteq \Diamond^{-1}(F)$. Thus $\Diamond^{-1}(F) = L$, and hence $F \notin N_\Diamond$, as required.

Proofs of (FHM3) and (FHM4) are straightforward.

(FHM5) Let $F \in X_L$ be such that $F \notin M_\Box$. Then for every $a \in L$, $\Box a \notin F$, that is $a \notin \Box^{-1}(F)$, thus $\Box^{-1}(F) = \emptyset$. Therefore $\Box^{-1}(F) \subseteq G$ for any $G \in X_L$, so $FR_\Box G$ holds.

(FHM6) Take $F \in X_L$ such that $F \notin N_\Diamond$. Then $\Diamond^{-1}(F) = L$, so $G \subseteq \Diamond^{-1}(F)$ for any $G \in X_L$. Thus $FS_\Diamond G$ holds. $\Box$

We now show that the lattice embedding $h : L \to L_{X_L}$, used in Section 2.5, preserves $\Box$ and $\Diamond$.

Lemma 11.3.5 *Let L be an HM-algebra. For any $a \in L$,*

(a) $h(\Box a) = \Box_{R_\Box}(h(a))$
(b) $h(\Diamond a) = \Diamond_{S_\Diamond}(h(a))$.

Proof:
(a) Let $F \in h(\Box a)$. Then $\Box a \in F$, so $F \in M_\Box$. Now, take any $G \in X_L$ such that $\Box^{-1}(F) \subseteq G$. Since $a \in \Box^{-1}(F)$, $a \in G$. Thus $G \in h(a)$. Hence $F \in [R_\Box]h(a)$. Therefore $F \in M_\Box \cap [R_\Box]h(a)$, which means that $F \in \Box_{R_\Box}(h(a))$. On the other hand, assume that $F \in \Box_{R_\Box}(h(a))$. Then $F \in M_\Box$ and $F \in [R_\Box]h(a)$. Suppose that $F \notin h(\Box a)$, that is $\Box a \notin F$, which means $a \notin \Box^{-1}(F)$. Since $F \in M_\Box$, there is some $b \in L$ such that $\Box b \in F$. Since $b \leq 1$ and $\Box$ is monotone, $\Box 1 \in F$. By Lemma 11.3.1(a), $\Box^{-1}(F)$ is a filter of L. Then there exists a prime filter of L, say G, such that $\Box^{-1}(F) \subseteq G$ and $a \notin G$. Then $F \notin [R_\Box]h(a)$, a contradiction.

(b) Assume $F \in h(\Diamond a)$ and $F \in N_\Diamond$. Then $\Diamond a \in F$ and for some $b \in L$, $\Diamond b \notin F$. Then $a \in \Diamond^{-1}(F)$ and since $0 \leq b$ and by (HM4) $\Diamond$ is monotone, $\Diamond 0 \notin F$. By Lemma 11.3.1(c), F is a co-filter of L, and furthermore, by Lemma 11.3.1(b), $\Diamond^{-1}(F)$ is also a co-filter of L. Consider the principal filter $\uparrow_\leq a$. We show that $\uparrow_\leq a \subseteq \Diamond^{-1}(F)$. Let $b \in \uparrow_\leq a$, that is $a \leq b$. Then, by (HM4), $\Diamond a \leq \Diamond b$. Hence $\Diamond b \in F$, so $b \in \Diamond^{-1}(F)$. By Lemma 11.3.2 there is a prime filter H such that $\uparrow_\leq a \subseteq H \subseteq \Diamond^{-1}(F)$. This means that $FS_\Diamond H$. We also have $a \in H$, so $H \in h(a)$.

We conclude that $F \in \Diamond_{S_\Diamond}(h(a))$. On the other hand, assume that if $F \in N_\Diamond$, then $F \in \langle S_\Diamond \rangle h(a)$, and suppose $F \notin h(\Diamond a)$. Then $\Diamond a \notin F$, and hence $F \in N_\Diamond$. It follows that there is some $G \in X_L$ such that $F S_\Diamond G$ and $G \in h(a)$. Thus $G \subseteq \Diamond^{-1}(F)$ and $a \in G$. Since $\Diamond a \notin F$, $a \notin \Diamond^{-1}(F)$, so $a \notin G$, a contradiction. □

We now show that the order-embedding $k : X \to X_{L_X}$, used in Section 2.5, preserves the sets M and N, and the relations R and S.

Lemma 11.3.6 *Let X be an HM-frame. For any $x, y \in X$,*

(a) $x \in M \;\Leftrightarrow\; k(x) \in M_{\Box_R}$

(b) $x \in N \;\Leftrightarrow\; k(x) \in N_{\Diamond_S}$

(c) $xRy \;\Leftrightarrow\; k(x) R_{\Box_R} k(y)$

(d) $xSy \;\Leftrightarrow\; k(x) S_{\Diamond_S} k(y)$.

Proof:
(a) Note that the following equivalences hold for any $x \in X$:

$$\begin{aligned} k(x) \in M_{\Box_R} &\Leftrightarrow \exists A \in L_X,\ \Box_R(A) \in k(x) \\ &\Leftrightarrow \exists A \in {}_X,\ x \in M \cap [R]A \\ &\Leftrightarrow \exists A \in L_X,\ x \in M \;\&\; x \in [R]A. \end{aligned}$$

If $x \in M$, then taking X for A we clearly have $k(x) \in M_{\Box_R}$. Conversely, if $k(x) \in M_{\Box_R}$, then $x \in M$.

(b) Note also that

$$\begin{aligned} k(x) \in N_{\Diamond_S} &\Leftrightarrow \exists A \in L_X,\ \Diamond_S(A) \notin k(x) \\ &\Leftrightarrow \exists A \in L_X,\ -N \cup \langle S \rangle A \notin k(x) \\ &\Leftrightarrow \exists A \in L_X,\ x \notin -N \cup \langle S \rangle A \\ &\Leftrightarrow \exists A \in L_X,\ x \in N \;\&\; x \in [S](-A). \end{aligned}$$

If $x \in N$, then taking $\emptyset$ for A we clearly have $k(x) \in N_{\Diamond_S}$. On the other hand, if $k(x) \in N_{\Diamond_S}$, then $x \in N$.

(c) Observe that

$$\begin{aligned} k(x) R_{\Box_R} k(y) &\Leftrightarrow \Box_R^{-1}(k(x)) \subseteq k(y) \\ &\Leftrightarrow \forall A \in L_X,\ \Box_R(A) \in k(x) \Rightarrow A \in k(y) \\ &\Leftrightarrow \forall A \in L_X,\ x \in M \cap [R]A \Rightarrow y \in A. \end{aligned}$$

Assume xRy and take $A \in L_X$ such that $x \in M$ and $x \in [R]A$. Thus $\forall z,\ xRz \Rightarrow z \in A$. Taking y for z we get $y \in A$, as required. On the other hand, take A to be $-\langle \leqslant \rangle \{y\}$. Clearly, $y \notin A$. Also, $A \in L_X$. Indeed, we have

$$[\leqslant](-\langle \leqslant \rangle \{y\}) = [\leqslant][\leqslant](-\{y\}) = [\leqslant](-\{y\}) = -\langle \leqslant \rangle \{y\}.$$

Now by assumption, $x \notin M$ or $x \notin [R](-\langle\leqslant\rangle\{y\})$. If $x \notin M$, then by (FHM5), xRy. If $x \notin [R](-\langle\leqslant\rangle\{y\})$, then for some $z \in X$, xRz and $z \leqslant y$. By (FHM3), xRy.

(d) Observe that

$$\begin{aligned} k(x) S_{\Diamond_S} k(y) &\Leftrightarrow k(y) \subseteq \Diamond_S^{-1} k(x) \\ &\Leftrightarrow \forall A \in L_X,\ A \in k(y) \Rightarrow A \in \Diamond_S(A) \in k(x) \\ &\Leftrightarrow \forall A \in L_X,\ y \in A \Rightarrow x \in \Diamond_S(A) \\ &\Leftrightarrow \forall A \in L_X,\ y \in A \Rightarrow (x \in -N \text{ or } x \in \langle S\rangle A) \\ &\Leftrightarrow \forall A \in L_X,\ y \in A \ \&\ x \in N \Rightarrow x \in \langle S\rangle A. \end{aligned}$$

Assume that xSy and take any $A \in L_X$ such that $y \in A$ and $x \in N$. Then $x \in \langle S\rangle A$, and hence $k(x) \in S_{\Diamond_S} k(y)$. For the reverse implication assume that $\forall A \in L_X,\ y \in A \Rightarrow (x \notin N$ or $x \in \langle S\rangle A)$. Take A to be $\uparrow_\leqslant y$. Then $y \in A$ and $A \in L_X$ since we have $[\leqslant]\uparrow_\leqslant y = [\leqslant]\langle\geqslant\rangle\{y\} = [\leqslant][\leqslant]\{y\} = [\leqslant]\{y\} = \uparrow_\leqslant y$. By assumption, $x \notin N$ or $x \in \langle S\rangle(\uparrow_\leqslant y)$. If $x \in N$, then by (FHM6), xSy. If $x \in \langle S\rangle\uparrow_\leqslant y$, then for some $z \in X$, xSz and $y \leqslant z$. By (FHM4) and reflexivity of $\geqslant$, xSy, as required. □

The developments of this section yield the following discrete representations for HM-algebras and HM-frames.

Theorem 11.3.7

(a) *Every HM-algebra is embeddable into the complex algebra of its canonical frame;*

(b) *Every HM-frame is embeddable into the canonical frame of its complex algebra.*

11.4 Axiomatic extensions of HM-algebras

To obtain a duality for Heyting algebras with normal modal operators we consider algebras, called *HK-algebras*, that are HM-algebras satisfying

(HM5) $\Box 1 = 1$

(HM6) $\Diamond 0 - 0$.

Recall the logic K is a basic classical modal logic based on Boolean algebras, here we have K based on Heyting algebras.

The appropriate frames, called *HK-frames*, are HM-frames satisfying

(FHM5') $N = X$

(FHM6') $M = X$.

It is easy to see that in the presence of (FHM5') the modal operator $\Box_R$ of the complex algebra satisfies $\Box_R(A) = [R]A$, and since $[R]X = X$, (HM5) is satisfied. Similarly, in the presence of (FHM6') $\Diamond_S(A) = \langle S\rangle A$, and since $\langle S\rangle\emptyset = \emptyset$, we have (HM6) satisfied. It is also easy to see that in the canonical frame $N_\Box$ and $M_\Diamond$ satisfy the conditions (FHM5') and (FHM6'), respectively.

The question arises which axioms guarantee that we may assume $R = S$ in every HK-frame. For distributive lattices with normal modal operators satisfying (HM1), (HM2), (HM3'), and (HM4') Dunn ([48]) showed that $\Diamond a \wedge \Box b \leq \Diamond(a \wedge b)$ is true in all such lattices if and only if $S \subseteq R$. Similarly, $\Box(a \vee b) \leq \Box a \vee \Diamond b$ corresponds to $R \subseteq S$.

Following [82] we present an axiomatization of Heyting algebras with normal modal operators such that the corresponding frames have a single relation adequate for both modalities $\Box$ and $\Diamond$. We consider a class of HK-algebras satisfying:

(HM7) $\Diamond(a \to b) \leq (\Box a \to \Diamond b)$

(HM8) $\Diamond a \to \Box b \leq \Box(a \to b)$.

Lemma 11.4.1 *For any HK1-algebra $(L, \Box, \Diamond)$ and for any $a \in L$,*

(a) $\Diamond \neg a \leq \neg \Box a$;

(b) $\neg \Diamond a \leq \Box \neg a$.

Proof:
(a) From (HM7) we get $\Diamond(a \to 0) \leq \Box a \to \Diamond 0 = \Box a \to 0 = \neg \Box a$.

(b) Since $0 \leq \Box 0$ and $\to$ is monotone in the second argument, $\Diamond a \to 0 \leq \Diamond a \to \Box 0$. Applying (HM8) we get the required result. □

An *HK1-frame* is a relational structure (X, R) such that X is a H-frame and $R \subseteq X \times X$ satisfies

(FH1) $(R; \leqslant) \subseteq (\leqslant; R)$

(FH2) $(\geqslant; R) \subseteq (R; \geqslant)$

(FH3) $(\leqslant; R) \cap (R; \geqslant) \subseteq R$.

Observe that if we assume that relation R satisfies (FHM3) and (FHM4) from Section 11.3, then (FH1) and (FH2) follow. Namely, assume (FHM3) holds, that is $(\leqslant; R; \leqslant) \subseteq R$. Since $R \subseteq (\leqslant; R)$, we get $(R; \leqslant) \subseteq (\leqslant; R; \leqslant) \subseteq (\leqslant; R)$. Similarly, if R satisfies (FHM4), then (FH2) holds.

The *complex algebra of an HK1-frame* X is the structure $(L_X, \Box_L, \Diamond_L)$ where L_X is the complex algebra of its H-frame reduct and for every $A \subseteq X$,

$$\Box_X A \stackrel{\text{def}}{=} [\leqslant; R]A$$
$$\Diamond_X A \stackrel{\text{def}}{=} \langle R \rangle A.$$

Lemma 11.4.2 *For any HK1-frame X and for any $A \subseteq X$,*

$$\langle \leqslant \rangle [R] A \subseteq [R] \langle \leqslant \rangle A.$$

Proof:
Let $x \in \langle\leqslant\rangle[R]A$. Thus there is some y such that $x \leqslant y$ and $y \in [R]A$. Suppose $x \notin [R]\langle\leqslant\rangle A$. This means that there exists z such that xRz and $z \notin \langle\leqslant\rangle A$. From $x \leqslant y$ and xRz it follows $y(\geqslant; R)z$, hence by (FH2) we get $y(R;\geqslant)z$. Then there is some u such that yRu and $u \geqslant z$. Now, $y \in [R]A$ with yRu yield $u \in A$, but $z \notin \langle\leqslant\rangle A$ and $u \geqslant z$ give $u \notin A$, a contradiction. □

Proposition 11.4.3 *The complex algebra of an HK1-frame is an HK1-algebra.*

Proof:
Note that L_X is closed on the operations $\Box_X$ and $\Diamond_X$, that is $[R]A = [\leqslant][R]A$ and $\langle R\rangle A = [\leqslant]\langle R\rangle A$ for any $A \in L_X$. In both cases the inclusion $\supseteq$ follows from reflexivity of $\leqslant$. Clearly, $A \subseteq [\leqslant]A$ by Lemma 1.10.1(g). Now assume $x \in \langle R\rangle A$. Then, for some y, xRy and $y \in A$. Take any z such that $x \leqslant z$. Now, by (FH2), $z(R;\geqslant)y$, that is, for some z', zRz' and $y \leqslant z'$. Since $A \in L_X$, from $y \in A$ and $y \leqslant z'$ we get $z' \in A$ and hence $z \in \langle R\rangle A$, as required.

(HM7) We show $\langle R\rangle[\leqslant](-A \cup B) \subseteq [\leqslant](-[\leqslant; R]A \cup \langle R\rangle B)$ for all $A, B \in L_X$. We prove this inclusion in the contrapositive form. First, note that by Lemma 1.8.1(a) and (b)

$$-\langle R\rangle[\leqslant](-A \cup B) = [R]\langle\leqslant\rangle(A \cap (-B))$$
$$-[\leqslant](-[\leqslant; R]A \cup \langle R\rangle B) = \langle\leqslant\rangle([\leqslant; R]A \cap \langle R\rangle B).$$

Now, we have

$\langle\leqslant\rangle([\leqslant; R]A \cap -\langle R\rangle B)$	
$\subseteq \langle\leqslant\rangle([R;\leqslant]A \cap [R](-B))$	(FH1), Lemma 1.8.4(i), Lemma 1.8.1(a)
$= \langle\leqslant\rangle([R][\leqslant]A \cap [R](-B))$	Lemma 1.8.4(o)
$= \langle\leqslant\rangle[R]([\leqslant]A \cap (-B))$	Lemma 1.8.4(a)
$= \langle\leqslant\rangle[R](A \cap (-B))$	since $A \in L_X$
$\subseteq [R]\langle\leqslant\rangle(A \cap (-B))$.	Lemma 11.4.2.

(HM8) We show, for all $A, B \in L_X$, that

$$[\leqslant](-\langle R\rangle A \cup [\leqslant; R]B) \subseteq [\leqslant; R][\leqslant](-A \cup B).$$

Assume $x \in [\leqslant](-\langle R\rangle A \cup [\leqslant; R]B)$ which means

$$\forall y \in X,\ x \leqslant y \Rightarrow (y \in -\langle R\rangle A \text{ or } y \in [\leqslant; R]B),$$

or equivalently,

$$\forall y \in X,\ x \leqslant y \ \&\ y \in \langle R\rangle A \Rightarrow y \in [\leqslant; R]B.$$

Take any $t \in X$ such that $x(\leqslant; R)t$ and any z such that $t \leqslant z$ and $z \in A$. Since $x(\leqslant; R)t$, there is some $v \in X$ such that $x \leqslant v$ and vRt. From vRt and $t \leqslant z$ we get $v(R;\leqslant)z$. By (FH1), $v(\leqslant; R)z$ which yields for some $w \in X$, $v \leqslant w$ and vRz. Furthermore, $x \leqslant v$ and $v \leqslant w$ imply $x \leqslant w$. Now, vRz and $z \in A$ yield

$w \in \langle R \rangle A$. Thus taking w for y we get $w \in [\leqslant; R]B$. Since wRz and $R \subseteq (\leqslant; R)$, by Lemma 1.10.1(j) we get $w(\leqslant; R)z$. Hence $z \in B$, as required. □

The *canonical frame of an HK1-algebra* L is the structure (X_L, R_L) such that X_L is the canonical frame of its H-algebra reduct and R_L is defined, for any $F, G \in X_L$, by

$$FR_LG \overset{\text{def}}{\Leftrightarrow} \Box^{-1}(F) \subseteq G \subseteq \Diamond^{-1}(F).$$

Proposition 11.4.4 *The canonical frame of an HK1-algebra is an HK1-frame.*

Proof:
(FH1) We need to show that $F(R_L; \subseteq)G$ implies $F(\subseteq; R_L)G$ for any $F, G \in X_L$. Assume that $F(R_L; \subseteq)G$. Then there is some $H \in X_L$ such that $\Box^{-1}(F) \subseteq H$, $H \subseteq \Diamond^{-1}(F)$, and $H \subseteq G$. Hence $\Box^{-1}(F) \subseteq G$. In order to show that $F(\subseteq; R_L)G$ we need to find some $H \in X_L$ such that $F \subseteq H$, $G \subseteq \Diamond^{-1}(H)$, and $\Box^{-1}(H) \subseteq G$. Consider the filter K generated by $F \cup \Diamond(G)$. We show that $K \subseteq \Box(G)$. Assume $\Box a \in K$. Then there is some $c \in F$ and some $d \in G$ such that $c \wedge \Diamond d \leq \Box a$. Hence $c \leq \Diamond d \rightarrow \Box a$ and thus, by (HM8), $c \leq \Box(d \rightarrow a)$. Since $c \in F$ and F is up-closed, we get $\Box(d \rightarrow a) \in F$, that is, $d \rightarrow a \in \Box^{-1}(F)$. Hence $d \rightarrow a \in G$, since $\Box^{-1}(F) \subseteq G$. Now $d \in G$ and $d \rightarrow a \in G$, so $d \wedge a = d \wedge (d \rightarrow a) \in G$ and hence, since G is up-closed, $a \in G$, so $\Box a \in \Box G$, as required. Thus $K \cap -\Box G = \emptyset$ and there is some $H \in X_L$ such that $K \subseteq H$ and $-\Box(G) \cap H = \emptyset$. Then $F \subseteq H$, $\Diamond(G) \subseteq H$, and $H \subseteq \Box(G)$. Hence there is some $H \in X_L$ such that $F \subseteq H$, $G \subseteq \Diamond^{-1}(H)$, and $\Box^{-1}(H) \subseteq G$.

(FH2) We need to show that $F(\supseteq; R_L)G$ implies $F(R_L; \supseteq)G$ for any $F, G \in X_L$. Assume $F(\supseteq; R_L)G$. Then there exists $H \in X_L$ such that $F \supseteq H$, $\Box^{-1}(H) \subseteq G$, and $G \subseteq \Diamond^{-1}(H)$. Hence $G \subseteq \Diamond^{-1}(F)$. In order to show that $F(R_L; \supseteq)G$ we need to find some $H \in X_L$ such that $H \subseteq \Diamond^{-1}(F)$, $\Box^{-1}(F) \subseteq G$, and $H \supseteq G$. Consider the filter K generated by $G \cup \Box^{-1}(F)$. We show that $K \subseteq \Diamond^{-1}(F)$. Assume $a \in K$. Then, there is some $c \in G$ and some $d \in \Box^{-1}(F)$ such that $c \wedge d \leq a$. Hence $c \leq d \rightarrow a$, so $\Diamond c \leq \Diamond(d \rightarrow a)$ and thus, by (HM7), $\Diamond c \leq \Box d \rightarrow \Diamond a$. Since $c \in G$ and $G \subseteq \Diamond^{-1}(F)$, $\Diamond c \in F$ and hence $\Box d \rightarrow \Diamond a \in F$. Now, $\Box d \in F$ and $\Box d \rightarrow \Diamond a \in F$, so $\Box d \wedge \Diamond a = \Box d \wedge (\Box d \rightarrow \Diamond a) \in F$ and hence, since F is up-closed, $\Diamond a \in F$, that is, $a \in \Diamond^{-1}(F)$, as required. Therefore we have $K \cap -\Diamond^{-1}(F) = \emptyset$, so there is some $H \in X_L$ such that $K \subseteq H$ and $H \cap -\Diamond^{-1}(F) = \emptyset$. Hence there is some $H \in X_L$ such that $H \subseteq \Diamond^{-1}(F)$, $\Box^{-1}(F) \subseteq H$, and $H \supseteq G$.

(FH3) We show $(\subseteq; R_L) \cap (R_L; \supseteq) \subseteq R_L$. Assume $F(\subseteq; R_L)G$ and $F(R_L; \supseteq)G$ for $F, G \in X_L$. Thus by the first assumption there is some $H \in X_L$ such that $F \subseteq H$ and $\Box^{-1}(H) \subseteq G \subseteq \Diamond^{-1}(H)$, and by the second assumption, there is some $J \in X_L$ such that $G \subseteq J$ and $\Box^{-1}(F) \subseteq J \subseteq \Diamond^{-1}(F)$. Since $F \subseteq H$, by Lemma 8.3.1(b), $\Box^{-1}(F) \subseteq \Box^{-1}(H)$. Hence $\Box^{-1}(F) \subseteq G$. Since $G \subseteq J$ and $J \subseteq \Diamond^{-1}(F)$, $G \subseteq \Diamond^{-1}(F)$. It follows that FR_LG, as required. □

Lemma 11.4.5 *Let (X_L, R_L) be the canonical frame of an HK1-algebra L. For every $F, G \in X_L$,*

(a) $F(\subseteq; R_L)G \Leftrightarrow \Box^{-1}(F)\subseteq G$;

(b) $F(R_L; \supseteq)G \Leftrightarrow G\subseteq \Diamond^{-1}(F)$.

Proof:
(a) Assume that $F(\subseteq; R_L)G$. Then there is some $H\in X_L$ such that $F\subseteq H$ and $\Box^{-1}(H)\subseteq G\subseteq \Diamond^{-1}(H)$. Since $F\subseteq H$ implies $\Box^{-1}(F)\subseteq \Box^{-1}(H)$, we get $\Box^{-1}(F)\subseteq G$. For the reverse implication, assume that $\Box^{-1}(F)\subseteq G$. Let H' be the filter of L generated by $F \cup \Diamond(G)$. Since G is a prime filter, $-G$ is a prime ideal, so $-G\neq\emptyset$, and hence $\Box(-G)\neq\emptyset$. Let I be an ideal generated by $\Box(-G)$. Suppose that $H' \cap I\neq\emptyset$. Then there are $a\in F$, $b\in G$, and $c'\in I$ such that $a\wedge\Diamond b\leq c'$. Next, there are some natural number n and $c_1,\ldots,c_n\notin G$ such that $c'\leq\Box c_1\vee\ldots\vee\Box c_n\leq\Box(c_1\vee\ldots\vee c_n)$. By primeness of G, $c=c_1\vee\ldots\vee c_n\notin G$. Since $a\wedge\Diamond b\leq\Box c$, $a\leq\Diamond b\to\Box c\leq\Box(b\to c)$ by axiom (H8). Since $a\in F$, $\Box(b\to c)\in F$, hence $b\to c\in G$. Now, $b\in G$ together with $b\to c\in G$ gives $c\in G$, a contradiction. Therefore $H'\cap I=\emptyset$. By the prime filter theorem for distributive lattices, there is $H\in X_L$ such that $H'\subseteq H$ and $H\cap I=\emptyset$. Hence $F\subseteq H$, $\Diamond(G)\subseteq H$, and $H\subseteq -\Box(-G)$. It follows that $\Box^{-1}(H)\subseteq\Box^{-1}(-(\Box(-G)))=-\Box^{-1}(\Box(-G))\subseteq G$. Since $\Diamond(G)\subseteq H$, $G\subseteq\Diamond^{-1}(H)$. Therefore $F(\subseteq; R_L)G$, as required.

(b) The left-to-right implication can be proved as in (a). Assume $G\subseteq\Diamond^{-1}(F)$. Consider the filter K of L generated by $G\cup\Box^{-1}(F)$. Suppose $K\cap-\Diamond^{-1}(F)\neq\emptyset$. Then there are $a\notin\Diamond^{-1}(F)$, $b\in G$, and $c\in\Box^{-1}(F)$ such that $b\wedge c\leq a$. Hence $b\leq c\to a$, so $\Diamond b\leq\Diamond(c\to a)\leq(\Box c\to\Diamond a)$ by axiom (HM7). Since $b\in G$ and $G\subseteq\Diamond^{-1}(F)$, we get $\Diamond b\in F$, and furthermore, $\Box c\to\Diamond a\in F$. Now, $\Box c\in F$ and $\Box c\to\Diamond a\in F$ imply $\Diamond a\in F$, a contradiction. Therefore $K\cap-\Diamond^{-1}(F)=\emptyset$. By Lemma 8.2.1(c), $-\Diamond^{-1}(F)$ is an ideal of L. Hence, by the prime filter theorem for distributive lattices, there is $H\in X_L$ such that $K\subseteq H$ and $H\cap-\Diamond^{-1}(F)=\emptyset$, that is, $H\subseteq\Diamond^{-1}(F)$. Clearly, $G\subseteq K$, so $G\subseteq H$. Also, $\Box^{-1}(F)\subseteq K$, thus $\Box^{-1}(F)\subseteq H$. Hence $F(R_L,\supseteq)G$. □

Let L be an HK1-algebra. We now show that the lattice embedding $h: L\to L_{X_L}$, used in Section 2.5, preserves $\Box$ and $\Diamond$.

Lemma 11.4.6 *For any $a\in L$,*

(a) $h(\Box a)=\Box_{X_L}h(a)$;

(b) $h(\Diamond a)=\Diamond_{X_L}h(a)$.

Proof:
(a) Note that, for every $F\in X_L$ and every $a\in L$,

$$\begin{aligned} F\in\Box_{X_L}h(a) &\Leftrightarrow F\in[\subseteq; R_L]h(a)\\ &\Leftrightarrow \forall G\in X_L,\ F(\subseteq; R_L)G\Rightarrow G\in h(a)\\ &\Leftrightarrow \forall G\in X_L,\ \Box^{-1}F\subseteq G\Rightarrow a\in G \end{aligned}$$

by Lemma 11.4.5(a). Assume that $F\in h(\Box a)$ and take any $G\in X_L$ such that $\Box^{-1}F\subseteq G$. Since $\Box a\in F$, $a\in G$. On the other hand, assume that $F\notin h(\Box a)$.

Then $a \notin \Box^{-1}(F)$, so $\langle\leq\rangle\{a\} \cap \Box^{-1}(F) = \emptyset$. Since $\langle\leq\rangle\{a\}$ is an ideal of L and, by Lemma 8.3.1(c), $\Box^{-1}(F)$ is a filter of L, by Theorem 1.6.8 there is some $G \in X_L$ such that $\Box^{-1}(F) \subseteq G$ and $\langle\leq\rangle\{a\} \cap G = \emptyset$. Thus $a \notin G$, and hence $F \notin \Box_{X_L} h(a)$.

(b) Observe that, for every $F \in X_L$ and every $a \in L$,

$$F \in \Diamond_{X_L} h(a) \Leftrightarrow F \in \langle R_L \rangle h(a) \Leftrightarrow \exists G \in X_L,\ FR_L G \ \&\ a \in G.$$

Assume that $F \in \Diamond_{X_L} h(a)$. Then, for some $G \in X_L$, $G \subseteq \Diamond^{-1}(F)$ and $a \in G$, thus $\Diamond a \in F$, that is $F \in h(\Diamond a)$. On the other hand, assume $F \in h(\Diamond a)$, so $\Diamond a \in F$. Suppose that $\langle\geq\rangle\{a\} \cap -\Diamond^{-1}(F) \neq \emptyset$. Then there is some $b \in L$ such that $a \leq b$ and $b \notin \Diamond^{-1}(F)$. Hence $\Diamond a \leq \Diamond b$ and $\Diamond b \notin F$, thus $\Diamond a \notin F$, a contradiction. Therefore $\langle\geq\rangle\{a\} \cap -\Diamond^{-1}(F) = \emptyset$, so by Theorem 1.6.8, there exists $H \in X_L$ such that $\langle\geq\rangle\{a\} \subseteq H$ and $H \subseteq \Diamond^{-1}(F)$. Using Lemma 11.4.5(b), we get $F(R_L; \supseteq)H$, so there is some $G \in X_L$ such that $FR_L G$ and $H \subseteq G$. Since $a \in H$, we obtain $a \in G$. Hence $F \in \langle R_L \rangle h(a)$, as required. □

Let X be an HK1-frame. We now show that the order-embedding $k : X \to X_{L_X}$, used in Section 2.5, preserves the relation R.

Lemma 11.4.7 *For any $x, y \in X$, $xRy \Leftrightarrow k(x) R_{L_X} k(y)$.*

Proof:
Note, for any $x, y \in X$, that

$$\begin{array}{ll} & k(x) R_{L_X} k(y) \\ \Leftrightarrow & (\Box_X^{-1}(k(x)) \subseteq k(y)) \ \&\ (k(y) \subseteq \Diamond_X^{-1}(k(x))) \\ \Leftrightarrow & (\forall A \in L_X, \Box_X A \in k(x) \Rightarrow A \in k(y)) \ \& \\ & (\forall B \in L_X, B \in k(y) \Rightarrow \Diamond_X B \in k(x)) \\ \Leftrightarrow & (\forall A \in X_L, x \in [\leqslant; R]A \Rightarrow y \in A) \ \& \\ & (\forall B \in X_L, y \in B \Rightarrow x \in \langle R \rangle B). \end{array}$$

Assume xRy and take any $A \in L_X$ such that $x \in [\leqslant; R]A$. Since xRy, we get $y \in A$. Now, take any $B \in L_X$ such that $y \in B$, which together with xRy gives $x \in \langle R \rangle B$. On the other hand, assume $k(x) R_{L_X} k(y)$. Let $A = [\leqslant]\langle\geqslant; R^{-1}; \geqslant\rangle\{x\}$. Clearly, $A \in L_X$ by Lemma 1.10.1(b) and (g). Since $(\leqslant; R; \leqslant)^{-1} = (\geqslant; R^{-1}; \geqslant)$, by Lemma 1.8.2(b) and Lemma 1.8.4(m) we have

$$\begin{aligned} \{x\} &\subseteq [\leqslant; R; \leqslant]\langle\geqslant; R^{-1}; \geqslant\rangle\{x\} \\ &= [\leqslant; R][\leqslant]\langle\geqslant; R^{-1}; \geqslant\rangle\{x\} \\ &= [\leqslant; R]A. \end{aligned}$$

Thus by the assumption $y \in [\leqslant]\langle\geqslant; R^{-1}; \geqslant\rangle\{x\}$. Since $\leqslant$ is reflexive, by Lemma 1.10.1(b) we get $y \in \langle\geqslant; R^{-1}; \geqslant\rangle\{x\}$ which implies there are $z, u \in X$ such that $y \geqslant z$, uRz, and $u \geqslant x$. Now, uRz and $z \leqslant y$ give $u(R; \leqslant)y$, which by (FH1) implies $u(\leqslant; R)y$. Then there is some t such that $u \leqslant t$ and tRy. Since $x \leqslant u$ and $u \leqslant t$, $x \leqslant t$, which together with tRy yields $x(\leqslant; R)y$.

Now consider the set $B = \langle\geqslant\rangle\{y\}$. Clearly, $B \in L_X$ and $y \in B$. By the assumption we obtain $x \in \langle R\rangle\langle\geq\rangle\{y\}$. Thus there is some $v \in X$ such that xRv and $y \leqslant v$, thus $x(R;\geqslant)y$.

Now, by (FH3), $x(\leqslant;R)y$ and $x(R;\geqslant)y$ imply xRy, as required. □

Thus we have the following discrete representations for HK1-algebras and HK1-frames.

Theorem 11.4.8

(a) *Every HK1-algebra is embeddable into the complex algebra of its canonical frame;*

(b) *Every HK1-frame is embeddable into the canonical frame of its complex algebra.*

11.5 Frontal Heyting algebras

A *frontal Heyting algebra, HF-algebra,* is a structure $(L, \Box)$ such that L is a Heyting algebra and $\Box$ is a unary operator satisfying, for all $a, b \in L$,

(HF1) $\Box(a \wedge b) = \Box a \wedge \Box b$

(HF2) $a \leq \Box a$

(HF3) $\Box a \leq b \vee (b \rightarrow a)$.

An *HF-frame* $(X, <)$ is a non-empty set X together with a transitive order $<$ such that the reflexive closure of $<$, that is, the relation $\leq$ such that for all $x, y \in X$, $x \leq y$ if and only if $x = y$ or $x < y$, is a partial order.

The *complex algebra of an HF-frame* X is a structure $(L_X, \Box_<)$ where L_X is the complex algebra of its H-frame reduct and for any $A \in L_X$,

$$\Box_< A \stackrel{\text{def}}{=} [<]A.$$

Proposition 11.5.1 *The complex algebra of an HF-frame is an HF-algebra.*

Proof:
Since the necessity operator $[<]$ distributes over $\wedge$, we get (HF1). The axiom (HF2) is obvious. To show (HF3) assume that, for any y, $x < y \Rightarrow y \in A$ and suppose that $x \notin B \cup (B \rightarrow_{\leqslant} A)$. It follows that $x \notin B$ and there is some y such that $x \leqslant y$ and $y \in B$ and $y \notin A$. If $x = y$, then we get a contradiction with $x \notin B$. If $x < y$, we get a contradiction with the assumption. □

The *canonical frame of an HF-algebra* L is a structure $(X_L, <_\Box)$ where X_L is the set of all prime filters of L and $<_\Box$ is a binary relation on X_L defined, for all $F, G \in X_L$, as

$$F <_\Box G \stackrel{\text{def}}{\Leftrightarrow} \Box^{-1}(F) \subseteq G.$$

Proposition 11.5.2 *The canonical frame of an HF-algebra is an HF-frame.*

Proof:
We have to show that the relation $\leqslant_\Box$ defined for all $F, G \in X_L$,

$$F \leqslant_\Box G \Leftrightarrow F = G \text{ or } F <_\Box G,$$

is a partial order. Reflexivity of $\leqslant_\Box$ is obvious. Transitivity and antisymmetry follow easily from the axiom (HF2). □

As observed in [77], $\Box_< A = A \cup \max(X - A)$, where for any $B \subseteq X$, $\max(B)$ is the set of all maximal elements of B, that is $\max B = \{x \in X : \neg(\exists y \in B \wedge x < y)\}$. Since elements of L_X can be seen as cones, the operator $\Box_<$ assigns to a cone its fronton, that is the set of maximal points of the complement of the cone. This explains the name of the class of HF-algebras.

The following lemma is adapted from [77].

Lemma 11.5.3 *Let L be an HF-algebra. For any $F, G \in X_L$,*

(a) *If $F <_\Box G$, then $F \subseteq G$;*
(b) *If $F \subset G$, then $F <_\Box G$;*
(c) $\leqslant_\Box = \subseteq$.

Proof:
(a) Let $F <_\Box G$ and $a \in F$. By (HF2) we get $\Box a \in F$, so $a \in G$.

(b) Let $F \subset G$, that is, $F \subseteq G$ and $F \neq G$. So there is some $b \in L$ such that $b \in G$ and $b \notin F$. Suppose not $F <_\Box G$. Hence, there is some $a \in L$ such that $\Box a \in F$ and $a \notin G$. Since $b \notin F$, we have $b \to a \notin F$. For otherwise, we would have $b \to a \in G$ and since $b \in G$ this would imply $a \in G$ which is a contradiction. Thus we get $b \notin F$ and $b \to a \notin F$. Hence, since F is prime, $b \vee (b \to a) \notin F$. By axiom (HF3), $\Box a \notin F$, a contradiction.

(c) Follows from (a) and (b). □

Let L be an HF-algebra. We now show that the lattice-embedding $h : L \to L_{X_L}$, used in Section 2.5, preserves the operation $\Box$.

Lemma 11.5.4 *For any $a \in L$, $h(\Box a) = \Box_{<_\Box}(h(a))$.*

Proof:
It is obvious that $h(\Box a) \subseteq \Box_{<_\Box}(h(a))$. For the other inclusion, take any $F \in X_L$ and assume $F \notin h(\Box a)$. Then $\Box a \notin F$ and hence, $a \notin \Box^{-1}(F)$. Now $\Box^{-1}(F)$ is a filter. Therefore by the prime filter theorem for bounded distributive lattices, there is some $G \in X_L$ such that $\Box^{-1}(F) \subseteq G$ and $a \notin G$. That is, there is some $G \in X_L$ such that $F <_\Box G$ and $a \notin G$. Thus, $F \notin [<_\Box]h(a) = \Box_{<_\Box}(h(a))$, as required. □

Let X be an HF-frame. We now show that the order-embedding $k : X \to X_{L_X}$, used in Section 2.5, is $<$-preserving.

Lemma 11.5.5 *For every* $x, y \in X$, $x < y \Leftrightarrow k(x) <_{\Box_<} k(y)$.

Proof:
Observe, for any $x, y \in X$,

$$k(x) \leq_{\Box_<} k(y) \Leftrightarrow \forall A \in L_X,\ \Box_< A \in k(x) \Rightarrow A \in k(y)$$
$$\Leftrightarrow \forall A \in L_X,\ x \in [<]A \Rightarrow y \in A.$$

Assume $x < y$ and take any $A \in L_X$ such that $x \in [<]A$. Then, since $[<]A$ is $<$-increasing, $y \in [<]A$ so $y \in A$ since $A \in L_X$. On the other hand, assume $k(x) <_{\Box_<} k(y)$. Then, for every $A \in L_X$, $x \in [<]A \Rightarrow y \in A$. Hence, for every $A \in L_X$, $x \in A \Rightarrow y \in A$. Thus, since $[<]\{x\} \in L_X$ and, by (HF2), $\{x\} \subseteq \Box_<\{x\}$, that is, $x \in [<]\{x\}$, $y \in [<]\{x\}$, and hence $x < y$. □

Therefore we get the following discrete representations for HF-algebras and HF-frames.

Theorem 11.5.6

(a) *Every HF-algebra is embeddable into the complex algebra of its canonical frame;*

(b) *Every HF-frame is embeddable into the canonical frame of its complex algebra.*

11.6 Symmetric Heyting algebras

A *symmetric Heyting algebra* of order $n \geq 2$, *SH_n-algebra*, is a structure $(L, \sim, \{s_i : i = 1, \ldots, n-1\})$ where L is an Heyting algebra and the unary operations $\sim$ and s_i, $i = 1, \ldots, n-1$ satisfy, for any $a, b \in L$,

(SH_n1) $\sim\sim a = a$

(SH_n2) $s_i(\sim a) = \sim(s_{n-i}(a))$, $\quad i = 1, \ldots, n-1$

(SH_n3) $s_i s_j(a) = s_j(a) \qquad i, j = 1, \ldots, n-1$

(SH_n4) $s_1(a) \leq a$

(SH_n5) $s_i(a \wedge b) = s_i(a) \wedge s_i(b)$

(SH_n6) $s_i(a \to b) = (s_i(a) \to s_i(b)) \wedge \ldots \wedge (s_{n-1}(a) \to s_{n-1}(b))$

(SH_n7) $s_1(a) \vee \neg s_1(a) = 1$, where $\neg a = a \to 0$.

An *SH_n-frame* is a structure $(X, N, \{S_i : i = 1, \ldots, n-1\})$ where X is an H-frame and N and S_i, $i = 1, \ldots, n-1$, are functions on X satisfying, for any $x, y \in X$,

(FSH_n1) $x \leqslant y \Rightarrow N(y) \leqslant N(x)$

(FSH_n2) $N(N(x)) = x$

(FSH_n3) $N(S_i(x)) = S_{i-1}(N(x))$, $\quad i = 1, \ldots, n-1$

(FSH_n4) $S_j S_i(x) = S_j(x)$, $\quad i, j = 1, \ldots, n-1$

(FSH$_n$5) $S_1(x) \leqslant x$
(FSH$_n$6) $x \leqslant S_{n-1}(x)$
(FSH$_n$7) $S_i(x) \leqslant S_j(x)$ for $i \leq j$
(FSH$_n$8) $x \leqslant y$ implies $S_i(x) \leqslant S_i(y)$ and $S_i(y) \leqslant S_i(x)$, $i = 1, \ldots, n-1$
(FSH$_n$9) $S_i(y) \leqslant y$ and $y \leqslant S_i(y)$ imply $S_i(y) = y$, $i = 1, \ldots, n-1$
(FSH$_n$10) $x \leqslant S_i(x)$ or $S_{i+1}(x) \leqslant x$, $i = 1, \ldots, n-2$, $n \geq 3$.

In the next lemma we collect some properties of S_i, $i = 1, \ldots, n-1$, that will be needed in the proof of Proposition 11.6.2.

Lemma 11.6.1 *Let $(X, N, \{S_i : i = 1, \ldots, n-1\})$ be an SH_n-frame. Then*

(a) *For every $x \in X$ there is some $i \in \{1, \ldots, n-1\}$ such that $x = S_i(x)$;*
(b) *If $y = S_i(y)$, then for every $z \in X$ and for every $j \in \{1, \ldots, n-1\}$,*

$$z = S_j(y) \text{ implies } y = S_i(z);$$

(c) *If $S_i(x) \leqslant y$, then $S_i(y) \leqslant y$;*
(d) *If $S_i(x) \leqslant y$, then for some $j \geq i$ and for some $z \in X$,*

$$x \leqslant z \text{ and } y = S_j(z).$$

Proof:
(a) Let $M_x = \{j \in \{1, \ldots, n-1\} : S_j(x) \leqslant x\}$. The set M_x is non-empty, since by (FSH$_n$5) we have $S_1(x) \leqslant x$. The set M_x is finite, so there is a maximal element i in M_x. If $i = n-1$, then since $i \in M_x$ we have $S_{n-1}(x) \leqslant x$. By (FSH$_n$6) and (FSH$_n$9) we obtain $x = S_i(x)$. If $i < n-1$, then since i is maximal, we have $i+1 \notin M_x$. Hence not $S_{i+1}(x) \leqslant x$. From (FSH$_n$10) it must be $x \leqslant S_i(x)$. Since $i \in M_x$, we have $S_i(x) \leqslant x$. From (FSH$_n$9) we get $x = S_i(x)$.

(b) Assume $z = S_j(y)$. Then by the assumption and (FSH$_n$4),

$$S_i(z) = S_i S_j(y) = S_i(y) = y.$$

(c) Assume $S_i(x) \leqslant y$. Then by (FSH$_n$8), $S_i(y) \leqslant S_i S_i(x)$ and by (FSH$_n$4), $S_i(y) \leqslant S_i(x)$. By transitivity of $\leqslant$ we get $S_i(y) \leqslant y$, as required.

(d) By (a), there exist h and k such that $x = S_h(x)$ and $y = S_k(y)$. Since $S_i(x) \leqslant y$, so by (FSH$_n$8) we have $S_h S_i(x) \leqslant S_h(y)$. From (FSH$_n$4) we obtain $S_h(x) \leqslant S_h(y)$. We conclude that $x \leqslant S_h(y)$. Put $z = S_h(y)$. Then $x \leqslant z$. Using (FSH$_n$4) we obtain $S_k(z) = S_k S_h(y) = S_k(y) = y$. Hence we have $y = S_k(z)$. If $k \geq i$, then we put $j = k$ and the required condition holds. If $k < i$, then by (FSH$_n$7) we have $S_k(y) \leqslant S_i(y)$. Hence we have $y \leqslant S_i(y)$. On the other hand, by the assumption and statement (c) we obtain $S_i(y) \leqslant y$. Using (FSH$_n$9) we obtain $y = S_i(y)$. By (b) we get $y = S_i(z)$. Choosing $j = i$ the required result follows. □

The *complex algebra of a* SH_n*-frame* $(X, N, \{S_i : i=1, \ldots, n-\}$ is a structure $(L_X, \sim_N, \{s_{S_i} : i=1, \ldots, n-1\})$ where L_X is the complex algebra of its H-frame reduct and for every $A \in L_X$,

$$\sim_N(A) \stackrel{\text{def}}{=} \{x \in X : N(x) \notin A\}$$
$$s_{S_i}(A) \stackrel{\text{def}}{=} \{x \in X : S_i(x) \in A\}.$$

Proposition 11.6.2 *The complex algebra of an* SH_n*-frame is an* SH_n*-algebra.*

Proof:
We need to show closure under the operations of $\sim$ and s_{S_i} for $i = 1, \ldots, n-1$, that is, $\sim_N(A) = [\leqslant]\sim_N A$ and $s_{S_i}(A) = [\leqslant]s_{S_i}(A)$. In both cases the $\supseteq$ inclusion follows from reflexivity of $\leqslant$.
Assume $x \in \sim_N A$. Take any $y \in X$ such that $x \leqslant y$. Then $N(x) \notin A$ and, by (N1), $N(y) \leqslant N(x)$. So, since $A = [\leqslant]A$, $N(y) \notin A$, as required.

Assume $x \in s_{S_i}(A)$. Take any y such that $x \leqslant y$. Then $S_i(x) \in A$ and, by (S6), $S_i(x) \leqslant S_i(y)$. Hence, since $A = [\leqslant]A$, $S_i(y) \in A$, as required.

(SH_n2), (SH_n3), (SH_n4) and (SH_n5) follow easily from the frame conditions (FSH_n3), (FSH_n4), (FSH_n5), and the fact that every $A \in L_X$ is $\leq$-increasing.

(SH_n6) We prove that the following statements are equivalent

(i) $\forall y \in X,\ S_i(x) \leqslant y \Rightarrow y \in -A \cup B$;
(ii) $\forall j \geq i,\ \forall y \in X,\ x \leqslant y \Rightarrow S_j(y) \in -A \cup B$.

From Lemma 11.6.1(d) it follows that (ii) implies (i). In order to show (i) implies (ii), we prove the contrapositive. Assume that there is some $j \geq i$ and some $y \in X$ such that $x \leqslant y$ and $S_j(y) \notin -A \cup B$. Then, by (FSH_n7), $S_i(x) \leqslant S_j(x)$ and, by (FSH_n8), $S_j(x) \leqslant S_j(y)$. Hence $S_i(x) \leqslant S_j(y)$. Hence (i) does not hold.

(SH_n7) Suppose that for every $x \in X$, $x \notin s_{S_1}(A) \cup \sim_N(s_{S_1}(A))$. So $S_1(x) \notin A$ and there exists some y such that $x \leqslant y$ and $S_1(y) \in A$. Since $x \leqslant y$, from (FSH_n8) we have $S_1(x) \leqslant S_1(y)$ and $S_1(y) \leqslant S_1(x)$. Since A is $\leqslant$-increasing, $S_1(y) \notin A$, a contradiction. □

The *canonical frame of an* SH_n*-algebra* L is a relational structure

$$(X_L, N_\sim, \{S_{s_i} : i=1, \ldots, n-1\})$$

where X_L is the canonical frame of its H-algebra reduct and for every $F \in X_L$

$$N_\sim(F) \stackrel{\text{def}}{=} \{a \in L : \sim a \notin F\}$$
$$S_{s_i}(F) \stackrel{\text{def}}{=} \{a \in L : s_i(a) \in F\}, \quad i=1, \ldots, n-1.$$

Lemma 11.6.3 *If* $F \in X_L$ *then* $N_\sim(F) \in X_L$ *and, for all* $i=1, \ldots, n-1$, $S_{s_i}(F) \in X_L$.

Proof:
The proof for $N_\sim$ follows from the properties of the De Morgan negation $\sim$. We show that $S_{s_i}(F)$ is a prime filter. Note that $a, b \in S_{s_i}(F)$ if and only if $s_i(a)$, $s_i(b) \in F$. Then $s_i(a) \wedge s_i(b) \in F$. By axiom (s4) $s_i(a \wedge b) \in F$. Hence $a \wedge b \in S_{s_i}(F)$.

Now let $s_i(a) \in F$ and $a \leq b$. The latter means that $a \to b = 1$, and hence $a \to b \in F$. Since $a \to b \leq s_{n-1}(a \to b)$, we have $s_{n-1}(a \to b) \in F$. By axiom (s5) for all $j \leq n-1$ we have $s_j(a) \to s_j(b) \in F$. In particular, it holds if we put $j = i$. So we get $s_i(a) \wedge (s_i(a) \to s_i(b)) \in F$. Since for any $c, d \in L$, $c \wedge (c \to d) \leq d$, we have $s_i(b) \in F$, so $b \in S_{s_i}(F)$.

To show that $S_{s_i}(F)$ is a prime filter, assume $a \vee b \in S_{s_i}(F)$. Then $s_i(a \vee b) \in F$. Since s_i distributes over $\vee$, we get $s_i(a) \in F$ or $s_i(b) \in F$, as required. □

Proposition 11.6.4 *The canonical frame of an SH_n-algebra is an SH_n-frame.*

Proof:
By Lemma 11.6.3, X_L is closed under $N_\sim$ and S_{s_i}, $i = 1, \ldots, n-1$.
Now we check that the frame axioms (FSH_n3) to (FSH_n10) are satisfied. Axioms (FSH_n4) and (FSH_n5) follow easily from the axioms (SH_n3) and (SH_n4), respectively. Axioms (FSH_n8) and (FSH_n9) follow from the properties of set inclusion.

(FSH_n3) Note that

$$a \in N_\sim(S_{s_i}(F)) \Leftrightarrow {\sim}a \notin S_{s_i}(F) \Leftrightarrow s_i({\sim}a) \notin F \Leftrightarrow {\sim}s_{n-i}(a) \notin F$$

by axiom (SH_n2). Moreover,

$${\sim}s_{n-i}(a) \notin F \Leftrightarrow s_{n-i}(a) \in N_\sim(F) \Leftrightarrow a \in S_{s_{n-i}}(N_\sim(F)).$$

(FSH_n6) Let $a \in F$ and suppose that $a \notin S_{s_{n-1}}(F)$. Hence $s_{n-1}(a) \notin F$. Since for every $a \in L$, $a \leq s_{n-1}(a)$ we get $a \notin F$, a contradiction.

(FSH_n7) Let $i \leq j$ and let $a \in S_{s_i}(F)$. Then $s_i(a) \in F$. Since for every $a \in L$, if $i \leq j$, then $s_i(a) \leq s_j(a)$, we have $s_j(a) \in F$. Hence $a \in S^c_j(F)$.

(FSH_n8) Let $a \in S_{s_i}(G)$. Then $s_i(a) \in G$, and hence $\neg s_i(a) \notin G$, for otherwise we have $0 = s_i(a) \wedge \neg s_i(a) \in G$, a contradiction. Suppose $s_i(a) \notin F$. Since for all i, $s_i(a) \vee \neg s_i(a) = 1$, we have $s_i(a) \vee \neg s_i(a) \in F$. But F is prime, so it follows that $\neg s_i(a) \in F$. By the assumption, $\neg s_i(a) \in G$, a contradiction.

(FSH_n10) Suppose that there exists some a such that $a \in F$ and $s_i(a) \notin F$ and there exists some b such that $s_{i+1}(b) \in F$ and $b \notin F$. Hence $a \wedge s_{i+1}(b) \in F$. Since for all $a, b \in L$, $a \wedge s_{i+1}(b) \leq b \vee s_i(a)$, we get $b \vee s_i(a) \in F$. Since F is a prime filter, $b \in F$ or $s_i(a) \in F$, a contradiction. □

We now show that the lattice-embedding $h : L \to L_{X_L}$, used in Section 2.5, preserves the unary operators $\sim$ and s_i, $i = 1, \ldots, n-1$.

Lemma 11.6.5 *Let L be SH$_n$-algebra. For every $a \in L$,*

(a) $h(\sim a) = \sim_{N_\sim}(h(a))$;

(b) $h(s_i(a)) = s_{S_{s_i}}(h(a))$, $i = 1, \ldots, n-1$.

Proof:

(a) We have

$$F \in \sim_{N_\sim}(h(a)) \Leftrightarrow N_\sim(F) \notin h(a) \Leftrightarrow a \notin N_\sim(F) \Leftrightarrow \sim a \in F \Leftrightarrow F \in h(\sim a).$$

(b) The following equivalences hold:

$$F \in s_{S_i}(h(a)) \Leftrightarrow S_{s_i}(F) \in h(a) \Leftrightarrow a \in S_{s_i}(F)m \Leftrightarrow s_i(a) \in F.$$

□

We now show that the order-embedding $k : X \to X_{L_X}$, used in Section 2.5, preserves the functions N and S_i for $i = 1, \ldots, n-1$.

Lemma 11.6.6 *Let X be an SH$_n$-frame. For all $x, y \in X$*

(a) $N(x) = y \Leftrightarrow N_{\sim_N}(k(x)) = k(y)$;

(b) $S_i(x) = y \Leftrightarrow S_{s_{S_i}}(k(x)) = k(y)$, *for* $i = 1, \ldots, n-1$.

Proof:

(a) Note that, for all $x, y \in X$,

$$\begin{aligned} N_{\sim_N}(k(x)) = k(y) &\Leftrightarrow \forall A \in L_X \sim_N A \notin k(x) \Leftrightarrow y \in A \\ &\Leftrightarrow \forall A \in L_X\ x \notin \sim_N A \Leftrightarrow y \in A \\ &\Leftrightarrow \forall A \in L_X\ N(x) \in A \Leftrightarrow y \in A. \end{aligned}$$

Assume that $N(x) = y$. Then clearly, for every $A \in L_X$, $N(x) \in A$ if and only if $y \in A$. On the other hand, assume that $N_{\sim_N}(k(x)) = k(y)$. Now $\uparrow_\leqslant N(x) \in L_X$ and $N(x) \in \uparrow_\leqslant N(x)$, so $y \in \uparrow_\leqslant N(x)$, that is, $N(x) \leqslant y$. Also $\uparrow_\leqslant y \in L_X$ and $y \in \uparrow_\leqslant y$, thus $N(x) \in \uparrow_\leqslant y$, that is, $y \leqslant N(x)$. Hence $N(x) = y$.

(b) Note that, for any $x, y \in X$,

$$\begin{aligned} S_{s_{S_i}}(k(x)) = k(y) &\Leftrightarrow \forall A \in L_X\ s_{S_i}(A) \in k(x) \Leftrightarrow y \in A \\ &\Leftrightarrow \forall A \in L_X\ x \in s_{S_i}(A) \Leftrightarrow y \in A \\ &\Leftrightarrow \forall A \in L_X\ S_i(x) \in A \Leftrightarrow y \in A. \end{aligned}$$

Assume that $S_i(x) = y$. Then clearly, for every $A \in L_X$, $S_i(x) \in A$ if and only if $y \in A$. On the other hand, assume that $s_{S_{s_i}}(k(x)) - k(y)$. Now $\uparrow_\leqslant S_i(x) \in L_X$ and $S_i(x) \in \uparrow_\leqslant S_i(x)$, so $y \in \uparrow_\leqslant S_i(x)$, that is, $S_i(x) \leqslant y$. Also $\uparrow_\leqslant y \in L_X$ and $y \in \uparrow_\leqslant y$, thus $S_i(x) \in \uparrow_\leqslant y$, that is, $y \leqslant S_i(x)$. Hence $S_i(x) = y$. □

Therefore we get the following discrete representations for SH$_n$-algebras and SH$_n$-frames.

Theorem 11.6.7

(a) *Every SH$_n$-algebra is embeddable into the complex algebra of its canonical frame;*

(b) *Every SH$_n$-frame is embeddable into the canonical frame of its complex algebra.*

11.7 Heyting-Brouwer algebras

A *Heyting-Brouwer algebra*, *HB-algebra*, $(L, \div)$ is a Heyting algebra L together with the binary operator $\div$, called a *difference* which is a residuum of $\vee$, that is, for every $a, b, c \in L$,

(HB) $a \div b \leq c \Leftrightarrow a \leq b \vee c$.

Note that $\div$ is not a symmetric difference. We can define a Brouwer negation in terms of $\div$ by

$$\ulcorner a \stackrel{\text{def}}{=} 1 \div a.$$

An *HB-frame* is just an H-frame $(X, \leqslant)$.

The *complex algebra of an HB-frame* $(X, \leqslant)$ is the structure $(L_X, \div_{\leqslant})$ where L_X is the complex algebra of $(X, \leqslant)$ considered as an H-frame and the binary operator $\div_{\leqslant}$ is defined, for all $A, B \in L_X$,

$$A \div_{\leqslant} B \stackrel{\text{def}}{=} \langle \geqslant \rangle (A \cap -B).$$

Proposition 11.7.1 *The complex algebra of an HB-frame is an HB-algebra.*

Proof:
It is sufficient to show that L_X is closed on $\div_{\leqslant}$, that is, $A \div_{\leqslant} B = [\leqslant](A \div_{\leqslant} B)$. From reflexivity of $\leqslant$, $[\leqslant](A \div_{\leqslant} B) \subseteq A \div_{\leqslant} B$. For the reverse inclusion, first observe that for every $A \subseteq X$, $\langle \geqslant \rangle A$ is $\leqslant$-increasing. Assume that, for some $z \in X$, $x \geqslant z$ and $z \in A$ and $x \leqslant y$. Suppose $y \notin \langle \geqslant \rangle A$. Thus, for every t, if $y \geqslant t$, then $t \notin A$. Take t to be z. Since $z \leqslant x \leqslant y$, we have $z \leqslant y$. Hence $z \notin A$, a contradiction. Now assume that $x \in \langle \geqslant \rangle (A \cap -B)$. Suppose that, for some y, $x \leqslant y$ and $y \notin \langle \geqslant \rangle (A \cap -B)$. Since $\langle \geqslant \rangle A$ is $\leqslant$-increasing, $x \notin \langle \geqslant \rangle (A \cap -B)$, a contradiction.

Now we show that, for all $A, B, C \in L_X$, $A \div_{\leqslant} B \subseteq C \Leftrightarrow A \subseteq B \cup C$. Assume $A \div_{\leqslant} B \subseteq C$, that is, $\langle \geqslant \rangle (A \cap -B) \subseteq C$. Hence

$$\forall x, \ (\exists y, \ x \geqslant y \ \& \ y \in A \ \& \ y \notin B) \Rightarrow x \in C.$$

Equivalently,

$$\forall x, \forall y, \ (x \geqslant y \ \& \ y \in A \ \& \ y \notin B) \Rightarrow x \in C.$$

In particular, if x is y, then we have

$$\forall y, \ y \in A \Rightarrow y \in B \text{ or } y \in C.$$

On the other hand, assume $A \subseteq B \cup C$. Let $x \in A \div_{\leqslant} B$. Hence, for some y, $x \geqslant y$ and $y \in A$ and $y \notin B$. Since $y \in A$, we have $y \in B$ or $y \in C$. Since $y \notin B$, necessarily $y \in C$. But $C \in L_X$, so $x \in C$, since $y \leqslant x$ and $y \in C$. □

The *canonical frame of an HB-algebra* is defined in the same way as the canonical frame of an H-algebra. Therefore Proposition 11.2.3 carries over to HB-frames.

Proposition 11.7.2 *The canonical frame of any HB-algebra is an HB-frame.*

Let L be an HB-algebra. We now show that the lattice-embedding $h : L \to L_{X_L}$, used in Section 2.5, preserves the binary operator $\div$.

Lemma 11.7.3 *For any $a, b \in L$, $h(a \div b) = \langle\supseteq\rangle(h(a) \cap -h(b))$.*

Proof:
Let $F \in X_L$ be such that $F \in h(a \div b)$. Then $a \div b \in F$. Consider an ideal of L generated by $-F \cup \{b\}$, say I'. Clearly, $-F \subseteq I'$ and $b \in I'$. Suppose $a \div b \in I'$. Then there exists some $c \in -F$ such that $a \div b \leq b \vee c$. Now we have $b \vee c = b \vee (b \vee c) \geq b \vee (a \div b) \geq a$. By (HB), $c \geq a \div b$. Since $c \in -F$ and $-F$ is an ideal of L, $a \div b \in -F$, a contradiction. Hence $a \div b \notin I'$. By the prime ideal theorem there exists a prime ideal, say I, such that $I' \subseteq I$ and $a \div b \notin I$. But $a \div b \leq a$, thus $a \notin I$. Moreover, since $-F \subseteq I'$, we get $-F \subseteq I$, and since $b \in I'$, it holds $b \in I$. Therefore there is a prime filter of L, namely $G = -I$, such that $G \subseteq F$, $a \in G$, and $b \notin G$. Hence $F \in \langle\supseteq\rangle(h(a) \cap -h(b))$, as required.
On the other hand, note that $a \vee b = b \vee (a \div b)$. Then $h(a) \cup h(b) = h(b) \cup h(a \div b)$. We intersect both sides of this equality with $-h(b)$ and we obtain $h(a) \cap -h(b) \subseteq h(a \div b)$. □

Since HB-frames are H-frames, the order-embedding $k : X \to X_{L_X}$, used in Section 2.5, is an embedding of HB-frames. Thus we have the following discrete representations for HB-algebras and HB-frames.

Theorem 11.7.4

(a) *Every HB-algebra is embeddable into the complex algebra of its canonical frame;*

(b) *Every HB-frame is embeddable into the canonical frame of its complex algebra.*

11.8 Apartness algebras

Let L be a Heyting algebra and let the ordering on L be denoted by $\leq$. An element a of L is called *complemented*, if $a \vee \neg a = 1$, where $\neg a$ is the pseudocomplement of a defined in Section 11.2. The set $B(L)$ of complemented elements of L is a subalgebra of L and a Boolean algebra. If $A \subseteq L$, we set

$$B(A) \stackrel{\text{def}}{=} A \cap B(L).$$

An *apartness algebra*, *A-algebra*, is a structure (L, ρ) where L is a Heyting algebra and a unary operator ρ on L satisfying, for all $a, b \in L$,

(A1) $\rho(0) = 1$

(A2) $\rho(a \vee b) = \rho(a) \wedge \rho(b)$

(A3) $\rho(a) \leq \neg a$

(A4) $a \leq \rho\rho(a)$

(A5) $\rho(a) \vee \rho\rho(a) = 1$.

The operator ρ is an abstract counterpart to an apartness complement discussed in [212]. An operator ρ is called *trivial*, if $\rho(a) = 0$ for every $a \neq 0$. An apartness algebra is *trivial* if its operator ρ is trivial. The axiom (A5) tells us that the image of ρ consists of complemented elements $\rho(a)$, in particular, that $\rho\rho(a) = \neg\rho(a)$, since $\rho(a) \wedge \rho\rho(a) = 0$ by (A3).

If L is a Stone algebra (see Section 10.2), then ρ obviously satisfies (A5). On the other hand, not every apartness algebra is a Stone algebra, as the example in Figure 11.1 shows. There, $\rho(\neg a) = 0$, and $\rho(b) = \rho(a)$.

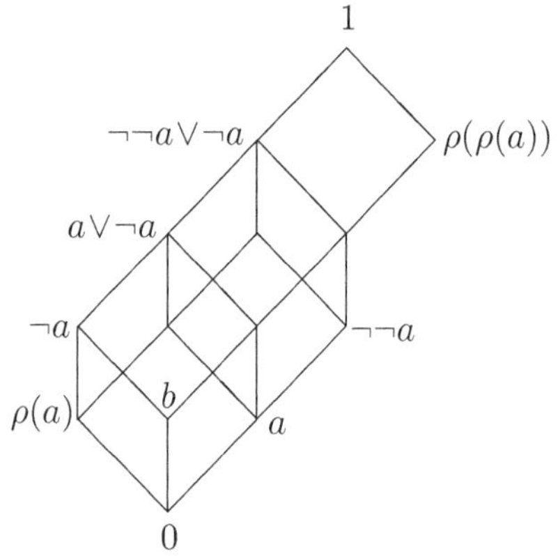

Figure 11.1: A non–Stone apartness algebra

Some simple properties of A-algebras are listed in the following lemma.

Lemma 11.8.1 *Let* (L, ρ) *be an A-algebra. For any* $a, b \in L$,

(a) ρ *is a sufficiency operator;*

(b) $\rho(a) = \rho\rho\rho(a)$;

(c) $a \leq b \leq \rho(\rho(a)) \Rightarrow \rho(b) = \rho(a)$;

(d) *if* $F, G \subseteq L$, $G = [\leqslant]G$ *and* $F \cap \rho(G) \neq \emptyset$ *then* $B(F) \cap \rho(B(G)) \neq \emptyset$.

Proof:
(a) Follows from (A1) and (A2).

(b) By (A4), $\rho(a) \leq \rho\rho\rho(a)$. Conversely, since $a \leq \rho\rho(a)$ by (A4) and ρ is antitone, we have $\rho\rho\rho(a) \leq \rho(a)$.

(c) Follows immediately from (A2) and (b).

(d) Assume $F \cap \rho(G) \neq \emptyset$, say, $a \in G$ and $\rho(a) \in F$. By (A5), $\rho(a), \rho\rho(a) \in B(L)$. Since $G = [\leqslant]G$ and $a \leq \rho\rho(a)$ by (A4), $\rho\rho(a) \in B(G)$, and $\rho\rho\rho(a) = \rho(a) \in F$. $\square$

An *apartness frame*, *A-frame*, is a structure (X, R) where X is a H-frame and R is a non-empty binary relation on X such that

(FA1) $\forall x, y \in X,\ x \leqslant y$ implies $R(x) \subseteq R(y)$

(FA2) R is irreflexive

(FA3) R is symmetric

(FA4) R is co-transitive.

Relations satisfying (FA2), (FA3), and (FA4) are considered in theories of incomplete information, where they are called diversity relations, see, for example, [40].

The *complex algebra of an A-frame* X is the structure (L_X, ρ_R) where L_X is the complex algebra of its H-frame reduct and, for every $A \subseteq X$,

$$\rho_R(A) \stackrel{\text{def}}{=} [\![R]\!](A)$$

where the operator $[\![\,]\!]$ is defined in Section 1.8.

Proposition 11.8.2 *The complex algebra of an A-frame is an A-algebra.*

Proof:
It was shown in Section 11.2 that L_X is a Heyting algebra. We show that ρ_R is well defined. Let $A \in L_X$, $x \in \rho_R(A)$, and $x \leqslant y$. By the definition of ρ_R we get $A \subseteq R(x)$, and (FA1) shows that $A \subseteq R(y)$. Thus $y \in \rho_R(A)$, and it follows that $\rho_R(A) = [\leqslant]\rho_R(A)$.

(A1) and (A2) follow from the fact that ρ_R is a sufficiency operator.

(A3) Let $x \in \rho(A)$. Then $A \subseteq R(x)$, and the irreflexivity of R shows that $x \notin A$. Hence, $\rho(A) \cap A = \emptyset$.

(A4) Let $x \in A$. We need to show that $x \in \rho_R\rho_R(A)$, that is, $\rho_R(A) \subseteq R(x)$. Assume that $y \in \rho_R(A)$, thus $A \subseteq R(y)$. Then $x \in A$ implies yRx and the symmetry of R implies that $y \in R(x)$.

(A5) Suppose there is some $x \in X$ such that $x \notin \rho_R(A)$ and $x \notin \rho_R\rho_R(A)$. Then there is some $y \in A$ such that $x(-R)y$ and there is some $z \in \rho_R(A)$ such that $x(-R)z$. Since R is symmetric, so is $-R$. Hence we have $z(-R)x$ and by co-transitivity of R we get $z(-R)y$. But since $z \in \rho_R(A)$ and $y \in A$, we have zRy, a contradiction. □

The *canonical frame of an A-algebra* L is a structure (X_L, R_ρ) where X_L is the canonical frame of its Heyting algebra reduct and, for all $F, G \in X_L$,

$$(F, G) \in R_\rho \stackrel{\text{def}}{\Leftrightarrow} F \cap \rho(G) \neq \emptyset.$$

Consider the following property of the operator ρ:

(ρ) $\rho(a) \wedge b = 0 \Rightarrow \rho(a) \leq \rho(b)$.

The property (ρ) can be expressed equationally in various ways.

Lemma 11.8.3 *The following are equivalent, for all $a, b \in L$,*

(a) $\rho(a) \wedge b = 0 \Rightarrow \rho(a) \leq \rho(b)$;
(b) $\rho(a) = \rho(\neg\rho(a))$;
(c) $\neg\rho(a) = \rho\rho(a)$.

Proof:
Assume that (a) holds. Then, since $\rho(a) \wedge \neg\rho(a) = 0$, we get $\rho(a) \leq \rho(\neg\rho(a))$. Conversely, since $\neg$ is antitone, by(A3) we have $\neg\neg a \leq \neg\rho(a)$. Since $a \leq \neg\neg a$, $a \leq \neg\rho(a)$. Since ρ is antitone, $\rho(\neg\rho(a)) \leq \rho(a)$. Hence (b) holds.

Assume that (b) holds. By (A3), we have $\rho\rho(a) \leq \neg\rho(a)$. Conversely, (A4) implies $\neg\rho(a) \leq \rho\rho(\neg\rho(a))$. Thus by (b) we get $\neg\rho(a) \leq \rho\rho(a)$, and hence (c) holds.

Assume that (c) holds and $\rho(a) \wedge b = 0$. Since $\neg\rho(a)$ is the greatest element such that $\rho(a) \wedge \neg\rho(a) = 0$, $b \leq \neg\rho(a)$. By (c), $b \leq \rho\rho(a)$. Since by Lemma 11.8.1(b), $\rho\rho(a) = \rho(a)$ and ρ is antitone, $\rho(a) \leq \rho(b)$. Hence (a) holds. □

Observe that (ρ) tells us that $\rho(L) \subseteq L$, and (A5) says that $\rho(L) \subseteq B(L)$. Thus, we see that the axiom (A5) implies the property (ρ). The converse, however, is not true. Consider the Heyting algebra shown in Figure 11.2.

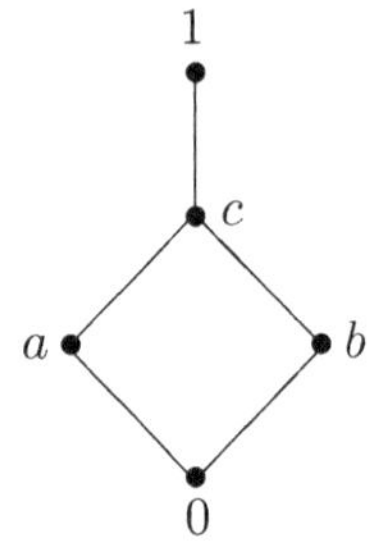

Figure 11.2: An algebra that satisfies (A1)–(A4) and (ρ), but not (A5)

If $\rho(a) = \neg a$, then L satisfies (A1)–(A4) and (ρ), but not (A5).

Proposition 11.8.4 *The canonical frame of a nontrivial A-algebra is an A-frame.*

Proof:
Since ρ is nontrivial, there is some $a \in L$, $a \neq 0$ such that $\rho(a) \neq 0$. Then there are prime filters F and G such that $a \in G$ and $\rho(a) \in F$, showing that $R_\rho \neq \emptyset$.

(FA1) Let $F \subseteq G$, and $(F, H) \in R_\rho$. Then, $F \cap \rho(H) \neq \emptyset$, thus, $G \cap \rho(H) \neq \emptyset$.

(FA2) Since $\rho(a) \wedge a = 0$ by (A3), $F \cap \rho(F) = \emptyset$.

(FA3) Let $F \cap \rho(G) \neq \emptyset$, say, $a \in G$ and $\rho(a) \in F$. Then, $\rho\rho(a) \in \rho(F)$, and since $a \leq \rho\rho(a)$ by (A4), $\rho\rho(a) \in G$.

(FA4) Let $(F, G) \in R_\rho$. We need to show that $(H, G) \notin R_\rho$ implies $(H, F) \in R_\rho$ for every $H \in X_L$. Now, $(F, G) \in R_\rho$ implies that $F \cap \rho(G) \neq \emptyset$, and so there is some $a \in G$ with $\rho(a) \in F$. Suppose that $H \cap \rho(G) = \emptyset$. In particular, $\rho(a) \notin H$. Since $\rho(a) \vee \neg\rho(a) = 1$ by (A5), and H is a prime filter, it follows that $\neg\rho(a) \in H$. Now, $\rho(a) \in F$ implies that $\rho\rho(a) \in \rho(F)$, and, by (ρ) and Lemma 11.8.3(c), $\rho\rho(a) = \neg\rho(a)$. This shows that $H \cap \rho(F) \neq \emptyset$, that is, $(H, F) \in R_\rho$. Hence, R_ρ is co-transitive. □

Let L be an A-algebra. It is shown in Section 11.2 that $h : L \to L_{X_L}$ is an embedding of Heyting algebras, hence it is sufficient to show that h preserves the operator ρ.

Lemma 11.8.5 *For any $a \in L$, $h(\rho(a)) = \rho_{R_\rho}(h(a))$.*

Proof:
Let $F \in X_L$ and $\rho(a) \in F$. We need to prove that $h(a) \subseteq R_\rho(F)$. Take G such that $a \in G$. Then $\rho(a) \in F \cap \rho(G)$, so $(F, G) \in R_\rho$. For the converse inclusion we show the contrapositive. Suppose that $\rho(a) \notin F$. To show that $F \notin \rho(h(a))$ we need to find a prime filter G such that $a \in G$ and $F \cap \rho(G) = \emptyset$.

Let $M = \{b \in B(L) : \rho(\neg b) \in F\}$. Suppose that $b_0, \ldots, b_n \in M$. Then

$$\rho(\neg b_0) \wedge \ldots \wedge \rho(\neg b_n) \in F$$

$$\rho(\neg b_0) \wedge \ldots \wedge \rho(\neg b_n) \overset{(A2)}{=} \rho(\neg b_0 \vee \ldots \vee \neg b_n) = \rho(\neg(b_0 \wedge \ldots \wedge b_n)),$$

thus M is closed under finite meets. If $b_0 \wedge \ldots \wedge b_n = 0$ then

$$\neg b_0 \vee \ldots \vee \neg b_n = 1,$$

since $b_i \in B(L)$, and thus, $\neg b_i \in F$ for some $i \in \{1, \ldots, n\}$. This contradicts (A3), since $\rho(\neg b_i) \in F$ by definition of M.

Next, we show that $M \cup \{a\}$ has the finite intersection property. By the preceding argument it is sufficient to show that $b \wedge a \neq 0$ for every $b \in M$. If $b \in M$ and $b \wedge a = 0$, then $a \leq \neg b$, and therefore, $\rho(\neg b) \leq \rho(a)$, since ρ is antitone. Now, $b \in M$ implies that $\rho(\neg b) \in F$, and furthermore, $\rho(a) \in F$. This contradicts our hypothesis $\rho(a) \notin F$. Whence $M \cup \{a\}$ has the finite intersection property and therefore the filter M' generated by $M \cup \{a\}$ is proper. Hence, by the prime filter theorem, there is a prime filter G of L containing M'. Assume that $F \cap \rho(G) \neq \emptyset$. Then there is some $c \in G$ such that $\rho(c) \in F$, and by Lemma 11.8.1(d) we conclude that $c \in B(L)$. Now, $c = \neg\neg c$, and it follows that $\neg c \in M$. Since $M \subseteq G$ and $c \in G$, we arrive at a contradiction. □

Let X be an A-frame. It is shown in Section 11.2 that the mapping $k : X \to X_{L_X}$ is an embedding of H-frames. We need to show that k preserves the relation R.

Lemma 11.8.6 *For any $x, y \in X$, $xRy \Leftrightarrow k(x) R_{\rho_R} k(y)$.*

Proof:
Clearly, $k(x)$ is a prime filter of L_X. Let xRy and set $A = \uparrow_{\leqslant} y$. Then $A \in k(y)$ and $A \subseteq R(x)$ by (FA1). Hence $x \in [[R]]A$ and therefore, $[[R]](A) \in k(x)$, since $[[R]]A \in L_X$. It follows that $k(x) \cap [[R]]k(y) \neq \emptyset$, and thus $k(x) R_\rho k(y)$. □

Therefore we get the following discrete representations for A-algebras and A-frames.

Theorem 11.8.7

(a) *Every A-algebra is embeddable into the complex algebra of its canonical frame;*

(b) *Every A-frame is embeddable into the canonical frame of its complex algebra.*

Chapter 12

Distributive residuated lattices

12.1 Introduction

Residuated lattices and, in particular, distributive residuated lattices play an important role in several branches of logic and algebra. Among the non-classical logics with semantics determined by residuated lattices there are substructural logics which include Lambek calculus and its various signature and/or axiomatic extensions and linear logic, multiple-valued logics, and fuzzy logics, among others. For recent developments in residuated lattices see [94].

In this chapter we consider distributive residuated lattices and some of their axiomatic extensions whose logics are grouped into two out of several hierarchies of substructural and fuzzy logics appearing in the literature, namely, those from [79] and [131].

The Esteva-Godo-Ono hierarchy [79] of substructural and fuzzy logics and their corresponding algebras starts with the full Lambek calculus with exchange and weakening which in the field of fuzzy logic is referred to as a monoidal logic [130]. An algebra L' is above an algebra L in the hierarchy whenever L' is an axiomatic extension of L. We consider algebras of logics from an upper part of that hierarchy consisting of some of the axiomatic extensions of the monoidal t-norm fuzzy logic MTL. Logic MTL introduced in [78] plays an important role in the field of fuzzy logics. Since its origin it is a subject of extensive study motivated by the following facts. It is complete with respect to the class of standard MTL-algebras, that is the MTL-algebras whose universe is the real unit interval [0,1] and the product is a left-continuous t-norm [137]. Furthermore, the necessary and sufficient condition for a t-norm to be residuated is left-continuity. It follows that logic MTL is the adequate tool for reasoning about left-continuous t-norms and their residua. In Section 12.5 we present a discrete duality for MTL-algebras based on [182, 175]. In Section 12.7, Section 12.8, and Section 12.9 discrete dualities for algebras SMTL, IMTL, and CMTL, respectively, are presented based on [175]. A discrete duality for ΠMTL algebras has not been approached yet.

In Section 12.9 the class of CMTL algebras was considered obtained from the MTL algebras by endowing them with the contraction axiom $a \leq a \odot a$. Since in MTL algebras also the reverse inequality $a \odot a \leq a$ holds, the product operator in these algebras is idempotent, $a = a \odot a$. Generalizations of contraction and idempotence to n-contraction law $a^n \leq a^{n+1}$ and n-potent law $a^n = a^{n+1}$, respectively, are studied in the field of residuated lattices and are shown to be related to important methodological problems. In this connection, the second hierarchy considered in this chapter is the hierarchy of n-contractive, $n \geq 2$, axiomatic extensions of logic MTL presented in [131]. An interest in the logics of that hierarchy stems from the result in [17] which shed a light on the problem of constructing a completion for axiomatic extensions of MTL satisfying the divisibility axiom. We say that a class of lattice-based algebras admits completion if for every algebra L from the class there is a complete algebra L' in that class and an embedding from L into L', where a lattice-based algebra is complete whenever its lattice reduct is complete. In [17] it is proved that any axiomatic extension of logic MTL with the axiom of divisibility added admits completion if and only if it satisfies n-potent law for some $n \geq 2$. In view of that theorem in the present chapter we approached the problem of developing a discrete duality for n-potent MTL algebras and for 2-potent BL algebras. The dualities we obtained are presented in Section 12.10 and Section 12.11, respectively, based on [67].

12.2 Distributive residuated lattices

A *residuated lattice*, *R-lattice*, is a structure $(L, \odot, \rightarrow, \leftarrow, 1')$ where L is a distributive lattice, $(L, \odot, 1')$ is a monoid, and $\rightarrow$ and $\leftarrow$ are the right residuum of $\odot$ and the left residuum of $\odot$, respectively, that is, for all $a, b, c \in L$

(R1) $a \odot b \leq c \Leftrightarrow b \leq a \rightarrow c$

(R2) $a \odot b \leq c \Leftrightarrow a \leq b \leftarrow c$

Given an R-lattice L and $A, B \subseteq L$, we define

$$A \odot B = \{c \in L : \exists a \in A \exists b \in B,\ a \odot b \leq c\}$$

Lemma 12.2.1 *Let L be an R-lattice and let $A, B, C \subseteq L$. Then*

(a) $A \odot (B \odot C) = (A \odot B) \odot C$;

(b) *If* $B \subseteq C$, *then* $(A \odot B) \subseteq (A \odot C)$.

Proof:
(a) Let $e \in A \odot (B \odot C)$. Then there is some $a \in A$ and there is some $d \in B \odot C$ such that $a \odot d \leq e$. Also, there is some $b \in B$ and there is some $c \in C$ such that $b \odot c \leq d$. By monotonicity of $\odot$ we get $a \odot (b \odot c) \leq a \odot d \leq e$. Hence, by associativity of $\odot$, $(a \odot b) \odot c \leq e$. Since $a \odot b \in A \odot B$, we get $e \in (A \odot B) \odot C$. The reverse inclusion can be proved in a similar way.

(b) Assume that $B \subseteq C$ and take $d \in A \odot B$. Then there is $a \in A$ and there is $b \in B$ such that $a \odot b \leq d$. Since $b \in B$, by assumption $b \in C$. Hence $d \in A \odot C$. □

An *R-frame* is a structure (X, R, I) where X is a bounded distributive lattice frame, R is a ternary relation on X, $I \subseteq X$, and, for all $x, x', y, y', z, z' \in X$,

(FR1) $I(x)$ & $x \leqslant x' \Rightarrow I(x')$
(FR2) $R(x,y,z)$ & $x' \leqslant x$ & $y' \leqslant y$ & $z \leqslant z' \Rightarrow R(x',y',z')$
(FR3) $R(x,y,z)$ & $R(z,y',z') \Rightarrow \exists u \in X,\ R(y,y',u)$ & $R(x,u,z')$
(FR4) $R(x,y,z)$ & $R(x',z,z') \Rightarrow \exists u \in X,\ R(x',x,u)$ & $R(u,y,z')$
(FR5) $I(x)$ & $(R(x,y,z)$ or $R(y,x,z)) \Rightarrow y \leqslant z$
(FR6) $\exists u \in X,\ I(x)$ & $R(x,u,x)$
(FR7) $\exists u \in X,\ I(x)$ & $R(u,x,x)$.

The *complex algebra of an R-frame* X is a structure

$$(L_X, \odot_R, \rightarrow_R, \leftarrow_R, 1'_I)$$

where L_X is the complex algebra of its distributive lattice frame reduct and, for all $A, B \in L_X$,

$$\begin{aligned}
1'_I &\stackrel{\text{def}}{=} I \\
A \odot_R B &\stackrel{\text{def}}{=} \{z \in X : \exists x \in A\ \exists y \in B\ R(x,y,z)\} \\
A \rightarrow_R B &\stackrel{\text{def}}{=} \{x \in X : \forall y, z \in X\ R(y,x,z) \ \&\ y \in A \Rightarrow z \in B\} \\
A \leftarrow_R B &\stackrel{\text{def}}{=} \{x \in X : \forall y, z \in X\ R(x,y,z) \ \&\ y \in A \Rightarrow z \in B\}.
\end{aligned}$$

Proposition 12.2.2 *The complex algebra of an R-frame is an R-lattice.*

Proof:
L_X is closed under $\cup$ and $\cap$. We show that it is also closed under $\odot_R$, $\rightarrow_R$, and $\leftarrow_R$, that is for all $A, B \subseteq X$, $A \odot_R B = [\leqslant](A \odot_R B)$, $A \rightarrow_R B = [\leqslant](A \rightarrow_R B)$, and $A \leftarrow_R B = [\leqslant](A \leftarrow_R B)$. In all cases the right-to-left inclusion follows from reflexivity of $\leqslant$. To show that $A \odot_R B \subseteq [\leqslant](A \odot_R B)$, assume that $x \in A \odot_R B$. Then there are $y, z \in X$ such that $y \in A$, $z \in B$, and $R(y,z,x)$. Take an $x' \in X$ such that $x \leqslant x'$. By (FR2), we get $R(y,z,x')$. Hence $x' \in A \odot_R B$. Now, we show that $A \rightarrow_R B \subseteq [\leqslant](A \rightarrow_R B)$. Let $x \notin [\leqslant] A \rightarrow_R B$. Then there is some $x' \in X$ such that $x \leqslant x'$ and $x' \notin A \odot_R B$. Hence there are $y, z \in X$ such that $R(y,x',z)$, $y \in A$, and $z \notin B$. By (FR2) it follows that $R(y,x,z)$, which gives that $x \notin A \rightarrow_R B$. In the similar way one can show that $A \leftarrow_R B \subseteq [\leqslant](A \leftarrow_R B)$. Using (FR1) it is easy to see that $1'_I \in L_X$.

Now, we show that $\odot_R$ is associative. Let $A, B, C \in L_X$, $x \in X$, and assume that $x \in A \odot_R (B \odot_R C)$. Then there are $y, z \in X$ such that $y \in A$, $z \in B \odot_R C$, and $R(y,z,x)$. Furthermore, there are also $u, t \in X$ satisfying $u \in B$, $t \in C$, and $R(u,t,z)$. By (FR4), $R(u,t,z)$ and $R(y,z,x)$ imply that there is some $v \in X$ such that $R(y,u,v)$ and $R(v,t,x)$. Now, $y \in A$, $u \in B$, and $R(y,u,v)$ yield $v \in A \odot_R B$ which together with $t \in C$ and $R(v,t,x)$ gives $x \in (A \odot_R B) \odot_R C$. Hence $A \odot_R (B \odot_R C) \subseteq (A \odot_R B) \odot_R C)$. The reverse inclusion can be shown in the similar way using (FR3).

To show that $1'_I$ is the unit element of $\odot_R$, take $A \in L_X$ and $x \in X$ such that $x \in 1'_I \odot_R A$, that is, $x \in I \odot_R A$. Then there are $y, z \in X$ such that $y \in I$, $z \in A$, and $R(y, z, x)$. Hence, by (FR5) we get $z \leqslant x$. Since $A = [\leqslant]A$, it follows that $x \in A$. For the reverse inclusion, assume that $x \in A$. By (FR7), there exists some $u \in X$ such that $u \in I$ and $R(u, x, x)$. Then $x \in I \odot_R A$. Proceeding in the similar way and using (FR6) we get $A \odot_R I = A$.

(R1) Let $A, B, C \in L_X$. We show that $A \odot_R B \subseteq X \Leftrightarrow B \subseteq A \rightarrow_R C$. Assume that $A \odot B \subseteq C$ and take $x \in X$ such that $x \notin A \rightarrow_R C$. This means that there are $y, z \in X$ such that $R(y, x, z)$, $y \in A$, and $z \notin C$. Hence, by the assumption, $z \notin A \odot_R B$, so for all $t, u \in X$, $R(u, t, z)$ and $u \in A$ imply $t \notin B$. Putting y for u and x for t we get $x \notin B$. Hence $B \subseteq A \rightarrow_R C$. For the reverse implication, assume that $B \subseteq A \rightarrow_R C$. Take $x \in A \odot_R B$. Then there are $y, z \in X$ such that $y \in A$, $z \in B$, and $R(y, z, x)$. By the assumption we get $z \in A \rightarrow_R C$. Hence, since $y \in A$ and $R(y, z, x)$, it follows that $x \in C$. Therefore $A \odot_R B \subseteq C$, as required.
The proof of (R2) is similar. □

The *canonical frame of an R-lattice* L is a structure $(X_L, R_\odot, I_{1'})$ such that X_L is the canonical frame of its lattice reduct and

$$I_{1'} \stackrel{\text{def}}{=} \{F \in X_L : 1' \in F\}$$

$$R_\odot(F, G, H) \stackrel{\text{def}}{\Leftrightarrow} F \odot G \subseteq H \text{ for all } F, G, H \in X_L.$$

Proposition 12.2.3 *The canonical frame of an R-lattice is an R-frame.*

Proof:
(FR1) The proof is straightforward.

(FR2) Let $F, F', G, G', H, H' \in X_L$ be such that $F \odot G \subseteq H$, $F' \subseteq F$, $G' \subseteq G$, and $H \subseteq H'$. Assume that $a \in F' \odot G'$. Then there are $b \in F'$, $c \in G'$ such that $b \odot c \leq a$. Hence $b \in F$ and $c \in G$, so we get $a \in F \odot G$, thus $a \in H$, and furthermore, $a \in H'$. Therefore $F' \odot G' \subseteq H'$ which means that $R_\odot(F', G', H')$.

(FR3) Let $F, G, G', H, H' \in X_L$ be such that $R_\odot(F, G, H)$ and $R_\odot(H, G', H')$, that is $F \odot G \subseteq H$ and $H \odot G' \subseteq H'$. Then, by Lemma 12.2.1(a)

$$(F \odot G) \odot G' = F \odot (G \odot G') \text{ and } (F \odot G) \odot G' \subseteq H \odot G'.$$

Since $H \odot G' \subseteq H'$, $F \odot (G \odot G') \subseteq H'$. Next, by Lemma 1.6.10, $G \odot G'$ is a filter of L. Hence, by applying theorem 1.6.11, there is a prime filter of L, say P, such that $G \odot G' \subseteq P$ and $F \odot P \subseteq H'$. Therefore $R_\odot(G, G', P)$ and $R_\odot(F, P, H')$ hold, as required.

(FR4) can be proved in the similar way.

(FR5) Let $F, G, H \in X_L$ be such that $F \in I_{1'}$ and $R_\odot(F, G, H)$ or $R_\odot(G, F, H)$ hold. This means that $1' \in F$ and $F \odot G \subseteq H$ or $G \odot F \subseteq H$. Let $a \in G$. If $F \odot G \subseteq H$, then $a = 1' \odot a \in F \odot G$, so $a \in H$. Similarly, if $G \odot F \subseteq H$, then $a = a \odot 1' \in G \odot F$, so again $a \in H$. Therefore $G \subseteq H$, as required.

(FR6) Let $F \in X_L$ and consider the principal filter generated by $1'$, $\uparrow_{\leq} 1'$. By Lemma 1.6.10 $F \odot \uparrow_{\leq} 1'$ is a filter of L. Then $a \in F \odot \uparrow_{\leq} 1'$ means that $b \odot 1' \leq a$ for some $b \in F$. But $b \odot 1' = b$, so $b \leq a$, thus $a \in F$. Therefore $F \odot \uparrow_{\leq} 1' \subseteq F$, so by Theorem 1.6.11, there is a prime filter of L, say G, such that $\uparrow_{\leq} 1' \subseteq G$ and $F \odot G \subseteq F$. Then $1' \in G$, so $G \in I_{1'}$ and $R(F, G, F)$.

The proof of (FR7) is similar. □

We now show that the lattice-embedding $h : L \to L_{X_L}$, used in Section 2.5, preserves the operations $\odot$, $\to$, and $\leftarrow$.

Lemma 12.2.4 *Let L be an R-lattice. For all $a, b \in L$,*

(a) $h(a \odot b) = h(a) \odot_{R_\odot} h(b)$;
(b) $h(a \to b) = h(a) \to_{R_\odot} h(b)$;
(c) $h(a \leftarrow b) = h(a) \leftarrow_{R_\odot} h(b)$.

Proof:
(a) Let $F \in h(a \odot b)$. Then $a \odot b \in F$. Consider the set $\uparrow_{\leq} a \odot \uparrow_{\leq} b$. By Lemma 1.6.10, it is a filter of L. Moreover, for any $c \in L$, if $c \in \uparrow_{\leq} a \odot \uparrow_{\leq} b$, then there are $a' \in \uparrow_{\leq} a$ and $b' \in \uparrow_{\leq} b$ such that $a' \odot b' \leq c$. Then $a \leq a'$ and $b \leq b'$. By monotonicity of $\odot$, we get $a \odot b \leq a' \odot b' \leq c$. Since $a \odot b \in F$, we have $c \in F$. Then $\uparrow_{\leq} a \odot \uparrow_{\leq} b \subseteq F$. Hence, by Theorem 1.6.11 there are prime filters G and H of L such that $\uparrow_{\leq} a \subseteq G$, $\uparrow_{\leq} b \subseteq H$, and $G \odot G \subseteq F$. Since $a \in \uparrow_{\leq} a$, $a \in G$ and similarly $b \in H$. Then $G \in h(a)$, $H \in h(b)$, and $R_\odot(G, H, F)$ which means that $F \in h(a) \odot_{R_\odot} h(b)$, as required.
Conversely, assume that $F \in h(a) \odot_{R_\odot} h(b)$. Then for some $G, H \in X_L$, $G \in h(a)$, $H \in h(b)$, and $R_\odot(G, H, F)$, that is, $G \odot H \subseteq F$. Hence $a \in G$ and $b \in H$, so $a \odot b \in G \odot H$, thus $a \odot b \in F$, which means that $F \in h(a \odot b)$, as required.

(b) Let $F \in h(a \to b)$. Then $a \to b \in F$. Take any $G, H \in X_L$ with $G \odot F \subseteq H$ and $a \in G$. Then $a \odot (a \to b) \in G \odot F$, so $a \odot (a \to b) \in H$. Since $a \to b \leq a \to b$, by (R1) we get $a \odot (a \to b) \leq b$, and whence $b \in H$. Therefore for all $G, H \in X_L$, $R_\odot(G, F, H)$ and $G \in h(a)$ imply $H \in h(b)$, which means that $F \in h(a) \to_{R_\odot} h(b)$, as required.
Conversely, assume that for all $G, H \in X_L$, $F \odot G \subseteq H$ and $a \in G$ implies $b \in H$ and let $F \notin h(a \to b)$, that is $a \to b \notin F$. By Lemma 1.6.10, $F \odot \uparrow_{\leq} a$ is a filter. We show that $b \notin F \odot \uparrow_{\leq} a$. For suppose $b \in F \odot \uparrow_{\leq} a$, then there exists $c, a' \in L$ such that $c \in F$, $a' \in \uparrow_{\leq} a$, $a \leq a'$ and $c \odot a \leq b$. Hence $c \leq a \to b$. Since $c \in F$, $a \to b \in F$ which is a contradiction. Hence $F \odot \uparrow_{\leq} a$ is a proper filter. By the prime filter theorem, there is some prime filter $H \in X_L$ such that $F \odot \uparrow_{\leq} a \subseteq H$ and $b \notin H$. By Theorem 1.6.11 there is some $G \in X_L$ such that $\uparrow_{\leq} a \subseteq G$ and $F \odot G \subseteq H$. Since $a \in G$, by the assumption $b \in H$ which is a contradiction.
The proof of (c) is similar. □

We now show that the order-embedding $k : X \to X_{L_X}$, used in Section 2.5, preserves relation R and the set I.

Lemma 12.2.5 *Let X be an R-frame. For all $x, y, z \in X$,*

(a) $R(x,y,z) \Leftrightarrow R_{\odot_R}(k(x),k(y),k(z))$;

(b) $x \in I \Leftrightarrow k(x) \in I_{1'_I}$.

Proof:
Note that the following equivalences hold:

$$\begin{aligned} &R_{\odot_R}(k(x),k(y),k(z) \\ &\quad\Leftrightarrow k(x) \odot_R k(y) \subseteq k(z) \\ &\quad\Leftrightarrow \forall A \in L_X,\ A \in k(x) \odot_R k(y) \Rightarrow A \in k(z) \\ &\quad\Leftrightarrow \forall A \in L_X\ (\exists B \in k(x)\ \&\ C \in k(y),\ B \odot_R C \subseteq A) \Rightarrow z \in A \\ &\quad\Leftrightarrow \forall A,B,C \in L_X,\ x \in B\ \&\ y \in C\ \&\ B \odot_R C \subseteq A \Rightarrow z \in A. \end{aligned}$$

Assume that $R(x,y,z)$ holds and consider any $A \in L_X$ such that for some $B, D \in L_X$, $x \in B$, $y \in D$, and $B \odot_R D \subseteq A$. Then $z \in B \odot_R D$, thus $z \in A$, so $R_{\odot_R}(k(x),k(y),k(z))$ holds. On the other hand, take any $x,y,z \in X$ such that $R_{\odot_R}(k(x),k(y),k(z))$. Consider $B = \langle \geqslant \rangle \{x\}$ and $C = \langle \geqslant \rangle \{y\}$. Then, clearly, $B, C \in L_X$ and $A \in L_X$ by (FR2). Furthermore, we have $B \odot_R C \subseteq A$, and hence by the assumption $z \in A$. Thus we have $R(x,y,z)$, as required.

(b) Follows easily from the corresponding definitions. □

We now obtain the discrete representations for R-lattices and R-frames.

Theorem 12.2.6

(a) *Every R-lattice is embeddable into the complex algebra of its canonical frame;*

(b) *Every R-frame is embeddable into the canonical frame of its complex algebra.*

In the next sections we consider some specific classes of R-lattices.

12.3 Bounded integral residuated lattices

A bounded R-lattice $(L, \odot, \to, \leftarrow, 1')$ is called *integral*, if

(R3) $1' = 1$.

An integral R-lattice will be written $(L, \odot, \to, \leftarrow)$.

An *integral R-frame* is a structure (X, R) satisfying (FR2), (FR3), (FR4) and, for all $x, y, z \in X$,

(FR8) $R(x,y,z)$ or $R(y,x,z) \Rightarrow y \leqslant z$.

(FR9) $\exists u \in X,\ R(x,u,x)$

(FR10) $\exists u \in X,\ R(u,x,x)$.

The *complex algebra of an integral R-frame* X is a structure

$$(L_X, \odot_R, \to_R, \leftarrow_R)$$

where L_X is a distributive lattice frame and $\odot_R$, $\to_R$, and $\leftarrow_R$ are defined by as in Section 12.2.

Proposition 12.3.1 *The complex algebra of an integral R-frame is an integral R-lattice.*

Proof:
It suffices to show that X is the unit element of $\odot_R$. Let $A \in L_X$ and let $x \in A \odot_R X$. Then there are $y \in A$ and $z \in X$ such that $R(y, z, x)$. Hence, by (FR8), $y \leqslant x$. Since $y \in A$ and $A = [\leqslant]A$, we get $x \in A$. On the other hand, let $x \in A$. By (FR9), there is some $u \in X$ such that $R(x, u, x)$. Hence $x \in A \odot_R X$. The equality $A = X \odot_R A$ can be proved in the similar way using (FR10). □

The *canonical frame of an integral R-lattice* L is the relational structure $(X_L, R_\odot)$ where X_L is the canonical frame of its lattice reduct and $R_\odot$ is defined as in Section 12.2.

Proposition 12.3.2 *The canonical frame of an integral R-lattice is an integral R-frame.*

Proof:
(FR8) Let $F, G, H \in X_L$ and let $a \in G$. If $F \odot G \subseteq H$, then, since $1 \in F$ and $a = 1 \odot a \in F \odot G$, $a \in H$. If $G \odot F \subseteq H$, then since $a = a \odot 1 \in G \odot F$, again $G \subseteq H$.

(FR9) Let $F \in X_L$. Since $\{1\}$ is a filter of L, by Lemma 1.6.10, $F \odot \{1\}$ is also a filter of L. Hence, if $a \in F \odot \{1\}$, then there is some $b \in F$ such that $b \odot 1 \leq a$. But $b \odot 1 = b$, thus $b \leq a$, so $a \in F$. Then $F \odot \{1\} \subseteq F$. By Theorem 1.6.11 it follows that there is a prime filter of L, say G, such that $F \odot G \subseteq F$. Then $R_\odot(F, G, F)$ holds, as required.

The proof of (FR10) is similar. □

We now obtain the discrete representations for integral R-lattices and integral R-frames.

Theorem 12.3.3

(a) *Every integral R-lattice is embeddable into the complex algebra of its canonical frame.*

(b) *Every integral R-frame is embeddable into the canonical frame of its complex algebra.*

12.4 Commutative residuated lattices

Now, we consider R-lattices L such that their product operator $\odot$ is commutative, that is, for all $a, b \in L$,

(R4) $a \odot b = b \odot a$.

They are referred to as *commutative R-lattices*. Note that commutativity of $\odot$ implies that $\rightarrow = \leftarrow$. Hence commutative R-lattices are written $(L, \odot, \rightarrow, 1')$.

A *commutative R-frame* is an R-frame (X, R, I) which satisfies axioms (FR1) – (FR4), (FR6), (FR7), and the following additional axiom, for all $x, y, z \in X$,

(FR11) $R(x,y,z) \Rightarrow R(y,x,z)$.

The *complex algebra of a commutative R-frame* X is a structure

$$(L_X, \odot_R, \rightarrow_R, 1'_I)$$

where L_X is the complex algebra of its distributive lattice frame reduct and $\odot_R$, $\rightarrow_R$, and $1'_I$ are defined as in Section 12.2.

Proposition 12.4.1 *The complex algebra of a commutative R-frame is a commutative R-lattice.*

Proof:
It suffices to show that for all $A, B \in L_X$, $A \odot_R B = B \odot_R A$. Let $x \in A \odot_R B$. Then there is some $y \in A$ and there is some $z \in B$ such that $R(y,z,x)$. By (FR11), this implies that $R(z,y,x)$. Hence $x \in B \odot_R A$. The reverse inclusion can be proved in the similar way. □

The *canonical frame of a commutative R-lattice* is defined in the same way as the canonical frame of an R-lattice.

Proposition 12.4.2 *The canonical frame of a commutative R-lattice is a commutative R-frame.*

Proof:
We only need to show (FR11). Let $F, G, H \in X_L$ be such that $R_\odot(F,G,H)$ holds. Then $F \odot G \subseteq H$ and since $F \odot G = G \odot F$, $R_\odot(G,F,H)$ holds. □

We now obtain the discrete representations for commutative R-lattice and commutative R-frames.

Theorem 12.4.3

(a) *Every commutative R-lattice is embeddable into the complex algebra of its canonical frame;*

(b) *Every commutative R-frame is embeddable into the canonical frame of it complex algebra.*

12.5 MTL-algebras

An *MTL-algebra* is a bounded, integral, and commutative R-lattice $(L, \odot, \rightarrow)$ satisfying the following *prelinearity axiom*, for all $a, b \in L$,

(MTL1) $(a \to b) \vee (b \to a) = 1$.

Since $\odot$ is commutative, the residuation axioms reduce to the following axiom, for all $a, b, c \in L$,

(MTL2) $a \odot b \leq c \Leftrightarrow a \leq b \to c \Leftrightarrow b \leq a \to c$

It is useful to have the following arithmetic properties of the operations $\odot$ and $\to$ of an MTL-algebra.

Lemma 12.5.1 *Let $(L, \odot, \to)$ be an MTL-algebra. For any $a, b, c \in L$,*

(a) *If $a \leq b$ then $a \odot c \leq b \odot c$;*
(b) $b \leq a \to b$;
(c) $a \odot (a \to b) \leq b$;
(d) $a \odot b \leq a \wedge b$;
(e) $a \odot (b \vee c) = (a \odot b) \vee (a \odot c)$;
(f) $a \odot (b \wedge c) = (a \odot b) \wedge (a \odot c)$.

An *MTL-frame* is a structure (X, R) such that X is its lattice frame reduct and for all $x, x', y, y', z, z' \in X$,

(FMTL1) $R(x, y, z)$ & $(x' \leqslant x)$ & $(y' \leqslant y)$ & $(z \leqslant z') \Rightarrow R(x', y', z')$
(FMTL2) $R(x, y, z)$ & $R(z, y', z') \Rightarrow \exists u \in X\ R(y, y', u)$ & $R(x, u, z')$
(FMTL3) $R(x, y, z)$ & $R(x', z, z') \Rightarrow \exists u \in X\ R(x', x, u)$ & $R(u, y, z')$
(FMTL4) $R(x, y, z) \Rightarrow y \leqslant z$
(FMTL5) $\exists u \in X,\ R(u, x, x)$
(FMTL6) $R(x, y, z) \Rightarrow R(y, x, z)$

and the following additional axiom for all $x, y, z, t, w \in X$,

(FMTL7) $R(x, y, z)$ & $R(x, t, w) \Rightarrow y \leqslant w$ or $t \leqslant z$.

Since MTL-algebras are integral residuated lattices, in their frames the set I is not needed. The following is proved in [78].

Lemma 12.5.2 *Every MTL-algebra is isomorphic to a subdirect product of linearly ordered MTL-algebras.*

As a consequence, an identity holds in all MTL–algebras iff it holds in all linearly ordered MTL algebras. Furthermore, for every MTL-algebra its lattice reduct is a distributive lattice.

The *complex algebra of an MTL-frame* X is the structure $(L_X, \odot_R, \to_R)$ where L_X is the complex algebra of its distributive lattice frame reduct and $\odot_R$ and $\to_R$ are defined as in Section 12.2.

Proposition 12.5.3 *The complex algebra of an MTL-frame is an MTL-algebra.*

Proof:
In view of Proposition 12.3.1 and Proposition 12.4.1 it suffices to show that (MTL1) holds. Let $A, B \in L_X$ and suppose that there is some $x \in X$ such that $x \notin (A \to_R B) \cup (B \to_R A)$. Then $x \notin A \to_R B$ and $x \notin B \to_R A$, and hence there are $y, y', z, z' \in X$ such that $R(x, y, z)$, $y \in A$, $z \notin B$, $R(x, y', z')$, $y' \in B$, and $z' \notin A$. Then, by (FMTL7), $y \leqslant z'$ or $y' \leqslant z$. If $y \leqslant z'$, then, since $y \in A$ and $A = [\leqslant]A$, we get $z' \in A$, a contradiction. If $y' \leqslant z$, then since $y' \in b$, we get $z \in B$ which is also a contradiction. □

The *canonical frame of an MTL-algebra* L is a structure $(X_L, R_\odot)$ where X_L is the canonical frame of its distributive lattice reduct and $R_\odot$ is a ternary relation on X_L defined as in Section 12.2, that is $R_\odot(F, G, H) \Leftrightarrow F \odot G \subseteq H$

Proposition 12.5.4 *The canonical frame of an MTL-algebra is an MTL-frame.*

Proof:
In view of Proposition 12.2.3, Proposition 12.3.2, and Proposition 12.4.2 it suffices to show that (FMTL7) holds. Let $F, G, G', H, H' \in X_L$ be such that $F \odot G \subseteq H$ and $F \odot G' \subseteq H'$. Suppose that neither $G \subseteq H'$ nor $G' \subseteq H$. Then there are $a, b \in L$ such that $a \in G$, $a \notin H'$, $b \in G'$, and $b \notin H$. By (MTL1), since F is a prime filter, $a \to b \in F$ or $b \to a \in F$. If $a \to b \in F$, then since $a \in G$, $(a \to b) \odot a \in F \odot G$, and hence $(a \to b) \odot a \in H$. But $a \to b \leq a \to b$, so by (MTL2), $(a \to b) \odot a \leq b$ and hence $b \in H$ which gives the required contradiction. In the case where $b \to a \in F$ the proof is similar. □

We now obtain the discrete representations for MTL-algebras and MTL-frames.

Theorem 12.5.5

(a) *Every MTL-algebra is embeddable into the complex algebra of its canonical frame;*

(b) *Every MTL-algebra is embeddable into the canonical frame of it complex algebra.*

12.6 Axiomatic extensions of MTL-algebras

Let an MTL-algebra $(L, \odot, \to)$ be given. A negation in L is defined, for any $a \in L$, by

$$\neg a \stackrel{\text{def}}{=} a \to 0.$$

Some properties of $\neg$ are listed in the following lemma.

Lemma 12.6.1 *For every MTL-algebra L and for all $a, b, c \in L$,*

(a) $a \leq \neg\neg a$;

(b) $(a \to b) \leq (\neg b \to \neg a)$;

(c) $a \odot \neg a = 0$.

In [79] a joint hierarchy of algebras of substructural logics and fuzzy logics is presented. In the hierarchy an algebra L' is above an algebra L whenever L' is an axiomatic extension of L. In the following sections we consider some of the logics from the upper part of this hierarchy consisting of the following axiomatic extensions of MTL-algebras

$$\text{SMTL} = \text{MTL} + a \wedge \neg a = 0 = \text{MTL} + \neg a \vee \neg\neg a = 1$$
$$\text{IMTL} = \text{MTL} + \neg\neg a \leq a$$
$$\text{CMTL} = \text{G} = \text{MTL} + a \leq a \odot a$$
$$\pi\text{MTL} = \text{SMTL} + (\pi 1)\ \neg\neg c \leq (((a \odot c) \to (b \odot c)) \to (a \to b))$$
$$\text{BL} = \text{MTL}+(\text{div})\ a \wedge b = a \odot (a \to b)$$
$$\mathfrak{L} = \text{MV} = \text{MTL} + a \vee b = (a \to b) \to b = \text{IMTL} + (\text{div}) = \text{BL} + \neg\neg a \leq a$$
$$\text{SBL} = \text{SMTL} + (\text{div})$$
$$\pi = \pi\text{MTL} + (\text{div}) = \text{SBL} + (\pi 1)$$
$$\text{Bool} = \text{G} + a \vee \neg a = 1$$

The axiom (div) is referred to as *divisibility.*

The algebras from the hierarchy and their logics are studied, among others, in [118, 119, 28].

The second hierarchy of logics based on distributive residuated lattices considered in this chapter is the one presented in [131], namely, a hierarchy of n-contractive fuzzy logics, whose semantics is determined by some axiomatic extensions of MTL-algebras. Since in logic MTL n-contraction is equivalent to n-potent law, the hierarchy may also be seen as a hierarchy of n-potent extensions of logic MTL. The hierarchy includes the families of logics C_nMTL and C_nIMTL obtained from logic MTL and IMTL, respectively, by adding to them the n-potent axiom. In the hierarchy, for $n \geqslant 1$, a logic C_nMTL is an axiomatic extension of logic C_{n+1}, and similarly for the logics C_nIMTL. Logics C_nMTL originated in [26], see also[13, 241]. Note that in this hierarchy the logic C_1MTL coincides with the logic CMTL to be considered in Section 12.9. Furthermore, the logic C_1IMTL coincides with the classical propositional logic Bool. The logic MTL is considered in Section 12.5 and logic IMTL in Section 12.8.

12.7 SMTL-algebras

An *SMTL-algebra* is an MTL-algebra $(L, \odot, \to)$ satisfying, for every $a \in L$,

(MTL3) $a \wedge \neg a = 0$.

An *SMTL-frame* is an MTL-frame (X, R) satisfying, for every $x \in X$,

(FMTL8) $\exists y, z \in X,\ R(x, y, z)\ \&\ x \leqslant y$.

Given an SMTL-frame (X, R), define, for every $A \subseteq X$,

$$\neg_R A \stackrel{\text{def}}{=} A \rightarrow_R \emptyset = \{x \in X : \forall y, z \in X,\ R(x, y, z) \Rightarrow y \notin A\}.$$

The *complex algebra of an SMTL-frame* is just the complex algebra of an MTL-frame.

Proposition 12.7.1 *The complex algebra of an SMTL-frame is an SMTL-algebra.*

Proof:
Let $x \in X$ be such that $x \in \neg_R A$. Then, by the definition of $\neg_R A$, for all $y, z \in X$, $R(x, y, z)$ implies $y \notin A$. By (FMTL8), there exist $y', z' \in X$ such that $R(x, y', z')$ and $x \leqslant y'$. Then $y' \notin A$, and furthermore $x \notin A$, since $A \in L_X$. Therefore $\neg_R A \subseteq -A$, that is $A \cap \neg_R A = \emptyset$, which completes the proof. □

The *canonical frame of an SMTL-algebra* is defined in the same way as the canonical frame of an MTL-algebra.

Proposition 12.7.2 *The canonical frame of an SMTL-algebra is an SMTL-frame.*

Proof:
We show that for any $F \in X_L$ there are $G, H \in X_L$ such that $R_\odot(F, G, H)$ and $F \subseteq G$. Let $F \in X_L$. We show that $F \odot F$ is a proper filter. Suppose otherwise, that is $0 \in F \odot F$. Then there exist $a, b \in F$ such that $a \odot b = 0$. Hence, since $\rightarrow$ is the residuum of $\odot$, $b \leq \neg a$, thus $a \wedge \neg a \in F$. But by axiom (MTL2), $a \wedge \neg a = 0$, whence $0 \in F$, a contradiction. By the prime filter theorem $F \odot F$ can be extended to a prime filter of L, say H, that is $F \odot F \subseteq H$. Hence we find two prime filters of L, namely $G = F$ and H, such that $R_\odot(F, G, H)$ and $F \subseteq G$, as required. □

Applying Theorem 12.5.5 we obtain the following discrete representations for SMTL-algebras and SMTL-frames.

Theorem 12.7.3

(a) *Every SMTL-algebra is embeddable into the complex algebra of its canonical frame;*

(b) *Every SMTL-frame is embeddable into the canonical frame of its complex algebra.*

Now, by Proposition 12.7.1, Proposition 12.7.2 and Theorem 12.7.3 we can establish discrete duality for SMTL-algebras and SMTL-frames.

12.8 IMTL-algebras

An *IMTL-algebra, MTL-algebra with involution*, is an MTL-algebra $(L, \odot, \rightarrow)$ satisfying, for every $a \in L$,

(MTL4) $\neg\neg a \leq a$.

Note that in every MTL-algebra $a \leq \neg\neg a$ holds for every $a \in L$, see Lemma 12.6.1(a). Hence, in view of (MTL4), $\neg$ is an involution.

An *IMTL-frame* is an MTL-frame (X, R) satisfying, for all $x, y \in X$,

(FMTL10) $\forall z \in X,\ (\exists t \in X,\ R(x, z, t) \Rightarrow \exists u \in X,\ R(z, y, u)) \Rightarrow y \leqslant x$

The *complex algebra of an IMTL-frame* is defined in the same way as the complex algebra of an MTL-frame, see Section 12.5.

Proposition 12.8.1 *The complex algebra of an IMTL-frame is an IMTL-algebra.*

Proof:
First, recall that $\neg_R A = \{x \in X : \forall y, z \in X,\ R(x, y, z) \Rightarrow y \notin A\}$. We need to show that $\neg_R \neg_R A \subseteq A$ for every $A \in L_X$.
Let $x \in X$ be such that $x \in \neg_R \neg_R A$. Then $\forall y, z \in X$, we have:

$$
\begin{array}{ll}
 & R(x, y, z) \Rightarrow y \notin \neg_R A \\
\Leftrightarrow & R(x, y, z) \Rightarrow \exists y', z' \in X,\ R(y, y', z')\ \&\ y' \in A \\
\Leftrightarrow & \exists y', z' \in X,\ R(x, y, z) \Rightarrow R(y, y', z')\ \&\ y' \in A \\
\Leftrightarrow & \exists y', z' \in X,\ (R(x, y, z) \Rightarrow R(y, y', z'))\ \&\ (R(x, y, z) \Rightarrow y' \in A)),
\end{array}
$$

which implies that, for all $y, z \in X$,

$$(\exists y', z' \in X,\ R(x, y, z) \Rightarrow R(y, y', z'))\ \&\ (\exists y', z' \in X,\ R(x, y, z) \Rightarrow y' \in A),$$

or equivalently,

(1) $\forall y, z \in X,\ \exists y', z' \in X,\ R(x, y, z) \Rightarrow R(y, y', z')$ and

(2) $\forall y, z \in X,\ \exists y' \in X,\ R(x, y, z) \Rightarrow y' \in A$.

By (FMTL9), it follows from (1) that $y' \leqslant x$. Furthermore, by (FMTL5), $R(u, x, x)$ holds for some $u \in X$, whence, by (FMTL6), $R(x, u, x)$, which together with (2) gives $y' \in A$. Hence, since $y' \leqslant x$ and $A \in L_X$, we get $x \in A$, as required. ☐

The *canonical frame of an IMTL-algebra* is just the canonical frame of an MTL-algebra, see Section 12.5.

Proposition 12.8.2 *The canonical frame of an IMTL-algebra is an IMTL-frame.*

Proof:
Let $F, G \in X_L$ be such that

(1) $\forall H \in X_L,\ (\exists P \in X_L,\ F \odot H \subseteq P) \Rightarrow (\exists P' \in X_L,\ H \odot G \subseteq P')$.

Let $a \in G$ be such that $a \in G$. We have to show that $a \in F$. First, we show that $F \odot \uparrow_{\leq} \neg a = L$. Suppose otherwise, that is $F \odot \uparrow_{\leq} \neg a \neq L$. By Lemma 1.6.10 $F \odot \uparrow_{\leq} \neg a$ is a filter of L and, by the assumption, it is a proper filter of L, so $e \notin F \odot \uparrow_{\leq} \neg a$ for some $e \in L$. Then, by the prime filter theorem, there is $P \in X_L$ such that $F \odot \uparrow_{\leq} \neg a \subseteq P$ and $e \notin P$. Hence, by Theorem 1.6.11, there is some $P' \in X_L$ such that $F \odot P' \subseteq P$ and $\uparrow_{\leq} \neg a \subseteq P'$. Now, putting P' for H in (1), by $F \odot P' \subseteq P$ we get $P' \odot G \subseteq P'$ for some $P' \in X_L$. Since $\neg a \in P'$ and $a \in G$ we get $\neg a \odot a \in P' \odot G \subseteq P'$. But $\neg a \odot a = 0$, whence $0 \in P'$, a contradiction. Therefore $0 \in F \odot \uparrow_{\leq} \neg a$. Then there exist $b, c \in L$ such that $b \in F$, $\neg a \leq c$, and $b \odot c = 0$. By the residuation condition we get $b \leq \neg c$, which together with $b \in F$ yields $\neg c \in F$. Furthermore, $\neg a \leq c$ implies $\neg c \leq \neg\neg a$. But $\neg\neg a \leq a$ by axiom (MTL4). Then $\neg c \leq a$, so since $\neg c \in F$, $a \in F$, as required. □

We now obtain the discrete representations for IMTL-algebras and IMTL-frames.

Theorem 12.8.3

(a) *Every IMTL-algebra is embeddable into the complex algebra of its canonical frame;*

(b) *Every IMTL-frame is embeddable into the canonical frame of its complex algebra.*

12.9 CMTL-algebras

A *CMTL-algebra* is an MTL-algebra $(L, \odot, \rightarrow)$ satisfying, for every $a \in L$,

(MTL3) $a \leq a \odot a$.

A *CMTL-frame* is an MTL frame (X, R) such that, for every $x \in X$,

(FMTL9) $R(x, x, x)$.

The *complex algebra of a CMTL-frame* (resp. the *canonical frame of a CMTL-algebra*) is defined as for an MTL-frame (resp. an MTL-algebra), see Section 12.5.

Proposition 12.9.1 *The complex algebra of a CMTL-frame is a CMTL-algebra.*

Proof:
We only need to show that $A \subseteq A \odot_R A$ for every $A \in L_X$. Let $x \in X$ be such that $x \notin A \odot_R A$. Then for all $y, z \in X$, $R(y, z, x)$ and $y \in A$ imply $z \notin A$. In particular, $R(x, x, x)$ implies $x \notin A$. By (FMTL9), $R(x, x, x)$ holds, whence $x \notin A$. □

Proposition 12.9.2 *The canonical frame of a CMTL-algebra is a CMTL-frame.*

Proof:
We only need to prove that (FMTL9) is satisfied, that is for every $F \in X_L$, $R_\odot(F, F, F)$ holds, so that $F \odot F \subseteq F$. Let $a \in F \odot F$. Then there are $b, c \in F$ such that $b \odot c \leq a$. Clearly, $b \wedge c \in F$, $b \wedge c \leq b$ and $b \wedge c \leq c$, so by monotonicity of $\odot$, $(b \wedge c) \odot (b \wedge c) \leq b \odot c$. By axiom (MTL3), $b \wedge c \leq (b \wedge c) \odot (b \wedge c)$. Therefore $(b \wedge c) \leq (b \odot c)$, and hence $b \odot c \in F$. Furthermore, since $b \odot c \leq a$, $a \in F$, as required. □

We now obtain the discrete representations for CMTL-algebras and CMTL-frames.

Theorem 12.9.3

(a) *Every CMTL-algebra is embeddable into the complex algebra of its canonical frame;*

(b) *Every CMTL-frame is embeddable into the canonical frame of its complex algebra.*

It is shown in [119] that the logic CMTL coincides with the logic G.

12.10 n-potent MTL-algebras

Let L be an MTL-algebra. For $a \in L$ and for $n \geq 1$, define

$$a^1 = a, \quad a^{n+1} = a \odot a^n.$$

Similarly, for $A, B \subseteq L$, let $A \odot B = \{c \in L : \exists a \in A, \exists b \in B, a \odot b \leq c\}$ and

$$A^1 = A, \quad A^{n+1} = A \odot A^n.$$

Lemma 12.10.1 *Let L be an MTL-algebra and let $F, G \in L$ be prime filters of L. Then $F \odot G$ is a prime filter of L.*

Proof:
Let $F, G \in L$ be prime filters of L. By Lemma 1.6.10, $F \odot G$ is a filter. Let $d \vee e \in F \odot G$, that is, for some $a \in F$ and some $b \in G$, $a \odot b \leq d \vee e$. By the residuation axiom (MTL2) from Section 12.5, $b \leq a \rightarrow (d \vee e) = (a \rightarrow d) \vee (a \rightarrow e)$. Thus $(a \rightarrow d) \vee (a \rightarrow e) \in G$. Since G is prime, either $a \rightarrow d \in G$ or $a \rightarrow e \in G$. Hence either $a \odot (a \rightarrow d) \in F \odot G$ or $a \odot (a \rightarrow e) \in F \odot G$. Since $a \odot (a \rightarrow d) \leq d$ and $a \odot (a \rightarrow e) \leq e$, either $d \in F \odot G$ or $e \in F \odot G$. □

Lemma 12.10.2 *Let L be an MTL-algebra. For for all $a, a_1, \ldots a_n \in L$ and all $n \geq 1$,*

(a) $a^{n+1} \leq a^n$;

(b) $a_1 \odot \ldots \odot a_n \leq a_1^n \vee \ldots \vee a_n^n$.

An *n-potent MTL-algebra, MTL^n-algebra,* is an MTL-algebra satisfying, for all $a \in L$,

(MTL^n) $a^n \leq a^{n+1}$.

An *MTL^n-frame* is an MTL-frame satisfying,

$$
\begin{array}{l}
(\mathrm{FMTL}^n)\ \forall x_1, \ldots x_n, y_1 \ldots y_{n-1}, z \in X, \\
\quad (R(x_1, x_2, y_1)\ \&\ R(y_1, x_3, y_2)\ \&\ \ldots\ \&\ R(y_{n-2}, x_n, z)) \\
\quad \Rightarrow \\
\quad \exists i \in \{1, \ldots, n\}, \exists u_1, \ldots u_{n-1}, \\
\quad (R(x_i, x_i, u_1)\ \&\ R(x_i, u_1, u_2)\ \&\ \ldots\ \&\ R(x_i, u_{n-1}, z)).
\end{array}
$$

The complex algebra of an MTL^n-frame and canonical frame of an MTL^n-algebra are defined as for an MTL-frame and an MTL-algebra, respectively, in Section 12.5.

Lemma 12.10.3 *Let L be an MTL-algebra and let $F_1, \ldots, F_n$, for $n \geq 1$, be prime filters of L . Then $(F_i)^{n+1} \subseteq F_1 \odot \ldots \odot F_n$ for some $i \in \{1, \ldots, n\}$.*

Proof:
Suppose that for all $i \in \{1, \ldots n\}$, $(F_i)^{n+1} \not\subseteq F_1 \odot \ldots \odot F_n$. Then there exist $a_1^i, \ldots, a_{n+1}^i \in F_i$ such that $a_1^i \odot \ldots a_{n+1}^i \notin F_1 \odot \ldots \odot F_n$. Since F_i is a filter, $d_i = a_1^i \wedge \ldots \wedge a_{n+1}^i \in F_i$ for all i. Since for each $j \in \{1, \ldots, n+1\}$, $d_i \leq a_j^i$ and $d_i^{n+1} \leq a_1^i \odot \ldots \odot a_{n+1}^i$. Hence $d_i^{n+1} \notin F_a \odot \ldots \odot F_n$. By the axiom ($\mathrm{MTL}^n$), $d_i^n \leq d_i^{n+1}$. Thus, for all i, $d_i^n \notin F_1 \odot \ldots \odot F_n$. Since $d_1 \odot \ldots \odot d_n \in F_1 \odot \ldots \odot F_n$ and, by Lemma 12.10.2,

$$d_2 \odot \ldots \odot d_n \leq d_1^n \vee \ldots \vee d_n^n \text{ and } d_1^n \vee \ldots \vee d_n^n \in F_1 \odot \ldots \odot F_n.$$

Since all F_i, for $i \in \{1, \ldots, n\}$, are prime filters, $F_1 \odot \ldots \odot F_n$ is a prime filter by Lemma 12.10.1. Thus, for some $i \in \{1, \ldots, n\}$, $d_i^n \in F_1 \odot \ldots \odot F_n$, a contradiction. □

Proposition 12.10.4 *The complex algebra of an MTL^n-frame is an MTL^n-algebra.*

Proof:
It suffices to show that the axiom (MTL^n) is satisfied. We show that for every $A \subseteq X$ and every $n \geq 1$, $A^n \subseteq A^{n+1}$. Let $z \in A^n$, that is,

$\exists x_1, \ldots x_n \in A, \exists y_1, \ldots y_{n-2} \in X,$
$R(x_1, x_2, y_1)\ \&\ R(y_1, x_3, y_2)\ \&\ \ldots\ \&\ R(y_{n-1}, x_n, z).$

By the frame axiom (FMTL^n),

$\exists i \in \{1, \ldots n\}, \exists u_1, \ldots u_{n-1} \in X,$
$R(x_i, u_1, u_2)\ \&\ R(x_i, u_2, u_3) \ldots\ \&\ R(x_i, u_{n-1}, z).$

Since $x_i \in A$, for every $i \in \{1, \ldots n\}$, we have $u_j \in A^{j+1}$ for every $j \in \{1, \ldots, n-1\}$. It follows that $z \in A^{n+1}$, as required. □

Proposition 12.10.5 *The canonical frame of an MTL^n-algebra is an MTL^n-frame.*

Proof:
Let $F_1, \ldots F_n, G_1, \ldots G_{n-2}, H$ be prime filters of L such that

$$F_1 \odot F_2 \subseteq G_1,\ G_1 \odot F_3 \subseteq G_2,\ \ldots,\ G_{n-2} \odot F_n \subseteq H.$$

It follows that $F_1 \odot \ldots \odot F_n \subseteq H$. By Lemma 12.10.3 there is $i \in \{1, \ldots, n\}$ such that $(F_i)^{n+1} \subseteq F_1 \odot \ldots \odot F_n \subseteq H$. Set $U_j = (F_i)^{j+1}$, for $j \in \{1, \ldots, n-1\}$. Then $F_i \odot F_i \subseteq U_1$, $F_i \odot U_1 \subseteq U_2, \ldots, F_i \odot U_{n-1} = F_i \odot F_i^n = (F_i)^{n+1} \subseteq H$, as required. □

The discrete representation theorems for MTL^n-algebras and MTL^n-frames can be proved analously to those for MTL-algebras and MTL-frames using the embeddings h and k, defined as for MTL-algebras.

Theorem 12.10.6

(a) *Every MTL^n-algebra is embeddable into the complex algebra of its canonical frame;*

(b) *Every MTL^n-frame is embeddable into the canonical frame of its complex algebra.*

Our results of this section together with a discrete duality for IMTL algebras presented in Section 12.8 provide the discrete representations for all C$_n$IMTL, $n \geq 1$, in the hierarchy of [131]. The corresponding frame axioms are those of n-potent MTL algebras together with (FMTL10) from Section 12.8. A discrete representation for n-potent MTL algebras with the axiom
(WNM) $((a \wedge b) \to 0) \vee ((a \wedge b) \to (a \odot b))$
has not been approached yet.

12.11 2-potent BL-algebras

In this section we consider the class of *2-potent BL-algebras*, *BL^2-algebras*, which are MTL-algebras satisfying, for all $a, b \in L$, the divisibility axiom and the 2-potent axiom

(Div) $b < a \Rightarrow A \odot (a \to b) = b$

(2-pot) $a^2 = a^3$.

There are several equivalent versions of the axiom (Div).

Lemma 12.11.1 *Let L be an MTL-algebra. The following are equivalent, for all $a, b \in L$,*

(a) $a \wedge b = a \odot (a \to b)$

(b) $a \odot (a \to b) = b \odot (b \to a)$

(c) $a < b \Rightarrow \exists c \in L, a = b \odot c$

(d) $(a \to b) \vee (b \to a \odot (a \to b)) = 1$

(e) (Div).

Proof:
Clearly, (a) is equivalent to (b) because $\wedge$ is commutative. Note that in (c) we may assume $a \leq b$ because if $a = b$ then $c = 1$ satisfies $a = a \odot 1$.

(a) $\Rightarrow$ (c) Take any $a, b \in L$ such that $a \leq b$. Then $a \wedge b = a$ and, by (b), $a = a \odot (a \to b) = b \odot (b \to a)$. Thus (c) holds with $c = b \to a$.

(c) $\Rightarrow$ (a) Note that, for every $c \in L$, $c \leq a$ and $c \leq b$ imply $c \leq a \odot (a \to b)$. Indeed, since $c \leq a$, by (c), there is some $d \in L$ such that $c = d \odot a$. Since $c = d \odot a \leq b$, by the residuation axiom (MTL2), $d \leq a \to b$. Thus $c = d \odot a \leq a \odot (a \to b)$. Since $a \wedge b \leq a, b$ we obtain $a \wedge b \leq a \odot (a \to b)$.

(a) $\Rightarrow$ (d) Note that

$$b \to a \odot (a \to b) = b \to (a \wedge b) = (b \to a) \wedge (b \to b) = (b \to a).$$

Now, by the prelinearity axiom,

$$(a \to b) \vee (b \to a \odot (a \to b)) = (a \to b) \vee (b \to a) = 1.$$

(d) $\Rightarrow$ (a) Since every MTL-algebra is decomposable as a product of linearly ordered MTL-algebras, it is sufficient to prove the implication in MTL-chains. If $a < b$ then $b \to a \neq 1$. By (d), $(b \to a) \vee (a \to b \odot (b \to a)) = 1$. By linearity, $a \to b \odot (b \to a) = 1$ and hence $a \leq b \odot (b \to a)$. Thus $b \wedge a \leq b \odot (b \to a)$. The reverse inequality holds in MTL. If $b \leq a$ then $b \to a = 1$. Hence $b \odot (b \to a) = b \odot 1 = b = a \wedge b$.

(a) $\Rightarrow$ (e) Let $b < a$. Then $a \wedge b = b$ and hence $a \odot (a \to b) = b$ as required.

(e) $\Rightarrow$ (a) Since (a) is equivalent to (c), the implication is obvious. □

Consider the equality

(V^2) $(a \to b) \vee (b \to a \odot b) = 1$.

Lemma 12.11.2 *Let L be a distributive, integral, commutative residuated lattice. Then the following are equivalent*

(a) *L satisfies (V^2);*

(b) *L is a 2-potent BL-algebra.*

Proof:
(a) $\Rightarrow$ (b) First, we show that L satisfies the axiom of prelinearity. Since $a \odot b \leq a$, we have $b \to (a \odot b) \leq b \to a$. By (V^2),

$$1 = (a \to b) \vee (b \to a \odot b) = (a \to b) \vee (b \to a).$$

Thus L is an MTL-algebra and hence, by Lemma 12.5.2, there is a family of linearly ordered MTL-algebras, $\{B_i\}_{i \in I}$ such that L is embeddable into the direct product ΠB_i and each B_i is a homomorphic image of L. It follow that each B_i satisfies (V^2). Now we show that each B_i satisfies divisibility (Div) and (2-pot).

For (Div), note that since $b \leq a \to b$, we have $a \odot b \leq a \odot (a \to b)$ which implies $b \to a \odot \leq b \to a \odot (a \to b)$. From ($V^2$)

$$1 = (a \to b) \vee (b \to a \odot b) \leq (a \to b) \vee (b \to a \odot (a \to b)).$$

By Lemma 12.11.2 the latter is equivalent to divisibility.

For (2-Pot) let $a \in B_i$. If $a^2 = a$ then $a^3 = a^2$. If $a^2 < a$ then $a \to a^2 \neq 1$. By (V^2) we have $(a \to a^2) \vee (a^2 \to a \odot a^2) = 1$. Since B_i is linearly ordered, $a^2 \to a^3 = 1$ and hence $a^2 \leq a^3$. Since in MTL-algebras $a^{n+1} \leq a^n$, for every $n \geq 1$, we also have $a^3 \leq a^2$.

(b) $\Rightarrow$ (a) The class of BL-algebras is generated by the set of algebras that are ordinal sums of copies of the 2-element BL-algebra and the 3-element BL-algebra, see [118]. Each such algebra is linearly ordered and hence, for any a and b in such an algebra, either $a \leq b$ or $b < a$. In the first case $a \to b = 1$ and hence (V^2) is satisfied. If $b < a$ and a, b are in the different components of the ordinal sum, then $a \odot b = b$. If a, b are in the same 2-element or 3-element chain then it is also easy to check that $a \odot b = b$. Thus $b \to (a \odot b) = 1$ and hence (V^2) holds. □

A *BL^2-frame* is an MTL-frame (X, R) satisfying, for all $x, y, z \in X$,

(FBL^2) $R(x, y, z) \wedge R(x, y', z') \Rightarrow (R(y, y', z) \vee y' \leq z)$.

The complex algebra of a BL^2-frame and the canonical frame of a BL^2-algebra are defined as for MTL-frames and MTL-algebras, respectively.

Proposition 12.11.3 *The complex algebra of a BL^2-frame is a BL^2-algebra.*

Proof:
In view of Lemma 12.11.2 and Proposition 12.10.4, it is sufficient to show that the complex algebra satisfies (V^2). Let $A, B \in L_X$. We show $(A \to_R b) \vee (B \to_R (A \odot_R B)) = X$. Suppose the contrary, that there is $x \in X$ such that $x \notin A \to_R B$ and $x \notin B \to_R (A \odot_R B)$. Then there are $y, z \in X$ such that $R(x, y, z)$ and $y \in A$ and $z \notin B$, and there are $y', z' \in X$ such that $R(x, y', z')$ and $y' \in B$ and $z' \notin A \odot_R B$. By (FBL^2) either $R(y, y', z')$ or $y' \leq z$. If $R(y, y', z')$ then, since $y \in A$ and $y' \in B$, we obtain $z' \in A \odot_R B$, a contradiction. If $y' \leq z$ then, since $y' \in B$, we obtain $z \in B$, a contradiction. □

Proposition 12.11.4 *The canonical frame of a BL^2-algebra is a BL^2-frame.*

Proof:
Let $F, G, H, G', H' \in X_L$. Assume that $F \odot G \subseteq H$ and $F \odot G' \subseteq H'$. Suppose that $G \odot G' \not\subseteq H'$ and $G' \not\subseteq H$. Then $G \odot G' \not\subseteq F \odot G'$ and $G' \not\subseteq F \odot G$. Let $a \in (G \odot G') - (F \odot G')$. It follows that there is some $b \in G$ and there is some

$d \in G'$ such that $b \odot d \leq a$ and $a \notin F \odot G'$. Let $c \in G' - (F \odot G)$. Consider $e = d \wedge c$. Since G' is a filter and $d, c \in G'$, we have $e \in G'$. Then $b \odot e \leq b \odot d \leq a$. Thus $b \odot e \notin F \odot G'$. Furthermore, $e \leq c \notin F \odot G$, so $e \notin F \odot G$. Using (V^2) we obtain $(b \to e) \vee (e \to b \odot e) = 1$. Since F is a prime filter we must have either $b \to e \in F$ or $e \to b \odot e \in F$. If $b \to e \in F$ then $b \odot (b \to e) \in G \odot F = F \odot G$. Since $b \odot (b \to e) \leq e$, $e \in G \odot F$, a contradiction. If $e \to b \odot e \in F$ then $e \odot (e \to (b \odot e)) \in G' \odot F = F \odot G'$. Since $e \odot (e \to (b \odot e)) \leq b \odot e$, $b \odot e \in F \odot G'$, again a contradiction. □

The discrete representation theorems for BL^2-algebras and BL^2-frames can be proved analogously to those for MTL-algebras and MTL-frames using the embeddings h and k, defined as for MTL-algebras.

Theorem 12.11.5

(a) *Every BL^2-algebra is embeddable into the complex algebra of its canonical frame;*

(b) *Every BL^2-frame is embeddable into the canonical frame of its complex algebra.*

It follows that based on the theorem on discrete duality for SMTL-algebras presented in Section 12.7 we also get a discrete duality for 2-potent SBL-algebras. The corresponding frame axioms are those of 2-potent BL-frame and (FMTL8) from Section 12.7. Similarly, based on the theorem on discrete duality for IMTL-algebras we get a discrete duality for 2-potent MV-algebras. The frame axioms are those of 2-potent BL-frames and (FMTL10) from Section 12.8. Furthermore, the theorem in [17] implies that no completion exists for SBL-algebras and MV–algebras alone.

Part III

General lattices with operators

Chapter 13

General lattices with modal operators

13.1 Introduction

In this chapter bounded general lattices with modal operators of possibility, $\Diamond$, necessity, $\Box$, sufficiency, $\boxminus$, and dual sufficiency, $\diamondminus$, respectively, are considered. In the classical modal algebras based on Boolean algebras discussed in Chapter 3 the operators of possibility and necessity, as well as the operators of sufficiency and dual sufficiency are mutually definable due to the presence of complement. In case of lattices all these operators are independent of each other. Discrete representation theorems for distributive lattices with these operators and for the corresponding frames are presented in Chapter 8. The basis for those representations are the discrete representation theorems for distributive lattices and distributive lattice frames considered in Section 2.5. In this chapter the basis for representations of general lattices with modal operators is the Urquhart representation for bounded general lattices presented in Section 2.8. An extension of the Urquhart method to a representation of general lattice frames is an open problem and therefore in this chapter we do not have representation theorems for frames. In Section 13.2, Section 13.4, Section 13.7, and Section 13.8 discrete representation theorems for the four classes of bounded general lattices with the modal operators mentioned above are presented based on [184], see also [65]. In each of these sections detailed proofs are given. In Section 13.2 a duality via truth for possibility general lattices and frames is presented. A duality via truth for necessity, sufficiency, and dual sufficiency general lattices can be obtained in a similar way based on developments in [184]. In Section 13.3 and Section 13.5 discrete representation theorems for some axiomatic extensions of possibility general lattices and necessity general lattices, respectively, are given, following [204]. In Section 13.6 bounded general lattices with both a possibility and a necessity operator and some of their axiomatic extensions are considered and the discrete representations are proved.

13.2 Possibility lattices

A *possibility (general) lattice* $(L, \Diamond)$, *PG-lattice*, is a general lattice L endowed with a unary operation $\Diamond$ on L satisfying, for all $a, b \in L$,

(PG1) $\Diamond(a \vee b) = \Diamond a \vee \Diamond b$ additive

(PG2) $\Diamond 0 = 0$ normal.

Observe, that any such possibility operator is isotone. For any $A \subseteq L$,

$$\Diamond^{-1}A = \{a \in L : \Diamond a \in A\}$$
$$\langle\geq\rangle\Diamond A = \{a \in L : \exists b \in A\,, \Diamond b \leq a\},$$

where $\leq$ is the natural ordering of the lattice L and $\geq = \leq^{-1}$. A filter (resp. ideal) of a possibility lattice $(L, \Diamond)$ is a filter (resp. ideal) of the underlying lattice L.

Lemma 13.2.1 *Let $(L, \Diamond)$ be a PG-lattice. For any $A, B \subseteq L$,*

(a) *If A is an ideal of L, then so is $\Diamond^{-1}(A)$;*
(b) *If A is a filter of L, then so is $\langle\geq\rangle\Diamond A$;*
(c) *If $\langle\geq\rangle\Diamond A \subseteq B$, then $A \subseteq \Diamond^{-1}(B)$;*
(d) *If B is a filter of L, then $A \subseteq \Diamond^{-1}(B)$ implies $\langle\geq\rangle\Diamond A \subseteq B$.*

Proof:
By way of example we prove (b), (c), and (d). The proofs of the remaining conditions are similar.

For (b), assume $A \subseteq L$ is a filter of L. Let $a, b \in \langle\geq\rangle\Diamond A$. By definition, there is some $c \in A$ such that $\Diamond c \leq a$, and there is some $d \in A$ such that $\Diamond d \leq b$. Since A is a filter, we have $c \wedge d \in A$. Since $c \wedge d \leq c$, we also have $\Diamond(c \wedge d) \leq \Diamond d$. Hence, $\Diamond(c \wedge d) \leq a$ and $\Diamond(c \wedge d) \leq b$. It follows that $\Diamond(c \wedge d) \leq a \wedge b$, and hence $a \wedge b \in \langle\geq\rangle\Diamond A$. Now assume that $a \leq b$ and $a \in \langle\geq\rangle\Diamond A$. It follows that there is some $c \in A$ such that $\Diamond c \leq a$. Hence $\Diamond c \leq b$, and therefore $b \in \langle\geq\rangle\Diamond A$.

For (c), assume $\langle\geq\rangle\Diamond A \subseteq B$. Take any $a \in A$. Since by reflexivity of $\leq$, $\Diamond a \leq \Diamond a$, so we get $\Diamond a \in \langle\geq\rangle\Diamond a$ and hence, by assumption, $\Diamond a \in B$. Therefore $a \in \Diamond^{-1}B$, as required.

For (d), assume $A \subseteq \Diamond^{-1}(B)$ and take $a \in \langle\geq\rangle\Diamond A$. Then there is some $b \in A$ such that $\Diamond b \leq a$. By the assumption $\Diamond b \in B$. But B is a filter of L, so we have $a \in B$, as required. □

A *possibility (general) frame*, *PG-frame*, is a structure (X, R, S) where X is a general lattice frame and R and S are binary relations on X satisfying,

(FPG1) $(\geqslant_1; R; \geqslant_1) \subseteq R$
(FPG2) $(\leqslant_2; S; \leqslant_2) \subseteq S$
(FPG3) $R \subseteq (S; \geqslant_1)$
(FPG4) $S \subseteq (\leqslant_2; R)$.

The *complex algebra of a PG-frame* X is $(L_X, \Diamond_S)$ where X_L is the complex algebra of a lattice frame as defined in Section 2.8 and $\Diamond_S$ is a unary operator defined for every $A \in L_X$ by

$$\Diamond_S A = l[S]r(A).$$

Lemma 13.2.2 *Let (X, R, S) be a PG-frame. For any $A \subseteq X$,*

(a) *$[S]A$ is a $\leqslant_2$-increasing set;*

(b) *$\langle R\rangle A$ is a $\leqslant_1$-increasing set.*

Proof:
We prove (a), the proof of (b) is similar. Let $x, y \in X$ be such that $x \in [S]A$ and $x \leqslant_2 y$. Suppose that $y \notin [S]A$, that is, there is some $z \in X$ such that ySz and $z \notin A$. By reflexivity of $\leqslant_2$ and (FPG2) we get xSz. Hence, since $x \in [S]A$, it follows that $z \in A$, a contradiction. □

Lemma 13.2.3 *Let (X, R, S) be a PG-lattice frame and let $A \subseteq X$.*

(a) *If A is r-stable, then $[S]A = r\langle R\rangle l(A)$ and $[S]A$ is r-stable;*

(b) *If A is l-stable, then $\Diamond_S A = lr\langle R\rangle A$ and $\Diamond_S A$ is l-stable.*

Proof:
(a) Let $x \notin r\langle R\rangle l(A)$. Then for some $y \in X$ such that $x \leqslant_2 y$ and $y \in \langle R\rangle l(A)$. Hence, there exists $z \in X$ such that yRz and $z \in l(A)$. By (FPG3), from yRz it follows that there is some $z' \in X$ such that ySz' and $z \leqslant_1 z'$. Now, $z \in l(A)$ and $z \leqslant_1 z'$ imply $z' \notin A$. Furthermore, by (FPG2), ySz', $x \leqslant_2 y$, and reflexivity of $\leqslant_2$ give xSz', which together with $z' \notin A$ yields $x \notin [S]A$. Hence $[S]A \subseteq r\langle R\rangle l(A)$. For the reverse inclusion, assume that $x \notin [S]A$, that is, there is some $y \in X$ such that xSy and $y \notin A$. Since A is r-stable, we have $A = rl(A)$. But $y \notin A$, so there is some $y' \in X$ such that $y \leqslant_2 y'$ and $y' \in l(A)$. By (FPG2), xSy and $y \leqslant_2 y'$ imply xSy'. Hence, by (FPG4), there is some $x' \in X$ such that $x \leqslant_2 x'$ and $x'Ry'$. Hence, since $y' \in l(A)$, we get $x' \in \langle R\rangle l(A)$, which together with $x \leqslant_2 x'$ yields $x \notin r\langle R\rangle l(A)$, a contradiction. Now, we show that $[S]A$ is r-stable. By Lemma 13.2.2(b), $\langle R\rangle l(A)$ is a $\leqslant_1$-increasing set. By Lemma 2.8.1(a), $r\langle R\rangle l(A)$ is r-stable. Hence, $[S]A$ is r-stable.

(b) First, we prove that $\Diamond_S A = lr(\langle R\rangle A)$, that is $l([S]r(A)) = lr(\langle R\rangle A)$. We show that $[S]r(A) = r(\langle R\rangle A)$. Then the required result will follow.
Let $x \in X$ be such that $x \notin r(\langle R\rangle A)$. Then there is some $y \in X$ such that $x \leqslant_2 y$ and $y \in \langle R\rangle A$, and furthermore, there is some $z \in X$ such that yRz and $z \in A$. By (FPG3), from yRz we get some $u \subset X$ such that ySu and $z \leqslant_1 u$. Now, by (FPG2), $x \leqslant_2 y$ and ySu imply xSu. Since A is l-stable, by Lemma 2.8.1(a) it is $\leqslant_1$-increasing, so $z \leqslant_1 u$ and $z \in A$ imply $u \in A$, that is $u \in lr(A)$. Hence, by Lemma 2.8.1(b), $u \notin r(A)$, which together with xSu gives $x \notin [S]r(A)$. For the reverse inclusion, take $x \in r(\langle R\rangle A)$ and suppose that $x \notin [S]r(A)$. Then there are $y, z \in X$ such that xSy, $y \leqslant_2 z$, and $z \in A$. By (FPG2), from xSy and $y \leqslant_2 z$ we get xSz. Hence, by (FPG4), there exists $u \in X$ such that $x \leqslant_2 u$ and uRz.

Now, $x \in r(\langle R\rangle A)$ and $x \leqslant_2 u$ imply $u \notin \langle R\rangle A$, which together with uRz gives $z \notin A$, a contradiction. By Lemma 2.8.1 (a) and Lemma 2.8.3(b), $lr(\langle R\rangle A)$ is l-stable; hence $\Diamond_S A$ is also l-stable. □

An alternative definition of the possibility operator in the complex algebra is possible such that the operator is defined in terms of relation R by $\Diamond_R A = lr\langle R\rangle A$ for every $A \in L_X$. Then in view of Lemma 13.2.3(b) we have $\Diamond_S A = \Diamond_R A$.

Proposition 13.2.4 *The complex algebra of a PG-frame is a PG-lattice.*

Proof:
We show that $\Diamond_S(A \vee_X B) = \Diamond_S A \vee_X \Diamond_S B$ and $\Diamond_S \emptyset = \emptyset$.

	$\Diamond_S(A \vee_X B)$	
$=$	$l[S]r(A \vee_X B)$	definition of $\Diamond_S$
$=$	$lr\langle R\rangle lr(A \vee_X B)$	Lemma 13.2.3(a)
$=$	$lr\langle R\rangle l(r(A) \cap r(B))$	Lemma 2.8.4(b), definition of $\vee_X$
$=$	$l[S](r(A) \cap r(B))$	Lemma 13.2.3(a)
$=$	$l([S]r(A) \cap [S]r(B))$	distributivity of $[S]$ over $\cap$.

Since A is l-stable, by Lemma 2.8.3(c), $r(A)$ is r-stable. By Lemma 13.2.3(a), $[S]r(A)$ is r-stable. Hence, $[S]r(A) = rl[S]r(A)$. By the definition of $\Diamond_S$, $rl[S]r(A) = r(\Diamond_S A)$. The similar reasoning shows that $rl[S]r(B) = r(\Diamond_S B)$. Hence $\Diamond_S(A \vee_X B) = l(r\Diamond_S A \cap r(\Diamond_S B)) = \Diamond_S A \vee_X \Diamond_S B$.
We also have $\Diamond_S \emptyset = l[S]r(\emptyset) = l[S]X = l(X) = \emptyset$. □

The *canonical frame of a PG-lattice* L is a structure $(X_L, R_\Diamond, S_\Diamond)$ where X_L is the canonical frame of its general lattice reduct (see Section 2.8) and $R_\Diamond$ and $S_\Diamond$ are defined, for all $F = (F_1, F_2)$ and $G = (G_1, G_2) \in X_L$, by

$$FS_\Diamond G \overset{\text{def}}{\Leftrightarrow} \Diamond^{-1}(F_2) \subseteq G_2$$

$$FR_\Diamond G \overset{\text{def}}{\Leftrightarrow} G_1 \subseteq \Diamond^{-1}(F_1).$$

Lemma 13.2.5 *Let $(L, \Diamond)$ be a PG-lattice. For every $F \in X_L$ and for every $a \in L$, the following conditions are equivalent,*

(a) $\Diamond a \in F_2$;

(b) *For every $G \in X_L$, if $FS_\Diamond G$, then $a \in G_2$.*

Proof:
By definition of $S_\Diamond$, (a) implies (b). On the other hand, suppose $\Diamond a \notin F_2$. We show that there is some $G \in X_L$ such that $FS_\Diamond G$ and $a \notin G_2$. Let $\uparrow_\leq a$ be the principal filter of L generated by a. Since $a \notin \Diamond^{-1}(F_2)$, $\uparrow_\leq a \cap \Diamond^{-1}(F_2) = \emptyset$. It follows that $(\uparrow_\leq a, \Diamond^{-1}(F_2))$ is a filter-ideal pair which, by Theorem 2.8.6, can be extended to a maximal filter-ideal pair, say, (G_1, G_2) of L. Then $\uparrow_\leq a \subseteq G_1$, and hence $a \in G_1$. Consequently, $a \notin G_2$. Since $\Diamond^{-1}(F_2) \subseteq G_2$, we have $FS_\Diamond G$, which completes the proof. □

Proposition 13.2.6 *The canonical frame of a PG-lattice is a PG-frame.*

Proof:
By way of example we prove (FPG2) and (FPG4)

For (FPG2), assume that $F_2' \subseteq F_2$, $\Diamond^{-1}(F_2) \subseteq G_2$, and $G_2 \subseteq G_2'$. Then, by Lemma 13.2.1(b), $F_2' \subseteq F_2$ implies $\Diamond^{-1}(F_2') \subseteq \Diamond^{-1}(F_2)$. Thus $\Diamond^{-1}(F_2') \subseteq G_2$, and hence $\Diamond^{-1}(F_2') \subseteq G_2'$, that is $F' S_\Diamond G'$ holds.

For (FPG4), assume that $\Diamond^{-1}(F_2) \subseteq G_2$. We have to show that there is some $F' \in X_L$ such that $F_2 \subseteq F_2'$ and $G_1 \subseteq \Diamond^{-1}(F_1')$. We show that $\langle \geq \rangle \Diamond G_1 \cap F_2 = \emptyset$. Assume the contrary, that is, there is some $a \in L$ such that $a \in \langle \geq \rangle \Diamond G_1$ and $a \in F_2$. Then there is some $b \in G_1$ such that $\Diamond b \leq a$. Since F_2 is an ideal of L, $\Diamond b \in F_2$, hence $b \in \Diamond^{-1}(F_2)$. By the assumption we obtain $b \in G_2$. It follows that $G_1 \cap G_2 \neq \emptyset$, a contradiction. By Lemma 13.2.1(b), $\langle \geq \rangle \Diamond G_1$ is a filter, so $(\langle \geq \rangle \Diamond G_1, F_2)$ is a filter-ideal pair which, by Theorem 2.8.6, can be extended to a maximal filter-ideal pair, say, $F' = (F_1', F_2')$ of L. Then we have $\langle \geq \rangle \Diamond G_1 \subseteq F_1'$ and $F_2 \subseteq F_2'$. From Lemma 13.2.1(c) we obtain $G_1 \subseteq \Diamond^{-1}(F_1')$, which completes the proof. □

Now we extend the representation of bounded lattices in Theorem 2.8.10 to PG-lattices. Let $(L, \Diamond)$ be a PG-lattice and let $h : L \to L_{X_L}$ be the mapping defined, for any $a \in L$, by $h(a) = \{F \in X_L : a \in F_1\}$. We now show that h preserves the possibility operator $\Diamond$ on L.

Lemma 13.2.7 *For every $a \in L$, $h(\Diamond a) = \Diamond_{S_\Diamond} h(a)$.*

Proof:
Let $F = (F_1, F_2) \in X_L$. Note that

$$\begin{aligned}
&F \in \Diamond_{S_\Diamond} h(a) \\
&\Leftrightarrow F \in l([S_\Diamond] r h(a)) \\
&\Leftrightarrow \forall G = (G_1, G_2) \in X_L,\ F_1 \subseteq G_1 \Rightarrow G \notin [S_\Diamond] r h(a) \\
&\Leftrightarrow \forall G = (G_1, G_2) \in X_L, F_1 \subseteq G_1 \Rightarrow (\exists H = (H_1, H_2) \in X_L,\ G S_\Diamond H \ \&\ H \notin r h(a)) \\
&\Leftrightarrow \forall G = (G_1, G_2) \in X_L,\ F_1 \subseteq G_1 \Rightarrow (\exists H = (H_1, H_2) \in X_L,\ G S_\Diamond H \ \&\ a \notin H_2)
\end{aligned}$$

where the last equivalence holds by Lemma 2.8.9(a). Let $F \in X_L$ be such that $F \in h(\Diamond a)$, that is $\Diamond a \in F_1$. Take $G \in X_L$ such that $F_1 \subseteq G_1$. Then $\Diamond a \in G_1$, hence $\Diamond a \notin G_2$. By Lemma 13.2.5 there is some $H \in X_L$ such that $G S_\Diamond H$ and $a \notin H_2$, whence $H \notin r h(a)$ by Lemma 2.8.9(a). On the other hand, assume $F \notin h(\Diamond a)$, that is $\Diamond a \notin F_1$. Let $\downarrow_\leq \Diamond a$ be the principal ideal of L generated by $\Diamond a$. Since $\Diamond a \notin F_1$, we have $F_1 \cap \downarrow_\leq \Diamond a = \emptyset$, thus $(F_1, \downarrow_\leq \Diamond a)$ is a filter-ideal pair of L which, by Theorem 2.8.6, can be extended to a maximal filter-ideal pair, say, (G_1, G_2) of L. Then $F_1 \subseteq G_1$ and $\downarrow_\leq \Diamond a \subseteq G_2$, so $\Diamond a \in G_2$. From Lemma 13.2.5 the required result follows. □

The following discrete representation for PG-lattices follows from Theorem 2.8.10 and Lemma 13.2.7.

Theorem 13.2.8 *Every PG-lattice is embeddable into the complex algebra of its canonical frame.*

Following [184] we now extend the discrete representation result of Theorem 13.2.8 to a duality via truth. The language of possibility logic $LATPG$ is obtained from the language $\mathcal{L}an_L$ of the general bounded lattice-based logic discussed in Section 2.9 by endowing it with a unary propositional connective $\Diamond$ of possibility and by a suitable extension of the notions of a formula and a sequent. The algebraic semantics of the language $\mathcal{L}an_{LPG}$ is an extension of the algebraic semantics of the logic in Section 2.9 obtained by taking PG-lattices $(L, \Diamond)$ as interpretation structures and by postulating that $v(\Diamond\alpha) = \Diamond v(\alpha)$, where on the right hand side at this equality the operator $\Diamond$ is the possibility operator of L. The frame semantics of the language $\mathcal{L}an_{LPG}$ is determined by the class of PG-frames (X, R, S). It is obtained from the frame semantics of the logic in Section 2.9 in the following way. Given a model $M = (X, R, S, m)$, we add the satisfaction clause, for any $x \in X$,

$$M, x \models \Diamond\alpha$$
$$\text{iff} \quad \text{for all } y \in X, \text{there are } z, t \in X \text{ such that } ySz,\ z \leq_2 t \text{ and } M, t \models \alpha.$$

Invoking the definition of $m(\alpha)$, namely, $m(\alpha) = \{x \in X : M, x, \models \alpha\}$, we have the following lemma.

Lemma 13.2.9 *For every model $M = (X, R, S, m)$ based on a PG-frame,*

(a) $m(\Diamond\alpha) = l[S]r(m(\alpha))$;

(b) $m(\Diamond\alpha)$ *is l-stable.*

Proof:
(a) follows easily from the satisfaction clause for formulas of the form $\Diamond\alpha$. For (b) required equality $m(\alpha) = lr(m(\alpha))$ follows from(a) and the fact that $[S]r(m(\alpha))$ is r-stable, as it is shown in the Proposition 13.2.4. □

Now it is easy to see that the complex algebra theorem holds.

Theorem 13.2.10 (Complex algebra theorem) *For all formulas α and β of $\mathcal{L}an_{LPG}$ and for every PG-frame $X \in \mathcal{F}rm_L$ the following conditions are equivalent:*

(a) *A sequent $\alpha \vdash \beta$ is true in X;*

(b) *A sequent $\alpha \vdash \beta$ is true in the complex algebra of X.*

Then the duality via truth theorem can be proved in a similar way as the Theorem 2.9.4.

Theorem 13.2.11 (Duality via truth) *For all formulas α and β of $\mathcal{L}an_{LPG}$ the following conditions are equivalent:*

(a) *A sequent $\alpha \vdash \beta$ is true in every PG-lattice;*

(b) *A sequent $\alpha \vdash \beta$ is true in every model based on a PG-frame.*

13.3 Axiomatic extensions of possibility lattices

Following [204], let us consider two possibility lattices endowed with additional axioms corresponding to well-known axioms in classical possibility algebras based on Boolean algebras. Namely, the following axioms are taken into account:

($\mathrm{T}_\Diamond$) $a \leq \Diamond a$

($4_\Diamond$) $\Diamond\Diamond a \leq \Diamond a$.

Let α stands for either of these axioms. A PG-lattice $(L, \Diamond)$ satisfying an axiom (α) is referred to as *PGα-lattice*.dex

Let (X, R, S) be a PG-frame. We consider the following additional axioms:

($\mathrm{FT}_\Diamond$) S is reflexive

($\mathrm{F4}_\Diamond$) S is transitive.

A PG-frame (X, R, S) is called a *PGα-frame* if it satisfies an axiom $(F\alpha)$ for $\alpha \in \{\mathrm{T}_\Diamond, 4_\Diamond\}$.

The *complex algebra of a PGα-frame* (resp. the *canonical frame of a PGα-lattice*) is defined in the same way as the complex algebra of a PG-frame (resp. canonical frame of a PG-lattice).

Proposition 13.3.1 *Let $\alpha \in \{T_\Diamond, 4_\Diamond\}$.*

(a) *The complex algebra of a PGα-frame is a PGα-lattice;*

(b) *The canonical frame of a PGα-lattice is a PGα-frame.*

Proof:
(a) For $(\mathrm{T}\alpha)$, take any $x \notin l([S]r(A))$. Then there is some $y \in X$ such that $x \leqslant_1 y$ and $y \in [S]r(A)$. Since S is reflexive, $y \in r(A)$, and hence $x \notin lr(A)$. But A is l-stable, so $A = lr(A)$. Therefore $x \notin A$.
For $(4_\Diamond)$, note that since l is antitone, it suffices to show that $[S]r(A) \subseteq [S]rl([S]r(A))$. Take any $x \in X$ such that $x \notin [S]rl([S]r(A))$. Then there is some $y \in X$ such that xSy and $y \notin rl([S]r(A))$. Since rl is a closure operator, $[S]r(A) \subseteq rl([S]r(A))$. Hence, $y \notin [S]r(A)$, that is, for some $z \in X$, ySz and $z \notin r(A)$. By transitivity of S, xSz, and hence $x \notin [S]r(A)$.

(b) For $(\mathrm{FT}_\Diamond)$, take any $F \in X_L$ and suppose that $a \in \Diamond^{-1}(F_2)$. Then $\Diamond a \in F_2$. Since F_2 is an ideal of L, by $(\mathrm{T}_\Diamond)$ $a \in F_2$. Then $\Diamond^{-1}(F_2) \subseteq F_2$, that is, $FS_\Diamond F$.
For $(\mathrm{F4}_\Diamond)$, take $F, G, H \in X_L$ such that $FS_\Diamond G$ and $GS_\Diamond H$. Then

$$\Diamond^{-1}(F_2) \subseteq G_2 \text{ and } \Diamond^{-1}(G_2) \subseteq H_2.$$

Let $a \in \Diamond^{-1}(F_2)$, that is, $\Diamond a \in F_2$. Since F_2 is an ideal, by $(4_\Diamond)$ $\Diamond a \in \Diamond^{-1}(F_2)$, so $\Diamond a \in G_2$, that is, $a \in \Diamond^{-1}(G_2)$. Hence $a \in H_2$. Therefore $\Diamond^{-1}(F_2) \subseteq H_2$, that is, $FS_\Diamond H$. □

Invoking Lemma 13.2.7 and Proposition 13.3.1, we obtain the following discrete representation for PGα-lattices.

Theorem 13.3.2 *Let $\alpha \in \{T_\Diamond, 4_\Diamond\}$. Every PG$\alpha$-lattice is embeddable into the complex algebra of its canonical frame.*

13.4 Necessity lattices

A *necessity (general) lattice*, *NG-lattice*, is a structure $(L, \Box)$ where L is a general lattice and $\Box$ is a unary operator on L satisfying for any $a, b \in L$,

(NG1) $\Box(a \wedge b) = \Box a \wedge \Box b$ multiplicative

(NG2) $\Box 1 = 1$ dual normal.

Observe, that any such necessity operator is monotone. For any $A \subseteq L$,

$$\Box^{-1}(A) = \{a \in L : \Box a \in A\}$$
$$\langle\leq\rangle\Box(A) = \{a \in L : \exists b \in A,\ a \leq \Box b\},$$

where $\leq$ is the natural ordering of the lattice L. A *filter* (resp. *ideal*) of a necessity lattice is a filter (resp. ideal) of the underlying lattice L.

Lemma 13.4.1 *Let $(L, \Box)$ be a necessity algebra. For any $A, B \subseteq L$,*

(a) *If A is a filter of L, then so is $\Box^{-1}(A)$;*
(b) *If A is an ideal of L, then so is $\langle\leq\rangle\Box A$;*
(c) *If $\langle\leq\rangle\Box A \subseteq B$, then $A \subseteq \Box^{-1}(B)$;*
(d) *If B is an ideal of L, then $A \subseteq \Box^{-1}(B)$ implies $\langle\leq\rangle\Box A \subseteq B$.*

Proof:
Similar to the proof of Lemma 13.2.1. □

A *necessity (general) frame*, *NG-frame*, is a structure (X, R, S) where X is a general lattice frame and R and S are binary relations on X satisfying,

(FNG1) $(\leqslant_1; R; \leqslant_1) \subseteq R$
(FNG2) $(\geqslant_2; S; \geqslant_2) \subseteq S$
(FNG3) $R \subseteq (\leqslant_1; S)$
(FNG4) $S \subseteq (R; \geqslant_2)$.

Lemma 13.4.2 *Let (X, R, S) be an NG-frame. Then, for every $A \subseteq X$,*

(a) *$[R]A$ is $\leqslant_1$-increasing;*
(b) *$\langle S\rangle A$ is $\leqslant_2$-increasing.*

Proof:
We prove (a); the proof of (b) is similar. Assume that $x \in [R]A$ and take $y \in X$ such $x \leqslant_1 y$. Suppose that there is some $z \in X$ such that yRz and $z \notin A$. By (FNG1), from $x \leqslant_1 y$ and yRz we get xRz, which together with $x \in [R]A$ gives $z \in A$, a contradiction. □

The *complex algebra of an NG-frame* X is a structure $(L_X, \Box_R)$ where L_X is the complex algebra of its general lattice frame reduct and $\Box_R$ is a unary operator defined, for every $A \in L_X$, by

$$\Box_R A \stackrel{\text{def}}{=} [R](A).$$

Lemma 13.4.3 *Let (X, R, S) be any NG-frame. For every l-stable set $A \subseteq X$,*

(a) $\Box_R A = l\langle S\rangle r(A)$;

(b) $\Box_R A$ *is l-stable.*

Proof:
(a) Suppose that there is some $x \in X$ such that $x \in [R]A$ and $x \notin l(\langle S\rangle r(A))$. Then there exist $y, z \in X$ such that $x \leqslant_1 y$, ySz, and $z \in r(A)$. By (FNG4), ySz implies that there is some $u \in X$ such that yRu and $z \leqslant_2 u$. Now, $z \in r(A)$ and $z \leqslant_2 u$ yield $u \notin A$. Furthermore, using (FNG1), from $x \leqslant_1 y$ and yRu we get xRu, which together with $x \in [R]A$ gives $u \in A$, a contradiction. Therefore $[R]A \subseteq l(\langle S\rangle r(A))$. For the reverse inclusion, suppose that there is some $x \in X$ such that $x \in l(\langle S\rangle r(A))$ and $x \notin [R]A$. Then there exists $y \in X$ such that xRy and $y \notin A$, so $y \notin lr(A)$. Hence there is some $z \in X$ such that $y \leqslant_1 z$ and $z \in r(A)$. Since xRy and $y \leqslant_1 z$, by (FNG1) we get xRz. So, by (FNG3), there exists $u \in X$ such that $x \leqslant_1 u$ and uSz. Now, $x \in l(\langle S\rangle r(A))$ and $x \leqslant_1 u$ imply $u \notin \langle S\rangle r(A)$. But $z \in r(A)$ and uSz yield $u \in \langle S\rangle r(A)$, a contradiction.

(b) By Lemma 13.4.2(b), $\langle S\rangle r(A)$ is $\leqslant_2$-increasing; so by Lemma 2.8.3(a), $l(\langle S\rangle r(A))$ is l-stable. Hence, by (a), $\Box_R A$ is l-stable. $\Box$

Proposition 13.4.4 *The complex algebra of an NG-frame is an NG-lattice.*

Proof:
In view of Proposition 2.8.5 we only have to show that that $\Box_R$ satisfies axioms (NG1) and (NG2). This easily follows from the fact that $[R]$ distributes over $\cap$ and preserves 1. $\Box$

The *canonical frame of an NG-lattice* L is a structure $(X_L, R_\Box, S_\Box)$ where X_L is a general lattice frame as defined in Section 2.8 and the binary relations $R_\Box$ and $S_\Box$ on X_L are defined, for any $F, G \in X_L$, by

$$FR_\Box G \stackrel{\text{def}}{\Leftrightarrow} \Box^{-1}(F_1) \subseteq G_1$$
$$FS_\Box G \stackrel{\text{def}}{\Leftrightarrow} G_2 \subseteq \Box^{-1}(F_2).$$

Proposition 13.4.5 *The canonical frame of an NG-lattice is an NG-frame.*

Proof:
By straightforward verification it is easy to show that the conditions (FNG1) and (FNG2) hold.

(FNG3) Let $F, G \in X_L$ be such that $FR_\Box G$ holds, that is $\Box^{-1}(F_1) \subseteq G_1$. First, we show that $F_1 \cap \langle\leq\rangle\Box G_2 = \emptyset$. Suppose otherwise, that is there is $a \in F_1$ such that $a \in \langle\leq\rangle\Box G_2$. Then there is some $b \in G_2$ such that $a \leq \Box b$. Since F_1 is a filter of L, $\Box b \in F_1$, that is $b \in \Box^{-1}(F_1)$, whence $b \in G_1$, which gives the required contradiction. Now, by Lemma 13.4.1(b), $\langle\leq\rangle\Box G_2$ is an ideal of L, so $(F_1, \langle\leq\rangle\Box G_2)$ is a filter-ideal pair and by Theorem 2.8.6 it can be extended to a maximal filter-ideal pair, say $H=(H_1,H_2)$. Then $F_1 \subseteq H_1$ and $\langle\leq\rangle\Box G_2 \subseteq H_2$. Applying Lemma 13.4.1(c) we obtain $G_2 \subseteq \Box^{-1}(H_2)$, that is $HS_\Box G$ holds.

(FNG4) Let $F, G \in X_L$ be such that $FS_\Box H$ holds, that is $G_2 \subseteq \Box^{-1}(F_2)$. We show that $\Box^{-1}(F_1) \cap G_2 = \emptyset$. For suppose otherwise, there is $a \in G_2$ such that $a \in \Box^{-1}(F_1)$. Thus $\Box a \in F_1$, so $\Box a \notin F_2$, whence, by the assumption, $a \notin G_2$, a contradiction. Since F_1 is a filter of L, so is $\Box^{-1}(F_1)$ by Lemma 13.4.1(a). Then $(\Box^{-1}(F_1), G_2)$ is a filter-ideal pair. Let $H=(H_1,H_2)$ be its extension to a maximal filter-ideal pair. Then $\Box^{-1}(F_1) \subseteq H_1$, that is $FR_\Box G$ holds, and $G_2 \subseteq H_2$, as required. □

Lemma 13.4.6 *For any* $a \in L$, $h(\Box a) = \Box_{R_\Box} h(a)$.

Proof:
Let $F \in X_L$ be such $F \notin h(\Box a)$, that is $\Box a \notin F_1$, so $a \notin \Box^{-1}(F_1)$. Hence

$$\Box^{-1}(F_1) \cap {\downarrow_\leq} a = \emptyset.$$

By Lemma 13.4.1(a), $\Box^{-1}(F_1)$ is a filter of L, therefore $(\Box^{-1}(F_1), {\downarrow_\leq} a)$ is a filter-ideal pair and by Theorem 2.8.6 it can be extended to a maximal filter-ideal pair, say G. Hence $\Box^{-1}(F_1) \subseteq G_1$, that is $FR_\Box G$, and ${\downarrow_\leq} a \subseteq G_2$, so $a \in G_2$, whence $a \notin G_1$, and furthermore, $G \notin h(a)$. Therefore $F \notin \Box_{R_\Box} h(a)$. On the other hand, take any $F \in X_L$ such that $F \notin \Box_{R_\Box} h(a)$. Then there exists $G \in X_L$ such that $FR_\Box G$, that is $\Box^{-1}(F_1) \subseteq G_1$, and $G \notin h(a)$, so $a \notin G_1$. Hence $a \notin \Box^{-1}(F_1)$, so $\Box a \notin F_1$, whence $F \notin h(\Box a)$. □

We conclude this section with the following discrete representation for NG-lattices.

Theorem 13.4.7 *Every NG-lattice is embeddable into the complex algebra of its canonical frame.*

13.5 Axiomatic extensions of necessity lattices

In this section we present two axiomatic extensions of necessity lattices ([204]) by endowing them with the axioms well-known in the classical modal algebras based on Boolean algebras. The following axioms will be considered:

($T_\Box$) $\Box a \leq a$

($4_\Box$) $\Box a \leq \Box\Box a$.

For $\alpha \in \{T_\Box, 4_\Box\}$, the necessity lattice endowed with either of these axioms (α) is called an *NGα-lattice.*

Let (X, R, S) be an NG-frame. We will consider the following frame axioms:

($FT_\Box$) R is reflexive

($F4_\Box$) R is transitive.

The *complex algebra of an NGα-frame* (resp. the *canonical frame of an NGα-lattice*) is just the complex algebra of an NG-frame (resp. the canonical frame of an NG-lattice).

Proposition 13.5.1 *Let* $\alpha \in \{T_\Box, 4_\Box\}$.

(a) *The complex algebra of an NG*α*-frame is an NG*α*-lattice;*

(b) *The canonical frame of an NG*α*-lattice is an NG*α*-frame.*

Proof:
(a) For ($T_\Box$), take any $x \in [R]A$. By reflexivity of R, $x \in A$.
For ($4_\Box$), take any $x \in X$ such that $x \notin [R][R]A$. Then there exist $y, z \in X$ such that xRy, yRz, and $z \notin A$. By ($FT_\Box$), xRz. Therefore $x \notin [R]A$.

(b) For ($T_\Box$), take any $F \in X_L$. Suppose that $a \in \Box^{-1}(F_1)$. Then $\Box a \in F_1$. Since F_1 is a filter of L, by ($T_\Box$) we get $a \in F_1$. Therefore $\Box^{-1}(F_1) \subseteq F_1$, that is, $FR_\Box F$.
For ($4_\Box$), take any $F, G, H \in X_L$ such that $FR_\Box G$ and $GR_\Box H$. Then

$$\Box^{-1}(F_1) \subseteq G_1 \text{ and } \Box^{-1}(G_1) \subseteq H_1.$$

Suppose that $a \in \Box^{-1}(F_1)$, that is, $\Box a \in F_1$. Since F_1 is a filter of L, by ($4_\Box$) we get $\Box\Box a \in F_1$, and hence $\Box a \in \Box^{-1}(F_1)$. So $\Box a \in G_1$, that is, $a \in \Box^{-1}(F_1)$, and hence $a \in H_1$. Then $\Box^{-1}(F_1) \subseteq H_1$, that is, $FR_\Box H$. ◻

We conclude this section by the following discrete representation result for NGα-lattices.

Theorem 13.5.2 *Let* $\alpha \in \{T_\Box, 4_\Box\}$. *Every NG*$\alpha$*-lattice is embeddable into the complex algebra of its canonical frame.*

13.6 Lattices with possibility and necessity operators

In this section we consider bounded lattices with possibility operator $\Diamond$ and necessity operator $\Box$ endowed with some connecting axioms analogous to those in the classical modal algebras.

First, let us define a *PNG-lattice* as a structure $(L, \Diamond, \Box)$ where $(L, \Diamond)$ is a PG-lattice and $(L, \Box)$ is an NG-lattice. Accordingly, by a *PNG-frame* we mean a structure (X, R, S, P, Q) such that (X, R, S) is a PG-frame and (X, P, Q) is an NG-frame. In view of Proposition 13.2.4, Proposition 13.4.4, Proposition 13.2.6, and Proposition 13.4.5 we get the following fact.

Proposition 13.6.1

(a) *The complex algebra of a PNG-frame is a PNG-lattice;*

(b) *The canonical frame of a PNG-lattice is a PNG-frame.*

In view of Theorem 13.2.8 and Theorem 13.4.7 we get the following discrete representation of PNG-lattices.

Theorem 13.6.2 *Every PNG-lattice is embeddable into the complex algebra of its canonical frame.*

Now consider the following additional axioms. For any $a \in L$,

(D) $\Box a \leq \Diamond a$
($E_\Diamond$) $\Diamond a \leq \Box\Diamond a$
($E_\Box$) $\Diamond\Box a \leq \Box a$
($B_\Diamond$) $\Diamond\Box a \leq a$
($B_\Box$) $a \leq \Box\Diamond a$.

If α stands for either of these axioms, then the corresponding PNG-lattice will be refereed to as *PNGα-lattice.*

Let (X, R, S, P, Q) be a PNG-frame. The following additional axioms will be considered:

(FD) $Id_X \cap (P^{-1}; S) \neq \emptyset$
($FE_\Diamond$) $(R^{-1}; P) \subseteq (\leqslant_1 ; S^{-1})$
($FE_\Box$) $(Q^{-1}; S) \subseteq (\leqslant_2 ; P^{-1})$
($FB_\Diamond$) $S \subseteq Q^{-1}$ & $R \subseteq P^{-1}$
($FB_\Box$) $Q^{-1} \subseteq S$ & $P^{-1} \subseteq R$.

Let ($F\alpha$) stands for either of the above frame axioms. A PNG-frame is called a *PNGα-frame* if it satisfies axiom ($F\alpha$).

The *complex algebra of a PNGα-frame* (resp. the *canonical frame of a PNGα-lattice*) is defined in the same way as the complex algebra of a PNG-frame (resp. the canonical frame of a PNG-lattice).

Proposition 13.6.3 *Let $\alpha \in \{D, E_\Diamond, E_\Box, B_\Diamond, B_\Box\}$. The complex algebra of a PNGα-frame is a PNGα-lattice.*

Proof:
In view of Proposition 13.2.4 and Proposition 13.4.4 it suffices to show that the additional axioms hold.

(D) Take any $x \notin \Diamond_S A$, that is, $x \notin l([S]r(A))$. Then there exists $z \in X$ such that $x \leqslant_1 z$ and $z \in [S]r(A)$. By (FD) there is $y \in X$ such that zPy and zSy. Since $z \in [S]r(A)$ and zSy, we get $y \in r(A)$, which by reflexivity of $\leqslant_2$ yields $y \notin A$. Next, by (FNG1), xPy, which together with $y \notin A$ gives $x \notin [P]A = \Box_P A$.

($E_\Diamond$) We show that for any $A \in L_X$, $\Diamond_R(A) \subseteq \Box_P \Diamond_S(A)$. By Lemma 13.2.3(b) and Lemma 13.4.3, $\Box_P \Diamond_S(A) \in L_X$. Since lr is a closure operator, it suffices to show that $\langle R\rangle A \subseteq [P]l([S]r(A))$. Suppose otherwise, that is, there exist $A \in L_X$ and $x \in X$ such that $x \in \langle R\rangle A$ and $x \notin [P]l([S]r(A))$. Then there exists $t \in X$ such that xRt and $t \in A$. Also, there exists $y \in X$ such that xPy and $y \notin l([S]r(A))$, that is, there exists $z \in X$ such that $y \leqslant_1 z$ and $z \in [S]r(A)$. Then, by (FNG1), xPz. By ($FE_\Diamond$), there exists $u \in X$ such that $t \leqslant_1 u$ and zSu. Hence $u \in r(A)$, and thus $t \notin lr(A)$. But A is l-stable, so $A = lr(A)$, and hence $t \notin A$, which gives the required contradiction.

($E_\Box$) We need to show that $l([S]r([P]A)) \subseteq l(\langle Q\rangle r(A))$. Since l is antitone, it suffices to show that $\langle Q\rangle r(A) \subseteq [S]r([P]A)$. Suppose that for some $A \in L_X$ and for some $x \in X$, $x \in \langle Q\rangle r(A)$ and $x \notin [S]r([P]A)$. Then there exists $y \in X$ such that xQy and $y \in r(A)$. Also, there exists $z \in X$ such that xSz and $z \notin r([P]A)$, that is, there exists $t \in X$ such that $z \leqslant_2 t$ and $t \in [P]A$. Furthermore, by (FNG1), xSz and $z \leqslant_2 t$ imply xSt. Now, using (FE$_\Box$), from xSt and xQy we get some $u \in X$ such that $y \leqslant_2 u$ and tPu. Hence $u \in A$ and thus $y \notin r(A)$ which gives the required contradiction.

(B$_\Diamond$) We show that $l([S]r([P]A)) \subseteq A$ for every $A \in L_X$. Let $x \in X$ be such that $x \in l([S]r([P]A))$ and suppose that $x \notin A$. Since $A = lr(A)$, there is some $y \in X$ such that $x \leqslant_1 y$ and $y \in r(A)$. From $x \in l([S]r([P]A))$ and $x \leqslant_1 y$ it follows that $y \notin [S]r([P]A)$, so there is some $z \in X$ such that ySz and $z \notin r([P]A)$, and furthermore, there exists $u \in X$ such that $z \leqslant_2 u$ and $u \in [P]A$. By (FPG2), ySz and $z \leqslant_2 u$ imply ySu, so by (FB$_\Diamond$) we get uQy, which by (FNG4) implies that there is some $t \in X$ such that uPt and $y \leqslant_2 t$. Now, $u \in [P]A$ and uPt give $t \in A$. But from $y \in r(A)$ and $y \leqslant_2 t$ we obtain $t \notin A$, a contradiction.

(B$_\Box$) We show that $A \subseteq [P]l([S]r(A))$ for any $A \in L_X$. Take $x \in X$ such that $x \notin [P]l([S]r(A))$. Then there is some $y \in X$ such that xPy and $y \notin l([S]r(A))$, so there exists $z \in X$ such that $y \leqslant_1 z$ and $z \in [S]r(A)$. By (FNG1), xPy and $y \leqslant_1 z$ imply xPz, that is $zP^{-1}x$, which by (FB$_\Box$) gives zRx. Hence, using (FPG3) we get $u \in X$ such that zSu and $x \leqslant_1 u$. Since $z \in [S]r(A)$ and zSu, we get $u \in r(A)$. But $x \leqslant_1 u$, so $x \notin lr(A)$, thus $x \notin A$. □

Proposition 13.6.4 *For $\alpha \in \{\mathrm{D}, \mathrm{E}_\Diamond, \mathrm{E}_\Box, \mathrm{B}_\Diamond, \mathrm{B}_\Box\}$, the canonical frame of a PNGα-lattice is a PNGα-frame.*

Proof:
(FD) Let $F = (F_1, F_2) \in X_L$. Suppose that $\Box^{-1}(F_1) \cap \Diamond^{-1}(F_2) \neq \emptyset$. Then there exists $a \in L$ such that $a \in \Box^{-1}F_1$ and $a \in \Diamond^{-1}F_2$, so $\Box a \in F_1$ and $\Diamond a \in F_2$. Since $\Box a \leq \Diamond a$ and F_1 is a filter of L, $\Diamond a \in F_1$, which gives the required contradiction. Hence $\Box^{-1}(F_1) \cap \Diamond^{-1}(F_2) = \emptyset$, so $(\Box^{-1}(F_1), \Diamond^{-1}(F_2))$ is a filter-ideal pair and by Theorem 2.8.6 it can be extended to a maximal filter-ideal pair, say $G = (G_1, G_2)$. Therefore $\Box^{-1}(F_1) \subseteq G_1$ and $\Diamond^{-1}(F_2) \subseteq G_2$, that is, $FP_\Box G$ and $F\, S_\Diamond G$.

(FE$_\Diamond$) Take $F, G, H \in X_L$ such that $FP_\Box G$ and $FR_\Diamond H$, that is $\Box^{-1}(F_1) \subseteq G_1$ and $H_1 \subseteq \Diamond^{-1}(F_1)$. We show that $H_1 \cap \Diamond^{-1}(G_2) = \emptyset$. Take any $a \in \Diamond^{-1}(G_2)$, that is, $\Diamond a \in G_2$. Then, since $y \subset X_L$, $\Diamond a \notin G_1$, and hence $\Diamond a \notin \Box^{-1}(F_1)$, that is, $\Box\Diamond a \notin F_1$. Thus, by (E$_\Diamond$), $\Diamond a \notin F_1$, so $a \notin \Diamond^{-1}(F_1)$ and hence $a \notin H_1$. Therefore we get $\Diamond^{-1}(G_2) \subseteq -H_1$, so $H_1 \cap \Diamond^{-1}(G_2) = \emptyset$. Since $\Diamond^{-1}(G_2)$ is an ideal of L, $(H_1, \Diamond^{-1}(G_2))$ is a filter-ideal pair and by Theorem 2.8.6 it can be extended to a maximal filter–ideal pair, say $H' = (H'_1, H'_2)$. Hence we get $H_1 \subseteq H'_1$ and $\Diamond^{-1}(G_2) \subseteq H_2$, that is $GS_\Diamond H'$.

(FE$_\Box$) Let $F, G, H \in X_L$ be such that $FS_\Diamond G$ and $FQ_\Box H$. Then $\Diamond^{-1}(F_2) \subseteq G_2$ and $H_2 \subseteq \Box^{-1}(F_2)$. We show that $\Box^{-1}(G_1) \cap H_2 = \emptyset$. Let $a \in L$ be such that $a \in \Box^{-1}(G_1)$, that is $\Box a \in G_1$. Then $\Box a \notin G_2$, so $\Box a \notin \Diamond^{-1}(F_2)$, that is, $\Diamond\Box a \notin F_2$ where F_2 is an ideal. Hence, by (E$_\Box$), $\Box a \notin F_2$, that is, $a \notin \Box^{-1}(F_2)$ and hence $a \notin H_2$. Therefore $\Box^{-1}(G_1) \subseteq -H_2$, so $\Box^{-1}(G_1) \cap H_2 \neq \emptyset$ where

$\Box^{-1}(G_1)$ is a filter of L. Then $(\Box^{-1}(G_1), H_2)$ is a filter-ideal pair and by Theorem 2.8.6 it can be extended to a maximal filter-ideal pair, say $H = (H'_1, H'_2)$. Therefore $\Box^{-1}(G_1) \subseteq H'_1$, that is, $GP_\Box H'$, and $H_2 \subseteq H'_2$.

(FB$_\Diamond$) Let $F, G \in X_L$ be such that $FR_\Diamond^{-1}G$ holds, that is $F_1 \subseteq \Diamond^{-1}(G_1)$. Take $a \in \Box^{-1}(F_1)$. Thus $\Box a \in F_1$, so $\Box a \in \Diamond^{-1}(G_1)$, and hence $\Diamond\Box a \in G_1$. By (B$_\Diamond$), $a \in G_1$, since G_1 is a filter of L. Therefore $\Box^{-1}(F_1) \subseteq G_1$, that is $FP_\Box G$ holds. Assume now that $FS_\Diamond G$ holds, that is $\Diamond^{-1}(F_2) \subseteq G_2$. Take $a \notin \Box^{-1}(G_2)$, thus $\Box a \notin G_2$. Hence $\Box a \notin \Diamond^{-1}(F_2)$, so $\Diamond\Box a \notin F_2$. Thus $a \notin F_2$, since F_2 is an ideal of L. Therefore $F_2 \subseteq \Box^{-1}(G_2)$, that is $FQ_\Box^{-1}G$ holds.

(FB$_\Box$) We show that $P_\Box \subseteq R_\Diamond^{-1}$ and $Q_\Box^{-1} \subseteq S_\Diamond$. Let $F, G \in X_L$ and assume $FP_\Box G$, that is $\Box^{-1}(F_1) \subseteq G_1$. Take any $a \in L$ such that $a \notin \Diamond^{-1}(G_1)$, thus $\Diamond a \notin G_1$. Hence $\Diamond a \notin \Box^{-1}(F_1)$, that is $\Box\Diamond a \notin F_1$. Since F_1 is a filter of L, by (B$_\Box$) we infer $a \notin F_1$. Hence $F_1 \subseteq \Diamond^{-1}(G_1)$, so $FR_\Diamond^{-1}G$ holds.
Assume now that $FQ_\Box^{-1}G$, so $F_2 \subseteq \Box^{-1}(G_2)$. Let $a \in \Diamond^{-1}(F_2)$. Then $\Diamond a \in F_2$, and hence $\Diamond a \in \Box^{-1}(G_2)$, that is $\Box\Diamond a \in G_2$. Since G_2 is an ideal of L, by (B$_\Box$) it follows that $a \in G_2$. Therefore $\Diamond^{-1}(F_2) \subseteq G_2$, that is $FS_\Diamond G$ holds. $\Box$

Now, in view of Theorem 13.6.2 we obtain the following discrete representation theorem for PNGα-lattices.

Theorem 13.6.5 *Let $\alpha \in \{\mathrm{D}, \mathrm{E}_\Diamond, \mathrm{E}_\Box\}$. Then every PNG$\alpha$-lattice is embeddable into the complex algebra of its canonical frame.*

Now, let us consider PNG-lattices where both (B$_\Diamond$) and (B$_\Box$) are satisfied. These structures are referred to as *PNG$_B$-lattices*.

Let a PNG-frame (X, R, S, P, Q) be such that (FB$_\Diamond$) and (FB$_\Box$) hold. Then $S = Q^{-1}$ and $P = R^{-1}$. Hence in these PNG-frames the number of relations characterizing modal operators may be reduced to two relations only, S and P, each one characterizes the operator $\Diamond$ and $\Box$, respectively.

More specifically, by a *PNG$_B$-frame* we mean a structure (X, S, P) where X is a general lattice frame and S and P are binary relations on X satisfying

(PNG$_B$1) $(\leqslant_2; S; \leqslant_2) \subseteq S$

(PNG$_B$2) $(\leqslant_1; P; \leqslant_1) \subseteq P$

(PNG$_B$3) $S \subseteq (\leqslant_2; P^{-1})$

(PNG$_B$4) $P \subseteq (\leqslant_1; S^{-1})$.

The *complex algebra of a PNG$_B$-frame* (X, S, P) is a structure $(L_X, \Diamond_S, \Box_P)$ where L_X is the complex algebra of its general lattice frame reduct and the unary operators $\Diamond_S$ and $\Box_P$ are defined as in Section 13.2 and Section 13.4, respectively, that is $\Diamond_S A = l([S]r(A))$ and $\Box_P A = [P]A$.

Proposition 13.6.6 *The complex algebra of a PNG$_B$-frame is a PNG$_B$-lattice.*

Proof:
Using similar arguments as in the proof of Lemma 13.2.3(b) we can show that for every l-stable set A, $\Diamond_S A = lr(\langle P^{-1}\rangle A)$ and hence $\Diamond_S A$ is l-stable. Similarly, proceeding as in the proof of Lemma 13.4.3 we can obtain that for any l-stable set A, $\Box_P A = l(\langle S^{-1}\rangle r(A))$ and $\Box_P A$ is also l-stable. Next, $\Diamond_S(A \vee_X B) = \Diamond_S A \vee_X \Diamond_S B$ can be shown as in the proof of Proposition 13.2.4. Also, $[P](A \wedge_X B) = [P](A \cap B) = [P]A \cap [P]B = \Box_P A \wedge_X \Box_P B$. Furthermore, $\Diamond_S \emptyset = l([S]r(\emptyset)) = l([S]X) = l(X) = \emptyset$ and $\Box_P X = [P]X = X$. Finally, $(B_\Diamond)$ and $(B_\Box)$ can be shown as in the proof of Proposition 13.6.3. □

The *canonical frame of a PNG$_B$-lattice* $(L, \Diamond, \Box)$ is a structure $(X_L, S_\Diamond, P_\Box)$ where X_L is the canonical frame of its lattice reduct and $S_\Diamond$ and $P_\Box$ are binary relations on X_L defined, for any $F, G \in X_L$, by

$$FS_\Diamond G \overset{\text{def}}{\Leftrightarrow} \Diamond^{-1}(F_2) \subseteq G_2$$
$$FP_\Box G \overset{\text{def}}{\Leftrightarrow} \Box^{-1}(F_1) \subseteq G_1.$$

Proposition 13.6.7 *The canonical frame of a PNG$_B$-lattice is a PNG$_B$-frame.*

Proof:
Verification of axioms (FPNG$_B$1) and (FPNG$_B$2) is straighforward.
For (FPNG$_B$3), we have to show that for any $F, G \in X_L$, if $FS_\Diamond G$, then there is some $H \in_L$ such that $F_2 \subseteq H_2$ and $\Box^{-1}(G_1) \subseteq H_1$. Assume that $FS_\Diamond G$ holds, that is $\Diamond^{-1}(F_2) \subseteq G_2$. Let $a \in F_2$. Since F_2 is an ideal of L, by $(B_\Diamond)$ we get $\Diamond\Box a \in F_2$, so $\Box a \in \Diamond^{-1}(F_2)$, whence $\Box a \in G_2$. Thus $a \notin \Box^{-1}(G_1)$. Therefore $\Box^{-1}(G_1) \cap F_2 = \emptyset$. Since $\Box^{-1}(G_1)$ is a filter of L, $(\Box^{-1}(G_1), F_2)$ is a filter-ideal pair of L and by Theorem 2.8.6 it can be extended to a maximal filter-ideal pair, say H. Then $\Box^{-1}(G_1) \subseteq H_1$ and $F_2 \subseteq H_2$, as required.
(FPNB$_B$4) can be proved in a similar way. □

We conclude with the following discrete representation for PNG$_B$-lattices.

Theorem 13.6.8 *Every PNG$_B$-lattice is embeddable into the complex algebra of its canonical frame.*

13.7 Sufficiency lattices

A *sufficiency (general) lattice, SG-lattice*, is a structure $(L, ◫)$ where L is a general lattice and ◫ is a unary operator on L satisfying, for any $a, b \in L$,

(SG1) $◫(a \vee b) = ◫a \wedge ◫b$ coadditive
(SG2) $◫0 = 1$ conormal.

Observe that any such sufficiency operator is antitone. For any $A \subseteq L$,

$$◫^{-1}(A) = \{a \in L : ◫a \in A\}$$
$$\langle\leq\rangle◫(A) = \{a \in L : \exists b \in A,\ a \leq ◫b\},$$

where $\leq$ is the natural ordering of the lattice L. A filter (resp. ideal) of a sufficiency bounded lattice $(L, \boxminus)$ is a filter (resp. ideal) of the underlying lattice L.

Lemma 13.7.1 *Let $(L, \boxminus)$ be an SG-lattice. For any $A, B \subseteq L$,*

(a) *If A is a filter of L then $\boxminus^{-1}(A)$ is an ideal of L;*

(b) *If A is a filter of L, then $\langle\leq\rangle\boxminus(A)$ is an ideal of L;*

(c) *If $\langle\leq\rangle\boxminus(A) \subseteq B$, then $A \subseteq \boxminus^{-1}(B)$;*

(d) *If A is an ideal of L, then $A \subseteq \boxminus^{-1}(B)$ implies $\langle\leq\rangle\boxminus(A) \subseteq B$.*

Proof:
The statements are easily verified using the relevant definitions. □

A *sufficiency (general) frame, SG-frame,* is a structure (X, R, S) where X is a general lattice frame and R and S are two binary relations satisfying

(FSG1) $(\leqslant_1; R; \leqslant_2) \subseteq R$

(FSG2) $(\geqslant_2; S; \geqslant_1) \subseteq S$

(FSG3) $R \subseteq (\leqslant_1; S)$

(FSG4) $S \subseteq (R; \geqslant_1)$.

Lemma 13.7.2 *Let (X, R, S) be an SG-frame. For any $A \subseteq X$,*

(a) *$[R]A$ is a $\leqslant_1$-increasing set.*

(b) *$\langle S\rangle A$ is a $\leqslant_2$-increasing set.*

Proof:
(a) Assume that $x \in [R]A$ and $x \leqslant_1 y$. Suppose that $y \notin [R]A$. Then there is some $z \in X$ such that yRz and $z \notin A$. Hence, by (FSG1), xRz and thus $z \in A$, which gives the required contradiction.
The proof of (b) is similar. □

The *complex algebra of an SG-frame* X is $(L_X, \boxminus_R)$ where L_X is the complex algebra of a general lattice frame as defined in Section 2.8 and $\boxminus_R$ is a unary operator defined, for any $A \in L_X$, by

$$\boxminus_R A \stackrel{\text{def}}{=} [R]r(A).$$

Lemma 13.7.3 *Let (X, R, S) be any SG-frame. For every l-stable set $A \in L_X$,*

(a) *$\boxminus_R A = l\langle S\rangle(A)$;*

(b) *$\boxminus_R A$ is l-stable.*

Proof:
(a) Assume that $x \in \boxminus_R A$, that is, $x \in [R]r(A)$. Suppose that $x \notin l\langle S\rangle A$, that is, there is some $y \in X$ such that $x \leqslant_1 y$ and $y \in \langle S\rangle(A)$. Then there is some $z \in X$ such that ySz and $z \in A$. Hence, by (FSG4) there is some $z' \in X$ such that

$z \leqslant_1 z'$ and yRz'. Since A is l-stable, by Lemma 2.8.1(a), A is $\leqslant_1$-increasing, and hence $z' \in A$. By (FSG1) $x \leqslant_1 y$ and yRz' imply xRz'. Hence, since $x \in [R]r(A)$, $z' \in r(A)$. It follows from Lemma 2.8.1(b) that $z' \notin A$, which gives the required contradiction. On the other hand, assume that $x \in l\langle S\rangle A$. Suppose that $x \notin [R]r(A)$, that is, there is some $y \in X$ such that xRy and $y \notin r(A)$. Hence there is some $z \in X$ such that $y \leqslant_2 z$ and $z \in A$. By (FSG3), xRy implies there is some $x' \in X$ such that $x \leqslant_1 x'$ and $x'Sy$. By (FSG2), $x'Sy$ and $y \leqslant_2 z$ imply $x'Sz$. So $x' \in \langle S\rangle(A)$. Hence, since $x \leqslant_1 x'$, $x \notin l\langle S\rangle A$, the required contradiction is obtained.

(b) By Lemma 13.7.2(b) $\langle S\rangle A$ is a $\leqslant_2$-increasing set, and by Lemma 2.8.3(b) $l\langle S\rangle A$ is an l-stable set. So by (a) $\boxdot_R A$ is an l-stable set. □

Proposition 13.7.4 *The complex algebra of an SG-frame (X, R, S) is an SG-lattice.*

Proof:
Invoking Proposition 2.8.5, it suffices to show that $\boxdot_R$ satisfies the axioms (SG1) and (SG2). For any $A, B \in L_X$,

$$\begin{aligned}\boxdot_R(A \vee_X B) &= [R]r(A \vee_X B)\\ &= [R]rl(r(A) \cap r(B))\\ &= [R](r(A) \cap r(B))\end{aligned}$$

since by Lemma 2.8.3(a) $r(A) \cap r(B)$ is r-stable. Furthermore, $[R](r(A) \cap r(B)) = [R]r(A) \cap [R]r(B) = \boxdot A \cap \boxdot B$. Similarly, $\boxdot_R \emptyset = l\langle S\rangle\emptyset = l(\emptyset) = X$. □

The *canonical frame of an SG-lattice* L is the structure $(X_L, R_\boxdot, S_\boxdot)$ where X_L is a general lattice frame as defined in Section 2.8 and the binary relations $R_\boxdot$ and $S_\boxdot$ on X_L are defined, for any $F, G \in X_L$, by

$$\begin{aligned}FR_\boxdot G &\overset{\text{def}}{\Leftrightarrow} \boxdot^{-1}(F_1) \subseteq G_2\\ FS_\boxdot G &\overset{\text{def}}{\Leftrightarrow} G_1 \subseteq \boxdot^{-1}(F_2).\end{aligned}$$

Lemma 13.7.5 *Let $(L, \boxdot)$ be any SG-lattice. For every $F \in X_L$ and for every $a \in L$, the following conditions are equivalent,*

(a) $\boxdot a \in F_1$;
(b) *For every $G \in X_L$, if $FR_\boxdot G$, then $a \in G_2$.*

Proof:
Assume that $\boxdot a \in F_1$ and $FR_\boxdot G$. From the corresponding definitions we get $a \in \boxdot^{-1}F_1$ and $\boxdot^{-1}F_1 \subseteq G_2$. Hence, $a \in G_2$. On the other hand, assume that $\boxdot a \notin F_1$. Let $\uparrow_\leq a = \{b \in L : a \leq b\}$ be the filter of L generated by a. By Lemma 13.7.1(a) $\boxdot^{-1}(F_1)$ is an ideal. Since $a \notin \boxdot^{-1}(F_1)$, $\boxdot^{-1}(F_1) \cap \uparrow_\leq a = \emptyset$. Hence $(\uparrow_\leq a, \boxdot^{-1}(F_1))$ is a filter-ideal pair which, by Theorem 2.8.6, can be

extended to a maximal filter-ideal pair (G_1, G_2) of L. So we have $\uparrow_{\leq} a \subseteq G_1$ and $\boxdot^{-1}(F_1) \subseteq G_2$. Thus $a \in G_1$ and $F R_{\boxdot} G$. So, since $G_1 \cap G_2 = \emptyset$, $a \notin G_2$, which completes the proof. □

Proposition 13.7.6 *The canonical frame of an SG-lattice is an SG-frame.*

Proof:
In view of Proposition 2.8.8 it remains to show that the axioms (FSG1) – (FSG4) are satisfied.

For (FSG1), assume that $F_1' \subseteq F_1$, $\boxdot^{-1}(F_1) \subseteq G_2$, and $G_2 \subseteq G_2'$. Then, by monotonicity of $\boxdot^{-1}$, $\boxdot^{-1}(F_1') \subseteq \boxdot^{-1}(F_1)$. Hence, $\boxdot^{-1}(F_1') \subseteq G_2'$.
The proof of (FSG2) is similar.

For (FSG3), assume that $\boxdot^{-1}(F_1) \subseteq G_2$. We will show that $F_1 \cap \langle \leq \rangle \boxdot G_1 = \emptyset$. Suppose there is some $a \in L$ such that $a \in F_1$ and $a \in \langle \leq \rangle \boxdot G_1$. Then there is some $b \in L$ such that $b \in G_1$ and $a \leq \boxdot b$. Since F_1 is a filter, $\boxdot b \in F_1$, and hence $b \in \boxdot^{-1}(F_1)$. Thus, by the assumption, $b \in G_2$ which gives the required contradiction. So $(F_1, \langle \leq \rangle \boxdot G_1)$ is a filter-ideal pair which, by Theorem 2.8.6, can be extended to a maximal filter-ideal pair (F_1', F_2') of L. Thus $F_1 \subseteq F_1'$ and $\langle \leq \rangle \boxdot G_1 \subseteq F_2'$ which means that $G_1 \subseteq \boxdot^{-1}(F_2')$ by Lemma 13.7.1(c). We conclude that $F_1 \subseteq F_1'$ and $F' S_{\boxdot} G$, as required.

For (FSG4), assume that $G_1 \subseteq \boxdot^{-1}(F_2)$. We show that $\boxdot^{-1}(F_1) \cap G_1 = \emptyset$. Suppose there is $a \in L$ such that $a \in \boxdot^{-1}(F_1)$ and $a \in G_1$. Then $\boxdot a \in F_1$. By the assumption $\boxdot a \in F_2$. Hence $F_1 \cap F_2 \neq \emptyset$, which gives a contradiction. Since by Lemma 13.7.1(a) $\boxdot^{-1}(F_1)$ is an ideal, $(G_1, \boxdot^{-1}(F_1))$ is a filter-ideal pair of L which, by Theorem 2.8.6, can be extended to a maximal filter-ideal pair (G_1', G_2') of L. It follows that $G_1 \subseteq G_1'$ and $\boxdot^{-1}(F_1) \subseteq G_2'$, and hence $F R_{\boxdot} G'$ and $G_1 \subseteq G_1'$, as required. □

Let $(L, \boxdot)$ be an SG-lattice and let h be the mapping defined, for any $a \in L$, by $h(a) = \{F \in X_L : a \in F_1\}$.

Lemma 13.7.7 *For every $a \in L$, $h(\boxdot a) = \boxdot_{R_{\boxdot}} h(a)$.*

Proof:
Note that, for any $F \in X_L$ and any $a \in L$,

$$
\begin{aligned}
& F \in h(\boxdot a) \\
& \quad \Leftrightarrow \boxdot a \in F_1 \\
& \quad \Leftrightarrow \forall G \in X_L,\ F R_{\boxdot} G \Rightarrow a \in G_2 && \text{Lemma 13.7.5} \\
& \quad \Leftrightarrow \forall G \in X_L,\ F R_{\boxdot} G \Rightarrow G \in r(h(a)) && \text{Theorem 2.8.9(a)} \\
& \quad \Leftrightarrow F \in \boxdot_{R_{\boxdot}} h(a).
\end{aligned}
$$

□

Invoking Theorem 2.8.10 and Lemma 13.7.7 we obtain the following discrete representation for SG-lattices.

Theorem 13.7.8 *Every SG-lattice is embeddable into the complex algebra of its canonical frame.*

13.8 Dual sufficiency lattices

A *dual sufficiency (general) lattice, dSG-lattice*, is a structure $(L, \diamondsuit)$ where L is a general lattice and $\diamondsuit$ is a unary operator on L satisfying for any $a, b \in L$

(dSG1) $\diamondsuit(a \wedge b) = \diamondsuit a \vee \diamondsuit b$ co-multiplicative

(dSG2) $\diamondsuit 1 = 0$ co-dual normal.

Observe that any such dual sufficiency operator is antitone. For any $A \subseteq L$,

$$\diamondsuit^{-1} A = \{a \in L : \diamondsuit a \in A\}$$
$$\langle \geq \rangle \diamondsuit A = \{a \in K : \exists b \in A,\ \diamondsuit b \leq a\},$$

where $\leq$ is the natural ordering of the lattice L.

Lemma 13.8.1 *Let $(L, \diamondsuit)$ be a dSG-lattice. For all $A, B \subseteq L$,*

(a) *If A is an ideal of L, then $\diamondsuit^{-1}(A)$ and $\langle \geqslant \rangle \diamondsuit(A)$ are filters of L;*

(b) *If $\langle \geqslant \rangle \diamondsuit(A) \subseteq B$, then $A \subseteq \diamondsuit^{-1}(B)$;*

(c) *If B is a filter of L and $A \subseteq \diamondsuit^{-1}(B)$, then $\langle \geqslant \rangle \diamondsuit(A) \subseteq B$.*

Proof:
The proof is by straightforward verification. □

A *dual sufficiency (general) frame, dSG-frame*, is a structure (X, R, S) where X is a general lattice frame and R and S are binary relations on X satisfying

(FdSG1) $(\geqslant_1; R; \geqslant_2) \subseteq R$

(FdSG2) $(\leqslant_2; S; \leqslant_1) \subseteq S$

(FdSG3) $R \subseteq (S; \geqslant_2)$

(FdSG4) $S \subseteq (\leqslant_2; R)$

Lemma 13.8.2 *Let (X, R, S) be a dSG-frame. Then for every $A \subseteq X$,*

(a) *$\langle R \rangle A$ is a $\leqslant_1$-increasing set;*

(b) *$[S]A$ is a $\leqslant_2$-increasing set.*

Proof:
(a) Let $x, y \in X$ be such that $x \in \langle R \rangle A$ and $x \leqslant_1 y$. Then there is some $z \in X$ such that xRz and $z \in A$. By (FdSG1), xRz and $x \leqslant_1 y$ imply yRz, which together with $z \in A$ gives $y \in \langle R \rangle A$.

The proof of (b) is similar. □

The *complex algebra of a dSG-frame* X is a structure $(L_X, \diamondsuit_S)$ where L_X is the complex algebra of its general lattice frame reduct and $\diamondsuit_S$ is a unary operator in L_X defined, for any $A \in L_X$, by

$$\diamondsuit_S A \stackrel{\text{def}}{=} l[S]A.$$

Lemma 13.8.3 *Let (X, R, S) be a dSG-frame. Then for every l-stable set $A \subseteq X$,*

(a) $[S]A = r(\langle R \rangle r(A))$;

(b) $\diamondsuit_S A$ *is an l-stable set.*

Proof:
(a) We show that $l[S]A \subseteq r(\langle R \rangle r(A))$. Let $x \notin r(\langle R \rangle r(A))$. Then there is some $y \in X$ such that $x \leqslant_2 y$ and $y \in \langle R \rangle r(A)$, and furthermore, there exists $z \in X$ such that yRz and $z \in r(A)$. By (FdSG3), yRz implies that there is some $u \in X$ such that ySu and $z \leqslant_2 u$. Next, by (FdSG2), from ySu and $x \leqslant_2 y$ it follows that xSu. Since $z \in r(A)$ and $z \leqslant_2 u$, we get $u \notin A$, which together with xSu gives $x \notin [S]A$.
The proof of the reverse inclusion is similar.

(b) By Lemma 13.8.2(b), $[S]A$ is $\leqslant_2$-increasing, so by Lemma 2.8.3(b), $l([S]A)$ is l-stable. $\qquad\qquad\square$

Proposition 13.8.4 *The complex algebra of a dSG-frame is a dSG-lattice.*

Proof:
Invoking Proposition 2.8.5, it suffices to show that (dSG1) and (dSG2) are satisfied. For any $A, B \in L_X$,

$$
\begin{aligned}
\diamondsuit_S(A \wedge_X B) & \\
= l([S](A \cap B)) &\qquad \text{definitions of } \diamondsuit_S \text{ and } \wedge_X \\
= l([S]A \cap [S]B) &\qquad \text{distributivity of } [S] \text{ over } \cap \\
= l(r(\langle R \rangle r(A)) \cap r(\langle R \rangle r(B))) &\qquad \text{Lemma 13.8.3(a)} \\
= l(rlr(\langle R \rangle r(A)) \cap rlr(\langle R \rangle r(B))) &\qquad \text{Lemma 13.8.2(a), Lemma 2.8.3(a)} \\
= lr(\langle R \rangle r(A)) \vee_X lr(\langle R \rangle r(B)) &\qquad \text{definition of } \vee_X \\
= l([S]A) \vee_X l([S]B) &\qquad \text{Lemma 13.8.3(a)} \\
= \diamondsuit_S A \vee_X \diamondsuit_S B. &
\end{aligned}
$$

Furthermore, $\diamondsuit_S X = l([S]X) = l(X) = \emptyset$. $\qquad\qquad\square$

The *canonical frame of a dSG-lattice* L is the structure $(X_L, R_\diamond, S_\diamond)$ where X_L is a general lattice frame as defined in Section 2.8 and the binary relations $R_\diamond$ and $S_\diamond$ on X_L are defined, for any $F, G \in X_L$, by

$$
FR_\diamond G \overset{\text{def}}{\Leftrightarrow} G_2 \subseteq \diamondsuit^{-1}(F_1)
$$

$$
FS_\diamond G \overset{\text{def}}{\Leftrightarrow} \diamondsuit^{-1}(F_2) \subseteq G_1.
$$

Lemma 13.8.5 *Let $(X_L, \diamondsuit_S)$ be the canonical frame of a dSG-lattice $(L, \diamondsuit)$. For every $F \in X_L$ and for every $a \in L$, the following conditions are equivalent*

(a) $\diamondsuit a \in F_2$;

(b) *For every $G \in X_L$, if $FS_\diamond G$ then $a \in G_1$.*

Proof:
Assume that the condition (b) does not hold, that is for some $F \in X_L$ and for some $a \in L$ there exists $G \in X_L$ such that $\diamond^{-1}(F_2) \subseteq G_1$ and $a \notin G_1$. Hence $a \notin \diamond^{-1}(F_2)$, that is $\diamond a \notin F_2$. Conversely, assume that $\diamond a \notin F_2$. Let $\downarrow_{\leq} a$ be the principal ideal of L generated by a. By Lemma 13.8.1(a), $\diamond^{-1}(F_2)$ is a filter of L. Then $\downarrow_{\leq} a \cap \diamond^{-1}(F_2) = \emptyset$, so $(\diamond^{-1}(F_2), \downarrow_{\leq} a)$ is a filter-ideal pair of L which, by Theorem 2.8.6, can be extended to a maximal filter-ideal pair (G_1, G_2) of L. Thus $\diamond^{-1}(F_2) \subseteq G_1$, that is $FS_{\diamond}G$ holds, and $\downarrow_{\leq} \subseteq G_2$. Since $a \in G_2$, we get $a \notin G_1$. Therefore (b) does not hold. □

Proposition 13.8.6 *The canonical frame of a dSG-lattice is a dSG-frame.*

Proof:
(FdSG1) Let $F, F', G, G' \in X_L$ be such that $F_1 \subseteq F'_1$, $FR_{\diamond}G$, and $G'_2 \subseteq G_2$. By monotonicity of $\diamond^{-1}$, $\diamond^{-1}(F_1) \subseteq \diamond^{-1}(F'_1)$. Hence, since $G_2 \subseteq \diamond^{-1}(F_1)$, we get $G'_2 \subseteq \diamond^{-1}(F'_1)$, that is, $F'R_{\diamond}G'$ holds.
The proof of (FdSG2) is similar.

(FdSG3) Let $F, G \in X_L$ be such that $FR_{\diamond}G$, that is $G_2 \subseteq \diamond^{-1}(F_1)$. Suppose that $\diamond^{-1}(F_2) \cap G_2 \neq \emptyset$. Then there is some $a \in G_2$ such that $\diamond a \in F_2$. Hence, by the assumption, $\diamond a \in F_1$, so $F_1 \cap F_2 \neq \emptyset$, a contradiction. Therefore $\diamond^{-1}(F_2) \cap G_2 = \emptyset$. Since F_2 is an ideal of L, by Lemma 13.8.1(a) $\diamond^{-1}(F_2)$ is a filter of L, thus $(\diamond^{-1}(F_2), G_2)$ is a filter-ideal pair and by Theorem 2.8.6 it can be extended to a maximal filter-ideal pair, say H. Then $\diamond^{-1}(F_2) \subseteq H_1$, that is, $FS_{\diamond}H$, and $G_2 \subseteq H_2$. Hence $F(S_{\diamond}; \supseteq)G$ holds.

(FdSG4) Assume that $FS_{\diamond}G$ holds, that is $\diamond^{-1}(F_2) \subseteq G_1$. Suppose $\langle\geq\rangle\diamond G_2 \cap F_2 \neq \emptyset$. Then there is some $a \in F_2$ such that $a \in \langle\geq\rangle\diamond(G_2)$, so there exists $b \in G_2$ such that $\diamond b \leq a$. Since $a \in F_2$ and F_2 is an ideal of L, we get $\diamond b \in F_2$. Hence, from $\diamond^{-1}(F_2) \subseteq G_1$, it follows that $b \in G_1$, so $b \in G_1 \cap G_2$, a contradiction. Therefore $\langle\geq\rangle\diamond G_2 \cap F_2 = \emptyset$. By Lemma 13.8.1(a), $\langle\geq\rangle\diamond(G_2)$ is a filter of L, thus $(\langle\geq\rangle\diamond(G_2), F_2)$ is a filter-ideal pair of L which, by Theorem 2.8.6, can be extended to a maximal filter-ideal pair (H_1, H_2) of L. Then $\langle\geq\rangle\diamond(G_2) \subseteq H_1$ and $F_2 \subseteq H_2$. By Lemma 13.8.1(b) we get $G_2 \subseteq \diamond^{-1}(H_1)$, that is, $HR_{\diamond}G$. Hence $F(\subseteq; R_{\diamond})G$. □

Let $(L, \diamond)$ be a dSG-lattice and let the mapping $h : L \to L_{X_L}$ be defined, for any $a \in L$, by $h(a) = \{F \in X_L : a \in F_1\}$.

Lemma 13.8.7 *For any* $a \in L$, $h(\diamond a) = \diamond_{S_{\diamond}} h(a)$.

Proof:
For any $F \in X_L$ and any $a \in L$,

$$\begin{aligned} F \in \diamond_{S_{\diamond}} h(a) &\Leftrightarrow F \in l[S_{\diamond}]h(a) \\ &\Leftrightarrow \forall G \in X_L,\ F_1 \subseteq G_1 \Rightarrow G \notin [S_{\diamond}]h(a) \\ &\Leftrightarrow \forall G \in X_L,\ F_1 \subseteq G_1 \Rightarrow (\exists H \in X_L,\ GS_{\diamond}H \ \&\ H \notin h(a)) \\ &\Leftrightarrow \forall G \in X_L,\ F_1 \subseteq G_1 \Rightarrow (\exists H \in X_L,\ \diamond^{-1}(G_2) \subseteq H_1 \ \&\ a \notin H_1). \end{aligned}$$

Assume that $F \in h(\Diamond a)$, that is, $\Diamond a \in F_1$. Take any $G \in X_L$ such that $F_1 \subseteq G_1$. Then $\Diamond a \in G_1$, so $\Diamond a \notin G_2$. By Lemma 13.8.5, there exists $H \in X_L$ such that $GS_{\Diamond}H$ and $a \notin H_1$, so the required condition holds. On the other hand, assume $F \notin h(\Diamond a)$. Thus $\Diamond a \notin F_1$, so $(F_1, \downarrow_{\leq} \Diamond a)$ is a filter-ideal pair of L which, by Theorem 2.8.6, can be extended to a maximal filter-ideal pair (G_1, G_2) of L. Then $F_1 \subseteq G_1$ and $\downarrow_{\leq} \Diamond a \subseteq G_2$. Hence $\Diamond a \in G_2$, which by Lemma 13.8.5 yields for any $H \in X_L$, $\Diamond^{-1}(G_2) \subseteq H_1$ implies $a \in H_1$, thus by the above equivalences, $F \notin \Diamond_{S_{\Diamond}} h(a)$. □

The following discrete representation for dSG-lattices follows from Theorem 2.8.10 and Lemma 13.8.7.

Theorem 13.8.8 *Every dSG-lattice is embeddable into the complex algebra of its canonical frame.*

Chapter 14

General lattices with negations

14.1 Introduction

The Urquhart representation theorem presented in Section 2.8 was extended in [5] to general lattices with various additional operators, in particular with a De Morgan negation. Inspired by those developments, in [71] representation theorems for various classes of lattices with negations were presented. This chapter is based on those developments. In Section 14.2 the negation operator is a De Morgan negation satisfying the same axioms as in the case of distributive lattices discussed in Section 9.8. In Section 14.3 the negation is a weak pseudocomplement which satisfies only one implication from the specific axiom of pseudocomplemented distributive lattices described in Section 9.10. In Section 14.4 an orthonegation is discussed. The representation theorem presented here differs from the topological representation theorem given in [112]. A discrete version of the latter theorem is discussed in Section 18.9 and Section 18.10. Finally, in Section 14.5 general lattices with pseudocomplement are considered. The theorem of [112] is extended to various axiomatic extensions of ortholattices in [45].

14.2 General lattices with De Morgan negation

A *De Morgan lattice* $(L, \neg)$ is a general lattice L with a unary operation $\neg$ satisfying, for any $a, b \in L$,

(DeMG1) $a \leq \neg b \Rightarrow b \leq \neg a$

(DeMG2) $\neg\neg a \leq a$.

In order to distinguish this class from DeM-distributive lattices presented in Section 9.8 we denote not necessarily distributive general De Morgan lattices by *DeMG-lattices*.

From (DeMG1) and (DeMG2) the following properties may be derived.

Lemma 14.2.1 *Let L be a DeMG-lattice. For any $a, b \in L$,*

(a) $\neg 1 = 0$;
(b) $\neg 0 = 1$;
(c) $a \leq b \Rightarrow \neg b \leq \neg a$;
(d) $a \leq \neg\neg a$;
(e) $\neg\neg a = a$;
(f) $\neg(a \vee b) = \neg a \wedge \neg b$;
(g) $\neg(a \wedge b) = \neg a \vee \neg b$;
(h) $\neg a = \neg b \;\Rightarrow\; a = b$.

Lemma 14.2.2 *Let L be a DeMG-lattice with the natural order $\leq$ and let F be a proper filter of L. Then*

(a) $\langle\leq\rangle\neg F$ *is an ideal of* L;
(b) $\neg a \in F \;\Leftrightarrow\; a \in \langle\leq\rangle\neg F$ *for every* $a \in L$.

Proof:
(a) By definition, $\langle\leq\rangle\neg F$ is down-closed and $0 \in \langle\leq\rangle\neg F$. Take any $a, b \in \langle\leq\rangle\neg F$. Then $a \leq \neg c$ and $b \leq \neg d$ for some $c, d \in F$. Since F is a filter, $c \wedge d \in F$ so $\neg(c \wedge d) \in \neg F$. From Lemma 14.2.1(c) one easily derives $(\neg c) \vee (\neg d) \leq \neg(c \wedge d)$ hence $a \vee b \leq \neg(c \wedge d)$ so $a \vee b \in \langle\leq\rangle\neg F$. Thus $\langle\leqslant\rangle\neg F$ is an ideal.

(b) If $\neg a \in F$, then $\neg\neg a \in \langle\leq\rangle\neg F$ hence $a \in \langle\leq\rangle\neg F$ by Lemma 14.2.1(d). If $a \in \langle\leq\rangle\neg F$, then $a \leq \neg b$ for some $b \in F$, so $b \leq \neg a$ by (DeMG1). Hence $\neg a \in F$. □

A *DeMG-frame* is a structure (X, N) where X is a general lattice frame and $N : X \to X$ is a unary operator satisfying, for all $x, y \in X$,

(FDeMG1) $N(N(x)) = x$
(FDeMG2) $x \leqslant_1 y \Rightarrow N(x) \leqslant_2 N(y)$
(FDeMG3) $x \leqslant_2 y \Rightarrow N(x) \leqslant_1 N(y)$.

Note that these axioms may be reduced to the axioms, for all $x, y \in X$,

(FDeMG1) $N(N(x)) = x$
(FDeMG2) $x \leqslant_1 y \Leftrightarrow N(x) \leqslant_2 N(y)$

The representation in this section essentially comes from [5], where the function N is called a *generalized Routley-Meyer star operator*. Here we give full details and in Section 14.4 extend the method to ortholattices.

The *complex algebra of a DeMG-frame* X is $(L_X, \neg_N)$ where L_X is the complex algebra of a lattice frame X and $\neg_N$ is a unary operator defined, for any $A \in L_X$, by

$$\neg_N A \stackrel{\text{def}}{=} \{x \in X : N(x) \in r(A)\}.$$

Proposition 14.2.3 *The complex algebra of a DeMG-frame is a DeMG-lattice.*

Proof:
We need to show that $\neg_N A$ is l-stable, that is, $lr(\neg_N A) = \neg_N A$, and that L_X satisfies (DeMG1) and (DeMG2). Since l and r form a Galois connection, by Lemma 2.8.2 we have $\neg_N A \subseteq lr(\neg_N A)$ if and only if $r(\neg_N A) \subseteq r(\neg_N A)$. For the converse, suppose that for every y, if $x \leqslant_1 y$, then $y \notin r(\neg_N A)$ and assume, to the contrary, that $x \notin \neg_N A$. Then $N(x) \notin r(A)$ and there is some z such that $N(x) \leqslant_2 z$ and $z \in A$. It follows by (FDeMG3) and (FDeMG1) that $x \leqslant_1 N(z)$ and hence, by the above assumption, $N(z) \notin r(\neg_N A)$. Thus, there is some t such that $N(z) \leqslant_2 t$ and $t \in \neg_N A$. By application of N and (FDeMG3) and (FDeMG1), we have that $z \leqslant_1 N(t)$ and $N(t) \in r(A)$, in particular $N(t) \notin A$. But $z \in A$ and A is $\leqslant_1$-increasing, as $A = lr(A)$, hence $N(t) \in A$, a contradiction.

To prove (DeMG1), suppose that $A \subseteq \neg_N B$. Then, for every $x \in X$, if $x \in A$, then $N(x) \in r(B)$. Suppose that $x \in B$ and, to the contrary, that $x \notin \neg_N A$, that is, $N(x) \notin r(A)$, in which case $N(x) \leqslant_2 y$ and $y \in A$ for some $y \in X$. By (FDeMG3) and (FDeMG1), $x \leqslant_1 N(y)$ hence $N(y) \in B$ since $B = lr(B)$ is $\leqslant_1$-increasing. But also $y \in \neg_N B$, by the assumption, and $N(y) \in r(B)$, a contradiction since $B \cap r(B) = \emptyset$.

To prove (DeMG2), let $x \in \neg_N \neg_N A$, hence $N(x) \in r(\neg_N A)$. We show that $x \in l(r(A))$ which equals A since A is l-closed. Let $x \leqslant_1 y$. Then $N(x) \leq_2 N(y)$, by (FDeMG2), hence $N(y) \in r(\neg_N A)$ since $r(\neg_N A)$ is $\leqslant_2$-increasing. Thus, $N(y) \notin \neg_N A$, that is, $y = N(N(y)) \notin r(A)$. Thus, $x \in l(r(A)) = A$. □

The *canonical frame of a DeMG-lattice* L is the structure $(X_L, N_\neg)$ where X_L is a lattice frame and, for all $F, G \in X_L$,

$$N_\neg(F) \stackrel{\text{def}}{=} (\neg F_2, \neg F_1).$$

Proposition 14.2.4 *The canonical frame of a DeMG-lattice is a DeMG-frame.*

Proof:
Axiom (FDeMG1) follows from (DeMG2); axioms (FDeMG2) and (FDeMG3) are immediate. Thus we need only show that $N_\neg$ is a function from X_L to X_L. That is, if $F \in X_L$, we have to show that $N_\neg(F) = (\neg F_2, \neg F_1)$ is a maximal filter-ideal pair. Let $a_1, a_2 \in F_2$. Then $\neg a_1, \neg a_2 \in \neg F_2$. Since $a_1 \vee a_2 \in F_2$ and $\neg a_1 \wedge \neg a_2 = \neg(a_1 \vee a_2)$, $\neg F_2$ is closed under $\wedge$. If $\neg a_1 \leq b$, then $\neg b \leq \neg\neg a_1 = a_1$, so $\neg b \in F_2$. Then $b = \neg\neg b \in \neg F_2$, so $\neg F_2$ is up-closed and $1 \in \neg F_2$. Thus, $\neg F_2$ is a filter. Similarly, $\neg F_1$ is an ideal. Also, $\neg F_1$ and $\neg F_2$ can be shown disjoint using the implication: $\neg b = \neg c \Rightarrow b = c$ and the fact that F_1 and F_2 are disjoint. To show maximality, suppose $G \in X_L$ and $\neg F_1 \subseteq G_2$ and $\neg F_2 \subseteq G_1$. Then $\neg\neg F_1 \subseteq \neg G_2$, that is, $F_1 \subseteq \neg G_2$ and also $F_2 \subseteq \neg G_1$. Since $(\neg G_2, \neg G_1)$ is a filter-ideal pair, the maximality of F implies $F_1 = \neg G_2$ and $F_2 = \neg G_1$. Thus, $\neg F_1 = G_2$ and $\neg F_2 = G_1$ so $N_\neg(F)$ is maximal. □

By Theorem 2.8.10, the mapping $h : L \to L_{X_L}$ defined, for any $a \in L$, by $h(a) = \{F \in X_L : a \in F_1\}$ is an embedding of the general lattice reduct of a DeMG-lattice into the complex algebra of its canonical frame. We now show that h preserves the De Morgan negation operator $\neg$ on L.

Lemma 14.2.5 *For any $a \in L$, $h(\neg a) = \neg_{N_\neg} h(a)$.*

Proof:
Take any $F \in h(\neg a)$. Then $\neg a \in F_1$, and hence $a = \neg\neg a \in \neg F_1$. Suppose that $\neg F_1 \subseteq G_2$. Then $a \notin G_1$, since G_1 and G_2 are disjoint. On the other hand, take any $F \in \neg_{N_\neg} h(a)$. Suppose, to the contrary, that $F \notin h(\neg_{N_\neg} a)$, that is, $\neg a \notin F_1$. By Lemma 14.2.2(b), $a \notin \langle\leq\rangle\neg F_1$ and hence $(\uparrow_\leq a, \langle\leq\rangle\neg F_1)$ is a filter-ideal pair, which may be extended to a maximal one, say G. Thus, $\neg F_1 \subseteq G_2$, so $a \notin G_1$, but $\uparrow_\leq a \subseteq G_1$, a contradiction. □

Hence, we have the following discrete representation for DeMG-lattices.

Theorem 14.2.6 *Every DeMG-lattice is embeddable into the complex algebra of its canonical frame.*

14.3 General lattices with a weak pseudocomplement

In this section we consider general lattices with a negation which is weaker than the pseudo-complement in that it satisfies only one implication from its definition, namely, $a \leq \neg b \Rightarrow a \wedge b = 0$.

A *weakly pseudocomplemented general lattice, wpG-lattice*, is a structure $(L, \neg)$ where L is a general lattice with a unary operation $\neg$ satisfying, for any $a, b \in L$,

(wpG1) $a \leq b \Rightarrow \neg b \leq \neg a$
(wpG2) $a \leq \neg\neg a$
(wpG3) $a \leq b \ \&\ a \leq \neg b \Rightarrow \forall c \in L\ , a \leq c$.

Note that for any weakly pseudo-complemented lattice, condition (DeMG1) and Lemma 14.2.1(a) and (b) also hold and, due to the presence of $\wedge$, (wpG3) is equivalent to $a \wedge \neg a = 0$. Furthermore, (wpG3) implies $a \leq \neg b \Rightarrow a \wedge b = 0$, but the converse implication does not necessarily hold, see the example in the proof of Theorem 14.5.1(a).

Lemma 14.3.1 *Let $(L, \neg)$ be a wpG-lattice. Then, for any proper filter $F \subseteq L$, $F \cap \langle\leq\rangle(\neg F) = \emptyset$.*

Proof:
Suppose there is some $a \in F \cap \langle\leq\rangle(\neg F)$. Then $a \leq \neg b$ for some $b \in F$, so $b \leq \neg a$. Thus, $\neg a \in F$ hence $0 = a \wedge (\neg a) \in F$ which is a contradiction. □

A *wpG-frame* is a structure (X, R) where X is a lattice frame and R is a binary relation on X satisfying,

(FwpG1) $\leqslant_1; R \subseteq R$
(FwpG2) $R; \leqslant_2 \subseteq R$
(FwpG3) $\forall x \in X,\ \exists y \in X,\ xRy \ \&\ x \leqslant_1 y$
(FwpG4) $R \subseteq \ \leqslant_1; R^{-1}$.

The *complex algebra of a wpG-frame* X is $(L_X, \neg_R)$ where L_X is the complex algebra of a lattice frame X and the unary operation $\neg_R$ on L_X is defined for any $A \in L_X$ by

$$\neg_R A \stackrel{\text{def}}{=} [R](-A).$$

Lemma 14.3.2 *If $A \in L_X$, then $\neg_R A \in L_X$.*

Proof:
We have to show that $[R](-A) = lr([R](-A))$. Note that

$$\begin{aligned} lr(\neg_R A) &= [\leqslant_1]\langle\leqslant_2\rangle[R](-A) \\ &= \{x : \forall s\ (x \leqslant_1 s \Rightarrow \exists t,\ s \leqslant_2 t \ \&\ \forall u,\ tRu \Rightarrow u \notin A)\}. \end{aligned}$$

Let $x \in \neg_R A$ and suppose that $x \leqslant_1 s$ for some s. We claim that $t = s$ satisfies the required properties. Clearly, $s \leqslant_2 s$. If sRu, then, since $x \leqslant_1 s$, it follows by (FwpG1) that xRu; hence $u \notin A$. Thus, $x \in lr(\neg_R A)$ so $\neg_R A \subseteq lr(\neg_R A)$.

For the reverse inclusion, let $x \in lr(\neg_R A)$ and suppose that, for some y, xRy and $y \in A$. By (FwpG4), there exists s such that $x \leqslant_1 s$ and yRs. Then, since $x \in lr(\neg_R A)$ and $x \leqslant_1 s$, there exists t such that $s \leqslant_2 t$ and $\forall u,\ tRu \Rightarrow u \notin A$. Since yRs and $s \leqslant_2 t$, by (FwpG2), we have yRt. By (FwpG4), there is some w such that $y \leqslant_1 w$ and tRw. Since A is $\leqslant_1$-increasing, $w \in A$. Now taking w for u we obtain a contradiction. □

Proposition 14.3.3 *The complex algebra of a wpG-frame is a wpG-lattice.*

Proof:
Invoking Lemma 14.3.2, it suffices to show that the axioms (wpG1), (wpG2), and (wpG3) are satisfied in L_X.

For (wpG1), take any $A, B \in L_X$ such that $A \subseteq B$. Let $x \in \neg_R B$. Then, xRy implies $y \notin B$ hence also $y \notin A$, so $x \in \neg_R A$.

For (wpG2), note that

$$\neg_R \neg_R A = [R]\langle R\rangle A = \{x : \forall y,\ xRy \Rightarrow \exists z\ (yRz \ \&\ z \in A)\}.$$

Let $x \in A$ and suppose that xRy for some y. By (FwpG4), there exists z such that yRz and $x \leqslant_1 z$. Since A is $\leqslant_1$-increasing and $x \in A$, we have $z \in A$. Thus, the required z exists, showing that $x \in \neg_R \neg_R A$.

For (wpG3), we prove its equivalent form $A \cap \neg_A A = \emptyset$. Take any $A \in L_X$ and suppose there exists $x \in A \cap \neg_R A$. By (FwpG3), there exists a y such that xRy and $x \leqslant_1 y$. Since $x \in \neg_R A$ and xRy we have $y \notin A$. But $x \in A$ and $A \in L_X$, hence A is $\leqslant_1$-increasing, so $x \leqslant_1 y$ implies $y \in A$, a contradiction. □

The *canonical frame of a wpG-lattice* L is a structure $(X_L, R_\neg)$ where X_L is a general lattice frame and $R_\neg$ is a binary relation on X_L defined, for any $F, G \in X_L$, by

$$F R_\neg G \overset{\text{def}}{\Leftrightarrow} \forall a,\ \neg a \in F_1 \Rightarrow a \in G_2.$$

Proposition 14.3.4 *The canonical frame of a wpG-lattice is a wpG-frame.*

Proof:
Properties (FwpG1) and (FwpG2) are straightforward to prove. For (FwpG3), suppose $F \in X_L$. By Lemmas 14.2.2(a) and 14.3.1, $(F_1, \langle\leq\rangle\neg F_1)$ is a filter-ideal pair, so we can extend it to a maximal one, say G. If $\neg a \in F_1$, then $a \in \langle\leq\rangle\neg F_1$ by Lemma 14.2.2(b). Hence $a \in G_2$. Thus, $F R_\neg G$. Also, $F_1 \subseteq G_1$, that is $F \leqslant_{1L} G$, so we have found the required G.

For (FwpG4), suppose $F, G \in X_L$ and $F R_\neg G$. By Lemma 14.2.2(a), $\langle\leq\rangle\neg G_1$ is an ideal. If $a \in F_1 \cap \langle\leq\rangle\neg G_1$, then $a \in F_1$ implies $\neg\neg a \in F_1$, which implies $\neg a \in G_2$. But $a \in \langle\leq\rangle\neg G_1$ implies $\neg a \in G_1$ by Lemma 14.2.2(b), which contradicts the fact that $G_1 \cap G_2 = \emptyset$. Thus, $F_1 \cap \langle\leq\rangle\neg G_1 = \emptyset$ and we can extend $(F_1, \langle\leq\rangle\neg G_1)$ to a maximal filter-ideal pair, say H. If $\neg a \in G_1$ then $a \in \langle\leq\rangle\neg G_1$ hence $a \in H_2$, so $G R_\neg H$. Also, $F \leqslant_{1L} H$, so we have proved (FwpG4).

For (FwpG5), suppose that $F, G, H \in X_L$ such that $H R_\neg F$ and $F \leqslant_{2L} G$. First, we show that $H_1 \cap \langle\leq\rangle\neg G_1 = \emptyset$. Suppose $a \in y_1 \cap \langle\leq\rangle\neg G_1$. Then, $\neg\neg a \in H_1$ hence $\neg a \in F_2$. Since $F \leqslant_{2L} G$ we have $\neg a \in G_2$. Also, $a \leq \neg b$ for some $b \in G_1$, so $\neg a \geq \neg\neg b \geq b$ hence $\neg a \in G_1$. This contradicts the fact that G_1 and G_2 are disjoint. We therefore have that $(H_1, \langle\leq\rangle\neg G_1)$ is a filter-ideal pair, so we may extend it to a maximal one, say H'. Then, $H_1 \subseteq H'_1$, that is $H \leqslant_{1L} H'$. Suppose $H' \leqslant_{2L} G'$ and $\neg a \in G_1$. Then $\neg\neg a \in \neg G_1$ so $a \in \langle\leq\rangle\neg G_1 \subseteq H'_2 \subseteq G'_2$ hence $a \in G'_2$. Thus, we have proved (FwpG5). □

Let $(L, \neg)$ be a wpG-lattice and let $h : L \to L_{X_L}$ be the general lattice embedding defined, for any $a \in L$, by $h(a) = \{F \in X_L : a \in F_1\}$. We now show that h preserves the weak pseudocomplement $\neg$.

Theorem 14.3.5 *For any* $a \in L$, $h(\neg a) = \neg_{R_\neg} h(a)$.

Proof:
Note that for any $a \in L$,

$$h(\neg a) = \{F : \neg a \in F_1\}$$
$$\neg_{R_\neg} h(a) = \{F : \forall G,\ F R_\neg G \Rightarrow a \notin G_1\}.$$

First, take any $F, G \in X_L$ such that $F \in h(\neg a)$ and $F R_\neg G$. Then $\neg a \in F_1$, so $a \in G_2$ and hence $a \notin G_1$, as required. On the other hand, suppose that $\neg a \notin F_1$. Then, by Lemma 14.2.2(b), $a \notin \langle\leq\rangle\neg F_1$. So $(\uparrow_\leq a, \langle\leq\rangle\neg F_1)$ forms a filter-ideal pair which can be extended to a maximal filter-ideal pair, say G. Now $\uparrow_\leq a \subseteq G_1$ and hence $a \in G_1$. Also, $\langle\leq\rangle\neg F_1 \subseteq G_2$ and hence, for any $c \in L$, $\neg c \in F_1$ implies, by Lemma 14.2.2(b), $c \in \langle\leq\rangle\neg F_1$ which in turn implies

$c \in G_2$. Therefore, there is some $G \in X_L$ such that $F R_\neg G$ and $a \in G_1$, that is, $F \notin \neg_{R_\neg} h(a)$, as required. □

Hence, we obtain the following discrete representation for wpG-lattices.

Theorem 14.3.6 *Every wpG-lattice is embeddable into the complex algebra of its canonical frame.*

14.4 General lattices with an orthonegation

An *ortholattice*, *O-lattice*, is a general lattice which is both a DeMG-lattice and a wpG-lattice. Since (wpG1) and (wpG2) hold in DeMG-lattices as mentioned in Lemma 14.2.1(c) and (d), respectively, it is sufficient to define O-lattices as DeMG-lattices satisfying (wpG3), that is, for any $a, b \in L$,

(O) $a \leq b \;\&\; a \leq \neg b \Rightarrow \forall c \in L,\, a \leq c$.

Recall that in the presence of a lattice meet operation $\wedge$, this condition can be restated as

(O') $a \wedge \neg a = 0$.

Theorem 14.4.1

(a) *The class of O-lattices is a proper subclass of the class of DeMG-lattices;*

(b) *The class of O-lattices is a proper subclass of the class of wpG-lattices.*

Proof:
(a) As a counterexample for the reverse inclusion, consider the lattice $\mathfrak{L}_3$ of 3-valued Łukasiewicz logic with the elements $1, 0, a$, where $\neg a = a$. Then $\mathfrak{L}_3$ is a DeMG-lattice, but $\neg a \wedge a = a \neq 0$, so $\mathfrak{L}_3$ is not an O-lattice.

(b) As a counterexample for the converse consider the pentagon $\mathcal{N}_5$ that is a lattice with 5 elements $1, 0, a, b, c$, where $a < b$ and c is incomparable with a, b. Let $\neg a = \neg b = c$ and $\neg c = b$. Then $\mathcal{N}_5$ is a wpG-lattice but, since (DeMG2) is false, $a < b = \neg\neg a$, it is not an O-lattice. □

An *orthoframe*, an *O-frame*, is a DeMG-frame (X, N) where $N : X \to X$ also satisfies

(FO) $\forall x \in X\; \exists y\; x \leqslant_1 y \;\&\; N(x) \leqslant_2 y$.

The *complex algebra of a DeMG-lattice* L is defined as for DeMG-frames.

Proposition 14.4.2 *The complex algebra of an O-frame is an O-lattice.*

Proof:
It is sufficient to show that (O') holds. Suppose, to the contrary, that there exists $A \in L_X$ such that $A \cap (\neg_N A) \neq \emptyset$, and let $x \in A \cap (\neg_N A)$. Then there exists y such that $x \leqslant_1 y$ and $N(x) \leqslant_2 y$. Since A is $\leqslant_1$-increasing, $y \in A$. Since $x \in \neg_N A$, $N(x) \in r(A)$. But then $N(x) \leqslant_2 y$ implies $y \notin A$, a contradiction. □

The *canonical frame of an O-lattice* L is defined as for DeMG-lattices.

Proposition 14.4.3 *The canonical frame of an O-lattice is an O-frame.*

Proof:
By Lemma 14.2.4, $(X_L, N_\neg)$ is a DeMG-frame. Take any $F \in X_L$. Then, by Lemma 14.2.2, $(F_1, \neg F_1)$ is a filter-ideal pair, so we may extend it to a maximal filter-ideal pair, say $G = (G_1, G_2)$. Then $F_1 \subseteq G_1$ and $\neg F_1 \subseteq G_2$, so we have found a G that satisfies the required conditions. □

Thus, the above propositions imply that given any O-lattice, the complex algebra of its canonical frame is an O-lattice as well. Since, by Theorem 14.2.6, the mapping $h : L \to L_{X_L}$ is an embedding of DeMG-lattices, we have the following discrete representation.

Theorem 14.4.4 *Every O-lattice is embeddable into the complex algebra of its canonical frame.*

The class of O-lattices is the intersection of the class of DeMG-lattices and the class of wpG-lattices. In this section we obtained a representation for O-lattices by extending the representation for DeMG-lattices presented in Section 14.2 to include the condition (wpG3). However, since O-lattices may also be considered as an axiomatic extension of wpG-lattices by the axiom (DeMG2), one would expect a connection between the two representations.

Let (X, N) be an O-frame. We construct from it a wpG-frame (X, R') where relation R' is defined in terms of N, $\leqslant_1$, and $\leqslant_2$, and the complex algebras obtained from either of these relational structures are the same. To find the connection, consider an O-lattice L. Its canonical O-frame is $(X_L, N_\neg)$ where, for any $F \in X_L$,

$$N_\neg(F) = (\neg F_2, \neg F_1).$$

Since L is also a wpG-lattice, its canonical wpG-frame is $(X_L, R_\neg)$ where, for any $F, G \in X_L$,

$$F\, R_\neg G \;\Leftrightarrow\; \forall a \in L,\ \neg a \in F_1 \Rightarrow a \in G_2.$$

We claim that in the presence of both (wpG3) and (DeMG2) the following relationship holds between $N_\neg$ and $R_\neg$. For any $F, G \in X_L$,

$$F R_\neg G \Leftrightarrow N_\neg(F) \leqslant_{2L} G.$$

Take any $F, G \in X_L$. Assume $F\, R_\neg G$. Then

$$\neg a \in \neg F_1 \Leftrightarrow a \in F_1 \Leftrightarrow \neg\neg a \in F_1 \Rightarrow \neg a \in G_2,$$

so $\neg F_1 \subseteq G_2$, that is $N_\neg(F) \leqslant_{2L} G$. Conversely, assume $N_\neg(F) \leqslant_{2L} G$. Take any $a \in L$ such that $\neg a \in F_1$. Then $a = \neg\neg a \in \neg F_1$, so $a \in G_2$, hence $F\, R_\neg G$.

Given an O-frame (X, N), a binary relation R' on X is defined, for any $x, y \in X$, by

$$xR'y \overset{\text{def}}{\Leftrightarrow} N(x) \leqslant_2 y.$$

Then (X, R') is a wpG-frame. In particular, (FwpG1) and (FwpG2) are straightforward and (FwpG3) is simply (FO). For (FwpG4), take $z = N(y)$ and, for (FwpG5), take $z = N(t)$.

Moreover, the complex algebra obtained from either of these relational structures is the same. For this it suffices to check that the two definitions of $\neg_R$ coincide.

$$\begin{aligned}\{x \in X : N(x) \in r(A)\} &= \{x \in X : \forall y \in X,\ N(x) \leqslant_2 y \Rightarrow y \notin A\} \\ &= \{x \in X : \forall y \in X,\ xR'y \Rightarrow y \notin A\}.\end{aligned}$$

Thus, the definition of $\neg_N$ in the DeMG-lattices coincides with the definition of $\neg_{R'}$ in the wpG-lattices.

A natural question arises whether a class of frames for O-lattices can be defined in the style of wpG-frames. That is, what conditions should be added to (FwpG1)–(FwpG5) in order to ensure that the complex algebra of such a structure also satisfies (DeMG2), so that it is an O-lattice.

14.5 Pseudocomplemented general lattices

A *pseudocomplemented general lattice*, *pG-lattice*, is a wpG-lattice $(L, \neg)$ with a unary operation $\neg$ satisfying, for any $a, b \in L$,

(pG) $a \wedge b = 0 \Rightarrow a \leq \neg b$.

Since any pG-lattice satisfies (wpG3) which implies $a \leq \neg b \Rightarrow a \wedge b = 0$, as mentioned in Section 14.3, the implication in (pG) may be replaced by the equivalence. Furthermore, it is easy to see that Lemma 14.2.1(a), (b), (c), (d), and $a \wedge \neg a = 0$ hold in pG-lattices.

Theorem 14.5.1

(a) *The class of pG-lattices is propely included in the class of wpG-latices;*

(b) *The class of pG-lattices is different from the class of O-lattices;*

(c) *The class of pG-lattices is different from the class of DeMG-lattices.*

Proof:
(a) As a counterexample for the converse, consider the lattice $\mathcal{L}$ with 6 elements $1, 0, a, b, c, d,$, where a, b, c, d are incomparable. Let $\neg a = b$, $\neg b = a$, $\neg c = d$ and $\neg d = c$. Then $\mathcal{L}$ is a DeMG-lattice, a wpG-lattices, and an O-lattice, but not a HG-lattice as the implication $a \wedge b = 0 \Rightarrow a \leq \neg b$ fails.

(b) The lattice $\mathcal{L}$ mentioned above is an O-lattice but not a pG-lattice, and the pentagon $\mathcal{N}_5$ defined in Theorem 14.4.1 is a pG-lattice but not an O-lattice.

(c) The lattice $\mathcal{L}$ is a DeMG-lattice but not a pG-lattice, and the pentagon $\mathcal{N}_5$ is a pG-lattice but not a DeMG-lattice. □

A *pG-frame* (X, R) is a wpG-frame satisfying

(FpG) $\forall x, y \in X,\ xRy \Rightarrow \exists z \in X\ (x \leqslant_1 z\ \&\ y \leqslant_1 z)$.

The *complex algebra of a pG-frame* X is defined as for wpG-lattices.

Proposition 14.5.2 *The complex algebra of a pG-frame is a pG-lattice.*

Proof:
Take any $A, B \in L_X$, such that $A \cap B = \emptyset$. Let $x \in A$ and let $y \in X$ be such that xRy. By (FpG), there exists $z \in X$ such that $x \leqslant_1 z$ and $y \leqslant_1 z$. Since $x \in A$ and A is $\leqslant_1$-increasing, we have $z \in A$ as well. If $y \in B$ then, since B is $\leqslant_1$-increasing, it would follow that $z \in B$ and hence that $z \in A \cap B$, contradicting our assumption that $A \cap B = \emptyset$. Thus, $y \notin B$ hence $x \in \neg_R B$, as required. □

The *canonical frame of a pG-lattice* L is defined as for wpG-lattices.

Proposition 14.5.3 *The canonical frame of a pG-lattice is a pG-frame.*

Proof:
We need only show that (FpG) holds. So, let $F, G \in X_L$ be such that $F\, R_\neg G$. Consider the filter generated by $F_1 \cup G_1$, denoted Z_1. We claim that $0 \notin Z_1$. If we suppose otherwise, then there exist $a_1, \ldots, a_n \in F_1$ and $b_1, \ldots, b_m \in G_1$ such that

$$(\textstyle\bigwedge_{i=1}^{n} a_i) \wedge (\bigwedge_{j=1}^{m} b_j) = 0.$$

If we set $a = \bigwedge_{i=1}^{n} a_i$ and $b = \bigwedge_{j=1}^{m} b_j$, then $a \in F_1$ and $b \in G_1$ and $a \wedge b = 0$. But this implies that $a \leq \neg b$, hence $\neg b \in F_1$. Finally, since $F\, R_\neg G$ and $\neg b \in F_1$, we have $b \in G_2$. Thus, $b \in G_1 \cap G_2$, a contradiction.

This shows that $0 \notin Z_1$ so $(Z_1, \{0\})$ is a filter-ideal pair. This can be extended to a maximal filter-ideal pair, say Z'. Clearly, $F \subseteq_1 Z'$ and $G \subseteq_1 Z'$, as required. □

Thus, we have shown that if L is a pG-lattice, then so is the complex algebra of its canonical frame. Moreover, from Section 14.3 we know that the mapping h is an embedding of wpG-lattices. Hence we have the following discrete representation for pG-lattices.

Theorem 14.5.4 *Every pG-lattice is embeddable into the complex algebra of its canonical frame.*

Chapter 15

Residuated and doubly residuated general lattices

15.1 Introduction

Residuated lattices evolved from the study of ideal lattices of rings, see [150]. In the development of their theory the papers by Ward and Dilworth, e.g., [251] have played an important role. Since then the theory has been a subject of investigations of many authors, see for example [14], [86], and [16]. Residuated lattices have a great variety of applications in algebra, logic, and computer science, in particular in lattice-ordered groups, substructural and fuzzy logics, and formal language theory. While in Chapter 12 the distributive residuated lattices and their various axiomatic extensions are studied, in the present chapter the focus is on general residuated lattices. In Section 15.2 a discrete representation theorem for them is developed. Then in analogy to Heyting algebras, where the relative pseudocomplement is the residual of meet, and Heyting-Brouwer agebras obtained from Heyting algebras by endowing them with the difference operator, which is the residual of join, in Section 15.3 we consider doubly residuated general lattices. A doubly residuated lattice is constructed from a residuated lattice, say L, as follows. We consider the opposite residuated lattice L^{op} and we extend the signature of L with the monoid of L^{op} and the residual of the product of L^{op}, which then is referred to as a sum. That is, a doubly residuated lattice is a general lattice with two monoids and with the residuals of their products. We present a discrete representation theorem for them. The doubly residuated lattices are the bases for the lattice-ordered relation algebras discussed in Chapter 16.

15.2 Residuated general lattices

A *bounded residuated (general) lattice*, *RG-lattice*, is a structure $(L, \odot, \rightarrow, \leftarrow, 1')$ such that

(RG1) L is a general lattice

(RG2) $(L, \odot, 1')$ is a monoid

(RG3) $\leftarrow$ is the *left residual of* $\odot$ and $\rightarrow$ is the *right residual of* $\odot$, defined, for all $a, b, c \in L$, by

$$a \odot b \leq c \Leftrightarrow a \leq b \leftarrow c$$
$$a \odot b \leq c \Leftrightarrow b \leq a \rightarrow c$$

Note that the existence of the residuals implies, for every $a, b \in L$, that

$$a \rightarrow b = \bigvee\{c \in L : c \odot a \leq b\}$$
$$a \leftarrow b = \bigvee\{c \in L : a \odot c \leq b\}.$$

A residuated lattice is called

- *integral* if it is bounded and $1 = 1'$
- *commutative* if $\odot$ is commutative
- *complete* if its lattice reduct is complete.

The following lemma collects the well-known properties of RG-lattices, see for example [15].

Lemma 15.2.1 *Let $(L, \odot, \rightarrow, \leftarrow, 1')$ be an RG-lattice. For all $a, b, c \in L$ and for every indexed family $(e_i)_{i \in I}$ of elements of L,*

(a) $a \odot (b \vee c) = (a \odot b) \vee (a \odot c)$;
$(b \vee c) \odot a = (b \odot a) \vee (c \odot a)$;

If $\bigvee_{i \in I} e_i$ exists, then
$a \odot \bigvee_{i \in I} e_i = \bigvee_{i \in I}(a \odot e_i)$;
$\bigvee_{i \in I} e_i \odot a = \bigvee_{i \in I}(e_i \odot a)$;

(b) $a \rightarrow (b \wedge c) = (a \rightarrow b) \wedge (a \rightarrow c)$;
$(b \wedge c) \leftarrow a = (b \leftarrow a) \wedge (c \leftarrow a)$;

If $\bigwedge_{i \in I} e_i$ exists, then
$a \rightarrow \bigwedge_{i \in I} e_i = \bigwedge_{i \in I}(a \rightarrow r_i)$;
$\bigwedge_{i \in I} e_i \leftarrow a = \bigwedge_{i \in I}(e_i \leftarrow a)$;

(c) $(b \vee c) \rightarrow a = (b \rightarrow a) \wedge (c \rightarrow a)$;
$a \leftarrow (b \vee c) = (a \leftarrow b) \wedge (a \leftarrow c)$;

If $\bigvee_{i \in I} e_i$ exists, then
$\bigvee_{i \in I} \rightarrow a = \bigwedge_{i \in I}(e_i \rightarrow a)$;
$a \leftarrow \bigvee_{i \in I} e_i = \bigwedge_{i \in I}(a \leftarrow e_i)$;

(d) *If $a \leq b$, then*
$a \odot c \leq b \odot c$ *and* $c \odot a \leq c \odot b$;
$b \rightarrow c \leq a \rightarrow c$ *and* $b \leftarrow c \leq a \leftarrow c$;
$c \rightarrow a \leq c \rightarrow b$ *and* $c \leftarrow a \leq c \leftarrow b$;

(e) $a \odot (a \rightarrow b) \leq b$;
$(a \leftarrow b) \odot b \leq a$;

(f) $(a \to b) \odot x \leq a \to (b \odot c)$;
$a \odot (b \leftarrow c) \leq (a \odot b) \leftarrow c$;

(g) $a \to (b \to c) = (b \odot a) \to c$;
$(a \leftarrow b) \leftarrow c = a \leftarrow (z \odot b)$;

(h) $a \to (b \leftarrow c) = (a \to b) \leftarrow c$;

(i) $1' \to a = a \leftarrow 1' = a$;

(j) $1' \leq a \to a$;
$1' \leq a \leftarrow a$;

(k) $b \leq a \to (a \odot b)$;
$a \leq (a \odot b) \leftarrow b$;

(l) $a \leq b \leftarrow (a \to b)$;
$a \leq (b \leftarrow a) \to b$;

(m) $a \leftarrow ((a \leftarrow b) \to a) = b \leftarrow a$;
$(a \leftarrow (b \to a)) \to a = b \to a$;

(n) $a \leftarrow (a \to a) = a$;
$(a \leftarrow a) \to a = a$;

(o) $(a \leftarrow b) \odot (b \leftarrow c) \leq a \leftarrow c$;
$(a \to b) \odot (b \to c) \leq a \to c$;

(p) $a \leq (a \to (a \odot b))$;

(q) $(a \to b) \odot c \leq a \to (b \odot c)$;
$a \odot (b \leftarrow c) \leq b \leftarrow (a \odot c)$;

(r) $(a \to b) \odot (b \to c) \leq (a \to c)$;
$(b \odot c) \odot (a \leftarrow b) \leq (a \leftarrow c)$;

(s) $(a \to c) \vee (b \to c) \leq (a \wedge b) \to c$.

By a *filter* (resp. *ideal*) of an RG-lattice L we mean a filter (resp. ideal) of its lattice reduct.

Given an RG-lattice L and a binary operation $*$ on L, it may be extended to an operation $*$ on 2^L defined, for any $A, B \subseteq L$, by

$$A * B \stackrel{\text{def}}{=} \{a * b : a \in A \ \&\ b \in B\}.$$

Lemma 15.2.2 *Let L be an RG-lattice, let F,F' be filters of L, and let I be an ideal of L. Then*

(a) *the following subsets of L are ideals of L*

$$I_1 = \{a \in L : F \cap (\{a\} \to I) \neq \emptyset\};$$
$$I_2 = \{a \in L : F \cap (I \leftarrow \{a\}) \neq \emptyset\};$$
$$I_3 = \{a \in L : I \cap (\{a\} \odot F) \neq \emptyset\};$$
$$I_4 = \{a \in L : I \cap (F \odot \{a\}) \neq \emptyset\};$$

(b) *the following subsets of L are filters of L*

$$F_1 = \{a \in L : F \cap (F' \to \{a\}) \neq \emptyset\};$$
$$F_2 = \{a \in L : F \cap (\{a\} \leftarrow F') \neq \emptyset\}.$$

Proof:
By way of example we show that I_1 is an ideal of L. Take $a, b \in L$ such that $a \in I_1$ and $b \leq a$. Then, by the definition of I_1, there is some $c \in I$ such that $a \to c \in F$. Since $b \leq a$, by Lemma 15.2.1(d) we get $a \to c \leq b \to c$. Thus $b \to c \in F$. Hence, for some $c \in I$, $b \to c \in F$, which means that $b \in I_1$. Furthermore, take $a, b \in L$ such that $a \in I_1$and $b \in I_1$. Then there are $a', b' \in I$ such that $a \to a' \in F$ and $b \to b' \in F$. Clearly, $a' \leq a' \vee b'$, so by Lemma 15.2.1(d) it follows that $a \to a' \leq a \to (a' \vee b')$. Since $a \to a' \in F$, we get $a \to (a' \vee b') \in F$. Analogously, $b \to (a' \vee b') \in F$. Thus $(a \to (a' \vee b')) \wedge (b \to (a' \vee b')) \in F$, so by Lemma 15.2.1(c), $(a \vee b) \to (a' \vee b') \in F$. Since I is an ideal of L and $a', b' \in I$, $a' \vee b' \in I$. Therefore for $c = a' \vee b' \in I$, $(a \vee b) \to c \in F$, hence $a \vee b \in I_1$. □

An *RG-frame* is a structure (X, R, S, Q, I) such that X is a general lattice frame as defined in Section 2.8, R, S, Q are ternary relations on X, and I is a unary relation on X such that, for all $x, x', y, y', z, z' \in X$,

(FRG1) $R(x, y, z)$ & $(x' \leqslant_1 x)$ & $(y' \leqslant_1 y)$ & $(z \leqslant_1 z') \Rightarrow R(x', y'z')$

(FRG2) $S(x, y, z)$ & $(x \leqslant_2 x')$ & $(y' \leqslant_1 y)$ & $(z' \leqslant_2 z) \Rightarrow S(x', y'z')$

(FRG3) $Q(x, y, z)$ & $(x' \leqslant_1 x)$ & $(y \leqslant_2 y')$ & $(z' \leqslant_2 z) \Rightarrow Q(x', y'z')$

(FRG4) $I(x)$ & $(x \leqslant_1 x') \Rightarrow I(x')$

(FRG5) $R(x, y, z) \Rightarrow \exists u \in X$, $(x \leqslant_1 u)$ & $S(u, y, z)$

(FRG6) $R(x, y, z) \Rightarrow \exists u \in X$, $(y \leqslant_1 u)$ & $Q(x, u, z)$

(FRG7) $S(x, y, z) \Rightarrow \exists u \in X$, $(z \leqslant_2 u)$ & $R(x, y, u)$

(FRG8) $Q(x, y, z) \Rightarrow \exists u \in X$, $(z \leqslant_2 u)$ & $R(x, y, u)$

(FRG9) $\exists u \in X$, $R(x, y, u)$ & $S(u, z, z')$
$\Rightarrow \exists t \in X$, $R(y, z, t)$ & $Q(x, t, z')$

(FRG10) $\exists u \in X$, $R(x, y, u)$ & $Q(x', u, z)$
$\Rightarrow \exists t \in X$, $R(x', x, t)$ & $S(t, y, z)$

(FRG11) $I(x)$ & $(R(x, y, z)$ or $R(y, x, z)) \Rightarrow y \leqslant_1 z$

(FRG12) $\exists u \in X$, $I(u)$ & $S(u, x, x)$

(FRG13) $\exists u \in X$, $I(u)$ & $Q(x, u, x)$.

Lemma 15.2.3 *For every RG-frame X, for every l-stable subset $A \subseteq X$, and for all $x, y, z \in X$,*

$$I(x) \;\&\; (y \in A) \;\&\; (R(x, y, z) \text{ or } R(y, x, z)) \Rightarrow z \in A.$$

Proof:
Take any $x, y, z \in X$ with $I(x)$, $y \in A$, and $R(x, y, z)$ or $R(y, x, z)$. By (FRG11) it follows that $y \leqslant_1 z$. Since A is l-stable, it is $\leqslant_1$-increasing, so $z \in A$. □

Let X be an RG-frame. Define two auxiliary operators $\circ_Q, \circ_S : 2^X \times 2^X \to 2^X$, for all $A, B \subseteq X$, by

$$A \circ_Q B \stackrel{\text{def}}{=} \{z \in X : \forall x, y \in X,\ Q(x, y, z)\ \&\ (x \in A) \Rightarrow (y \in r(B))\}$$
$$A \circ_S B \stackrel{\text{def}}{=} \{z \in X : \forall x, y \in X,\ S(x, y, z)\ \&\ (y \in B) \Rightarrow (x \in r(A))\}.$$

Lemma 15.2.4 *For any $A, B \subseteq X$,*

(a) *$A \circ_Q B$ and $A \circ_S B$ are $\leqslant_2$-increasing;*

(b) *If A and B are l-stable, then $A \circ_Q B$ and $A \circ_S B$ are r-stable;*

(c) *If A and B are l-stable, then $A \circ_Q B = A \circ_S B$.*

Proof:
(a) Let $A, B \subseteq X$ and suppose that $A \circ_Q B$ is not $\leqslant_2$-increasing. Then there exist $x, z \in X$ such that $x \in A \circ_Q B$, $x \leqslant_2 z$, and $z \notin A \circ_Q B$. By the definition of $\circ_Q$, $y \notin A \circ_Q B$ means that there exist $u, w \in X$ such that $Q(z, u, y)$, $z \in A$, and $u \notin r(B)$. However, by reflexivity of $\leqslant_1$ and $\leqslant_2$, from (FRG3), $x \leqslant_2 z$ and $Q(z, u, y)$ imply $Q(z, u, x)$. Hence, since $z \in A$ and $u \notin r(B)$, we get $x \notin A \circ_Q B$, a contradiction. The proof for $A \circ_S B$ is similar.

(b) Let $A, B \subseteq X$ be l-stable sets. By (a), $A \circ_Q B$ is $\leqslant_2$-increasing. Then $A \circ_Q B \subseteq rl(A \circ_Q B)$. It suffices to show that $rl(A \circ_Q B) \subseteq A \circ_Q B$.
Let $z \in X$ and assume that $z \notin A \circ_Q B$. We show that $z \notin rl(A \circ_Q B)$. By the definition of $\circ_Q$, there exist $x, y \in X$ such that $Q(x, y, z)$, $x \in A$, and $y \notin r(B)$. Then there is some $y' \in X$ such that $y \leqslant_2 y'$ and $y' \in B$. By (FRG3), $Q(x, y, z)$ and $y \leqslant_2 y'$ imply $Q(x, y', z)$. Hence, by (FRG8), there is some $z' \in X$ such that $z \leqslant_2 z'$ and $R(x, y', z')$.
Now we show that $z' \in l(A \circ_Q B)$. Let $z'' \in X$ and assume that $z' \leqslant_1 z''$. Hence, by (FRG1), from $R(x, y' z')$ we get $R(x, y', z'')$ which by (FRG6) implies that there is some $y'' \in X$ such that $y' \leqslant_1 y''$ and $Q(x, y'', z'')$. Hence, since B is l-stable and $y' \in B$, $y'' \in B$, so $y'' \notin r(B)$. Furthermore, since $x \in A$, we get $z'' \notin A \circ_Q B$. However, z'' is an arbitrary element satisfying $z' \leqslant_1 z''$, so we obtain that for any $z'' \in X$, $z' \leqslant_1 z''$ implies $z'' \notin A \circ_Q B$, so $z'' \in l(A \circ_Q B)$, as required. Proceeding in a similar way we can show that $A \circ_S B$ is r-stable.

(c) We show that $A \circ_S B \subseteq A \circ_Q B$. Let $z \in X$ and assume that $z \notin A \circ_Q B$. By the assumption there exist $x, y \in X$ such that $Q(x, y, z)$, $x \in A$, and $y \notin r(B)$. Since $y \notin r(B)$, there is some $y' \subset X$ such that $y \leqslant_2 y'$ and $y' \in B$. Then, by (FRG3), we get $Q(x, y', z)$. By (FRG8) there exists some $z' \in X$ such that $z \leqslant_2 z'$ and $R(x, y', z')$. Hence, by applying (FRG5), there exists some $x' \in X$ such that $x \leqslant_1 x'$ and $S(x', y', z')$. Now, since A is l-stable, the conditions $x \in A$ and $x \leqslant_1 x'$ imply $x' \notin r(A)$. Hence, $y' \in B$ and $S(x', y', z')$ imply $z' \notin A \circ_S B$. It follows from (a) that $A \circ_S B$ is $\leqslant_2$-increasing, so $z \leqslant_2 z'$ and $x' \notin r(A)$ imply $z \notin A \circ_S B$. The proof of the reverse inclusion is similar. □

The *complex algebra of an RG-frame* X is a structure $(L_X, \odot_Q, \to_R, \leftarrow_R, \emptyset, X, 1'_I)$, where L_X is the complex algebra of the lattice frame reduct of X and, for all $A, B \in L_X$,

$$
\begin{aligned}
A \odot_Q B &\overset{\text{def}}{=} l(A \circ_Q B)\\
A \to_R B &\overset{\text{def}}{=} \{x \in X : \forall y, z \in X,\ R(y,x,z)\ \&\ y \in A \Rightarrow z \in B\}\\
A \leftarrow_R B &\overset{\text{def}}{=} \{x \in X : \forall y, z \in X,\ R(x,y,z)\ \&\ y \in A \Rightarrow z \in B\}\\
1'_I &\overset{\text{def}}{=} lr(I).
\end{aligned}
$$

Lemma 15.2.5 *For any $A, B \subseteq X$,*

(a) *$A \to_R B$ and $A \leftarrow_R B$ are $\leqslant_1$-increasing;*
(b) *If A and B are l-stable, then so are $A \to_R B$, $A \leftarrow_R B$, and $A \odot_Q B$;*
(c) *$1'_I$ is l-stable.*

Proof:
(a) Suppose that for some $A, B \subseteq X$, $A \to_R B$ is not $\leqslant_1$-increasing. Then there are $x, y \in X$ such that $x \in A \to_R B$, $x \leqslant_1 y$, and $y \notin A \to_R B$. By the definition of $\to_R$, $y \notin A \to_R B$ means that there are $u, w \in X$ such that $R(u,y,w)$, $u \in A$, and $w \notin B$. By (FRG1), $R(u,y,w)$ and $x \leqslant_1 y$ imply $R(u,x,w)$. Then $x \notin A \to_R B$, a contradiction.
In a similar way one can show that $A \leftarrow_R B$ is $\leqslant_1$-increasing.

(b) Let $A, B \subseteq X$ be l-stable sets. We show that $A \leftarrow_R B$ is also l-stable. By (a), $A \leftarrow_R B$ is $\leqslant_1$-increasing, so by Lemma 2.8.1(c), $(A \leftarrow_R B) \subseteq lr(A \leftarrow_R B)$. It suffices to show that $lr(A \leftarrow_B B) \subseteq (A \leftarrow_R B)$.

Take $x \in X$ such that $x \notin A \leftarrow_R B$. Then, by the definition of $\leftarrow_R$, there exist $y, z \in X$ such that $R(x,y,z)$, $y \in B$, and $z \notin A$. Since $A = lr(A)$, $z \notin A$ means that there is some $z' \in X$ such that $z \leqslant_1 z'$ and $z' \in r(A)$. By (FRG1), $R(x,y,z)$ and $z \leqslant_1 z'$ imply $R(x,y,z')$. Hence, by (FRG5), there is some $x' \in X$ such that $x \leqslant_1 x'$ and $S(x',y,z')$. Let $x'' \in X$ be such that $x' \leqslant_2 x''$. By (FRG2), $S(x',y,z')$ implies $S(x'',y,z')$, so by (FRG7), there is some $z'' \in X$ such that $z' \leqslant_2 z''$ and $R(x'',y,z'')$. By Lemma 2.8.1(a'), $r(A)$ is $\leqslant_2$-increasing, so $z' \in r(A)$ and $z' \leqslant_2 z''$ imply $z'' \in r(A)$, and hence $z'' \notin A$. Now, by the definition of $\leftarrow_R$, from $y \in B$, $R(x'',y,z'')$, and $z'' \notin A$ it follows that $x'' \notin A \leftarrow_R B$. Hence for any x'', $x' \leqslant_2 x''$ implies $x'' \notin A \leftarrow_R B$, thus $x' \in r(A \leftarrow_R B)$. Since $x \leqslant_1 x'$, so finally we $x \notin lr(A \leftarrow_R B)$, as required. We can show that $A \to_R B$ is l-stable in a similar way.

Finally, we show that $A \odot_Q B$ is l-stable. By Lemma 15.2.4(b), $A \circ_Q B$ is r-stable. Then, from Lemma 2.8.3(b) it follows that $l(A \circ_Q B)$ is l-stable, that is, $A \odot_Q B$ is l-stable.

(c) By Lemma 2.8.3(c), $lr(I) = 1'_I$ is l-stable. □

The next three lemmas involve verifying that the RG-lattice axioms are satisfied in the complex algebra of any RG-frame X.

Lemma 15.2.6 *For any $A, B, C \in L_X$, $A \odot_Q (B \odot_Q C) = (A \odot_Q B) \odot_Q C$.*

Proof:
Let $A, B, C \subseteq L_X$. By Lemma 15.2.4(c) it suffices to show that $A \circ_Q (B \odot_Q C) = (A \odot_Q B) \circ_S C$.
For the left-to-right inclusion, take $z \in X$ such that $z \notin (A \odot_Q B) \circ_S C$. Thus, by the definition of $\circ_S$, there are $x, y \in X$ such that $S(x, y, z)$, $y \in C$, and $x \notin r(A \odot_Q B)$. But $r(A \odot_Q B) = rl(A \circ_Q B)$ by the definition of $\odot_Q$. Furthermore, by Lemma 15.2.4(c), $rl(A \circ_Q B) = rl(A \circ_S B)$. Moreover, $A \circ_S B$ is r-stable by Lemma 15.2.4(b), so $rl(A \circ_S B) = A \circ_S B$. Then $r(A \odot_Q B) = A \circ_S B$. Hence, since $x \notin r(A \odot_Q B)$, $x \notin A \circ_S B$. Then there are $u, v \in X$ such that $S(u, v, x)$, $v \in B$, and $u \notin r(A)$. Furthermore, $u \notin r(A)$ means that there exists some $u' \in X$ such that $u \leqslant_2 u'$ and $u' \in A$. By (FRG2), $S(u, v, x)$ and $u \leqslant_2 u'$ imply $S(u', v, x)$. Hence, by (FRG7), there exists some $x' \in X$ such that $x \leqslant_2 x'$ and $R(u', v, x')$. Applying again (FRG2), from $S(x, y, z)$ and $x \leqslant_2 x'$ we get $S(x', y, z)$. Therefore, for some x', $R(u'v, x')$ and $S(x', y, z)$ hold. By (FRG9), this implies that there is some $w \in X$ such that $R(v, y, w)$ and $Q(u', w, z)$. Hence, by (FRG), there is some $v' \in X$ such that $v \leqslant_1 v'$ and $S(v', y, w)$. Since B is l-stable, it is $\leqslant_1$-increasing, so $v \in B$ and $v \leqslant_1 v'$ imply $v' \in B$, thus $v' \notin r(B)$. Thus we have obtained that there are $y, v' \in X$ such that $y \in C$, $S(v', y, w)$, and $v' \notin r(B)$ hold which by the definition of $\circ_S$ means that $w \notin B \circ_S C$. From Lemma 15.2.4(b), $B \circ_S C$ is r-stable, so $B \circ_S C = rl(B \circ_S C) = rl(B \circ_Q C) = r(B \odot_Q C)$ by Lemma 15.2.4(c). Hence $w \notin B \circ_S C$ implies $w \notin r(B \odot_Q C)$. Therefore we finally get that there exist $u', w \in X$ such that $u' \in A$, $Q(u', w, z)$, and $w \notin r(B \odot_Q C)$ hold which by the definition of $\circ_Q$ means that $z \notin A \circ_Q (B \odot_Q C)$.
Proceeding in a similar way and using (FRG10), the reverse inclusion can be proved. □

Lemma 15.2.7 *For every $A \in L_X$,*

(a) $1'_I \odot_Q A = A$;
(b) $A \odot_Q 1'_I = A$.

Proof:
(a) Let $A \in L_X$. We show that $1'_I \circ_Q A = r(A)$ which will imply $1'_I \odot_Q A = l(1'_I \circ_Q A) = lr(A) = A$.
In order to show $1'_I \circ_Q A \subseteq r(A)$, take $z \in X$ such that $z \notin r(A)$. Then there exists some $w \in X$ such that $z \leqslant_2 w$ and $w \in A$. By (FRG12) there exists some $x \in X$ such that $x \in I$ and $S(x, w, w)$. By Lemma 2.8.3(c), $I \subseteq lr(I)$. Then $x \in lr(I)$, or equivalently, by the definition of $1'_I$, $x \in 1'_I$. Furthermore, by (FRG2), $z \leqslant_2 w$ and $S(x, w, w)$ imply $S(x, w, z)$. Hence, in view of (FRG7), there is some $z' \in X$ such that $z \leqslant_2 z'$ and $R(x, w, z')$. Then, by (FRG6), there exists some $y \in X$ such that $w \leqslant_1 y$ and $Q(x, y, z')$. Also, by (FRG3), $Q(x, y, z')$ and $z \leqslant_2 z'$ yield $Q(x, y, z)$. By the assumption, A is l-stable, so it is $\leqslant_1$-increasing, hence $w \in A$ and $w \leqslant_1 y$ imply $y \in A$. Thus $y \notin r(A)$ which together with $x \in 1'_I$ and $Q(x, y, z)$ gives $z \notin 1'_I \circ_Q A$.
For the reverse implication, take $z \in X$ such that $z \notin 1'_I \circ_Q A$. Then $Q(x, y, z)$, $x \in 1'_I$, and $y \notin r(A)$ for some $x, y \in X$. Next, $y \notin r(A)$ implies that there is some

$y' \in X$ such that $y \leqslant_2 y'$ and $y' \in A$. Then, by (FRG3), we get $Q(x, y', z)$. By (FRG8) there is some $z' \in X$ such that $z \leqslant_2 z'$ and $R(x, y', z')$. Now, it suffices to show that $z' \in A$. Then, in view of $z \leqslant_2 z'$, $z \notin r(A)$ will follow.
Suppose otherwise, that is $z' \notin A$, so $z' \notin lr(A)$. This means that there exists some $z'' \in X$ such that $z' \leqslant_1 z''$ and $z'' \in r(A)$. Applying (FRG1), $R(x, y', z')$ and $z' \leqslant_1 z''$ imply $R(x, y', z'')$. Hence, by (FRG5), there is some $x' \in X$ such that $x \leqslant_1 x'$ and $S(x', y', z'')$. Since $x \in 1'_I = lr(I)$, we get $x' \notin r(I)$, so there is some $x'' \in X$ such that $x' \leqslant_2 x''$ and $x'' \in I$. Also, $S(x', y', z'')$ and $x' \leqslant_2 x''$ give $S(x'', y', z'')$ by (FRG2). Hence, by (FRG7), $z'' \leqslant_2 w$ and $R(x'', y', w)$ for some $w \in X$. Now, by Lemma 15.2.3, $y' \in A$, $x'' \in I$, and $R(x'', y', z'')$ imply $w \in A$, and furthermore, $z'' \notin r(A)$, a contradiction.
(b) Let $A \in L_X$. As before, we first show that $1'_I \circ_Q A = r(A)$. Then we will get $1'_I \odot_Q A = l(1'_I \circ_Q A) = lr(A) = A$.
For the left-to-right inclusion, take $z \in X$ such that $z \notin r(A)$. Thus there is some $x \in X$ such that $z \leqslant_2 x$ and $x \in A$. By (FRG13), there exists $y \in X$ such that $y \in I$ and $Q(x, y, x)$. By Lemma 2.8.3(c) $I \subseteq lr(I) = 1'_I$. But $y \in I$, so $y \in 1'_I$, hence $y \notin r(1'_I)$. Next, by (FRG3), $z \leqslant_2 x$ and $Q(x, y, x)$ imply $Q(x, y, z)$, which together with $x \in A$ and $y \notin r(1'_I)$ yields $z \notin A \circ_Q 1'_I$. Now we show that $r(A) \subseteq 1'_I \circ_Q A$. Assume $y \notin 1'_I \circ_Q A$. Then, by the definition of $\circ_Q$, there are $x, y \in X$ such that $Q(x, y, z)$, $x \in A$, and $y \notin r(1'_I)$. Next, $y \notin r(1'_I)$ gives some y' such that $y \leqslant_2 y'$ and $y' \in 1'_I$. By (FRG3), $Q(x, y, z)$ and $y \leqslant_2 y'$ imply $Q(x, y', z)$. Hence, by (FRG8), there is some $z' \in X$ such that $z \leqslant_2 z'$ and $R(x, y', z')$. Now, proceeding as in the proof of (a) we get $z' \in A$, which together with $z \leqslant_2 z'$ gives $z \notin r(A)$, as required. □

Lemma 15.2.8 *For any $A, B, C \in L_X$,*

(a) $(A \odot_Q B) \subseteq C \Leftrightarrow B \subseteq (A \rightarrow_R C)$;
(b) $(A \odot_Q C) \subseteq B \Leftrightarrow A \subseteq (C \leftarrow_R B)$.

Proof:
Since both equivalences can be proved in the analogous way, we show the first one only.
For the left-to-right implication, assume that $A \odot_Q B \not\subseteq C$. Then $x \in A \odot_Q B$ and $x \notin C$ for some $x \in X$. By the assumption $x \notin lr(C)$, so there is some $x' \in X$ such that $x \leqslant_1 x'$ and $x' \in r(C)$. Since $A \odot_Q B$ is l-stable, we get $x' \notin A \circ_S B$, thus there exist $y, z \in X$ such that $S(y, z, x')$, $z \in B$, and $y \notin r(A)$. Next, $y \notin r(A)$ means that there is $y' \in X$ such that $y \leqslant_2 y'$ and $y' \in A$. By (FRG2), $S(y, z, x')$ and $y \leqslant_2 y'$ imply $S(y', z, x')$, which by (FRG5) yields that there is some $x'' \in X$ such that $x' \leqslant_2 x''$ and $R(y', z, x'')$. Since $x' \in r(C)$, we get $x'' \notin C$. Hence, since $y' \in A$ and $R(y', z, x'')$, $z' \notin A \rightarrow_R C$. But $z \in B$, so we finally get $B \not\subseteq A \rightarrow_R C$, as required.
For the right-to-left implication, assume that $B \not\subseteq A \rightarrow_R C$. Then there exists $x \in X$ such that $x \in B$ and $x \notin A \rightarrow C$. By the definition of $\rightarrow_Q$, there are $y, z \in X$ such that $R(y, x, z)$, $y \in A$, and $z \notin C$. Suppose that $A \odot_Q B \subseteq C$. Then $z \notin A \odot_Q B$. Since $A \odot_Q B = l(A \circ_Q B) = l(A \circ_S B)$, we have $z \notin A \circ_S B$. Thus there is some z' such that $z \leqslant_1 z'$ and $z' \in A \circ_S B$. Furthermore, by

(FRG3), $R(y,x,z)$ and $z \leqslant_1 z'$ imply $R(y,x,z')$. By (FRG5) there is some $y' \in X$ such that $y \leqslant_1 y'$ and $S(y',x,z')$. Hence, since $x \in B$ and $z' \in A \circ_S B$, we get $y' \in r(A)$, which together with $y \leqslant_1 y'$ yields $y \notin lr(A)$. Thus $y \notin A$, a contradiction. □

As a consequence of the preceding lemmas, we conclude that

Proposition 15.2.9 *The complex algebra of an RG-frame is an RG-lattice.*

The *canonical frame of an RG-lattice L* is the structure

$$(X_L, R_\odot, S_\odot, Q_\odot, I_{1'}),$$

where X_L is the canonical frame of the lattice reduct of L, $I_{1'} \subseteq X_L$, and $R_\odot$, $S_\odot$, and $Q_\odot$ are ternary relations on X_L defined, for all $F, G, H \in X_L$, by

$$\begin{aligned} I_{1'} &\stackrel{\text{def}}{=} \{F \in X_L : 1' \in F\} \\ R_\odot(F,G,H) &\stackrel{\text{def}}{\Leftrightarrow} \forall a,b \in L,\ a \in F_1\ \&\ b \in G_1 \Rightarrow a \odot b \in H_1 \\ S_\odot(F,G,H) &\stackrel{\text{def}}{\Leftrightarrow} \forall a,b \in L,\ a \odot b \in H_2\ \&\ b \in G_1 \Rightarrow a \in F_2 \\ Q_\odot(F,G,H) &\stackrel{\text{def}}{\Leftrightarrow} \forall a,b \in L,\ a \odot b \in H_2\ \&\ a \in F_1 \Rightarrow b \in G_2 \end{aligned}$$

It is easy to see that, for all $F, G, H \in X_L$,

$$\begin{aligned} R_\odot(F,G,H) &\Leftrightarrow F_1 \odot G_1 \subseteq H_1 \\ S_\odot(x,y,z) &\Leftrightarrow -F_2 \odot G_1 \subseteq -H_2 \\ Q_\odot(x,y,z) &\Leftrightarrow F_1 \odot -G_2 \subseteq -H_2. \end{aligned}$$

Proposition 15.2.10 *The canonical frame of an RG-lattice is an RG-frame.*

Proof:
We have to show that axioms (FRG1)–(FRG13) are satisfied in the canonical frame of L. Proofs of (FRG1)–(FRG4) are straightforward.

(FRG5) Take $F = (F_1, F_2), G = (G_1, G_2), H = (H_1, H_2) \in X_L$ such that $R_\odot(F,G,H)$ holds. Define the set

$$I = \{c \in L : H_2 \cap (\{c\} \odot G_1) \neq \emptyset\}.$$

Suppose $F_1 \cap I \neq \emptyset$. Then there are $a, b \in L$ such that $a \in F_1$, $b \in G_1$, and $a \odot b \in H_2$. By the assumption it follows that $a \odot b \in H_1$, so $H_1 \cap H_2 \neq \emptyset$, a contradiction. Hence $F_1 \cap I = \emptyset$. By Lemma 15.2.2, I is an ideal of L. Then (F_1, I) is a filter-ideal pair of L which, by Theorem 2.8.6, can be extended to a maximal filter-ideal pair, say (U_1, U_2). Then $F_1 \subseteq U_1$ and $I \subseteq U_2$. Let $a, b \in L$ be such that $a \odot b \in H_2$ and $b \in G_1$. Then $a \in I$, so $a \in U_2$. Therefore $S_\odot(F,G,H)$ holds and $F \leqslant_{1L} U$, hence (FRG5) holds.
(FRG6), (FRG7), and (FRG8) can be proved in a similar way.

(FRG9). Let $F, G, H, H' \in X_L$ and assume that there exists $U \in X_L$ such that $R_\odot(F, G, U)$ and $S_\odot(U, H, H')$. Define two subsets $J, I \subseteq L$ by

$$J = \uparrow_\leq (G_1 \odot H_1)$$
$$I = \{c \in L : H_2' \cap (F_1 \odot \{c\}) \neq \emptyset\}.$$

Suppose $J \cap I \neq \emptyset$. Then there is some $a \in L$ such that $a \in J$ and $a \in I$, so there are $b, c \in L$ such that $b \in G_1$, $c \in H_1$, and $b \odot c \leq a$. Also, since $a \in I$, there is $d \in L$ such that $d \in F_1$ and $d \odot a \in H_2'$. Since $b \odot c \leq a$, $(b \odot c) \vee a = a$, which by Lemma 15.2.1(a) gives $d \odot a = d \odot ((b \odot c) \vee a) = (d \odot (b \odot c)) \vee (d \odot a)$. Thus $d \odot (b \odot c) \leq d \odot a$. Since H_2' is and ideal of L and $d \odot a \in H_2'$, we get $d \odot (b \odot c) \in H_2'$. Hence, by axiom (RG1), $(d \odot b) \odot c \in H_2'$, which together with $S_\odot(U, H, H')$ and $c \in H_1$ yields $d \odot b \in U_2$. On the other hand, $R_\odot(F, G, U)$, $b \in G_1$, and $d \in F_1$ give $d \odot b \in U_1$. Hence $U_1 \cap U_2 \neq \emptyset$, a contradiction. Therefore $J \cap I = \emptyset$. Note that J is a filter of L and, by Lemma 15.2.2, I is an ideal of L. So (J, I) is a filter-ideal pair of L and by Theorem 2.8.6 it can be extended to a maximal filter-ideal pair, say (W_1, W_2). Then $J \subseteq W_1$ and $I \subseteq W_2$. Finally, take $a, b \in L$ satisfying $a \odot b \in H_2'$ and $a \in F_1$. By the definition of I, $b \in I$, thus $b \in W_2$. This means that $Q_\odot(F, W, H')$ holds. Furthermore, assume that $a \in G_1$ and $b \in H_1$. By the definition of J, $a \odot b \in J$, so $a \odot b \subseteq W_1$. Hence $R_\odot(G, H, W)$ also holds.
(FRG10) can be shown in a similar way.

(FRG11) Take any $F, G, H \in X_L$ such that $F \in I_{1'}$ and

$$R_\odot(F, G, H) \text{ and } R_\odot(G, F, H).$$

Then $1' \in F$. If $R_\odot(F, G, H)$, then for any $b \in G_1$, $1' \odot b = b \in H_1$, so $G_1 \subseteq H_1$. If $R_\odot(G, F, H)$, then for any $b \in G_1$, $b \odot 1' = b \in H_1$, thus $G_1 \subseteq H_1$.

(FRG12) Let $F \in X_L$ and consider the set

$$I = \{a \in L : (\{a\} \odot F_1) \cap F_2 \neq \emptyset\}.$$

By Lemma 15.2.2, I is an ideal of L. Let $\uparrow_\leq 1'$ be the principal filter generated by $1'$. Suppose that $\uparrow_\leq 1' \cap I \neq \emptyset$. Then there is $a \in L$ such that $a \in \uparrow_\leq 1'$, that is, $1' \leq a$, and $a \in I$, that is for some $b \in F_1$, $a \odot b \in F_2$. By monotonicity of $\odot$, since $1' \leq a$, $b = 1' \odot b \leq a \odot b$. Hence $b \in F_2$, so $F_1 \cap F_2 \neq \emptyset$, a contradiction. Therefore $\uparrow_\leq 1' \cap I = \emptyset$, thus $(\uparrow_\leq 1', I)$ is a filter-ideal pair of L and by Theorem 2.8.6 it can be extended to a maximal filter-ideal pair, say (U_1, U_2). Then $\uparrow_\leq 1' \subseteq U_1$ and $I \subseteq U_2$. Clearly, $1' \in \uparrow_\leq 1'$, so $1' \in U_1$, which means that $I_{1'}(U)$ holds. Moreover, let $a, b \in L$ be such that $a \odot b \in F_2$ and $b \in F_1$. Then $a \in I$, so $a \in U_2$, thus $S_\odot(U, F, F)$ holds as well.

(FRG13) Let $F \in X_L$ and consider the set

$$J = \{a \in L : (F_1 \odot \{a\}) \cap F_2 \neq \emptyset\}.$$

Using similar reasoning to that in (FRG12), $\uparrow_\leq 1' \cap J = \emptyset$. By Lemma 15.2.2, J is an ideal of L, so $(\uparrow_\leq 1', J)$ is a filter-ideal pair of L. Let (U_1, U_2) be its extension to a maximal filter-ideal pair. Then $\uparrow_{\leq} 1' \subseteq U_1$ and $J \subseteq U_2$. Thus $1' \in U_1$, so $I_{1'}(U)$ holds. Also, if $a, b \in L$ are such that $a \odot b \in F_2$ and $a \in F_1$, then $b \in J$, so $b \in U_2$. Hence $Q_{\odot}(F, U, F)$ holds, as required. □

Given a canonical frame $(X_L, R_{\odot}, S_{\odot}, Q_{\odot}, I_{1'})$ of an RG-lattice L, ternary relations $R_{\leftarrow}$ and $R_{\rightarrow}$ on X_L may be defined, for any $F, G, H \in X_L$, by

$$R_{\leftarrow}(F,G,H) \overset{\text{def}}{\Leftrightarrow} \forall a, b \in L,\ b \leftarrow a \in F_1 \ \&\ a \in G_1 \Rightarrow b \in H_1$$

$$R_{\rightarrow}(F,G,H) \overset{\text{def}}{\Leftrightarrow} \forall a, b \in L,\ a \in F_1 \ \&\ a \rightarrow b \in G_1 \Rightarrow b \in H_1.$$

It is interesting to note that

Lemma 15.2.11 *For any RG-lattice L, $R_{\odot} = R_{\rightarrow} = R_{\leftarrow}$.*

Proof:
We show that $R_{\odot} = R_{\leftarrow}$. Take any $F, G, H \in X_L$ such that $R_{\odot}(F, G, H)$. Suppose there exist $a, b \in L$ such that $b \leftarrow a \in F_1$, $a \in G_1$, and $b \notin H_1$. Then, by the definition of $R_{\odot}$, $(b \leftarrow a) \odot a \in H_1$. Next, by Lemma 15.2.1(d), $(b \leftarrow a) \odot a \leq b$. Hence $b \in H_1$, a contradiction. On the other hand, suppose that for some $F, G, H \in X_L$, $R_{\leftarrow}(F, G, H)$ holds and there are $a, b \in L$ such that $a \in F_1$, $b \in G_1$, but $a \odot b \notin H_1$. By Lemma 15.2.1(j), $a \leq (a \odot b) \leftarrow b$, so $(a \odot b) \leftarrow b \in F_1$. Hence, by $R_{\leftarrow}(F, G, H)$, we get $a \odot b \in H_1$, a contradiction.
The proof of $R_{\odot} = R_{\rightarrow}$ is similar. □

Let L be an RG-lattice, let $(X_L, R_{\odot}, S_{\odot}, Q_{\odot}, I_{1'})$ be its canonical frame, and let $(L_{X_L}, \vee_{X_L}, \wedge_{X_L}, \odot_{Q_{\odot}}, \rightarrow_{R_{\odot}}, \leftarrow_{R_{\odot}}, \emptyset, X_L, 1'_{I_{1'}})$ be the complex algebra of the canonical frame of L. Let $h : L \rightarrow L_{X_L}$ be the embedding of the underlying general lattices defined, for every $a \in L$, by $h(a) = \{F \in X_L : a \in F_1\}$.

Lemma 15.2.12 *For all $a, b \in L$,*

(a) $h(1') = 1'_{I_{1'}}$;

(b) $h(a \odot b) = h(a) \odot_{Q_{\odot}} h(b)$;

(c) $h(a \rightarrow b) = h(a) \rightarrow_{R_{\odot}} h(b)$;

(d) $h(a \leftarrow b) = h(a) \leftarrow_{R_{\odot}} h(b)$.

Proof:
(a) Note that $I_{1'} = h(1')$, so by Theorem 2.8.9(b), it is an l-stable set. Also, by the definition of $1'_I$, $1'_{I_{1'}} = lr(I_{1'})$, thus $1'_{I_{1'}} = I_{1'} = h(1')$.

(b) Take $a, b \in L$ and $H \in X_L$ such that $H \in h(a \odot b)$. Then $a \odot b \in H_1$. We have to show that $H \in h(a) \odot_{Q_{\odot}} h(b)$. By the definition of $\odot_Q$, this means that $H \in l(h(a) \circ_{Q_{\odot}} h(b))$, or equivalently, by Lemma 15.2.4(c), $H \in l(h(a) \circ_{S_{\odot}} h(b))$, that is for every $K \in X_L$, $H \leqslant_{1L} K$ implies $K \notin h(a) \circ_{S_{\odot}} h(b)$. Let $H \leqslant_{1L} K$, that is, $H_1 \subseteq K_1$. We show that $K \notin h(a) \circ_{S_{\odot}} h(b)$.
Consider the set

$$I = \{c \in L : (\uparrow_{\leq} a \odot \{c\}) \cap K_2 \neq \emptyset\}.$$

Suppose that $b \in I$. Then there exists $e \in {\uparrow_{\leq}} a$, that is, $a \leq e$, and $e \odot b \in K_2$. By monotonicity of $\odot$, $a \odot b \leq e \odot b$, so $a \odot b \in K_2$, thus $a \odot b \notin K_1$. Since $H_1 \subseteq K_1$, we get $a \odot b \notin H_1$, a contradiction. Therefore $b \notin I$. By Lemma 15.2.2, I is an ideal of L. Hence $({\uparrow_{\leq}} b, I)$ is a filter-ideal pair of L. Let (G_1, G_2) be its extension to a maximal filter-ideal pair. Then ${\uparrow_{\leq}} b \subseteq G_1$ and $I \subseteq G_2$, so $b \in G_1$ and thus $G \in h(b)$.
Now, consider the set

$$J = \{c \in L : (\{c\} \odot G_1) \cap K_2 \neq \emptyset\}.$$

Suppose that $a \in J$. Then there exists some $e \in G_1$ such that $a \odot e \in K_2$. Hence, by the definition of I, we get $e \in I$, so $e \in G_2$. Then $e \notin G_1$, a contradiction. Therefore $a \notin J$. By Lemma 15.2.2, J is an ideal of L. Then ${\uparrow_{\leq}} a \cap J = \emptyset$, and moreover, $({\uparrow_{\leq}} a, J)$ is a filter-ideal pair of L. Let (F_1, F_2) be its extension to a maximal filter-ideal pair. Hence ${\uparrow_{\leq}} a \subseteq F_1$ and $J \subseteq F_2$. Then $a \in F_1$, thus $F \in h(a)$, which implies $F \notin rh(a)$.
Finally, we show that $S_{\odot}(F, G, K)$ holds. By the definition of $S_{\odot}$ this means that for all $c, d \in L$, $c \notin F_2$ & $d \in G_1 \Rightarrow c \odot d \notin K_2$. Let $c \notin F_2$ and $d \in G_1$. Since $J \subseteq F_2$, we get $c \notin J$, so for every $e \in G_1$, $c \odot e \notin K_2$. In particular, $c \odot d \notin K_2$, hence $S_{\odot}(F, G, K)$ holds. Therefore, for some $F, G \in X_L$, $G \in h(b)$, $F \notin rh(a)$, and $S_{\odot}(F, G, K)$ hold; so $K \notin h(a) \circ_{S_{\odot}} h(b)$, as required. For the reverse inclusion, assume that $a, b \in L$ and $H \in X_L$ are such that $H \in h(a) \odot_{Q_{\odot}} h(b)$. Thus $H \in l(h(a) \circ_{Q_{\odot}} h(b))$. Let $G \in X_L$ be such that $H_1 \subseteq G_1$. Since $H \in l(h(a) \circ_{Q_{\odot}} h(b))$, we have $G \notin h(a) \circ_{Q_{\odot}} h(b)$, so there are $F, K \in X_L$ such that $Q_{\odot}(F, K, G)$, $F \in h(a)$, and $K \notin rh(b)$. Then $a \in F_1$, and moreover, by Theorem 2.8.9(a), $b \notin K_2$. Hence $a \odot b \notin G_2$, since $Q_{\odot}(F, K, G)$ holds. Again, by Theorem 2.8.9(a), $G \notin h(a \odot b)$. However, $H \subseteq_1 G$ implies $H \in lrh(a \odot b)$. Since $h(a \odot b)$ is l-stable, $H \in h(a \odot b)$.

(c) Take any $F \in X_L$ such that $F \in h(a \to b)$ and $F \notin h(a) \to_R h(b)$. Then there exist $G, H \in X_L$ such that $R_{\odot}(G, F, H)$, $G \in h(a)$, and $H \notin h(b)$. By Lemma 15.2.11, $R_{\odot}(G, F, H) = R_{\to}(G, F, H)$. Since $a \in G_1$ and $a \to b \in F_1$ we get $b \in H_1$, that is, $H \in h(b)$ which gives a contradiction. For the reverse inclusion, assume $F \notin h(a \to b)$. Thus $a \to b \notin F_1$. Consider the set

$$I = \{c \in L : F_1 \cap (\{c\} \to {\downarrow_{\leq}} b) \neq \emptyset\},$$

By Lemma 15.2.2, I is an ideal of L. Since $b \in {\downarrow_{\leq}} b$, $a \notin I$, so ${\uparrow_{\leq}} a \cap I = \emptyset$. Then $({\uparrow_{\leq}} a, I)$ is a filter-ideal pair and by Theorem 2.8.6 it can be extended to a maximal filter-ideal pair, say $G = (G_1, G_2)$. Then ${\uparrow_{\leq}} a \subseteq G_1$ and $I \subseteq G_2$. But $a \in G_1$, so $G \in h(a)$.
Now consider the set

$$K = \{c \in L : F_1 \cap (G_1 \to \{c\}) \neq \emptyset\}.$$

By Lemma 15.2.2, K is a filter of L. Take any $b \in K$. Then there is some $b' \in L$ such that $b' \in G_1$ and $b' \to b \in F_1$. Hence, by the definition of I, $b' \in I \subseteq G_2$, so $b' \notin G_1$, a contradiction. Therefore $b \notin K$, so $(F, {\uparrow_{\leq}} b)$ is a filter-ideal pair. Let (H_1, H_2) be its extension to a maximal filter-ideal pair. Thus $K \subseteq H_1$ and

$\downarrow_{\leq} b \subseteq H_2$. Since $b \in H_2$, $b \notin H_1$, that is $H \notin h(b)$.
Now, take any $c, d \in L$ such that $c \to d \in F_1$ and $c \in G_1$. Then $d \in K$ and, since $K \subseteq H_1$, $d \in H_1$. By the definition of $R_{\to}$, $R_{\to}(G, F, H)$ holds, and so $R_{\odot}(G, F, H)$ by Lemma 15.2.11. Therefore, for some $G, H \in X_L$, $G \in h(a)$, $H \notin h(b)$, and $R_{\odot}(G, F, H)$. Thus $F \notin h(a) \to_{R_{\odot}} h(b)$.
The proof of (d) is similar. □

From Lemma 15.2.12 and Theorem 2.8.10, the following discrete representation theorem for RG-lattices may be obtained.

Theorem 15.2.13 *Any RG-lattice is embeddable into the complex algebra of its canonical frame.*

15.3 Doubly residuated general lattices

A *doubly residuated lattice, DRG-lattice,* is a structure

$$(L, \odot, \to, \leftarrow, 1', \oplus, \rightarrowtriangle, \leftarrowtriangle, 0')$$

such that

(DRG1) $(L, \odot, \to, \leftarrow, 1')$ is a residuated lattice

(DRG2) $(L, \oplus, 0')$ is a monoid

(DRG3) $\rightarrowtriangle$ and $\leftarrowtriangle$ are the *right* and the *left residual of* $\oplus$, respectively, defined, for all $a, b \in L$, by

$$a \leq b \oplus c \overset{\text{def}}{\Leftrightarrow} b \rightarrowtriangle a \leq c$$
$$a \leq c \circ b \overset{\text{def}}{\Leftrightarrow} b \leftarrowtriangle a \leq c.$$

The operation $\oplus$ is called the *sum*. As in Section 15.2, we consider bounded doubly residuated lattices with 0 and 1 as the smallest and the greatest elements of its lattice reduct.

It is easily noted that the existence of the residuals of sum $\oplus$ implies, for any $a, b \in L$, that

$$a \rightarrowtriangle b = \bigwedge\{c \in L : b \leq a \oplus c\}$$
$$a \leftarrowtriangle b = \bigwedge\{c \in L : b \leq c \oplus a\}.$$

Lemma 15.3.1 *Let L be a general lattice. The structure*

$$(L, \odot, \to, \leftarrow, 1', \oplus, \rightarrowtriangle, \leftarrowtriangle, 0')$$

is a DRG-lattice if and only if $(L, \odot, \to, \leftarrow, 1')$ and $(L^{op}, \oplus, \rightarrowtriangle, \leftarrowtriangle, 0')$ are RG-lattices.

The above lemma implies that properties of operations $\oplus$, $\multimap$, and $\multimapinv$ can be easily obtained from the analogous properties of the operations $\odot$, $\rightarrow$, and $\leftarrow$, respectively.

A *DRG-frame* is a structure

$$(X, R, S, Q, I, \Theta, \Upsilon, \Omega, J)$$

such that (X, R, S, Q, I) is an RG-frame, Θ, Υ, and Ω are ternary relations on X, and $J \subseteq X$ such that, for all $x, x', y, y', z, z' \in X$,

(FDRG1) $\Theta(x,y,z)\ \&\ x' \leqslant_2 x\ \&\ y' \leqslant_2 y\ \&\ z \leqslant_2 z' \Rightarrow \Theta(x',y',z')$

(FDRG2) $\Upsilon(x,y,z)\ \&\ x \leqslant_1 x'\ \&\ y' \leqslant_2 y\ \&\ z' \leqslant_1 z \Rightarrow \Upsilon(x',y',z')$

(FDRG3) $\Omega(x,y,z)\ \&\ x' \leqslant_2 x\ \&\ y \leqslant_1 y'\ \&\ z' \leqslant_1 z \Rightarrow \Omega(x',y',z')$

(FDRG4) $J(x)\ \&\ x \leqslant_2 x' \Rightarrow J(x')$.

(FDRG5) $\Theta(x,y,z) \Rightarrow \exists u \in X\ (x \leqslant_2 u\ \&\ \Upsilon(u,y,z))$

(FDRG6) $\Theta(x,y,z) \Rightarrow \exists u \in X\ (y \leqslant_2 u\ \&\ Q(x,u,z))$

(FDRG7) $\Upsilon(x,y,z) \Rightarrow \exists u \in X\ (z \leqslant_1 u\ \&\ \Theta(x,y,u))$

(FDRG8) $\Omega(x,y,z) \Rightarrow \exists u \in X\ (z \leqslant_1 u\ \&\ \Theta(x,y,u))$

(FDRG9) $\exists u \in X (\Theta(x,y,u) \,\&\, \Upsilon(u,z,y))$
$\Rightarrow \exists w \in X (\Theta(y,z',w) \,\&\, \Omega(x,w,y))$

(FDRG10) $\exists u \in X (\Theta(x,y,u) \,\&\, \Omega(z,u,z'))$
$\Rightarrow \exists w \in X (\Theta(z,x,w) \,\&\, \Upsilon(w,y,z'))$

(FDRG11) $J(x)\ \&\ (\Theta(x,y,z) \text{ or } \Theta(y,x,z)) \Rightarrow y \leqslant_2 z$

(FDRG12) $\exists u \in X\ (J(u)\ \&\ \Upsilon(u,x,x))$

(FDRG13) $\exists u \in X\ (J(u)\ \&\ \Omega(x,u,x))$.

Let $(X, \leqslant_1, \leqslant_2)$ be a general lattice frame. By the *opposite lattice frame* we mean a structure $X^{op} = (X, \leqslant_1^{op}, \leqslant_2^{op})$, where $\leqslant_1^{op} = \leqslant_2$ and $\leqslant_2^{op} = \leqslant_1$. Note that a DRG-frame can be viewed as a join of the RG-frame based on a general lattice frame $(X, \leqslant_1, \leqslant_2)$ with the RG-frame based on the opposite general lattice frame $(X, \leqslant_2, \leqslant_1)$. Therefore we have the following fact.

Lemma 15.3.2 *If $(X, R, S, Q, I, \Theta, \Upsilon, \Omega, J)$ is a DRG-frame then the structure $(X^{op}, \Theta, \Upsilon, \Omega, J)$ is an RG-frame.*

Lemma 15.3.2 implies that the properties of relations Θ, Υ, Ω, and the set J can be easily obtained from the properties of relations R, S, Q and I, respectively, by interchanging the roles of orderings $\leqslant_1$ and $\leqslant_2$.

The *complex algebra of a DRG-frame* X is a structure

$$(L_X, \odot_R, \rightarrow_Q, \leftarrow_Q, 1'_I, \oplus_\Omega, \multimap_\Theta, \multimapinv_\Theta, 0'_J)$$

such that $(L_X, \odot_R, \rightarrow_Q, \leftarrow_Q, 1'_I)$ is the complex algebra of its RG-frame reduct and, for all $A, B \subseteq X$,

$$\begin{aligned}
0'_J &\stackrel{\text{def}}{=} l(J) \\
A \oplus_\Omega B &\stackrel{\text{def}}{=} \{z \in X : \forall x, y \in X,\ \Omega(x,y,z)\ \&\ x \in r(A) \Rightarrow y \in B\} \\
A \twoheadrightarrow_\Theta B &\stackrel{\text{def}}{=} \{x \in X : \forall y, z \in X,\ \Theta(y,x,z)\ \&\ y \in A \Rightarrow z \in B\} \\
A \twoheadleftarrow_\Theta B &\stackrel{\text{def}}{=} \{x \in X : \forall y, z \in X,\ \Theta(x,y,z)\ \&\ y \in A \Rightarrow z \in B\}.
\end{aligned}$$

Proposition 15.3.3 *The complex algebra of a DRG-frame is a DRG-lattice.*

Proof:
Since J is $\leqslant_2$-increasing, $l(J) = 0'_J$ is l-stable by (FDRG4). Hence, by Proposition 15.2.9 and Lemma 15.3.2 the result follows. □

The *canonical frame of a DRG-lattice* L is a structure $(X_L, \Theta_\oplus, \Upsilon_\oplus, \Omega_\oplus, J_{0'})$ such that X_L is the canonical frame of its RG-lattice reduct and $\Theta_\oplus$, $\Upsilon_\oplus$, $\Omega_\oplus$, and $J_{0'}$ are defined, for all $F, G, H \in X_L$, by

$$\begin{aligned}
J_{0'} &\stackrel{\text{def}}{=} \{F \in X_L : 0' \in F_2\} \\
\Theta_\oplus(F,G,H) &\stackrel{\text{def}}{\Leftrightarrow} \forall a, b \in L\ (a \in F_2\ \&\ b \in G_2 \Rightarrow a \oplus b \in H_2) \\
\Omega_\oplus(F,G,H) &\stackrel{\text{def}}{\Leftrightarrow} \forall a, b \in L\ (a \in F_2\ \&\ a \oplus b \in H_1 \Rightarrow b \in G_1) \\
\Upsilon_\oplus(F,G,H) &\stackrel{\text{def}}{\Leftrightarrow} \forall a, b \in L\ (b \in G_2\ \&\ a \oplus b \in H_1 \Rightarrow a \in F_1).
\end{aligned}$$

Proposition 15.3.4 *The canonical frame of a DRG-lattice is a DRG-frame.*

Proof:
The proof follows from Proposition 15.2.9 and Lemma 15.3.2. □

We have the following discrete representation for DRG-lattices.

Theorem 15.3.5 *Any DRG-lattice is embeddable into the complex algebra of its canonical frame.*

Chapter 16

General lattices with relational operators

16.1 Introduction

In Section 5.2 the Tarski relation algebras based on Boolean algebras are considered. In [63] a generalization of those algebras is introduced such that instead of a Boolean algebra a general lattice is endowed with the abstract counterparts of the relational composition and converse. Since relational composition behaves like a product in residuated lattices, the developments in Section 15.2 apply to it. To deal with the relational converse, in Section 16.2 residuated lattices are endowed with an involution and specific axioms are postulated, analogous to the axioms of the operator of converse in Boolean relation algebras, which say how the involution acts with lattice operations and product. In Boolean relation algebras the operation of relative sum is expressible in terms of the relational composition and complement, but it is not the case in lattice based relation algebras. Given a residuated lattice L the relative sum behaves like the product in the opposite lattice L^{op}. Thus an adequate basis for lattice based relation algebras is that of doubly residuated lattices, presented in Section 15.3, and endowed with involutions as presented in Section 16.3. Then the two specific axioms are added which show how the neutral elements of product and sum act with each other, as presented in Section 16.4. The content of this chapter is based on [64].

16.2 Residuated general lattices with involution

In this section we consider RG-lattices endowed with an involutive operator $\smile$. Specific axioms say how this operator acts with the lattice join $\vee$ and also with the product $\odot$ of the underlying RG-lattice. Intuitively, $\smile$ resembles relational converse operation. This fact will be useful in Section 16.4.

An *RG-lattice with involution*, *RGI-lattice*, is a structure $(L, \odot, \rightarrow, \leftarrow, \smile, 1')$

such that $(L, \odot, \rightarrow, \leftarrow, 1')$ is an RG-lattice and $^\smile$ is a unary operation in L satisfying for all $a, b \in L$,

(RGI1) $a^{\smile\smile} = a$

(RGI2) $(a \vee b)^\smile = a^\smile \vee b^\smile$

(RGI3) $(a \odot b)^\smile = b^\smile \odot a^\smile$.

The following lemma shows basic properties of the operation $^\smile$.

Lemma 16.2.1 *Let $(L, \odot, \rightarrow, \leftarrow, {}^\smile, 1')$ be an RGI-lattice. For all $a, b \in L$,*

(a) $0^\smile = 0$ *and* $1^\smile = 1$;

(b) $a \leq b$ *implies* $a^\smile \leq b^\smile$;

(c) $(a \wedge b)^\smile = a^\smile \wedge b^\smile$.

Proof:
(a) Note that

$$0^\smile = 0 \vee 0^\smile = 0^{\smile\smile} \vee 0^\smile = (0^\smile \vee 0)^\smile = 0^{\smile\smile} = 0$$
$$1 = 1 \vee 1^\smile = 1^{\smile\smile} \vee 1^\smile = (1^\smile \vee 1)^\smile = 1^\smile.$$

(b) Let $a \leq b$. Then $a \vee b = b$, so $(a \vee b)^\smile = b^\smile$. By (RGI2), $a^\smile \vee b^\smile = b^\smile$, thus $a^\smile \leq b^\smile$.

(c) Since $a \wedge b \leq a$, by (b) we get $(a \wedge b)^\smile \leq a^\smile$. Similarly, $(a \wedge b)^\smile \leq b^\smile$. Then $(a \wedge b)^\smile \leq a^\smile \wedge b^\smile$. On the other hand, since $a^\smile \wedge b^\smile \leq a^\smile$, $(a^\smile \wedge b^\smile)^\smile \leq a$ by (b) and (RGI1). Similarly, $(a^\smile \wedge b^\smile)^\smile \leq b$, so $(a^\smile \wedge b^\smile)^\smile \leq a \wedge b$. Applying again (RGI1) and (b) we get $a^\smile \wedge b^\smile = (a^\smile \wedge b^\smile)^{\smile\smile} \leq (a \wedge b)^\smile$. □

For any $A \subseteq L$, $A^\smile = \{a^\smile \in L : a \in A\}$.

Lemma 16.2.2 *Let $(L, \odot, \rightarrow, \leftarrow, {}^\smile, 1')$ be an RGI-lattice. For all $A, B \subseteq L$,*

(a) $A^\smile = \{a \in L : a^\smile \in A\}$;

(b) $A^{\smile\smile} = A$;

(c) $(-A)^\smile = -(A^\smile)$;

(d) $(A \cup B)^\smile = A^\smile \cup B^\smile$;

(e) $(A \cap B)^\smile = A^\smile \cap B^\smile$.

Proof:
By way of example we show (b) and (e).

(b) Note that for every $a \in L$,

$$a \in A \Leftrightarrow a^\smile \in A^\smile \Leftrightarrow a^{\smile\smile} \in A^{\smile\smile} \Leftrightarrow a \in A^{\smile\smile}.$$

(e) For every $a \in L$,

$$\begin{aligned} a \in (A \cap B)^{\smile} &\Leftrightarrow a^{\smile\smile} \in (A \cap B)^{\smile} \\ &\Leftrightarrow a^{\smile} \in A \cap B \\ &\Leftrightarrow a^{\smile} \in A \text{ and } a^{\smile} \in B \\ &\Leftrightarrow a \in A^{\smile} \text{ and } a \in B^{\smile} \\ &\Leftrightarrow a \in A^{\smile} \cap B^{\smile} \end{aligned}$$

which completes the proof. □

An *RGI-frame* is a structure (X, R, S, Q, I, C) such that (X, R, S, Q, I) is an RG-frame and mapping $C : X \to X$ satisfies, for all $x, y, z \in X$,

(FRGI1) $x \leqslant_1 y \Rightarrow C(x) \leqslant_1 C(y)$
(FRGI2) $x \leqslant_2 y \Rightarrow C(x) \leqslant_2 C(y)$
(FRGI3) $C(C(x)) = x$
(FRGI4) $Q(x, y, z) = S(C(y), C(x), C(z))$.

The *complex algebra of an RGI-frame* X is a structure $(L_X, \odot_Q, \to_R, \leftarrow_R, 1'_I, {}^{\smile c})$ such that $(L_X, \odot_Q, \to_R, \leftarrow_R, \emptyset, X, 1'_I)$ is the complex algebra of its RG-frame reduct and ${}^{\smile c} : 2^X \to 2^X$ is defined, for every $A \in L_X$, by

$$A^{\smile c} \stackrel{\text{def}}{=} \{C(x) \in X : x \in A\}.$$

The following lemma gives main properties of the operation ${}^{\smile c}$.

Lemma 16.2.3 *Let X be an RGI-frame. For all $A, B \subseteq X$,*

(a) $A^{\smile c} = \{x \in X : C(x) \in A\}$;
(b) $A^{\smile c \smile c} = A$;
(c) $A \subseteq B \Rightarrow A^{\smile c} \subseteq B^{\smile c}$;
(d) $(A \cap B)^{\smile c} = A^{\smile c} \cap B^{\smile c}$;
(e) $l(A^{\smile c}) = l(A)^{\smile c}$ *and* $r(A^{\smile c}) = r(A)^{\smile c}$;
(f) *If A is l-stable, then so is $A^{\smile c}$.*

Proof:
By way of example we show (b), (d), and (e).

(b) Note that by (FRGI3), for any $x \in X$ the following hold:

$$x \in A \Leftrightarrow C(C(x)) \in A \Leftrightarrow C(x) \in A^{\smile c} \Leftrightarrow x \in A^{\smile c \smile c}.$$

(d) By (c), $(A \cap B)^{\smile c} \subseteq A^{\smile c}$ and $(A \cap B)^{\smile c} \subseteq B^{\smile c}$, so $(A \cap B)^{\smile c} \subseteq A^{\smile c} \cap B^{\smile c}$. On the other hand, (b) and (c) imply $(A^{\smile c} \cap B^{\smile c})^{\smile c} \subseteq A^{\smile c \smile c} = A$. Similarly, $(A^{\smile c} \cap B^{\smile c})^{\smile c} \subseteq B$. Hence $(A^{\smile c} \cap B^{\smile c})^{\smile c} \subseteq A \cap B$, so using again (b) and (c) we get $A^{\smile c} \cap B^{\smile c} = (A^{\smile c} \cap B^{\smile c})^{\smile c \smile c} \subseteq (A \cap B)^{\smile c}$.

(e) Assume that $x \notin l(A)^{\smile c}$. Thus $C(x) \notin l(A)$, so there is some $y \in X$ such that $C(x) \leqslant_1 y$ and $y \in A$. By (FRGI1), $C(x) \leqslant_1 y$ implies $C(C(x)) \leqslant_1 C(y)$, so from (FRGI3) we get $x \leqslant_1 C(y)$. Next, by (FRGI3), it follows that $C(C(y)) \in A$, thus $C(y) \in A^{\smile c}$. Hence, since $x \leqslant_1 C(y)$, we obtain $x \notin l(A^{\smile c})$. In a similar way the other equality can be proved. □

Proposition 16.2.4 *The complex algebra of an RGI-frame is an RGI-lattice.*

Proof:
Let $(L_X, \odot_Q, \rightarrow_R, \leftarrow_R, 1'_I, {}^{\smile c})$ be the complex algebra of an RGI-frame. By Proposition 15.2.9, $(L_X, \odot_Q, \rightarrow_R, \leftarrow_R, 1'_I)$ is an RG-lattice. Also L_X is closed under the operation ${}^{\smile c}$ by Lemma 16.2.3. It remains to show that the axioms (RGI1), (RGI2), and (RGI3) are satisfied.

Note that (RGI1) immediately follows from Lemma 16.2.3(b).

(RGI2) Take any $A, B \subseteq X$ and any $x \in X$. Then we have:

$$
\begin{array}{rcll}
x \in (A \vee_X B)^{\smile c} & \Leftrightarrow & x \in l(r(A) \cap r(B))^{\smile c} & \text{definition } \vee_X \\
& \Leftrightarrow & x \in l((r(A) \cap r(B))^{\smile c}) & \text{Lemma 16.2.3(e)} \\
& \Leftrightarrow & x \in l(r(A)^{\smile c} \cap r(B)^{\smile c}) & \text{Lemma 16.2.3(d)} \\
& \Leftrightarrow & x \in l(r(A^{\smile c}) \cap r(A^{\smile c})) & \text{Lemma 16.2.3(e)} \\
& \Leftrightarrow & x \in A^{\smile c} \vee_X B^{\smile c}. &
\end{array}
$$

(RGI3) Take any $A, B \in L_X$ and $z \in X$ such that $z \in (A \odot B)^{\smile c}$. Then $C(z) \in A \odot_Q B$, that is $N(z) \in l(A \circ_Q B)$. Let $z' \in X$ be such that $N(z) \leqslant_1 z'$. Then $z' \notin A \circ_Q B$, so there are $x, y \in X$ such that $Q(x, y, z')$, $x \in A$, and $y \notin r(B)$. Hence there is some $y' \in X$ such that $y \leqslant_2 y'$ and $y' \in B$. Since $x \in A$ and $y' \in B$, we get $C(x) \in A^{\smile c}$ and $C(y') \in B^{\smile c}$. But B is l-stable, so by Lemma 16.2.3(f), $B^{\smile c}$ is also l-stable. Hence $C(y') \notin r(B)$. By (FRG3), $Q(x, y, z')$ and $y \leqslant_2 y'$ give $Q(x, y', z')$ which by (FRGI4) is equivalent to $S(C(y'), C(x), C(z'))$. Since $C(x) \in A^{\smile c}$ and $C(y') \notin r(B)$, we obtain $C(z') \notin B^{\smile c} \circ_S A^{\smile c}$. This means, by Lemma 15.2.4(b), that $C(z') \notin B^{\smile c} \circ_Q A^{\smile c}$. Furthermore, by (FRGI1) and (FRGI3), $C(z) \leqslant_1 z'$ means $z \leqslant_1 C(z')$. Then $z \in l(B^{\smile c} \odot_Q A^{\smile c})$, that is, $z \in B^{\smile c} \odot_Q A^{\smile c}$.
The reverse inclusion can be proved in a similar way. □

The *canonical frame of an RGI-lattice* L is a structure

$$(X_L, R_\odot, S_\odot, Q_\odot, I_{1'}, C_\smile),$$

where $(X_L, R_\odot, S_\odot, Q_\odot, I_{1'})$ is the canonical frame of the its RG-lattice reduct and $C_\smile : X_L \rightarrow X_L$ is defined, for every $(F_1, F_2) \in X_L$, by

$$C_\smile(F_1, F_2) \stackrel{\text{def}}{=} (F_1{}^\smile, F_2{}^\smile).$$

Lemma 16.2.5 *Let L be an RGI-lattice. Then*

(a) *If A is a filter of L, then so is $A^\smile$;*

(b) *If A is an ideal of L, then so is $A^\smile$;*

(c) (F_1, F_2) *is a filter-ideal pair of* L *if and only if* $(F_1{}^{\smile}, F_2{}^{\smile})$ *is a filter-ideal pair of* L*;*

(d) *If* $F \in X_L$*, then* $C_{\smile}(F) \in X_L$.

Proof:
(a) Let A be a filter of L and let $a, b \in L$ be such that $a \in A^{\smile}$ and $a \leq b$. Then, by Lemma 16.2.2(a), $a^{\smile} \in A$, and also $a^{\smile} \leq b^{\smile}$ by Lemma 16.2.1(b). Hence $b^{\smile} \in A$, thus $b \in A^{\smile}$. Moreover, assume that $a, b \in A^{\smile}$. Then, again by Lemma 16.2.2(a), $a^{\smile} \in A$ and $b^{\smile} \in A$, so $a^{\smile} \wedge b^{\smile} \in A$. By Lemma 16.2.1(c), $(a \wedge b)^{\smile} \in A$, thus $(a \wedge b)^{\smile\smile} \in A$ by the definition of $A^{\smile}$. Therefore, by axiom (RGI1), $a \wedge b \in A^{\smile}$.
The proof of (b) is similar.

(c) Let $F = (F_1, F_2)$ be a filter-ideal pair. By (a), $F_1{}^{\smile}$ is a filter of L, and by (b), $F_2{}^{\smile}$ is an ideal of L. Then $F_1{}^{\smile} \cap F_2{}^{\smile} = \emptyset$, since otherwise there is some $a \in L$ such that $a \in F_1{}^{\smile}$ and $a \in F_2{}^{\smile}$, so by Lemma 16.2.2(a), $a^{\smile} \in F_1 \cap F_2$, which contradicts the assumption. Hence $(F_1{}^{\smile}, F_2{}^{\smile})$ is a filter-ideal pair of L. On the other hand, take $F_1, F_2 \subseteq L$ and assume that $(F_1{}^{\smile}, F_2{}^{\smile})$ is a filter-ideal pair of L. Then, by Lemma 16.2.2(b) and (a), $F_1 = F_1{}^{\smile\smile}$ is a filter of L and $F_2 = F_2{}^{\smile\smile}$ is an ideal of L. Next, by Lemma 16.2.2(e), $(F_1 \cap F_2)^{\smile} = \emptyset$, so $F_1 \cap F_2 = (F_1 \cap F_2)^{\smile\smile} = \emptyset$. Therefore (F_1, F_2) is a filter-ideal pair of L.

(d) Let $F = (F_1, F_2)$ be a filter-ideal pair. Suppose that $(F_1{}^{\smile}, F_2{}^{\smile})$ is not maximal. Then, by Theorem 2.8.6, it can be extended to a maximal filter-ideal pair, say $G = (G_1, G_2)$. Hence $F_1{}^{\smile} \subseteq G_1$, $F_2{}^{\smile} \subseteq G_2$, and $(F_1{}^{\smile}, F_2{}^{\smile}) \neq (G_1, G_2)$. By Lemma 16.2.3(b) and (c), $F_1 \subseteq G_1{}^{\smile}$, $F_2 \subseteq G_2{}^{\smile}$, and $(F_1, F_2) \neq (G_1{}^{\smile}, G_2{}^{\smile})$. Thus (F_1, F_2) is not maximal, a contradiction. □

Proposition 16.2.6 *The canonical frame of an RGI-lattice is an RGI-frame.*

Proof:
Let $(X_L, R_{\odot}, S_{\odot}, Q_{\odot}, I_{1'}, C_{\smile})$ be the canonical frame of an RGI-lattice. From Proposition 15.2.10, $(X_L, R_{\odot}, S_{\odot}, Q_{\odot}, I_{1'})$ is an RG-frame. Now $C_{\smile}$ is well-defined by Lemma 16.2.5(d). It remains to show that axioms (FRGI1)–(FRGI4) are satisfied.

(FRGI1) Let $(F_1, F_2), (G_1, G_2) \in X_L$ be such that $F_1 \subseteq G_1$. Since $^{\smile}$ is monotone, $F_1{}^{\smile} \subseteq G_1{}^{\smile}$, hence $C_{\smile}(F) \subseteq_1 C_{\smile}(G)$.
The proof of (FRGI2) is smilar.

(FRGI3) For any $F = (F_1, F_2) \in X_L$, we have

$$C_{\smile}(C_{\smile}(F)) = (F_1{}^{\smile\smile}, F_2{}^{\smile\smile}) = (F_1, F_2) = F.$$

(FRGI4) Note that since $a = a^{\smile\smile}$, for any $a \in L$ there exists $a' \in L$, namely $a' = a^{\smile}$, such that $a = a'^{\smile}$. For all $F, G, H \in X_L$,

$$
\begin{aligned}
&S_{\odot}(C_{\smile}(G), C_{\smile}(F), C_{\smile}(H)) \\
&\quad \Leftrightarrow \forall a, b \in L\ (a \odot b \in H_2{}^{\smile}\ \&\ b \in F_1{}^{\smile} \Rightarrow a \in G_2{}^{\smile}) \\
&\quad \Leftrightarrow \forall a, b \in L\ ((a \odot b)^{\smile} \in H_2\ \&\ b^{\smile} \in F_1 \Rightarrow a^{\smile} \in G_2) \\
&\quad \Leftrightarrow \forall a, b \in L\ (b^{\smile} \odot a^{\smile} \in H_2\ \&\ b^{\smile} \in F_1 \Rightarrow a^{\smile} \in G_2) \\
&\quad \Leftrightarrow \forall a, b \in L\ (b \odot a \in H_2\ \&\ b \in F_1 \Rightarrow a \in G_2) \\
&\quad \Leftrightarrow Q_{\odot}(F, G, H)
\end{aligned}
$$

which completes the proof. □

We have the following discrete representation for RGI-lattices.

Theorem 16.2.7 *Every RGI-lattice is embeddable into the complex algebra of its canonical frame.*

Proof:
For every $F \in X_L$ and for every $a \in L$ we have:

$$F \in h(a^{\smile}) \Leftrightarrow a^{\smile} \in F_1 \Leftrightarrow a \in F_1{}^{\smile c} \Leftrightarrow C_{\smile}(F) \in h(a) \Leftrightarrow F \in h(a)^{\smile c_{\smile}}.$$

Hence $h(a^{\smile}) = h(a)^{\smile c_{\smile}}$. By Theorem 15.2.13 the required result follows. □

16.3 Doubly residuated general lattices with involution

A *doubly residuated general lattice with involution, DRGI-lattice*, is a structure $(L, \odot, \rightarrow, \leftarrow, \oplus, \twoheadrightarrow, \twoheadleftarrow, ^{\smile}, ^{\frown}, 1', 0')$ such that $(L, \odot, \rightarrow, \leftarrow, ^{\smile}, 1')$ an RGI-lattice, $(L, \odot, \rightarrow, \leftarrow, \oplus, \twoheadrightarrow, \twoheadleftarrow, 1', 0')$ is a DRG-lattice, and $^{\frown}$ is a unary operation in L satisfying, for all $a, b \in L$,

(DRGI1) $a^{\frown\frown} = a$

(DRGI2) $(a \wedge b)^{\frown} = a^{\frown} \wedge b^{\frown}$

(DRGI3) $(a \oplus b)^{\frown} = b^{\frown} \oplus a^{\frown}$.

Note that a DRGI-lattice can be viewed as a kind of join of two RGI-lattices, one based on L and the other based on L^{op}.

Lemma 16.3.1 *Let* $(L, \odot, \rightarrow, \leftarrow, \oplus, \twoheadrightarrow, \twoheadleftarrow, ^{\smile}, ^{\frown}, 1', 0')$ *be a DRGI-lattice. Then* $(L, \oplus, \twoheadrightarrow, \twoheadleftarrow, ^{\frown}, 0')$ *is a RGI-lattice.*

A *DRGI-frame* is a structure

$$(X, R, S, Q, I, C, \Theta, \Upsilon, \Omega, J, \Gamma)$$

such that (X, R, S, Q, I, C) is an RGI-frame, $(X, R, S, Q, \Theta, \Upsilon, \Omega, I, J)$ is a DRG-frame, and for all $x, y, z \in X$ the following assertions are satisfied:

(DRGI1) $x \leqslant_1 y \Rightarrow \Gamma(x) \leqslant_1 \Gamma(y)$

(DRGI2) $x \leqslant_2 y \Rightarrow \Gamma(x) \leqslant_2 \Gamma(y)$
(DRGI3) $\Gamma(\Gamma(x)) = x$
(DRGI4) $\Omega(x, y, z) = \Upsilon(\Gamma(y), \Gamma(x), \Gamma(z))$.

Observe that a DRGI-frame can be viewed as a kind of join of two RGI-frames: the one based on a lattice frame $(X, \leqslant_1, \leqslant_2)$ and the other other one based on the opposite lattice frame $(X, \leqslant_1^{op}, \leqslant_2^{op})$, defined in Section 15.3. Therefore we have the following fact.

Lemma 16.3.2 *Let $(X, R, S, Q, I, C, \Theta, \Upsilon, \Omega, J, \Gamma)$ be a DRGI-frame. Then $(X^{op}, \Theta, \Upsilon, \Omega, J, \Gamma)$ is an RGI-frame.*

Proof:
The proof follows from the definitions of an RGI-frame and a DRGI-frame. □

From the above lemma it follows that the properties of relations Θ, Υ, Ω, J, and Γ can be easily obtained from the properties of the relations R, S, Q, I, and C, respectively, by interchanging the roles of the orderings $\leqslant_1$ and $\leqslant_2$.

The *complex algebra of a DRGI-frame* X is a structure

$$(L_X, \odot_R, \rightarrow_R, \leftarrow_R, \oplus_\Theta, \twoheadrightarrow_\Theta, \twoheadleftarrow_\Theta, {}^{\smile_N}, {}^{\frown_\Gamma}, 1'_I, 0'_J)$$

such that $(L_X, \odot_R, \rightarrow_R, \leftarrow_R, {}^{\smile_N}, \emptyset, X, 1'_I)$ is the complex algebra of its RGI-frame reduct, $(L_X, \odot_R, \rightarrow_R, \leftarrow_R, \oplus_\Theta, \twoheadrightarrow_\Theta, \twoheadleftarrow_\Theta, 1'_I, 0'_J)$ is the complex algebra of its DRG-frame reduct, and ${}^{\frown_\Gamma} : X \rightarrow X$ is defined, for every $A \subseteq X$, by

$$A^{\frown_\Gamma} \stackrel{\text{def}}{=} \{\Gamma(x) : x \in A\}.$$

From Lemma 16.3.2 and Lemma 16.2.3 we have the following lemma.

Lemma 16.3.3 *Let X be an DRGI-frame. For all $A, B \subseteq X$,*

(a) $A^{\frown_\Gamma} = \{x \in X : C(x) \in A\}$;
(b) $A^{\frown_\Gamma \frown_\Gamma} = A$;
(c) $A \subseteq B \Rightarrow A^{\frown_\Gamma} \subseteq B^{\frown_\Gamma}$;
(d) $(A \cap B)^{\frown_\Gamma} = A^{\frown_\Gamma} \cap B^{\frown_\Gamma}$;
(e) $l(A^{\frown_\Gamma}) = l(A)^{\frown_\Gamma}$ *and* $r(A^{\frown_\Gamma}) = r(A)^{\frown_\Gamma}$;
(f) *If A is l-stuble, then so is $A^{\frown_\Gamma}$.*

Proposition 16.3.4 *The complex algebra of a DRGI-frame is a DRGI-lattice.*

Proof:
By Lemma 16.3.3(f), $A^{\frown_\Gamma} \in L_X$ for any $A \in L_X$. The required result follows from Proposition 16.2.4 and Lemma 16.3.2. □

The *canonical frame of a DRGI-lattice* L is a structure

$$(X_L, R_\odot, S_\odot, Q_\odot, I_{1'}, C_\smile, \Theta_\oplus, \Upsilon_\oplus, \Omega_\oplus, J_{0'}, \Gamma_\frown)$$

such that $(X_L, R_\odot, S_\odot, Q_\odot, I_{1'}, C_\smile)$ is the canonical frame of its RGI-lattice reduct, $(X_L, R_\odot, S_\odot, Q_\odot, I_{1'}, \Theta_\oplus, \Upsilon_\oplus, \Omega_\oplus, J_{0'})$ is the canonical frame of its DRG-lattice reduct, and $\Gamma_\frown : X_L \to X_L$ is defined for every $(F_1, F_2) \in X_L$ as

$$\Gamma_\frown(F_1, F_2) \stackrel{\text{def}}{=} (F_1{}^\frown, F_2{}^\frown),$$

where for any $A \subseteq L$, $A^\frown = \{a^\frown \in L : a \in A\}$.

From Lemma 16.2.5 and Lemma 16.3.1 the following facts follow.

Lemma 16.3.5 *Let $(L, \odot, \to, \leftarrow, \oplus, \twoheadrightarrow, \twoheadleftarrow, {}^\smile, {}^\frown, 1', 0')$ be a DRGI-lattice. For all $A, B \subseteq L$,*

(a) *If A is a filter of L, then so is $A^\frown$;*
(b) *If A is an ideal of L, then so is $A^\frown$;*
(c) *(F_1, F_2) is a filter-ideal pair of L if and only if $(F_1{}^\frown, F_2{}^\frown)$ is a filter-ideal pair of L;*
(d) *If $F \in X_L$, then $\Gamma_\frown(F) \in X_L$.*

Proposition 16.3.6 *The canonical frame of a DRGI-lattice is a DRGI-frame.*

Proof:
From Lemma 16.3.5 we have that $\Gamma_\frown$ is well-defined. The required result follows from Proposition 16.2.6 and Lemma 16.3.1. □

We conclude this section by the following discrete representation for DRGI-lattices.

Theorem 16.3.7 *Every DRGI-lattice is embeddable into the complex algebra of its canonical frame.*

Proof:
In view of Theorem 15.3.5 and Theorem 16.2.7 it remains to show that the mapping $h : X \to X_{LX}$, defined as $h(a) = \{(F_1, F_2) \in X_L : a \in F_1\}$, preserves the operation $^\frown$. Note that

$$F \in h(a^\frown) \Leftrightarrow a^\frown \in F_1 \Leftrightarrow a \in F_1^{\frown r} \Leftrightarrow \Gamma_\frown(F) \in h(a) \Leftrightarrow F \in h(a)^{\frown r \frown}$$

which completes the proof. □

16.4 Lattice-based relation algebras

A *lattice-based relation algebra, LRA-algebra,* is a DRGI-lattice

$$(L, \odot, \to, \leftarrow, \oplus, \twoheadrightarrow, \twoheadleftarrow, {}^\smile, {}^\frown, 1', 0')$$

satisfying

(LRA1) $0' \wedge 1' = 0$

(LRA2) $0' \vee 1' = 1$.

An *LRA-frame* is a DRGI-frame $(X, R, S, Q, I, C, \Theta, \Upsilon, \Omega, J, \Gamma)$ satisfying

(FLRA1) $lr(I) \cap l(J) = \emptyset$
(FLRA2) $r(I) \cap rl(J) = \emptyset$.

The *complex algebra of an LRA-algebra* and the *canonical frame of an LRA-algebra* are defined in the same way as the complex algebra of a DRGI-frame and the canonical frame of a DRGI-lattice, respectively.

Proposition 16.4.1 *The complex algebra of an LRA-algebra is an LRA-frame.*

Proof:
It suffices to show that the axioms (LRA1) and (LRA2) are satisfied.

For (LRA2), note that $0'_J \vee_X 1'_I = l(rlr(I) \cap rl(J))$. By Lemma 2.8.3(c) and since I is l-stable, $I \subseteq lr(I)$. Hence $rlr(I) \subseteq r(I)$. Furthermore, $rlr(I) \cap rl(J) \subseteq r(I) \cap rl(J) = \emptyset$ by (FLRA2). Therefore $rlr(I) \cap rl(J) = \emptyset$, so $l(rlr(I) \cap rl(J)) = l(\emptyset) = X_L$.
(LRA1) follows from (FLRA2). □

Proposition 16.4.2 *The canonical frame of an LRA-algebra is an LRA-frame.*

Proof:
It suffices to show that (FLRA1) and (FLRA2) hold.
Note first that

$$l(\{F \in X_L : a \in F_2\}) = \{F \in X_L : a \in F_1\}.$$

Indeed, take $a \notin F_1$. Then $F_1 \cap \downarrow_\leq a = \emptyset$, so $(F_1, \downarrow_\leq a)$ is a filter-ideal pair and by Theorem 2.8.6 it can be extended to the maximal filter-ideal pair, say G. Hence $F_1 \subseteq G_1$ and $a \in G_2$. Therefore $F \notin l(\{H \in X_L : a \in H_2\})$. For the reverse inclusion, let $a \in F_1$ and let $G \in X_L$ be such that $F_1 \subseteq G_1$. Then $a \in G_1$, thus $a \notin F_2$. Similarly, $r(\{F \in X_L : a \in F_1\}) = \{F \in X_L : a \in F_2\}$.

(FLRA1) Observe that

$$\begin{aligned} & lr(I_{1'}) \cap l(J_{0'}) \\ &= lr(\{F \in X_L : 1' \in F_1\}) \cap l(\{F \in X_L : 0' \in F_2\}) \\ &= l(\{F \in X_L : 1' \in F_2\}) \cap l(\{F \in X_L : 0' \in F_2\}) \\ &= \{F \in X_L : 1' \in F_1\} \cap \{F \in X_L : 0' \in F_1\} \\ &= \{F \in X_L : 1' \in F_1 \ \& \ 0' \in F_1\} \\ &\subseteq \{F \in X_L : 1' \wedge 0' \in F_1\}. \end{aligned}$$

However, by (LRA1), $1' \wedge 0' = 0$. Since F_1 is a proper filter of L, $0 \notin F_1$, so $\{F \in X_L : 1 \wedge 0' \in F_1\} = \emptyset$, and consequently, $lr(I_{1'}) \cap l(J_{0'}) = \emptyset$, as required.

(FLRA2) Note that

$$
\begin{aligned}
&r(I)\cap rl(J)\\
&\quad = r(\{F\in X_L : 1'\in F_1\})\cap rl(\{F\in X_L : 0'\in F_2\})\\
&\quad = r(\{F\in X_L : 1'\in F_1\})\cap r(\{F\in X_L : 0'\in F_1\})\\
&\quad = \{F\in X_L : 1'\in F_2\}\cap\{F\in X_L : 0'\in F_2\}\\
&\quad = \{F\in X_L : 1'\in F_2 \ \& \ 0'\in F_2\}\\
&\quad \subseteq \{F\in X_L : 1'\vee 0'\in F_2\}.
\end{aligned}
$$

Since $1'\vee 0' = 1 \neq F_2$, it follows that $\{F\in X_L : 1'\vee 0'\in F_2\} = \emptyset$. In conclusion, $r(I)\cap rl(J) = \emptyset$. □

As a consequence of the preceeding two results we obtain a discrete representation for LRA-algebras.

Theorem 16.4.3 *Any LRA-algebra is embeddable into the complex algebra of its canonical frame.*

Chapter 17

Algebras of substructural logics

17.1 Introduction

The basic substructural logics are the full Lambek calculus, FL, originated in [151], see also [152, 244], and its three axiomatic extensions obtained from FL by endowing it with the structural rules of exchange, contraction, and weakening, respectively. An algebraic approach to substructural logics presented in [170] is implicitly based on residuated lattices. The algebraic counterpart of FL, FL-algebras, is the class of residuated lattices endowed with a distinguished fixed element of the lattice which provides a means of extending the signature of FL with a negation. A survey of algebraic approaches to substructural logics can be found in [149]. The algebraic counterparts of the three axiomatic extensions of FL mentioned above are FLe-algebras, FLc-algebras, and FLw-algebras. They are considered in Section 17.3, Section 17.4, and Section 17.5, respectively, where the representation theorems are proved based on the Urquhart representation of lattices. Of special importance is the class of Flew-algebras obtained as a join of FLe-algebras and FLw-algebras. The corresponding logic is the logic of residuated lattices, it gives rise to the Esteva-Godo-Ono hierarchy of substructural and fuzzy logics mentioned in Section 12.1 and Section 12.6, see [79]. In the field of fuzzy logic it is referred to as a monoidal logic, see [130]. In the hierarchy the two immediate successors of FLew are obtained by endowing it with a negation and postulating the axioms

(1) $\neg\neg a = a$

(2) $a \wedge \neg a = 0$,

thus obtaining FLew+(1), and FLew+(2), respectively. Then the successive levels of the hierarchy are constructed by adding to those two axiomatic extensions of FLew the axioms

(3) $a \vee b = ((a \rightarrow b) \rightarrow b) \wedge ((b \rightarrow a) \rightarrow a)$

(4) distributivity of the underlying lattice.

The next level of the hierarchy is constructed by adding the axiom of prelinearity

(5) $(a \to b) \vee (b \to a) = 1$.

Then we obtain the logics whose algebras are

$$\text{FLew} + (3) + (4) + (5) = \text{MTL}$$

$$\text{FLew} + (1) + (3) + (4) + (5) = \text{IMTL}$$

$$\text{FLew} + (2) + (3) + (4) + (5) = \text{SMTL}.$$

They are discussed in Section 12.5, Section 12.9, and Section 12.8, respectively. In Section 17.6 an Urquhart-style representation theorem for algebras FLew+(2) is presented. The content of this chapter is based on [176, 177].

17.2 FL-algebras

An *FL-algebra* is a structure (L, δ) such that L is a residuated lattice and $\delta \in L$ is an arbitrary, fixed element of L.

In general, FL-algebras need not be bounded. In this section we will consider bounded FL-algebras in order to conform to the requirements of the Urquhart representation of lattices, which is the basis for our results. In other words, by an FL-algebra we mean a structure L, where L is an RG-lattice and $\delta \in L$.

Note that if the least element 0 exists in L, from the residuation condition it follows that $a \odot 0 = 0 \odot a = 0$. Then the greatest element can be defined as $1 = 0 \to a$. Moreover, we assume that $1' \neq 0$. This assumption is equivalent to non-triviality of the algebra and is essential for proving the representation theorems presented in this chapter. Since for all the classes of algebras considered in this chapter we always assume that in the complex algebras of the frames the least element is defined as the empty set and the greatest element is defined as the universe of the underlying frame, in the remaining sections we will not write explicitly the lattice bounds.

An *FL-frame* is a relational structure (X, D) where X is an RG-frame, $D \subseteq X$, and, for all $x, x' \in X$,

(FFL) $D(x)$ & $x \leqslant_1 x' \Rightarrow D(x')$.

The *complex algebra of an FL-frame* X is a structure (L_X, δ_D) where L_X is the complex algebra of its RG-frame reduct and δ_D is defined by

$$\delta_D = lr(D).$$

Proposition 17.2.1 *The complex algebra of an FL-frame is an FL-algebra such that δ_D is not a zero element.*

The *canonical frame of an FL-algebra* L is a structure

$$(X_L, D_\delta)$$

such that X_L is the canonical frame of an underlying RG-lattice and D_δ is defined by

$$D_\delta \stackrel{\text{def}}{=} \{F \in X_L : \delta \in F_1\}.$$

Proposition 17.2.2 *The canonical frame of an FL-algebra is an FL-frame.*

Let $h : L \to L_{X_L}$ be as in Section 15.2, that is, for any $a \in L$, $h(a) = \{(F_1, F_2) \in X_L : a \in F_1\}$. A discrete representation for FL-algebras is obtained.

Theorem 17.2.3 *Every FL-algebra is embeddable into the complex algebra of its canonical frame.*

Proof:
By straightforward verification it is easy to see that $h(\delta) = D_\delta$. Then result follows from Theorem 15.2.13. □

17.3 FLe-algebras

By an *FLe-algebra* we mean an FL-algebra L such that $\odot$ is commutative. An *FLe-frame* is an FL-frame (X, D) satisfying the following condition for all $x, y, z \in X$

(FFLe) $R(x, y, z) \Rightarrow R(y, x, z)$.

The *complex algebra of an FLe-frame* is just the complex algebra of an FL-frame.

Proposition 17.3.1 *The complex algebra of an FLe-frame is an FLe-algebra.*

Proof:
It suffices to show that $A \odot_S B = B \odot_S A$ for every $A \in L_X$. Then, in view of Proposition 17.2.1, the results follows.
Since $A \odot_Q B = l(A \circ_Q B)$, and, by Lemma 15.2.4(c), $A \circ_Q B = A \circ_S B$ for all l-stable sets $A, B \subseteq X$, we need to show that $A \circ_S B = B \circ_S A$.
($\subseteq$) Let $x \in X$ be such that $z \notin B \circ_S A$. Then there exist some $x, y \in X$ such that $S(x, y, z)$, $y \in A$, and $x \notin r(B)$. Next, there is some $x' \in X$ satisfying $x \leqslant_2 x'$ and $x' \in B$. Now, by (FRG2), we get $S(x', y, z)$, which by (FRG3) implies that there exists $z' \in X$ such that $z \leqslant_2 z'$ and $R(x', y, z')$. Hence, by (FRLe), we get $R(y, x', z')$, so in view of (FRG5) there exists some $y' \in X$ such that $y \leqslant_1 y'$ and $S(x', y', z')$. By the assumption, A is l-stable, so it is $\leqslant_1$-increasing. Then, since $y \in A$ and $y \leqslant_1 y'$, it follows that $y' \in A$, so $y' \notin r(A)$. Hence, by the definition of $\odot_S$ and the conditions $x' \in B$ and $S(y', x', z')$ we get $z' \notin A \circ_S B$. Now suppose that $z \in A \circ_S B$. From Lemma 15.2.4(b), $A \circ_S B$ is r-stable, so it is $\leqslant_2$-increasing. Whence, since $z \leqslant_2 z'$, we get $z' \in A \circ_S B$, a contradiction. Therefore $z \notin A \circ_S B$, as expected.
($\supseteq$) can be proved in the similar way. □

Proposition 17.3.2 *The canonical frame of an FLe-algebra is an FLe-frame.*

Proof:
We only need to show that (FFLe) is satisfied. Take $F, G, H \in X_L$ such that $R_\odot(F, G, H)$ holds. Then, for all $a, b \in L$, $a \in F_1$ and $b \in G_1$ imply $a \odot b \in H_1$. By commutativity of $\odot$, $b \odot a \in H_1$. Therefore $R_\odot(G, F, H)$ holds. □

From Theorem 17.2.3 and Proposition 17.3.1 and Proposition 17.3.2 we obtain the discrete representation for FLe-algebras.

Theorem 17.3.3 *Every FLe-algebra us embeddable into the complex algebra of its canonical frame.*

17.4 FLc-algebras

An *FLc-algebra* is an FL-algebra L satisfying the contraction condition, that is for every $a \in L$,

(FLc) $a \leq a \odot a$.

The following assertions can be easily shown.

Lemma 17.4.1 *For every FLc-algebra L and for all $a, b \in L$,*

(a) $a \leq \neg\neg a$;

(b) $a \leq \neg a \Leftrightarrow b \leq \neg b$.

An *FLc-frame* is an FL-frame (X, D) satisfying the following condition for every $x \in X$,

(FFLc) $R(x, x, x)$

The *complex algebra of an FLc-frame* is defined in the same way as the complex algebra of an FL-frame.

Proposition 17.4.2 *The complex algebra of an FLc-frame is an FLc-algebra.*

Proposition 17.4.3 *The canonical frame of an FLc-algebra is an FLc-frame.*

Proof:
It is sufficient to prove that $R_\odot(F, F, F)$ holds for every $F \in X_L$. First, note that since $a \wedge b \leq a$ and $a \wedge b \leq b$, by monotonicity of $\odot$ we have $(a \wedge b) \odot (a \wedge b) \leq a \odot a$. Also, by (FLc), $a \wedge b \leq (a \wedge b) \odot (a \wedge b)$. Then $a \wedge b \leq a \odot b$. Let $F \in X_L$ and let $a, b \in F_1$. Since F_1 is a filter of L, $a \wedge b \in F_1$, and furthermore, $a \odot b \in F_1$. Hence $R_\odot(F, F, F)$ holds. □

From Theorem 17.2.3 and Propositions 17.4.2 and 17.4.3 we get the following discrete representation for FLc-algebras.

Theorem 17.4.4 *Every FLc-algebra is embeddable into the complex algebra of its canonical frame.*

17.5 FLw-algebras

An *FLw-algebra* is an FL-algebra L satisfying, for all $a, b \in L$,

(FLw1) $a \odot b \leq a$

(FLw2) $a \odot b \leq b$

(FLw3) δ is the bottom element of the underlying lattice L.

An *FLw-frame* is an FL-frame (X, D) satisfying, for all $x, y, z \in X$,

(FFLw1) $S(x, y, z) \Rightarrow z \leqslant_2 x$

(FFLw2) $Q(x, y, z) \Rightarrow z \leqslant_2 y$

(FFLw3) $D = \emptyset$.

The *complex algebra of an FLw-frame* is defined in the same way as the complex algebra of an FL-frame.

Proposition 17.5.1 *The complex algebra of an FLw-frame is an FLw-algebra.*

Proof:
Due to (FFLw), the axiom (FLw3) holds in the complex algebra L_X. It remains to show that for all l-stable subsets $A, B \subseteq X$, $A \odot_Q B \subseteq A$ and $A \odot_Q B \subseteq B$. Let $x \in A \odot_Q B$ and let $x' \in X$ be an arbitrary element satisfying $x \leqslant_1 x'$. Since $A \odot_Q B = l(A \circ_Q B)$, we get $x' \notin A \circ_Q B$, so there exist $y, z \in X$ such that $Q(y, z, x')$, $y \in A$, and $z \notin r(B)$. Furthermore, there is some $z' \in X$ such that $z \leqslant_2 z'$ and $z' \in B$. Moreover, by (FFLw2) we get $x' \leqslant_2 z$. By transitivity of $\leqslant_2$ it follows that $x' \leqslant_2 z'$, so $x' \notin r(B)$, since $x' \in B$. Therefore we get that for any $x' \in X$, $x \leqslant_1 x'$ implies $x' \notin r(B)$, which means that $x \in lr(B)$. But by the assumption, B is l-stable. Hence $x \in B$.
By (FFLw3) and the fact that $A \circ_Q B = A \circ_S B$ the second inclusion can be proved in the analogous way. □

Proposition 17.5.2 *The canonical frame of an FLw-algebra is an FLw-frame.*

Proof:
(FFLw1) Let $F, G, H \in X_L$ and let $a \in L$ be such that $a \in H_2$ and $S_\odot(F, G, H)$. By (FLw1), $a \odot 1 \leq a$, we get $a \odot 1 \in H_2$. Moreover, $1 \in G_1$, since G_1 is a filter of L. Hence, by the definition of $S_\odot$ we obtain $a \in F_2$. Therefore $H \leqslant_{2L} F$.
(FFLw2) Let $F, G, H \in X_L$ and let $b \in L$ be such that $Q_\odot(F, G, H)$ and $b \in H_2$. Similarly, by (FLw2), $1 \odot b \leq b$, so $1 \odot b \in H_2$. Since $1 \in x_1$, by the definition of $Q_\odot$ we finally get $b \in G_2$. Hence $H \leqslant_{2L} G$.
(FFLw3) Since $\delta = 0$, by the definition of D_δ it directly follows that $D_\delta = \emptyset$. □

We conclude this section by the following discrete representation for FLw-algebras.

Theorem 17.5.3 *Every FLw-algebra is embeddable into the complex algebra of its canonical frame.*

17.6 An axiomatic extension of FLew-algebras

In this section we develop a representation result for FLew-algebras whose complement operation $\neg a = a \to 0$ satisfies the axiom (2) in Section 17.1.

An *FLew2-algebra* is a FLew-algebra L satisfying the following axiom for every $a \in L$,

(FLew2) $a \wedge \neg a = 0$

where $\neg a = a \to 0$.

Lemma 17.6.1 *Let L be an FLew2-algebra and let F be a proper filter of L.*

(a) *For every $a \in L$, if $a \in \neg F$, then $\neg a \in F$;*

(b) *$\langle\leq\rangle(\neg F)$ is an ideal of L;*

(c) *$F \cap \langle\leq\rangle(\neg F) = \emptyset$.*

Proof:
(a) Let $a \in \neg F$. Then $a = \neg b$ for some $b \in F$, so clearly $a \leq \neg b$. Hence, by Lemma 17.4.1(b), $b \leq \neg a$, which implies that $\neg a \in F$.

(b) Since $1 \in F$, we have $0 = \neg 1$, so $0 \in \langle\leq\rangle\neg F$. Let $a, b \in \langle\leq\rangle\neg F$. Then there exist $a', b' \in L$ such that $a', b' \in \neg F$, $a \leq a'$, and $b \leq b'$. By (a) we get $\neg a', \neg b' \in F$. Furthermore, by Lemma 17.4.1(b) it follows that $\neg a' \leq \neg a$. Analogously, $b \leq b'$ implies $\neg b' \leq \neg b$. Hence we get $c, d \in F$ such that $c = \neg a'$, $d = \neg b'$, $c \leq \neg a$, and $d \leq \neg b$. By Lemma 17.4.1(b), from $c \leq \neg a$ we get $a \leq \neg c$, and similarly, $d \leq \neg b$ implies $b \leq \neg d$. Then $a \vee b \leq \neg c \vee \neg d$. But $\neg c \vee \neg d \leq \neg(c \wedge d)$ by Lemma 15.2.1(s). Since $c \wedge d \in F$, it follows that $\neg(c \wedge d) \in \neg F$. Hence, $a \vee b \in \langle\leq\rangle(\neg F)$.

(c) Suppose that there is some $a \in F$ such that $a \in \langle\leq\rangle(\neg F)$. Then there exists $b \in L$ such that $b \in \neg F$ and $a \leq b$. By (a), from $b \in \neg F$ it follows that $\neg b \in F$. Moreover, by Lemma 15.2.1(d), $a \leq b$ implies $\neg b \leq \neg a$. Hence $\neg a \in F$, which together with $a \in F$ gives $a \wedge \neg a \in F$. By (FLew2), $a \wedge \neg a = 0$, so $0 \in F$, which contradicts the fact that F is proper. □

An *FLew2-frame* is an FLew-frame (X, R, S, Q) satisfying the following axiom

(FFew2) $\forall x \in X\ \exists y, z \in X,\ R(y, x, z)\ \&\ x \leqslant_1 y$.

Let (X, R, S, Q) be an FLew2-frame. For any $A \subseteq X$,

$$\neg_R A \stackrel{\text{def}}{=} A \to_R \emptyset \quad = \quad \{x \in X : (\forall y, z \in X)\ R(y, x, z) \Rightarrow y \notin A\}$$

From Lemma 15.2.5(a) we get the following fact.

Lemma 17.6.2 *Let (X, R, S, Q) be an FLew2-frame. Then for every l-stable set $A \subseteq X$, $\neg_R A$ is also l-stable.*

The *complex algebra of an FLew2-frame* is defined in the same way as the complex algebra of an FL-frame.

Proposition 17.6.3 *The complex algebra of an FLew2-frame is an FLew2-algebra.*

Proof:
We have to show that $A \wedge_X \neg_R A = \emptyset$, that is, $A \cap \neg_R A = \emptyset$. Let $x \in A$. By (FLew2), there exist $y, z \in X$ such that $R(y, x, z)$ and $x \leqslant_1 y$. Since A is l-stable, by Lemma 2.8.1(a) it is $\leqslant_1$-increasing, so $x \in A$ and $x \leqslant_1 y$ imply $y \in A$. Now, by the definition of $\neg_R$, from $R(y, x, z)$ and $y \in A$ we get $x \notin \neg_R A$. □

The *canonical frame of an FLew2-algebra* is defined in the same way as the canonical frame of an FL-algebra.

Proposition 17.6.4 *The canonical frame of an FLew2-algebra is an FLew2-frame.*

Proof:
In view of Proposition 17.3.2 and Proposition 17.5.2 it suffices to show that (FLew2) holds.
Let $F \in X_L$. By Lemma 17.6.1(b) and (c), $\langle\leq\rangle(\neg F_1)$ is an ideal of L and $F_1 \cap \langle\leq\rangle(\neg F_1) = \emptyset$. Then $(F_1, \langle\leq\rangle(\neg F_1))$ is a filter-ideal pair of L which by Theorem 2.8.6 can be extended to a maximal filter-ideal pair, say $G = (G_1, G_2)$. Hence $F_1 \subseteq G_1$ and $\langle\leq\rangle(\neg F_1) \subseteq G_2$. Now, consider the set

$$B \stackrel{\text{def}}{=} \{b \in L : (\exists a \in G_1)\ a \to b \in F_1\}.$$

We show that B is a filter of L. Since $a \to 1 = 1$, we have $a \to 1 \in F_1$ for any $a \in L$, in particular for $a \in G_1$. Hence $1 \in B$. Moreover, let $b \in B$ and consider an arbitrary $b' \in L$ such that $b \leq b'$. By the definition of B, there exists $a \in G_1$ such that $a \to b \in F_1$. But $a \to b \leq a \to b'$, so $a \to b' \in F_1$. Then $b' \in B$. By Lemma 17.6.1(b), $(B, \langle\leq\rangle(\neg B))$ is a filter-ideal pair of L and by Theorem 2.8.6 it can be extended to the maximal filter-ideal pair, say H.
Let $a, b \in L$ be such that $a \in G_1$ and $a \to b \in F_1$. Then $b \in B \subseteq H_1$. Hence, by Lemma 15.2.11, $R(G, F, H)$ holds. Thus we have proved that for an arbitrary $F \in X_L$ there exist $G, H \in X_L$ such that $R_{\odot}(G, F, H)$ and $F_1 \leqslant_{1L} G_1$, which means that (FFLew2) holds. □

Chapter 18

Beyond Urquhart representation

18.1 Introduction

In this chapter we present some discrete representations of general lattices different from the Urquhart-style representation described in Section 2.7 where one of the characteristics is the requirement of disjointness of the filter and the ideal in every filter-ideal pair which is an element of the universe of a lattice frame. A dual, in a sense, representation of lattices is obtained by replacing the assumption of disjointness by exhaustiveness, understood as the postulate that the union of the filter and the ideal in every pair in the universe of a lattice frame coincides with the whole lattice. Such an exhaustive representation is presented in Section 18.2. In [156] an exhaustive representation is developed for a paraconsistent logic. Dualizing the pair of disjoint and exhaustive representations, in Section 18.3 and Section 18.4 we consider the pair of non-disjoint and non-exhaustive representations, respectively.

In [122] a topological representation of general lattices is presented such that the dual spaces are two-sorted structures consisting of a set of filters, and a set of ideals of a lattice, together with a heterogenous binary relation included in the Cartesian product of these sets. In Section 18.5 we present a discrete version of that representation such that the lattice frames are abstract two-sorted structures (X, Y, R) where X and Y are non-empty sets and $R \subseteq X \times Y$. The appropriate constructions of complex algebras of the lattice frames and the canonical frames of lattices are given. The complex algebras are lattices whose universes consist of pairs of some subsets of X and Y, respectively. The two universes of the canonical frame of a lattice are the set of filters and the set of ideals of the lattice. As usual the representation theorem says that every lattice is embeddable into the complex algebra of its canonical frame. In Section 18.6 we present Dedekind-MacNeille representations of posets and lattices. In Section 18.7 we present canonical extensions of lattices as MacNeille representation algebras of structures presented in [108]. In [256] the representation of lattices presented in Section 18.6 is generalized to a topological representation of lattices. The representation algebra is obtained by a generalization of Dedekind cuts [159] used to construct the real numbers from the rational numbers [37]. In Section 18.8 a discrete version of that representation is developed. This representation is an instance of the representation from Section 18.5 such that some additional conditions are postulated on the universes of the complex algebra and the canonical frame. We also mention an interpretation which is given in formal concept analysis [256] to the notions used in the Dedekind-MacNeille-Wille representation. Next, in Section 18.9 we present a discrete version of the topological representation of ortholattices developed in [112]. The frames associated with ortholattices are those presented in [111]. In Section 18.10 we show that this representation can be made two-sorted following the developments in Section 18.5.

The Urquhart-style representations, in Section 2.8 and Sections 18.2–18.4 and the Hartonas-Dunn-style representations, in Section 18.5 and Section 18.10, are based on a similar idea. The one sorted universes of the canonical frames of a lattice L in the former representations are defined as certain subsets of the Cartesian product of the family of filters and the family of ideals of L, respectively. In the latter representations the universes of the two-sorted canonical frames of L are the family of filters and the family of ideals of L, respectively. Then an appropriate subset of the Cartesian product of those sets is identified in terms of the relation in the frame.

A remark on notation is in order. In Sections 18.2–18.4 of this chapter the notation for filter-ideal pairs in the canonical frames of the respective lattices is the same as in Chapters 13–17, that is $F = (F1, F2)$. Although this notation may be questioned, because usually we denote filters by F and ideals by I with indices if necessary, our notation enables us to assign a symbol to a pair and symbols to the components of the pair using the same letter of the alphabet. However, in the remaining sections of this chapter, where the canonical frames of algebras are two-sorted with the universes consisting of a set of filters and a set of ideals of lattices, we use the traditional notation F for elements of the first sort and I for the elements of the second sort.

The contents of the Sections 18.2–18.4 are from [179]. The contents of Sections 18.5 and 18.8 are based on [61].

18.2 Exhaustive representation of lattices

In Section 2.8 the Urquhart-style discrete representation of bounded general lattices is presented. It is a disjoint representation in the sense that the universes of canonical frames of lattices consist of maximal filter-ideal pairs such that the filter and the ideal in each pair are disjoint. Following [179] in this section we present a representation of general lattices which is dual, in a sense, to the disjoint representation, namely, the universes of canonical frames consist of filter-ideal pairs such that the union of the filter and the ideal in each pair equals the whole lattice. Such representation is referred to as the exhaustive representation. In order to uniformly describe the two representations and compare them we add the words disjoint or exhaustive and the superscripts d or e to the notions involved in their constructions. In particular, the disjoint canonical frame of a lattice L will be written $(X^d_L, \leq^d_1, \leq^d_2)$ and the exhaustive canonical frame of a lattice will be written $(X^e_L, \leq^e_1, \leq^e_2)$.

Let L be a general lattice. We recall from Section 2.8 that a general lattice frame is a structure $(X, \leqslant_1, \leqslant_2)$ such that $X \neq \emptyset$, $\leqslant_1$ and $\leqslant_2$ are preorders on X, and $\leqslant_1 \cap \leqslant_2 = Id_X$.

Given a general lattice frame X, two mappings $\lambda, \rho : 2^X \to 2^X$ may be defined, for any $A \subseteq X$, by

$$\lambda(A) \stackrel{\text{def}}{=} \langle \leqslant_1^{-1} \rangle (-A)$$
$$\rho(A) \stackrel{\text{def}}{=} \langle \leqslant_2^{-1} \rangle (-A).$$

A subset $A \subseteq X$ is called *λ-stable* (resp. *ρ-stable*) whenever $\lambda\rho(A) = A$ (resp. $\rho\lambda(A) = A$).

Theorem 18.2.1 *Let $(X, \leqslant_1, \leqslant_2)$ be a lattice frame. For every $A \subseteq X$,*

(a) *$-lr(A) = \lambda\rho(-A)$ and $-\lambda\rho(A) = lr(-A)$;*
(b) *$-r(A) = \rho(-A)$ and $-l(A) = \lambda(-A)$.*

By the above theorem, for every $A \subseteq X$, A is λ-stable if and only if $-A$ is l-stable. Moreover, the dual properties to those in Lemma 2.8.1 and Lemma 2.8.3 hold.

Lemma 18.2.2 *Let X be a general lattice frame. For every $A \subseteq X$,*

(a) *$\lambda(A)$ is $\leqslant_1$-decreasing and $\rho(A)$ is $\leqslant_2$-decreasing;*
(b) *$-A \subseteq \lambda(A)$ and $-A \subseteq \rho(A)$;*
(c) *λ and ρ are antitone.*

Lemma 18.2.3 *Let X be a general lattice frame. For all $A, B \subseteq X$,*

(a) *If A is $\leqslant_1$-decreasing, then $\rho(A)$ is ρ-stable;*
(b) *If A is $\leqslant_2$-decreasing, then $\lambda(A)$ is λ-stable;*
(c) *$\lambda\rho$ and $\rho\lambda$ are interior operators on 2^X.*

Lemma 18.2.4 *Let X be a general lattice frame. The antitone maps λ and ρ form a dual Galois connection between the lattice of $\leqslant_1$-decreasing subsets of X and the lattice of $\leqslant_2$-decreasing subsets of X.*

The *exhaustive complex algebra of a general lattice frame X* is a structure $(L^e_X, \vee^e_X, \wedge^e_X, \emptyset, X)$ where $L^e_X = \{A \subseteq X : A = \lambda\rho(A)\}$ and, for any $A, B \in L^e_X$,

$$A \vee^e_X B \stackrel{\text{def}}{=} A \cup B$$
$$A \wedge^e_X B \stackrel{\text{def}}{=} \lambda(\rho(A) \cup \rho(B)).$$

Lemma 18.2.5 *For all* $A, B \in L^e_X$,

$$A \vee^e_X B = -(-A \wedge^d_X -B)$$
$$A \wedge^e_X B = -(-A \vee^d_X -B).$$

Proposition 18.2.6 *The exhaustive complex algebra of a general lattice frame is a general lattice.*

Let L be a general lattice and let $F_1, F_2 \subseteq L$. We say that (F_1, F_2) is an *exhaustive pair of* L whenever $(-F_1, -F_2)$ is a maximal disjoint pair of the opposite lattice L^{op}. The pair $(-F_1, -F_2)$ will be denoted by $-F$.

Let $F = (F_1, F_2)$ and $G = (G_1, G_2)$ be exhaustive pairs. We define orderings $\leqslant^e_1$, $\leqslant^e_2$, and $\leqslant^e$ on the family of exhaustive pairs by

$$F \leqslant^e_{1L} G \overset{\text{def}}{\Leftrightarrow} -F_1 \subseteq -G_1$$
$$F \leqslant^e_{2L} G \overset{\text{def}}{\Leftrightarrow} -F_2 \subseteq -G_2$$
$$\leqslant^e_L \overset{\text{def}}{=} \leqslant^e_{1L} \cap \leqslant^e_{2L}.$$

Lemma 18.2.7 *If* (F_1, F_2) *is an exhaustive pair of* L*, then for all* $a, b \in L$,

(a) $0 \notin F_1$ *and* $1 \notin F_2$;
(b) $F_1 \cup F_2 = L$;
(c) $a \in F_1$ & $a \leq b \Rightarrow b \in F_1$;
(d) $b \in F_2$ & $a \leq b \Rightarrow a \in F_2$;
(e) $a \vee b \in F_1 \Rightarrow a \in F_1$ *or* $b \in F_1$;
(f) $a \wedge b \in F_2 \Rightarrow a \in F_2$ *or* $b \in F_2$.

Lemma 18.2.8 *Let* F *and* G *be exhaustive pairs of a general lattice* L*. Then*

$$(-F \leqslant^d_{1L} -G \ \&\ -F \leqslant^d_{2L} -G \Rightarrow -F = -G)$$
$$\Rightarrow \ (F \leqslant^e_{1L} G \ \&\ F \leqslant^e_{2L} G \Rightarrow F = G).$$

Lemma 18.2.9 *If* $(-F_1, -F_2)$ *is a maximal disjoint pair of* L^{op} *with respect to the ordering* $\leqslant^d_L$*, then the exhaustive pair* (F_1, F_2) *is maximal with respect to the ordering* $\leqslant^e_L$.

The *exhaustive canonical frame of a general lattice* L is a structure

$$(X^e_L, \leqslant^e_{1L}, \leqslant^e_{2L})$$

where X^e_L is the family of all exhaustive pairs of L and the orderings $\leqslant^e_{1L}$ and $\leqslant^e_{2L}$ are defined as above.

Let $(L, \vee, \wedge)$ be a general lattice. Let $f : L \to L^{op}$ be a mapping such that, for all $a, b \in L$,

$$f(a \vee b) = f(a) \wedge f(b)$$
$$f(a \wedge b) = f(a) \vee f(b)$$
$$f(f(a)) = a.$$

and let $g : X_L^e \to X_{L^{op}}^d$ be a mapping such that, for any $(F_1, F_2) \in X_L^e$,

$$g(F_1, F_2) = (-F_1, -F_2).$$

Then g is an isomorphism (of frames) from X_L^e onto $X_{L^{op}}^d$.

Let the mapping $h^e : L \to L^e_{X_L^e}$ be defined, for any $a \in L$, by

$$h^e(a) \stackrel{\text{def}}{=} g^{-1}(-h^d(f(a))).$$

Lemma 18.2.10

(a) *For every $a \in L$, $h^e(a)$ is λ-stable;*

(b) *h^e is a lattice embedding.*

Proof:
(a) By Theorem 2.8.9(b), for every $a \in L$ we have

$$\begin{aligned}\lambda\rho(h^e(a)) &= \lambda g^{-1}(-rh^d(f(a))) \\ &= g^{-1}(-lr(h^d(f(a)))) \\ &= g^{-1}(-h^d(f(a))) \\ &= h^e(a).\end{aligned}$$

(b) Since h^d, f, and g^{-1} are injective, h^e is also injective. We now show that h^e preserves operations. For all $a, b \in L$

$$\begin{aligned}h^e(a \vee b) &= g^{-1}(-h^d(f(a \vee b))) \\ &= g^{-1}(-h^d(f(a) \wedge f(b))) \\ &= g^{-1}(-(h^d(f(a)) \wedge^d_{X_L} h^d(f(b)))) \\ &= g^{-1}(-h^d(f(a)) \vee^e_{X_L} -h^d(f(b))) \\ &= g^{-1}(-h^d(f(a))) \vee^e_{X_L} g^{-1}(-h^d(f(b))) \\ &= h^e(a) \vee^e_{X_L} h^e(b).\end{aligned}$$

Similarly, $h^e(a \wedge b) = h^e(a) \wedge^e_{X_L} h^e(b)$. □

We now obtain the following (exhaustive) discrete representation for general lattices.

Theorem 18.2.11 *Every general lattice is embeddable into the exhaustive complex algebra of its exhaustive canonical frame.*

18.3 Non-disjoint representation of lattices

In [46] yet another representation of lattices is presented. In this section we present this result in the style we use for previous presentations, stressing a duality between lattices and some general frames.

Let L be a general lattice which is not necessarily bounded.

A *frame* is a structure $(X,<)$ where X is a non-empty set and $<$ is a transitive relation on X. Given a frame $(X,<)$, the two mappings $t,s: 2^X \to 2^X$ are defined, for any $A \subseteq X$, by

$$t(A) \stackrel{\text{def}}{=} [\![>]\!]A$$
$$s(A) \stackrel{\text{def}}{=} [\![<]\!]A$$

where $>$ stands for $<^{-1}$ and the sufficiency operator $[\![\,]\!]$ is defined as in Section 1.8. A set $A \subseteq X$ is called t-stable (resp. s-stable) whenever $ts(A)=A$ (resp. $st(A)=A$).

Lemma 18.3.1 *Let $(X,<)$ be a frame. For every $A \subseteq X$,*

(a) *$t(A)$ is $<$-decreasing and $s(A)$ is $<$-increasing;*
(b) *If $<$ is irreflexive, then $t(A) \subseteq -A$ and $s(A) \subseteq -A$;*
(c) *st and ts are closure operators on 2^X.*

Lemma 18.3.2 *Let X be a frame. The antitone maps s and t form a Galois connection between the lattice of $<$-increasing subsetsof X and the lattice of $<$-decreasing subsets of X.*

A *non-disjoint complex algebra of a frame X* is a structure

$$(L_X^{nd}, \vee_X^{nd}, \wedge_X^{nd})$$

where $L_X^{nd} = \{A \subseteq X : ts(A) = A\}$ and, for all $A, B \subseteq X$,

$$A \vee_X^{nd} B \stackrel{\text{def}}{=} t(s(A) \cap s(B))$$
$$A \wedge_X^{nd} B \stackrel{\text{def}}{=} A \cap B.$$

Lemma 18.3.3 *For all $A, B \subseteq X$,*

(a) *$A \vee_X^{nd} B$ is t-stable;*
(b) *If $A, B \in L_X^{nd}$ then so is $A \wedge_X^{nd} B$.*

Proposition 18.3.4 *The non-disjoint complex algebra of a frame is a general lattice.*

Let L be a lattice. By a *non-disjoint filter-ideal pair of L* we mean a pair (F_1, F_2) such that F_1 is filter of L, F_2 is an ideal of L, and $F_1 \cap F_2 \neq \emptyset$. The ordering on non-disjoint filter-ideal pairs is defined by

$$(F_1, F_2) <_L^{nd} (G_1, G_2) \stackrel{\text{def}}{\Leftrightarrow} F_1 \subseteq G_1.$$

A non-disjoint filter-ideal pair is called *principal* whenever it is a minimal element with respect to the partial order $\leqslant_L^{nd} = <_L^{nd} \cup =$.

Lemma 18.3.5 *Let $(L, \vee, \wedge)$ be a lattice. Then (F_1, F_2) is a principal filter-ideal pair of L if and only if there is some $a \in L$ such that $F_1 = \uparrow_{\leq} a$ and $F_2 = \downarrow_{\leq} a$.*

The *non-disjoint canonical frame of a general lattice* L is a structure

$$(X_L^{nd}, <_L^{nd})$$

where X_L^{nd} is a family of all principal non-disjoint filter-ideal pairs of a lattice L and $<_L^{nd}$ is defined as above.

Proposition 18.3.6 *The non-disjoint canonical frame of a general lattice is a frame.*

Let L be a lattice. Consider a mapping $h^{nd} : L \to L_{X_L^{nd}}^{nd}$ defined, for any $a \in L$, by

$$h^{nd}(a) \stackrel{\text{def}}{=} \{(F_1, F_2) \in X_L^{nd} : a \in F_1\}.$$

Lemma 18.3.7 *For every $a \in L$,*

(a) *$sh^{nd}(a) = \{(F_1, F_2) \in X_L^{nd} : a \in F_2\}$;*

(b) *$h^{nd}(a)$ is a t-stable set;*

(c) *h^{nd} is a lattice-embedding.*

Proof:
(a) First, note that for every $a \in L$,

$$F \in sh^{nd}(a) \Leftrightarrow F \in [\![<_L^{nd}]\!]h(a) \Leftrightarrow \forall G \in X_L^{nd},\ a \in G_1 \Rightarrow F_1 \subseteq G_1.$$

Take $F \in X_L^{nd}$. By Lemma 18.3.5, there is some $e \in L$ such that $F = (\uparrow_{\leq} e, \downarrow_{\leq} e)$. Assume that $F \in sh^{nd}(a)$. Then

$$\forall b \in L,\ a \in \uparrow_{\leq} b \Rightarrow \uparrow_{\leq} e \subseteq \uparrow_{\leq} b.$$

Taking b for a we get $a \leq e$, thus $a \in \downarrow_{\leq} e = F_2$. On the other hand, assume that $a \in F_2 = \downarrow_{\leq} e$. Thus $a \leq e$. Take any $b \in L$ such that $b \leq a$. Hence $b \leq e$, so $\uparrow_{\leq} e \subseteq \uparrow_{\leq} b$. Therefore, $a \in \uparrow_{\leq} b$ implies $\uparrow_{\leq} e \subseteq \uparrow_{\leq} b$ for any $b \in L$. By Lemma 18.3.5, b uniquely generates a principal filter-ideal pair, say G, we get $a \in G_1$ implies $F_1 \subseteq G_1$ for any $G \in X_L^{nd}$. Hence $F \in sh^{nd}(a)$.

(b) We have to show that $[\![>_L^{nd}]\!][\![<_L^{nd}]\!]h(a) = h(a)$. Note that for every $F \in X_L^{nd}$,

$$\begin{aligned} & F \in [\![>_L^{nd}]\!][\![<_L^{nd}]\!]h(a) \\ & \Leftrightarrow \forall G \in X_L^{nd},\ G \in [\![<_L^{nd}]\!]h(a) \Rightarrow G <_L^{nd} F \\ & \Leftrightarrow \forall G \in X_L^{nd},\ (\forall H \in X_L^{nd},\ a \in H_1 \Rightarrow G_1 \subseteq H_1) \Rightarrow G_1 \subseteq H_1. \end{aligned}$$

Let $F, G, H \in X_L^{nd}$. By Lemma 18.3.5, there exist $b \in L$, $c \in L$, and $e \in L$ such that $F = (\uparrow_{\leq} b, \downarrow_{\leq} b)$, $G = (\uparrow_{\leq} c, \downarrow_{\leq} c)$, and $H = (\uparrow_{\leq} e, \downarrow_{\leq} e)$. Hence

$$F \in [\![>_L^{nd}]\!][\![<_L^{nd}]\!]h(a) \Leftrightarrow \forall c \in L,\ (\forall e \in L,\ a \in \uparrow_{\leq} e \Rightarrow \uparrow_{\leq} c \subseteq \uparrow_{\leq} e) \Rightarrow \uparrow_{\leq} c \subseteq \uparrow_{\leq} b.$$

Assume $F \in [\![>_L^{nd}]\!][\![<_L^{nd}]\!]h(a)$. Then, taking a for c we get $\uparrow_{\leq} a \subseteq \uparrow_{\leq} b$, thus $a \in \uparrow_{\leq} b = F_1$. Hence $F \in h(a)$.

The reverse inclusion follows from Lemma 1.8.4(n).

(c) From Lemma 18.3.5 it follows that h^{nd} is injective. We show that h^{nd} preserves lattice operations. Note that for all $a, b \in L$,

$h^{nd}(a) \vee_{X_L^{nd}} h^{nd}(b)$	
$= t(s(\{F \in X_L^{nd} : a \in F_1\}) \cap s(\{F \in X_L^{nd} : b \in F_1\}))$	definition of $\vee^{nd}$
$= t(\{F \in X_L^{nd} : a \in F_2\} \cap \{F \in X_L^{nd} : b \in F_2\})$	Lemma 18.3.7(a)
$= t(\{F \in X_L^{nd} : a \vee b \in F_2\})$	F_2 is an ideal of L
$= t(s(\{F \in X_L^{nd} : a \vee b \in F_1\})$	Lemma 18.3.7(a)
$= ts(h^{nd}(a \vee b))$	definition of h^{nd}.
$= h^{nd}(a \vee b)$	Lemma 18.3.7(b).

Similarly, $h^{nd}(a) \vee_{X_L^{nd}} h^{nd}(b) = h^{nd}(a \vee b)$. □

We now obtain the following discrete representation of general lattices.

Theorem 18.3.8 *Every general lattice is embeddable into the non-disjoint complex algebra of its non-disjoint canonical frame.*

18.4 Non-exhaustive representation of lattices

In this section, as in Section 18.3, we consider general lattices which are not necessarily bounded.

A *frame* is a structure $(X, <)$ where $X \neq \emptyset$ and $<$ is a transitive relation on X. Given a frame $(X, <)$ the mappings $\tau : 2^X \to 2^X$ and $\sigma : 2^X \to 2^X$ are defined as follows,

$$\tau(A) \stackrel{\text{def}}{=} \langle\!\langle < \rangle\!\rangle A$$
$$\sigma(A) \stackrel{\text{def}}{=} \langle\!\langle > \rangle\!\rangle A.$$

A set $A \subseteq X$ is τ-stable (resp. σ-stable) whenever $\tau\sigma(A) = A$ (resp. $\sigma\tau(A) = A$).

Theorem 18.4.1 *For any $A \subseteq X$,*

$$\tau(A) = -t(-A)$$
$$\sigma(A) = -s(-A).$$

Therefore, dualising Lemma 18.3.1, we get the following facts.

Lemma 18.4.2 *Let $(X, <)$ be a frame. For all $A, B \subseteq X$*

(a) *$\tau(A)$ is $<$-increasing and $\sigma(A)$ is $<$-decreasing;*
(b) *τ and σ are antitone;*
(c) *$\sigma\tau$ and $\tau\sigma$ are interior operators on 2^X.*

Lemma 18.4.3 *Let X be a frame. The antitone maps σ and τ form a Galois connection between the lattice of $<$-increasing subsets of X and the lattice of $<$-decreasing subsets of X.*

The *non-exhaustive complex algebra of a frame* $(X, <)$ is a structure

$$(L_X^{ne}, \vee_X^{ne}, \wedge_X^{ne})$$

where $L_X^{ne} = \{A \subseteq X : \tau(\sigma(A)) = A\}$ and, for all $A, B \subseteq X$,

$$A \vee_X^{ne} B \stackrel{\text{def}}{=} A \cup B$$
$$A \wedge_X^{ne} B \stackrel{\text{def}}{=} \tau(\sigma(A) \cup \sigma(B)).$$

Since τ (resp. σ) and t (resp. s) are dual in the sense of Theorem 18.4.1, it follow that

Lemma 18.4.4 *For all $A, B \subseteq X$,*

$$-(A \vee_X^{ne} B) = -A \wedge_X^{nd} -B$$
$$-(A \wedge_X^{ne} B) = -A \vee_X^{nd} -B.$$

Hence, for any $A \subseteq X$, A is τ-stable if and only of $-A$ is t-stable.

Proposition 18.4.5 *The non-exhaustive complex algebra of a frame is a general lattice.*

A *non-exhaustive pair of general lattice* L is a pair (F_1, F_2) where $-F_1$ is a principal filter of L^{op} and $-F_2$ is a principal ideal of L^{op} such that $F_1 \cup F_2 \neq L$.

The *non-exhaustive canonical frame of a general lattice* L is a structure $(X_L^{ne}, <_L^{ne})$ where X_L^{ne} is the family of all non-exhaustive pairs of L and $<_L^{ne}$ is an ordering on X_L^{ne} defined, for non-exhaustive pairs (F_1, F_2) and (G_1, G_2) of L, by

$$(F_1, F_2) <_L^{ne} (G_1, G_2) \stackrel{\text{def}}{\Leftrightarrow} G_1 \subseteq F_1.$$

Proposition 18.4.6 *The non-exhaustive canonical frame of a general lattice is a frame.*

Consider any lattice L. Let $f : L \to L$ be defined as in Section 18.3 and let $j : X_L^{ne} \to X_L^{nd}$ be defined for $(F_1, F_2) \in X_L^{ne}$ as

$$j(F_1, F_2) \stackrel{\text{def}}{=} (-F_1, -F_2).$$

Also, define a mapping $h^{ne} : L \rightarrow L^{ne}_{X^{ne}_L}$ for every $a \in L$ by

$$h^{ne}(a) \stackrel{\text{def}}{=} j^{-1}(-h^{nd}(f(a))).$$

Since the definition of h^{ne} in terms of h^{nd} is analogous to that of h^e in terms of h^d, similar arguments to that in the proof of Lemma 18.2.10 show the following lemma.

Lemma 18.4.7 *For every $a \in L$,*

(a) *$h^{ne}(a)$ is a τ-stable set;*

(b) *h^{ne} is a lattice embedding.*

We conclude this section with the following non-exhaustive representation for general lattices.

Theorem 18.4.8 *Every general lattice is embeddable into the non-exhaustive complex algebra of its non-exhaustive canonical frame.*

18.5 Representation of lattices with two-sorted frames

Let $(L, \vee, \wedge, 0, 1)$ be a general lattice. A *lattice frame* is a structure (X, Y, R) such that X and Y are non-empty sets and $\emptyset \neq R \subseteq X \times Y$. Such a frame will be denoted by XY. We extend the definition of sufficiency operators (see Section 1.8) to the operators determined by a heterogenous relation R. For every $A \subseteq X$ and every $B \subseteq Y$, we define

$$[\![R^{-1}]\!]A \stackrel{\text{def}}{=} \{y \in Y : \forall x \in X,\ x \in A \Rightarrow xRy\}$$
$$[\![R]\!]B \stackrel{\text{def}}{=} \{x \in X : \forall y \in Y,\ y \in B \Rightarrow xRy\}.$$

In [14] these operators are referred to as *polarities.* Let

$$f_R \stackrel{\text{def}}{=} [\![R]\!][\![R^{-1}]\!]$$
$$g_R \stackrel{\text{def}}{=} [\![R^{-1}]\!][\![R]\!].$$

Lemma 18.5.1 *Let (X, Y, R) be a lattice frame. Then*

(a) *$[\![R]\!]$ and $[\![R^{-1}]\!]$ form a Galois connection between the posets $(2^X, \subseteq)$ and $(2^Y, \subseteq)$, that is, for any $A \subseteq X$ and for any $B \subseteq Y$, $A \subseteq [\![R]\!]B$ if and only if $B \subseteq [\![R^{-1}]\!]A$;*

(b) *$[\![R]\!]$ and $[\![R^{-1}]\!]$ are antitone, which implies that they transform join into meet;*

(c) *f_R and g_R are closure operators on 2^X and 2^Y, respectively, that is, they are extensive, monotone, and idempotent.*

Let $L_X = \{A \subseteq X : f_R(A) = A\}$ and $L_Y\{B \subseteq Y : g_R(B) = B\}$. We define the following operations and constants on L_X and L_Y. For $A, A' \in L_X$,

$$A \vee_X A' = f_R(A \cup A') \qquad 0_X = f_R(\emptyset)$$
$$A \wedge_X A' = A \cap A' \qquad 1_X = X\,.$$

For $B, B' \in L_Y$,

$$B \vee_Y B' = B \cap B' \qquad 0_Y = Y$$
$$B \wedge_Y B' = g_R(B \cup B') \qquad 1_Y = g_R(\emptyset)\,.$$

Lemma 18.5.2

(a) $(L_X, \vee_X, \wedge_X, 0_X, 1_X)$ *and* $(L_Y, \vee_Y, \wedge_Y, 0_Y, 1_Y)$ *are general lattices;*

(b) *The natural order on* L_X *and* L_Y *is set inclusion* $\subseteq$ *and its converse* $\supseteq$, *respectively.*

Proof:
By way of example we show (a) for L_X. First, we show that L_X is closed on the operations. Since f_R is idempotent, L_X is closed on $\vee_X$. Since f_R is extensive, $A \cap A' \subseteq f_R(A \cap A')$. Consequently, since $A \cap A' \subseteq A$, $A \cap A' \subseteq A'$, and f_R is monotone, $f_R(A \cap A') \subseteq f_R(A) \cap f_R(A') = A \cap A'$. Satisfaction of the lattice axioms in L_X is by an easy verification. For example, $A \vee_X (A \wedge_X A') = f_R(A \cup (A \cap A')) = f_R(A) = A$, and hence the absorption law holds. □

The *complex algebra of a lattice frame* X is a structure $(L_{XY}, \vee_{XY}, \wedge_{XY})$ where $L_{XY} = L_X \times L_Y$ and for all (A, B), $(A', B') \in L_X \times L_Y$,

$$(A, B) \vee_{XY} (A', B') = (A \vee_X A', B \vee_Y B')$$
$$(A, B) \wedge_{XY} (A', B') = (A \wedge_X A', B \wedge_Y B')$$
$$0_{XY} = (0_X, 0_Y)$$
$$1_{XY} = (1_X, 1_Y).$$

Lemma 18.5.3

(a) L_{XY} *is a general lattice;*

(b) *The natural order of* L_{XY} *is, for all* (A, B), $(A', B') \in L_X \times L_Y$,

$$(A, B) \leqslant_{XY} (A', B') \Leftrightarrow A \subseteq A' \text{ and } B' \subseteq B\,.$$

Example 18.5.4 *Let* $R = X \times Y$. *If* $A \subseteq X$ *and* $B \subseteq Y$, *then* $f_R(A) = X$ *and* $g_R(B) = Y$. *Thus* $L_X = \{X\}$, $L_Y = \{Y\}$, *and* $L_{XY} = \{(X, Y)\}$.

Example 18.5.5 [14] *Let* $X = Y$ *and let* R *be the diversity relation on* X, *that is* xRy *if and only if* $x \neq y$. *Then for every* $A \subseteq X$, $A = f_R(A)$. *Indeed, since* f_R *is extensive,* $A \subseteq f_R(A)$. *Conversely,*

$$\begin{aligned} x \in f_R(A) &\Leftrightarrow x \in [\![\neq]\!][\![\neq]\!]A \\ &\Leftrightarrow \forall y,\ y \in [\![\neq]\!]A \Rightarrow x \neq y \\ &\Leftrightarrow \forall y,\ ((\forall z,\ z \in A \Rightarrow y \neq z) \Rightarrow x \neq y) \\ &\Leftrightarrow \forall y(x = y \Rightarrow \exists z(y = z \ \&\ z \in A)). \end{aligned}$$

Taking x *for* y *we get* $x = z$ *and* $z \in A$ *for some* z, *which yields* $x \in A$.

The *canonical frame of a general lattice* L is a structure (X_L, Y_L, R_L) such that X_L is the set of filters of L, Y_L is the set of ideals of L, and for $F \in X_L$, $I \in Y_L$,

$$F\, R_L I \Leftrightarrow F \cap I \neq \emptyset .$$

Clearly, the canonical frame of a lattice is a lattice frame.

Let the mappings $h_1 : L \to 2^{X_L}$ and $h_2 : L \to 2^{Y_L}$ be defined, for every $a \in L$, by

$$h_1(a) = \{F \in X_L : a \in F\}$$
$$h_2(a) = \{I \in Y_L : a \in I\}.$$

We define $h : L \to L_{X_L Y_L}$, for every $a \in L$, by

$$h(a) \stackrel{\text{def}}{=} (h_1(a),\ h_2(a)).$$

Lemma 18.5.6 *For every* $a \in L$,

(a) $h_1(a) = f_{R_L}\big(h_1(a)\big)$;

(b) $h_2(a) = g_{R_L}\big(h_2(a)\big)$.

Proof:
For (a), note that

$$\begin{aligned} F \in f_{R_L}\big(h_1(a)\big) &\Leftrightarrow F \in [\![R_L]\!][\![R_L^{-1}]\!]h_1(a) \\ &\Leftrightarrow \forall I \in Y_L,\ I \in [\![R_L^{-1}]\!]h_1(a) \Rightarrow F\, R_L I \\ &\Leftrightarrow \forall I \in Y_L,\ (\forall H \in X_L,\ H \in h_1(a) \Rightarrow I\, R_L^{-1} H) \Rightarrow F\, R_L I \\ &\Leftrightarrow \forall I \in Y_L,\ (\forall H \in X_L,\ a \in H \Rightarrow I \cap H \neq \emptyset) \Rightarrow F \cap I \neq \emptyset . \end{aligned}$$

Since f_{R_L} is extensive, $h_1(a) \subseteq f_{R_L}\big(h_1(a)\big)$. Conversely, assume $F \in f_{R_L}\big(h_1(a)\big)$ and consider $I = {\downarrow_{\leq}}\, a$. Then for every $H \in X_L$, $a \in H$ implies ${\downarrow_{\leq}}\, a \cap H \neq \emptyset$ which yields $F \cap {\downarrow_{\leq}}\, a \neq \emptyset$. Thus there is some $b \in L$ such that $b \in F$ and $b \leq a$ which imply $a \in F$. Hence $F \in h_1(a)$. The proof of (b) is analogous. □

Lemma 18.5.7 *For all* $a, b \in L$,

(a) $h_1(a) \vee_{X_L} h_1(b) = h_1(a \vee b)$;

(b) $h_2(a) \vee_{Y_L} h_2(b) = h_2(a \vee b)$.

Proof:
For (a), note that

$$\begin{aligned} & F \in h_1(a) \vee_{X_L} h_1(b) \\ &\quad \Leftrightarrow F \in f_{R_L}\big(h_1(a) \cup h_1(b)\big) \\ &\quad \Leftrightarrow F \in [\![R_L]\!][\![R_L^{-1}]\!]\big(h_1(a) \cup h_1(b)\big) \\ &\quad \Leftrightarrow \forall J \in Y_L,\ J \in [\![R_L^{-1}]\!]\big(h_1(a) \cup h_1(b)\big) \Rightarrow F\, R_L J \\ &\quad \Leftrightarrow \forall J \in Y_L,\ \big(\forall H \in X_L,\ H \in h_1(a) \cup h_1(b) \Rightarrow J\, R_L^{-1} H\big) \Rightarrow F\, R_L J \\ &\quad \Leftrightarrow \forall J \in Y_L,\ (\forall H \in X_L,\ a \in H \text{ or } b \in H \Rightarrow J \cap H \neq \emptyset) \Rightarrow F \cap J \neq \emptyset . \end{aligned}$$

Assume $F \in h_1(a \vee b)$, that is, $a \vee b \in H$. Take $J \in Y_L$ such that

$$\forall H \in X_L,\ a \in H \ \text{ or } \ b \in H \Rightarrow H \cap J \neq \emptyset .$$

Consider $H = \uparrow_{\leq} a$. Then $a \in J$. Similarly, for $H = \uparrow_{\leq} b$ we get $b \in J$. Thus $a \vee b \in J$ and hence $F \cap J \neq \emptyset$. It follows that $F \in f_{R_L}\big(h_1(a) \cup h_1(b)\big)$. Conversely, assume $F \in f_{R_L}\big(h_1(a) \cup h_1(b)\big)$ and consider $J = \downarrow_{\leq} (a \vee b)$. Then

$$\big(\forall H \in X_L,\ a \in H \ \text{ or } \ b \in H \Rightarrow \downarrow_{\leq} (a \vee b) \cap H \neq \emptyset\big) \Rightarrow F \cap \downarrow_{\leq} (a \vee b) \neq \emptyset .$$

Thus there is some $c \in L$ such that $c \leq a \vee b$ which yields $a \vee b \in F$, as required.

For (b), note that

$$\begin{aligned} I \in h_2(a) \vee_{Y_L} h_2(b) &\Leftrightarrow I \in h_2(a) \cap h_2(b) \\ &\Leftrightarrow a \in I \ \&\ b \in I \\ &\Leftrightarrow a \vee b \in I \\ &\Leftrightarrow I \in h_2(a \vee b) . \end{aligned}$$

□

Analogously, we can prove the following lemma.

Lemma 18.5.8 *For any $a, b \in L$,*

(a) $h_1(a) \wedge_{X_L} h_1(b) = h_1(a \wedge b)$;

(b) $h_2(a) \wedge_{Y_L} h_2(b) = h_2(a \wedge b)$.

Lemma 18.5.9

(a) $h_1(0) = 0_{X_L}$, $h_1(1) = 1_{X_L}$;

(b) $h_2(0) = 0_{Y_L}$, $h_2(1) = 1_{Y_L}$.

Proof:
Note that for every $F \in X_L$, $F \in h_1(0)$ iff $F = L$ iff $\forall I \in Y_L$, $F \cap I \neq \emptyset$ iff $F \in f_{R_L}(\emptyset)$ iff $F \in 0_{X_L}$. Furthermore, $h(1) = \{F \in X_L : 1 \in F\} = X_L = 1_{X_L}$. Similarly, $h_2(0) = \{I \in Y_L : 0 \in I\} = Y_L = 0_{Y_L}$. Now, for every $I \in Y_L$, $I \in h_2(1)$ iff $I = L$ iff $\forall F \in X_L$, $I \cap F \neq \emptyset$ iff $I \in g_{R_L}(\emptyset)$ iff $i \in 1_{Y_L}$. □

Now we show the following facts.

Lemma 18.5.10

(a) $h(a \vee b) = h(a) \vee_{X_L Y_L} h(b)$;

(b) $h(a \wedge b) = h(a) \wedge_{X_L Y_L} h(b)$;

(c) $h(0) = 0_{X_L Y_L}$;

(d) $h(1) = 1_{Y_L Y_L}$;

(e) *h is injective.*

Proof:
(a) and (b) follow from Lemma 18.5.7 and Lemma 18.5.8, respectively, and (c), (d) from Lemma 18.5.9. For (e) assume $h(a) = h(b)$. Then for every $F \in X_L$ and for every $I \in Y_L$, $a \in F \cap I$ if and only if $b \in F \cap I$. In particular, $a \in \uparrow_{\leq} a \cap \downarrow_{\leq} a$ if and only if $b \in \downarrow_{\leq} a \cap \uparrow_{\leq} a$ and hence $a = b$ □

Theorem 18.5.11 *Every general lattice is embeddable into the complex algebra of its two-sorted canonical frame.*

Example 18.5.4 shows that, in general, a frame (X, Y, R) cannot be embedded into the canonical frame of its complex algebra.

18.6 Dedekind-MacNeille representations

Let $(X, \leqslant)$ be a poset and let $L_X = \{A \subseteq X : A = [\![\leqslant]\!][\![\geqslant]\!]A\}$. Following the notation established in Section 18.5, $L_X = \{A \subseteq X : A = f_{\leqslant}(A)\}$. Consider the poset $(L_X, \subseteq)$ and let $m : X \to L_X$ be a mapping defined for every $x \in L$ as $m(x) = \downarrow_{\leqslant} x = \{y \in X : y \leqslant x\}$.

Theorem 18.6.1 *The mapping m is an order-embedding.*

Proof:
First, we show that m is well defined, that is $m(x) = f_{\leqslant}(m(x))$. Since $f_{\leqslant}$ is a closure operator, $m(x) \subseteq f_{\leqslant}(m(x))$. Conversely, note that $[\![\geqslant]\!]\downarrow_{\leqslant} x = \uparrow_{\leqslant} x = \{y \in X : x \leqslant y\}$. Then $f_{\leqslant}(\uparrow_{\leqslant} x) = \uparrow_{\leqslant} x$ as required. It is easy to see that m is order preserving, $x \leqslant y$ iff $\downarrow_{\leqslant} x \subseteq \downarrow_{\leqslant} y$, and one-one. □

L_X can be made into a complete lattice with the operations of infinite meet and join defined, for every family $\{A_i : i \in I\}$ of subsets of X, by

$$\bigwedge_{i \in I} A_i = \bigcap_{i \in I} A_i \text{ and } \bigvee_{i \in I} A_i = f_{\leqslant}(\bigcup_{i \in I} A_i),$$

respectively.

An equivalent representation of posets can be constructed with Dedekind cuts. Given a poset $(X, \leqslant)$ and $A, B \subseteq X$, a *cut* is a pair (A, B) such that $A = [\![\leqslant]\!]B$ and $B = [\![\geqslant]\!]A$. In some papers, cuts are referred to as Galois-stable pairs of sets. This is motivated by the fact that for any binary relation R the operators $[\![R]\!]$ and $[\![R^{-1}]\!]$ are a Galois pair, see Section 1.4 and Section 1.8.

If (A, B) is a cut, then $A = f_{\leqslant}(A)$. Conversely, if $A = f_{\leqslant}(A)$ then $(A, [\![\leqslant]\!]A)$ is a cut. Let $Cut(X)$ be the family of cuts of X. Define an ordering relation $\leqslant_{Cut(X)}$ on the set $Cut(X)$ by

$$(A, B) \leqslant_{Cut(X)} (A', B') \text{ iff } A \subseteq A'.$$

Note that $(A, B) \leqslant_{Cut(X)} (A', B')$ iff $B' \subseteq B$.

Consider a poset $(Cut(X), \leqslant_{Cut(X)})$ and a mapping $l : X \to Cut(X)$ defined as $l(x) = (\downarrow_{\leqslant} x, \uparrow_{\leqslant} x)$. It is easy to see that

Theorem 18.6.2 *The mapping l is an order-embedding.*

Also in this case $Cut(X)$ can be made into a complete lattice. For any family $\{(A_i, B_i) : i \in I\}$ of cuts of X define

$$\bigwedge\{(A_i, B_i) : i \in I\} = \Big(\bigcap_{i\in I} A_i,\ [\![\geqslant]\!](\bigcap_{i\in I} A_i)\Big)$$

$$\bigvee\{(A_i, B_i) : i \in I\} = \Big([\![\leqslant]\!](\bigcap_{i\in I} B_i),\ \bigcap_{i\in I} B_i\Big).$$

This provides a means for the Dedekind-MacNeille representation of lattices. It is known that for every Boolean algebra its MacNeille representation algebra is a Boolean algebra, but it is not necessarily the case for distributive lattices, the MacNeille representation algebra of a distributive lattice may not be distributive.

Examples of MacNeille representations for some lattice-based structures can be found in [121, 27, 234, 102], among others.

18.7 Canonical Extensions

Canonical extensions have their origin in the papers [139, 140]. In discrete dualities the counterparts of canonical extensions are complex algebras of canonical frames. While canonical extensions are a means of comprehending the Stone duality purely algebraically, their counterparts in discrete dualities are built from two structures of algebraic and logical origin, respectively.

The canonical extension of a Boolean algebra B is the powerset algebra

$$(2^{Ult(B)}, -, \cup, \cap, \emptyset, Ult(B))$$

where $Ult(B)$ is the set of ultrafilters of B. It is easy to see that it coincides with the complex algebra of the canonical frame of B described in Section 2.3.

Canonical extensions of Boolean algebras with n-ary operators which are normal and additive in each argument originated in [139]. In particular, with the method presented there one can obtain the canonical extension of a possibility algebra $(B, \Diamond)$. It is shown in Section 3.6 that the canonical extension of a possibility Boolean algebra coincides with the complex algebra of its canonical frame. Similar result is proved in that section for Boolean algebras with a sufficiency operator $(B, \boxminus)$. Clearly, this property extends to the algebras with n-ary operators that behave as possibility or sufficiency uniformly in each of their arguments. Canonical extensions for Boolean algebras with operators which are neither monotone nor antitone, such as the algebras considered in Section 4.5 are not known in the literature. Similarly, the concept of canonical extension is not applicable to the Boolean algebras considered in Section 3.6.

Canonical extensions for distributive lattices are introduced and studied in [103, 104, 105]. They are presented in a universal algebra setting based on topological representations of distributive lattices. As in the case of Boolean algebras, canonical extensions of distributive lattices can be shown to coincide with the complex algebras of their canonical frames considered in Section 2.3.

The inspiration for the notion of canonical extension of a general lattice can be traced back to representation theorems for lattices presented in [123, 47] and to the notion of an intermediate structure derived from the developments in [108], see also [107, 229]. Canonical extensions of general lattices presented in [101] are MacNeille completions of intermediate structures, see [100, 107, 229].

In this section we present canonical extensions of general lattices as complex algebras of their canonical frames in accordance with the framework developed in the book.

Let $(L, \vee, \wedge)$ be a general lattice and let $(X, \leqslant)$ be a general lattice frame such that $X \neq \emptyset$ and $\leqslant$ is a partial order on X.

The *complex algebra of a general lattice frame* X is a structure $(L_X, \vee_X, \wedge_X)$ where

$$L_X = \{A \subseteq X : A = f_{\leqslant}(A)\}$$

and for all $A, B \in L_X$,

$$A \vee_X B = f_{\leqslant}(A \cup B)$$
$$A \wedge_X B = A \cap B.$$

We recall that according to the notation in Section 18.5 $f_{\leqslant} = [\![\leqslant]\!][\![\leqslant^{-1}]\!]$ and for any binary relation R the pair of operators $[\![R]\!]$ and $[\![R^{-1}]\!]$ is a Galois pair, as discussed in Section 1.4 and Section 1.8. Note that the natural order on L_X is set inclusion, thus $(L_X, \subseteq)$ is the MacNeille representation poset for $(X, \leqslant)$, as discussed in Section 18.6.

Let $\mathcal{F}_0(L)$ and $\mathcal{I}_0(L)$ be the set of filters of L and the set of ideals of L, respectively. Consider a relation $\preccurlyeq$ on a disjoint union $\mathcal{F}_0(L) \uplus \mathcal{I}_0(L)$ defined, for all $F, G \in \mathcal{F}_0(L)$ and for all $I, J \in \mathcal{I}_0(L)$, by

$$F \preccurlyeq G \text{ iff } G \subseteq F$$
$$I \preccurlyeq J \text{ iff } I \subseteq J$$
$$F \preccurlyeq I \text{ iff } F \cap I \neq \emptyset$$
$$I \preccurlyeq F \text{ iff } \forall a \in I, \forall b \in F, a \leq b,$$

where $\leq$ is the natural order on L. The relation $\preccurlyeq$ is a partial order on $\mathcal{F}_0(L)$ and on $\mathcal{I}_0(L)$. However, it is not antisymmetric, and hence not a partial order, on $\mathcal{F}_0(L) \uplus \mathcal{I}_0(L)$ since taking a principal filter $\uparrow_{\leq} a$ and a principal ideal $\downarrow_{\leq} a$ we have

$$\uparrow_{\leq} a \preccurlyeq \downarrow_{\leq} a \quad \text{and} \quad \downarrow_{\leq} a \preccurlyeq \uparrow_{\leq} a \quad \text{but} \quad \uparrow_{\leq} a \neq \downarrow_{\leq} a.$$

Thus $(\mathcal{F}_0(L) \uplus \mathcal{I}_0(L), \preccurlyeq)$ cannot be taken as a canonical frame of L. We construct the canonical frame by defining an equivalence relation $\sim$ on $\mathcal{F}_0(L) \uplus \mathcal{I}_0(L)$, for any $A, B \in \mathcal{F}_0(L) \uplus \mathcal{I}_0(L)$, by

$$A \sim B \quad \text{iff} \quad A \preccurlyeq B \text{ and } B \preccurlyeq A.$$

Then the *canonical frame* of L is a structure $(X_L, \leqslant_L)$ where $X_L = (\mathcal{F}_0(L) \uplus \mathcal{I}_0(L)) |_\sim$ and $\leqslant_L$ is $\preccurlyeq |_\sim$. It is easy to see that $\leqslant_L$ is a partial order on X_L and hence the canonical frame of L is a lattice frame, as required. In the literature [108, 107, 229] this structure is referred to as the intermediate structure.

Then the complex algebra of the canonical frame of L, $(L_{X_L}, \vee_{X_L}, \wedge_{X_L})$, is its MacNeille representation algebra.

18.8 Dedekind-MacNeille-Wille representation of lattices

In [256] the Dedekind-MacNeille representation of lattices with cuts is generalized. We present it as an instance of the representation of lattices with two-sorted frames developed in Section 18.5.

Let $(L, \vee, \wedge, 0, 1)$ be a general lattice. The corresponding frames are two-sorted structures (X, Y, R) where $X, Y \neq \emptyset$, and $R \subseteq X \times Y$. By a *cut* in a frame (X, Y, R) we mean a pair (A, B) such that $A \subseteq X$, $B \subseteq Y$, $A = [\![R]\!]B$, and $B = [\![R^{-1}]\!]A$.

Lemma 18.8.1 *Let* (A, B) *be a cut in a frame* (X, Y, R)*. Then*

(a) $f_R(A) = [\![R]\!]B$;

(b) $g_R(B) = [\![R^{-1}]\!]A$.

Proof:
Since the operators f_R, g_R are extensive, and the sufficiency operator is antitone, we have

$$f_R(A) = f_R([\![R]\!]B) \supseteq [\![R]\!]B \supseteq [\![R]\!]\, g_R(B) = f_R(A)\,.$$

The proof of (b) is similar. □

Consider the complex algebra

$$(L_{XY}, \vee_{XY}, \wedge_{XY}, 0_{XY}, 1_{XY})$$

of a frame (X, Y, R) defined as in Section 18.5.

Lemma 18.8.2 *If* (A, B) *and* (A', B') *are cuts in* (X, Y, R)*, then*

(a) $(A, B) \vee_{XY} (A', B') = \big([\![R]\!](B \cap B'),\ B \cap B'\big)$;

(b) $(A, B) \wedge_{XY} (A', B') = \big(A \cap A',\ [\![R^{-1}]\!](A \cap A')\big)$;

(c) $0_{XY} = \big([\![R]\!]Y,\ Y\big)$;

(d) $1_{XY} = \big(X,\ [\![R^{-1}]\!]\, X\big)$;

(e) $(A, B) \leq_{XY} (A', B')$ *iff* $A \subseteq A'$ *(equivalently,* $B' \subseteq B$*).*

Proof:
This is an easy verification based on Lemma 18.8.1. □

The *Dedekind-MacNeille-Wille complex algebra*, DMNW-complex algebra, of a lattice frame (X, Y, R) is the structure $(C_{XY}, \vee_{XY}, 0_{XY}, 1_{XY})$ where C_{XY} is the set of cuts in (X, Y, R) and the operations and constants are defined as in Lemma 18.8.2. They form a proper subclass of the class of complex algebras considered in Section 18.5.

Let F_a (resp. I_a) be the principal filter (resp. the principal ideal) of a lattice L generated by $a \in L$.

The *Dedekind-MacNeille-Wille canonical frame* of L, DMNW-canonical frame, is a structure (X_L, Y_L, R_L) such that

$$X_L = \{F_a : a \in L\},\ Y_L = \{I_a : a \in L\},\ \text{and } F_a R_L I_b \text{ iff } a \leq b.$$

Note that $F_a\ R_L\ I_b$ holds iff $F_a \cap I_b \neq \emptyset$. These frames form a proper subclass of the class of canonical frames of lattices considered in Section 18.6.

Define $h : L \to C_{X_L Y_L}$ by $h(a) = (h_1(a), h_2(a))$ where

$$h_1(a) = \{F_b \in X_L : b \leq a\} \quad h_2(a) = \{I_b \in Y_L : a \leq b\}.$$

Lemma 18.8.3 *For all $a, b \in L$,*

(a) *h is a cut in the canonical frame of L;*

(b) $h(a \vee b) = h(a) \vee_{XY} h(b)$;

(c) $h(a \wedge b) = h(a) \wedge_{XY} h(b)$;

(d) $h(0) = 0_{XY}$, $h(1) = 1_{XY}$;

(e) *h is injective.*

Proof:
(a) First, we show $h_1(a) = [\![R_L]\!]h_1(a)$. Assume $b \leq a$ and take $c \in L$ such that $a \leq c$. Then, clearly, $b \leq c$. Conversely, assume $a \leq c$ implies $b \leq c$ for every $c \in L$. Taking a for c we get $b \leq a$. The proof of $h_2(a) = [\![R_L^{-1}]\!]h_2(a)$ is similar.
(b) We show $h_1(a \vee b) = [\![R_L]\!](h_2(a) \cap h_2(b))$. Note that $F_c \in [\![R_L]\!](h_1(a) \cap h_2(b))$ iff $\forall d \in L$, $a \leq d$ & $b \leq d \Rightarrow c \leq d$. Assume $F_c \in h_1(a \vee b)$. Then $c \leq a \vee b$. If $a \leq d$ and $b \leq d$, then since $a \vee b \leq d$, we get $c \leq d$. Conversely, taking $a \vee b$ for d we have $c \leq a \vee b$ as required. The proof of $h_2(a \vee b) = h_2(a) \cap h_2(b)$ is similar.
(c) We show $h_2(a \wedge b) = [\![R_L^{-1}]\!](h_1(a) \cap h_1(b))$. Note that $I_c \in [\![R_L^{-1}]\!](h_1(a) \cap h_1(b))$ iff $\forall d \in L$, $d \leq a$ & $d \leq b \Rightarrow d \leq c$. If $I_c \in h_2(a \wedge b)$, that is $a \wedge b \leq c$, then taking d such that $d \leq a$ and $d \leq b$ we obtain $d \in a \wedge b$, and hence $d \leq c$. Conversely, taking $a \wedge b$ for d we get $a \wedge b \leq c$, and hence $I_c \in h_2(a \wedge b)$. The proof of $h_2(a \wedge b) = h_1(a) \cap h_1(b)$ is similar.
(d) It is easy to verify that $h_1(b) = [\![R_L]\!]Y_L$ and $h_2(0) = Y_L$. Similarly, $h_1(1) = X_L$ and $h_2(1) = [\![R_L^{-1}]\!]X_L$.

(e) Assume $h(a) = h(b)$. Then for all $c, d \in L$, $a \in F_c \cap I_d$ iff $b \in F_c \cap I_d$. In particular, taking a for c and d we get $a = b$. □

We conclude with the representation theorem for DMNW-algebras.

Theorem 18.8.4 *Every general lattice is embeddable into DMNW-complex algebra of its DMNW-canonical frame.*

In formal concept analysis [256, 96] a special intuitive interpretation is given to the notions involved in the representation of lattices presented in this section.

In a lattice frame (X, Y, R) sets X and Y are interpreted as a set of objects and a set of attributes of those objects, respectively, and relation R holds between an object o and attribute a whenever o has the attribute a. Here the attributes are 2-valued, an object may have an attribute or not. With this interpretation, a frame is referred to as context. In information systems with incomplete information [191] attributes are treated as maps which assign a subset of values to an object. Such attributes are multivalued.

Given a context, its DMNW-complex algebra is a concept lattice, where cuts are treated as concepts. The operators $[\![R]\!]$ and $[\![R^{-1}]\!]$ are referred to as an extent and an intent operators, respectively.

18.9 Ortholattices

In [111], see also [110], a frame semantics was developed for an orthologic whose algebraic semantics is given in terms of ortholattices presented in Section 14.4. In [112] a topological representation of ortholattices is developed. In the present section a discrete version of that representation of O-lattices is presented such that a construction of the representation algebra is based on the frames considered in [111]. We recall that an O-lattice is a structure $(L, \vee, \wedge, \neg, 0, 1)$ such that $(L, \vee, \wedge, 0, 1)$ is a general lattice and the following axioms are satisfied

(O1) $a \wedge \neg a = 0$

(O2) $\neg\neg a = a$

(O3) $a \leqslant b \Rightarrow \neg b \leqslant \neg a$.

In order to distinguish between the frames associated to O-lattices in Section 14.4, O-frames, the frames considered in this section will be referred to as *O'-frames.* Specifically, an *O'-frame* is a structure $(X, \perp)$ where $X \neq \emptyset$ and $\perp$ is a binary relation on X which is irreflexive and symmetric. These structures are named *orthogonality spaces* in [85].

A set $A \subseteq X$ is called $\perp$*-closed* whenever $[\![\perp]\!][\![\perp]\!]A \subseteq A$.

Lemma 18.9.1 *Let* $(X, \perp)$ *be an O'-frame. For every* $A \subseteq X$, $A \subseteq [\![\perp]\!][\![\perp]\!]A$.

Proof:
Since $\perp$ is symmetric, $\perp = \perp^{-1}$. By Lemma 1.8.5(b), $[\![\perp]\!][\![\perp]\!]$ is a closure operator, and hence it is expanding. □

The *complex algebra of an O'-frame* X is a structure $(L_X, \vee_\perp, \cap, \neg_\perp, \emptyset, X)$ such that

$$L_X \stackrel{\text{def}}{=} \{A \subseteq X : A \text{ is } \perp\text{-closed}\}$$
$$A \vee_\perp B \stackrel{\text{def}}{=} \neg_\perp(\neg_\perp A \cap \neg_\perp B)$$
$$\neg_\perp A \stackrel{\text{def}}{=} [\![\perp]\!]A.$$

Lemma 18.9.2 *Let $(X, \perp)$ be an O'-frame. Then for all $A, B \subseteq X$, if A and B are $\perp$-closed, then so are $A \cap B$, $\neg_\perp A$, and $A \vee_\perp A$.*

Proof:
$(\cap)$ We show $[\![\perp]\!][\![\perp]\!](A \cap B) \subseteq A \cap B$. Since $A \cap B \subseteq A$ and $[\![\perp]\!][\![\perp]\!]$ is monotone as a closure operator, $[\![\perp]\!][\![\perp]\!](A \cap B) \subseteq [\![\perp]\!][\![\perp]\!]A \subseteq A$. Similarly, $[\![\perp]\!][\![\perp]\!](A \cap B) \subseteq B$. Thus the required inclusion holds.
$(\neg_\perp)$ By Lemma 18.9.1, $A \subseteq [\![\perp]\!][\![\perp]\!]A$. Since the sufficiency operator is antitone, we have $[\![\perp]\!][\![\perp]\!][\![\perp]\!]A \subseteq [\![\perp]\!]A$, and hence $\neg_\perp A$ is $\perp$-closed.
$(\vee_\perp)$ Since $\vee_\perp$ is definable in terms of $\neg_\perp$ and $\cap$, and L_X is closed on these operators, L_X is also closed on $\vee_\perp$. □

Proposition 18.9.3 *The complex algebra of an O'-frame is an O-lattice.*

Proof:
In view of Lemma 18.9.2 it is sufficient to show that L_X satisfies the lattice axioms and the axioms (O1) and (O2). Axiom (O3) follows from the antitonicity of the sufficiency operator.

Since the intersection of $\perp$-closed sets is $\perp$-closed, the sufficiency operators transform join to meet, and by Lemma 18.9.1 the lattice axioms follow easily. By way of example we show associativity of $\vee_\perp$. Note that

$$\begin{aligned}
A \vee_\perp (B \vee_\perp C) &= \neg_\perp(\neg_\perp A \cap \neg_\perp \neg_\perp(\neg_\perp B \cap \neg_\perp C)) \\
&= \neg_\perp(\neg_\perp A \cap (\neg_\perp B \cap \neg_\perp C)) \\
&= \neg_\perp((\neg_\perp A \cap \neg_\perp B) \cap \neg_\perp C) \\
&= \neg_\perp(\neg_\perp \neg_\perp(\neg_\perp A \cap \neg_\perp B) \cap \neg_\perp C) \\
&= \neg_\perp(\neg_\perp(A \vee_\perp B) \cap \neg_\perp C) \\
&= (A \vee_\perp B) \vee_\perp C.
\end{aligned}$$

(O1) Since $\perp$ is irreflexive, by Lemma 1.10.1(c), $\neg_\perp A \subseteq -A$, whence $A \cap \neg_\perp A = \emptyset$, as required.

(O2) Since A is $\perp$-closed, we get $\neg_\perp \neg_\perp A = A$. □

The *canonical frame of an O-lattice* L is a structure $(X_L, \perp_L)$ where $X_L = \{F \subseteq L : F \text{ is a proper filter of } L\}$ and for all $F, G \in X_L$,

$$F \perp_\neg G \stackrel{\text{def}}{\Leftrightarrow} \exists a \in L,\ \neg a \in F \ \&\ a \in G.$$

Lemma 18.9.4 *Let F be a proper filter of L. Then for every $a \in L$, $a \vee \neg a \in F$.*

Proof:
Since $F \neq \emptyset$, there is some $b \in F$. By (O1), $0 = a \wedge \neg a \leq \neg b$, so $b \leq \neg(a \wedge \neg a) = \neg a \vee \neg\neg a = \neg a \vee a$. □

Lemma 18.9.5 *Let F be a proper filter of L and let $\neg a \notin F$. Then there is a proper filter G of L such that $a \in G$ and for every $b \in L$, either $\neg b \notin F$ or $b \notin G$.*

Proof:
Let $G = \{c \in L : a \leq c\}$. Clearly, G is closed on $\wedge$. Also, G is a proper filter of L, for suppose $0 = b \wedge \neg b \in G$. Then $a \leq b \wedge \neg b$. Hence $\neg(b \wedge \neg b) = \neg b \vee \neg\neg b = \neg b \vee b \leq \neg a$. By Lemma 18.9.4, $\neg b \vee b \in F$. Since $\neg a \notin F$, we get a contradiction. Now suppose that for some $b \in L$, $\neg b \in F$ and $b \in G$. Since $b \in G$ and $a \leq b$, we have $\neg b \leq \neg a$. Since $\neg b \in F$, $\neg a \in F$, a contradiction. □

Proposition 18.9.6 *The canonical frame of an O-lattice is an O'-frame.*

Proof:
We show that $\perp_L$ is irreflexive and symmetric. Let F be a proper filter of L. Hence, there is no $a \in L$ such that $a \in F$ and $\neg a \in F$, since otherwise $a \wedge \neg a = 0 \in F$. Hence $F \perp_L F$ does not hold for any $F \in X_L$.
For symmetry of $\perp_L$, take $F, G \in X_L$ such that $F \perp_L G$. Then there is some $a \in L$ such that $\neg a \in F$ and $a \in G$. Let $b = \neg a$. Hence $\neg b = \neg\neg a = a \in G$ and $b \in F$. Therefore $G \perp_L F$ holds. □

Consider the mapping $h : L \to L_{X_L}$ defined, for any $a \in L$, by $h(a) = \{F \in X_L : a \in F\}$.

Theorem 18.9.7 *For every $a \in L$,*

(a) *h is injective;*
(b) *$h(a)$ is $\perp_\neg$-closed;*
(c) *$h(\neg a) = \neg_{\perp_\neg} h(a)$;*
(d) *$h(a \wedge b) = h(a) \cap h(b)$;*
(e) *$h(0) = \emptyset$ and $h(1) = X_L$.*

Proof:
(a) Since the canonical frame of L is an O'-frame, Lemma 18.9.1 applies to it. Since the natural order in L_{X_L} is set inclusion, h is injective.
(b) We show that $[\![\perp]\!][\![\perp]\!]h(a) \subseteq h(a)$. Note that

$$\begin{aligned} F \in [\![\perp]\!][\![\perp]\!]h(a) &\Leftrightarrow \forall G \in X_L,\ G \in [\![\perp]\!]h(a) \Rightarrow F \perp_\neg G \\ &\Leftrightarrow \forall G \in X_L\ ((\forall H \in X_L,\ H \in h(a) \Rightarrow G \perp_\neg G) \Rightarrow F \perp_\neg G). \end{aligned}$$

Suppose $a \notin F$. Then $\neg\neg a \notin F$. By Lemma 18.9.5 there is some $G_0 \in X_L$ such that $\neg a \in G_0$ and for every $b \in L$, either $\neg b \notin F$ or $b \notin G$. Taking G_0 for G we get

$$(\forall H \in X_L, H \in h(a) \Rightarrow G_0 \perp_\neg H) \ \Rightarrow\ F \perp_\neg G_0.$$

If $H \in X_L$ is such that $a \in H$, then $G_0 \perp_\neg H$, since $\neg a \in G_0$. It follows that $F \perp_\neg G_0$, which means that for some $c \in L$, $\neg c \in F$ and $c \in G_0$. Taking c for b we get a contradiction.

(c) Observe that

$$\begin{aligned} F \in \neg_{\perp_\neg} h(a) &\Leftrightarrow F \in [\![\perp_\neg]\!] h(a) \\ &\Leftrightarrow \forall G \in X_L,\ G \in h(a) \Rightarrow F \perp_\neg G \\ &\Leftrightarrow \forall G \in X_L,\ a \in G \Rightarrow (\exists b \in L,\ \neg b \in F \ \&\ b \in G). \end{aligned}$$

Assume that $\neg a \in F$ and take $G \in X_L$ such that $a \in G$. Then $F \in \neg_{\perp_\neg} h(a)$. Conversely, take any $F \in X_L$ with $\neg a \notin F$. By Lemma 18.9.5, there is some $G \in X_L$ such that for every $b \in L$ either $\neg b \notin F$ or $b \notin G$. Hence $F \notin \neg_{\perp_\neg} h(a)$.

Proofs of (d) and (e) are immediate. Preservation of $\vee$ follows from (c) and (d). □

We conclude with the following discrete representation for O-lattices.

Theorem 18.9.8 *Every O-lattice is embeddable into the complex algebra of its canonical frame.*

Now, consider the mapping $k : X \to X_{L_X}$ defined, for any $x \in X$, by $k(x) = \{A \in L_X : x \in A\}$.

Lemma 18.9.9 *For every $x \in X$, $k(x)$ is a proper filter of L_X.*

Proof:
It is easy to see that $k(x)$ is a filter of L_X. We only show that $k(x)$ is a proper filter of L. Suppose there is some $y \in X$ such that for every $A \in L_X$, $y \in A$. It follows that for every $A \subseteq X$, $[\![\perp]\!][\![\perp]\!] \subseteq A$ implies $y \in A$. Consider $B = \{z \in X : z \perp y\}$. To show that B is $\perp$-closed, assume that $z \in [\![\perp]\!][\![\perp]\!]B$. Then we have

$$\forall t \in X, (\forall u \in X,\ u \perp y \Rightarrow t \perp u) \ \Rightarrow\ z \perp t.$$

Take y for t. Since $\perp$ is symmetric, we get $z \perp y$. From irreflexivity of $\perp$ it follows that $y \notin B$, a contradiction. □

Lemma 18.9.10 *The mapping k preserves the relation $\perp$.*

Proof:
We show $x \perp y$ if and only if $k(x) \perp_{\neg_\perp} k(y)$. Note that

$$\begin{aligned} k(x) \perp_{\neg_\perp} k(y) &\Leftrightarrow \exists A \in L_X,\ \neg_\perp A \in k(x)\ \&\ A \in k(y) \\ &\Leftrightarrow \exists A \in L_X,\ x \in [\![\perp]\!] A\ \&\ y \in A \\ &\Leftrightarrow \exists A \in L_X,\ \forall z \in X,\ (z \in A \Rightarrow x \perp z)\ \&\ y \in A. \end{aligned}$$

Assume $k(x) \perp_{\neg_\perp} k(y)$. Since $y \in A$, taking y for z we get $x \perp y$. On the other hand, assume $x \perp y$. Then clearly the set $\{t \in X : x \perp t\}$ satisfies the required conditions. □

However, k is not injective as the following example ([54]) shows.

Example 18.9.11 *Let* $X = \{x, y, z\}$ *and* $\perp = \{(x,z), (y,z), (z,x), (z,y)\}$. *Then* $[\![\perp]\!]\{x\} = \{z\} = [\![\perp]\!]\{y\}$ *which implies* $f_\perp(\{x\}) = f_\perp(\{y\})$. *Then* L_X *consists of the sets* $\emptyset$, $\{z\}, \{x,y\}, \{x,y,z\}$ *which yields* $k(x) = k(y)$ *but* $x \neq y$.

Consequently, we do not have a representation theorem for *O'-frames*.

18.10 Representation of ortholattices with two-sorted frames

In this section we show that the representation theorem for ortholattices developed in [112] can be viewed as an instance of the representation theorem for lattices with two-sorted frames presented in Section 18.5.

Let $(L, \vee, \wedge, 0, 1, \neg)$ be an ortholattice. We define $\neg A = \{\neg a : a \in L\}$.

Lemma 18.10.1 *For any ortholattice* L *and for* $F, I \subseteq L$,

(a) F *is a filter of* L *iff* $\neg F$ *is an ideal of* L;
(b) *Every ideal of* L *has the form* $\neg F$ *for some filter* F *of* L;
(c) *Let* F *be a filter of* L. *Then* F *is proper iff* $\neg F$ *is proper.*

A *two-sorted orthoframe* is a structure $(X, X, \perp)$ such that $X \neq \emptyset$ and $\perp$ is an irreflexive and symmetric relation on X. Let $f_\perp$ and $g_\perp$ be defined as in Section 18.6. Clearly, since $\perp$ is symmetric, $f_\perp = g_\perp$.

By the *complex algebra of a two sorted orthoframe* X we mean the structure $(L_{XX}, \vee_{XX}, \wedge_{XX}, 0_{XX}, 1_{XX}, \neg_\perp)$ where, for any $A, B, A', B' \subseteq X$,

$$\begin{aligned} L_{XX} &= \{(A,B) : A = f_\perp(a) \text{ and } B = f_\perp(B)\}, \\ (A,B) \vee_{XX} (A'B') &= (f_\perp(A \cup A'),\ B \cap B'), \\ (A,B) \wedge_{XX} (A'B') &= (A \cap A',\ f_\perp(B \cup B')), \\ 0_{XX} &= (f_\perp(\emptyset), X)\ (= (\emptyset, X)), \\ 1_{XX} &= (X, f_\perp(\emptyset))\ (= (X, \emptyset)), \\ \neg_\perp(A,B) &= ([\![\perp]\!]A,\ [\![\perp]\!]B). \end{aligned}$$

By Lemma 18.5.2, $(L_{XX}, \vee_{XX}, \wedge_{XX}, 0_{XX}, 1_{XX})$ is a general lattice.

Lemma 18.10.2 *$\neg_\perp$ is an orthocomplement on L_{XX}, that is for all $A, B \subseteq X$*

(a) $(A,B) \wedge_{XX} \neg_\perp(A,B) = 0_{XX}$;
(b) $\neg_\perp \neg_\perp(A,B) = (A,B)$;
(c) $\neg_\perp$ *is antitone.*

Proof:
(a) Since, for every $C \subseteq X$, $[\![\perp]\!]C \subseteq -C$, and operator $f_\perp$ is extensive, it follow that

$$(A,B) \wedge_{XX} ([\![\perp]\!]A,\ [\![\perp]\!]B) = \big(A \cap [\![\perp]\!]A,\ f_\perp(B \cup [\![\perp]\!]B)\big) = (\emptyset, X) = 0_{XX}.$$

(b) Since $A = f_\perp(A)$ and $B = f_\perp(B)$, the required equality holds.

(c) follows from antitonicity of the sufficiency operator. □

Proposition 18.10.3 *The complex algebra of an orthoframe $(X, X, \perp)$ is an ortholattice.*

The *two-sorted canonical frame* of an ortholattice L is the structure

$$(X_L, X_L, \perp_L)$$

such that X_L is the family of proper filters of L and, for all $F, G \in X_L$,

$$F \perp_L G \ \Leftrightarrow\ \exists a \in L,\ \neg a \in F \text{ and } a \in G.$$

By an easy verification of irreflexivity and symmetry of $\perp_L$, we obtain

Proposition 18.10.4 *The canonical frame of an ortholattice L is a two-sorted orthoframe.*

Let Y_L be the family of proper ideals of L and let $R_L \subseteq X_L \times Y_L$ be defined as the relation in the canonical frame of a lattice in Section 18.5, that is, for every $F \in X_L$ and every $I \in Y_L$,

$$F R_L I \ \Leftrightarrow\ F \cap I \neq \emptyset.$$

Lemma 18.10.5 *For all $F \in X_L$, $I \in Y_L$, $F R_L I$ iff $F \perp_L (\neg I)$.*

Proof:
Note that

$$F R_L I \Leftrightarrow \exists a \in L, a \in F \ \&\ a \in I \Leftrightarrow \exists a \in L, a \in F \ \&\ \neg a \in \neg I.$$

Hence, $F R_L I \Leftrightarrow (\neg I) \perp_L F \Leftrightarrow F \perp_L (\neg I)$. □

Lemma 18.10.6 *Let $A \subseteq X_L$ and $B \subseteq Y_L$. Then*

(a) $f_{R_L}(A) = f_{\perp_L}(A)$;

(b) $g_{R_L}(B) = \neg f_{\perp_L}(\neg B)$.

Proof:
(a) Using symmetry of $\perp_L$ and Lemma 18.10.5, we have for every $F \in X_L$,

$$
\begin{array}{rcl}
F \in f_{\perp_L}(A) & \Leftrightarrow & F \in [\![\perp_L]\!][\![\perp_L]\!]A \\
& \Leftrightarrow & \forall G \in X_L, G \in [\![\perp_L]\!]A \Rightarrow F \perp_L G \\
& \Leftrightarrow & \forall G \in X_L, (\forall H \in X_L,\ H \in A \Rightarrow G \perp_L H) \Rightarrow F \perp_L G \\
& \Leftrightarrow & \forall \neg G \in Y_L, (\forall H \in X_L,\ H \in A \Rightarrow \neg G R_L^{-1} H) \Rightarrow F R_L \neg G \\
& \Leftrightarrow & F \in [\![R_L]\!][\![R_L^{-1}]\!]A \\
& \Leftrightarrow & F \in f_{R_L}(A).
\end{array}
$$

(b) Let $I \in Y_L$.

$$
\begin{array}{rcl}
I \in g_{R_L}(B) & \Leftrightarrow & I \in [\![R_L^{-1}]\!][\![R_L]\!]B \\
& \Leftrightarrow & \forall H \in X_L, H \in [\![R_L]\!]B \Rightarrow I R_L^{-1} H \\
& \Leftrightarrow & \forall H \in X_L, (\forall J \in Y_L,\ J \in B \Rightarrow H R_L J) \Rightarrow H R_L I \\
& \Leftrightarrow & \forall H \in X_L, (\forall \neg J \in X_L,\ \neg J \in \neg B \Rightarrow H \perp_L (\neg J)) \Rightarrow H \perp_L (\neg I) \\
& \Leftrightarrow & \forall H \in X_L, H \in [\![\perp_L]\!](\neg B) \Rightarrow H \perp_L (\neg I) \\
& \Leftrightarrow & \neg I \in [\![\perp_L]\!][\![\perp_L]\!](\neg B) \\
& \Leftrightarrow & \neg I \in f_{\perp_L}(\neg B) \\
& \Leftrightarrow & I \in \neg f_{\perp_L}(\neg B).
\end{array}
$$

□

Let $h : L \to L_{X_L X_L}$ be defined, for any $a \in L$, by

$$h(a) = (h_1(a), h_1(a)), \text{ where } h_1(a) = \{F \in X_L : a \in F\}.$$

Lemma 18.10.7 *The mapping h is well defined.*

Proof:
Since $f_{\perp_L} = g_{\perp_L}$ it is sufficient to show that for every $a \in L$, $f_{\perp_L}(h_1(a)) = h_1(a)$. By Lemma 18.10.6(a), for every $A \subseteq X_L$, $A = f_{\perp_L}(A)$ iff $A = f_{R_L}(A)$. By Lemma 18.4.7, the required equality holds. □

Thus by Lemma 18.10.7 and Theorem 18.4.8 h is a lattice-embedding.

Lemma 18.10.8 *For every $a \in L$, $h(\neg a) = \neg_{\perp_L} h(a)$.*

Proof:
Note that $\neg_{\perp_L}(h(a) = ([\![\perp_L]\!]h_1(a),\ [\![\perp_L]\!]h_1(a))$, and for any $F \in X_L$,

$$F \in [\![\perp_L]\!]h_1(a) \text{ iff } \forall G \in X_L, a \in G \Rightarrow F \perp_L G.$$

Assume $\neg a \in F$ and take $G \in X_L$ such that $a \in G$. Then $F \perp_L G$. Conversely, consider $G = \uparrow_{\leq} a$. Clearly, $a \in G$. Thus $F \perp_L \uparrow_{\leq} a$ which means that for some $b \in L$, $\neg b \in F$ and $b \in \uparrow_{\leq} a$. It follows that $a \leq b$, and hence $\neg b \leq \neg a$. Since $\neg b \in F$, $\neg a \in F$, which yields $F \in h_1(\neg a)$. □

Theorem 18.10.9 *Every ortholattice is embeddable into the complex algebra of its two-sorted canonical frame.*

Bibliography

[1] A. Abramsky. A Cook's tour of the finitary non-well-founded sets. Invited Lecture at BCTCS, 1988.

[2] W. Ackermann. *Solvable Cases of the Decision Problem.* North Holland Publishing Company, 1954.

[3] M. Aiello, I. Pratt, and J. van Benthem, editors. *Handbook of Spatial Logic.* Springer, 2007.

[4] P.S. Alexandroff. Diskrete raüme. *Math Sbornik*, 2:501–518, 1937.

[5] G. Allwein and J. M. Dunn. Kripke models for linear logic. *J. Symb. Logic*, 58:514–545, 1993.

[6] G. Amati and F. Pirri. A uniform tableau method for intuitionistic modal logic. *Studia Logica*, 53:29–60, 1994.

[7] R. Balbes and P. Dwinger. *Distributive Lattices.* University of Missouri Press, Columbia, 1974.

[8] D. Becchio. Logique trivalente de Łukasiewicz. *Annales Scientifiques de l'Université de Clermont-Ferrand*, 66:33–83, 1978.

[9] G. Bezhanishvili. Varieties of monadic Heyting algebras, Part I. *Studia Logica*, 61(3):367–402, 1998.

[10] G. Bezhanishvili. Varieties of monadic Heyting algebras, Part II. *Studia Logica*, 62(1):21–48, 1999.

[11] G. Bezhanishvili, N. Bezhanishvili, D. Gabelaia, and A. Kurz. Bitopological duality for distributive lattices and Heyting algebras. *Mathematical Structures in Computer Science*, 20(3):359–393, 2010.

[12] A. Białynicki-Birula and H. Rasiowa. On constructible falsity in the constructive logic with strong negation. *Colloquium Mathematicum*, 6:287–310, 1958.

[13] M. Bianchi and F. Montagna. n-Contractive BL-logics. *Archive for Mathematical Logic*, 50(3):257–285, 2011.

[14] G. Birkhoff. *Lattice Theory*. American Mathematical Society Colloquium Publications 25 (3rd edition). American Mathematical Society, Providence, R.I., 1979. 1st edition in 1940.

[15] K. Blount and C. Tsinakis. The structure of residuated lattices. *Int. J. of Algebra Comput.*, 13(4):437–461, 2003.

[16] T. S. Blyth and M. F. Janowitz. *Residuation Theory*. Pergamon Press, New York, USA, 1972.

[17] M. Bousaniche and L. Cabrer. Completions in subvarieties of BL-algebras. In *Proceedings of the 40th IEEE International Symposium on Multiple-Valued Logic*, pages 89–92, 2010.

[18] D. Bridges and L. Vita. Apartness spaces as a framwork for constructive topology. *Annals of Pure and Applied Logic*, 119:61–83, 2003.

[19] C. Brink, D. Gabbay, and H. J. Ohlbach. Towards automating duality. *Computers and Mathematics with Applications*, 29(2):73–90, 1995.

[20] S. Celani. Notes on the representation of distributive modal algebras. *Miskolc Mathematical Notes*, 9:81–89, 2008.

[21] S. Celani. Topological duality for Boolean algebras with a normal n-ary monotonic operator. *Order*, 26:49–67, 2009.

[22] S. Celani and R. Jansana. A new semantics for positive modal logic. *Notre Dame Journal of Formal Logic*, 38(1):1–18, 1997.

[23] L. A. Chagrova. An undecidable problem in correspondence theory. *Journal of Symbolic Logic*, 56(4):1261–1272, 1991.

[24] C. C. Chen and G. Grätzer. Stone lattices II. Structure theorems. *Canadian Journal of Mathematics*, pages 895–903, 1969.

[25] L. Chin and A. Tarski. Distributive and modular laws in the arithmetic of relation algebras. *University of California Publications*, 1:341–381, 1951.

[26] A. Ciabattoni, F. Esteva, and L. Godo. T-norm based logics with n-contraction. *Neural Network World*, 12:453–460, 2002.

[27] A. Ciabattoni, N. Galatos, and K. Terui. MacNeille completions of FL-algebras. *Algebra Universalis*, 66:405–420, 2011.

[28] L. O. Cignoli, I. M. L. D'Ottaviano, and D. Mundici. *Algebraic Foundations of Many-Valued Reasoning*, volume 7 of *Trends in Logic*. Springer, Heidelberg, 1999.

[29] B. I. Clarke. A calculus of individuals based on 'connection'. *Notre Dame Journal of Formal Logic*, 22:204–218, 1981.

[30] A. Cohn and S. Hazarika. Qualitative spatial reasoning: An overview. *Fundamenta Informaticae*, 46(1–2):1–29, 2001.

[31] S. Comer. On connections between information systems, rough sets and algebraic logic. In C. Rauszer, editor, *Algebraic Methods in Logic and Computer Science*, volume 28, pages 117–124. Banach Center Publications, Warsaw, 1993.

[32] W. Conradie, V. Goranko, and D. Vakarelov. Algorithmic correspondence and completeness in modal logic i. the core algorithm sqema. *Logical Methods in Computer Science*, 2(1):1–26, 2006.

[33] B. A. Davey and H. Priestley. *Introduction to Lattices and Order*. Cambridge University Press, Cambridge, 1990.

[34] T. de Laguna. Point, line and surface as sets of solids. *Journal of Philosophy*, 19:449–461, 1922.

[35] A. de Morgan. On the syllogism: No. IV, and on the logic of relations. *Transactions of the Cambridge Philosophical Society*, 10:331–358, 1864.

[36] H. C. M. de Swart. *Game Theory and Social Choice*. Tilburg University Press, 1999.

[37] R. Dedekind. *Essays on the Theory of Numbers (contains English translations of Stetigkeit und irrationale Zahlen)*. Dover, (1963) 1901.

[38] S. Demri. A completeness proof for a logic with an alternative necessity operator. *Studia Logica*, 58:99–112, 1997.

[39] S. Demri. A logic with relative knowledge operators. *Journal of Logic, Language and Information*, 8:167–185, 1999.

[40] S. Demri and E. Orłowska. *Incomplete Information: Structure, Inference, Complexity*. EATCS Monographs in Teoretical Computer Science. Springer Verlag, Berlin, Heidelberg, 2002.

[41] G. Dimov and D. Vakarelov. Contact algebras and region-based theory of space: A proximity approach I. *Fundamenta Informaticae*, 74(2–3):209–249, 2006.

[42] K. Došen. Duality between modal algebras and neighbourhood frames. *Studia Logica*, 48(2):219–234, 1988.

[43] K. Došen. Negation in the light of modal logic. In D. M. Gabbay and H. Wansing, editors, *What is Negation?*, pages 77–86. Kluwer, Dordrecht, 1999.

[44] J. Dugundji. *Topology*. Allyn and Bacon, Boston, 1966.

[45] J. M. Dunn. Generalized orthonegation. In D. Gabbay and H. Wansing, editors, *Negation: A Notion in Focus*, pages 3–26. W. de Gruyter, Berlin, 1996.

[46] J. M. Dunn and G. M. Hardegree. *Algebraic Methods in Philosophical Logic*. Clarendon Press, Oxford, 2001.

[47] J. M. Dunn and C. Hartonas. Stone duality for lattices. *Algebra Universalis*, 37:391–401, 1997.

[48] M. Dunn. Positive modal logic. *Studia Logica*, 55(2):301–317, 1995.

[49] I. Düntsch. Rough relation algebras. *Fundamenta Informaticae*, 21:321–331, 1994.

[50] I. Düntsch. A logic for rough sets. *Theoretical Computer Science*, 179(1-2):427–436, 1997.

[51] I. Düntsch. Rough sets and algebras of relations. In E. Orłowska, editor, *Incomplete Information: Rough Set Analysis*, pages 95–108. Physica-Verlag, Heidelberg, New York, 1998.

[52] I. Düntsch. Contact relation algebras. In E. Orłowska and A. Szałas, editors, *Relational Methods for Computer Science Applications*, volume 65 of *Studies in Fuzziness and Soft Computing*, pages 113–133. Springer, Heidelberg, 2001.

[53] I. Düntsch, 2011. Private communication.

[54] I. Düntsch, 2013. Private communication.

[55] I. Düntsch and E. Orłowska. Beyond modalities: sufficiency and mixed algebras. In E. Orłowska and A. Szałas, editors, *Relational Methods in Algebra, Logic, and Computer Science*, pages 277–299. Physica Verlag, Heidelberg, 2001.

[56] I. Düntsch and E. Orłowska. A discrete duality between apartness algebras and apartness frames. *Journal of Applied Non-Classical Logics*, 18(2–3):209–223, 2008.

[57] I. Düntsch and E. Orłowska. An algebraic approach to preference relations. In H. C. M. de Swart, editor, *Relational and Algebraic Methods in Computer Science: 12th International Conference, RAMICS 2011 Rotterdam, The Netherlands, May 30 – June 3, 2011, Proceedings.*, volume 6663 of *Lecture Notes in Computer Science*, pages 141–147. Springer, 2011.

[58] I. Düntsch and E. Orłowska. Discrete dualities for double Stone algebras. *Studia Logica*, 99:127–142, 2011.

[59] I. Düntsch and E. Orłowska. Mixed algebras revisited, 2011. Preprint.

[60] I. Düntsch and E. Orłowska. Discrete duality for rough relation algebras. *Fundamenta Informaticae*, 127:35–47, 2013.

[61] I. Düntsch and E. Orłowska. Discrete representation of lattices with two-sorted frames. Manuscript, 2013.

[62] I. Düntsch and E. Orłowska. Discrete dualities for some algebras with relations. *Journal of Logical and Algebraic Methods in Programming*, 83(2):169–179, 2014.

[63] I. Düntsch, E. Orłowska, and A. M. Radzikowska. Lattice–based relation algebras and their representability. In H. C. M. de Swart et al, editor, *Theory and Applications of Relational Structures as Knowledge Instruments*, volume 2929 of *Lecture Notes in Computer Science*, pages 234–258. Springer–Verlag, 2003.

[64] I. Düntsch, E. Orłowska, and A. M. Radzikowska. Lattice–based relation algebras II. In H. C. M. de Swart et al, editor, *Theory and Applications of Relational Structures as Knowledge Instruments II*, volume 4342 of *Lecture Notes in Artificial Intelligence*, pages 267–289. Springer–Verlag, 2006.

[65] I. Düntsch, E. Orłowska, A. M. Radzikowska, and D. Vakarelov. Relational representation theorems for some lattice–based structures. *Journal of Relational Methods in Computer Science JoRMiCS*, 1:132–160, 2004.

[66] I. Düntsch, E. Orłowska, and I. Rewitzky. Structures with multirelations, their discrete dualities and applications. *Fundamenta Informaticae*, 100:77–98, 2010.

[67] I. Düntsch, E. Orłowska, and C. van Alten. Discrete dualities for n-potent MTLalgebras and 2-potent BLalgebras. Fuzzy Sets and Systems Available online 28 September 2014, ISSN 0165-0114, http://dx.doi.org/10.1016/j.fss.2014.09.014, 2014.

[68] I. Düntsch and D. Vakarelov. Region-based theory of discrete spaces: A proximity approach. *Annals of Mathematics and Artificial Intelligence*, 49:5–14, 2007.

[69] I. Düntsch and M. Winter. A representation theorem for Boolean contact algebras. *Theoretical Computer Science*, 347(3):498–512, 2005.

[70] I. Düntsch and M. Winter. Rough relation algebra revisited. *Fundamenta Informaticae*, 73:1–18, 2006.

[71] W. Dzik, E. Orłowska, and C. van Alten. Relational representation theorems for general lattices with negations. In R. Schmidt, editor, *Proceedings of 9th International Conference of Relational Methods in Computer Science*, volume 4136 of *Lecture Notes in Computer Science*, pages 162–176. Springer–Verlag, 2006.

[72] W. Dzik, E. Orłowska, and C. van Alten. Relational representation theorems for general lattices with negations: A survey. In H. C. M. de Swart, E. Orłowska, and G. Schmidt, editors, *Theory and Applications of Relational Structures as Knowledge Instruments II*, volume 4342 of *Lecture Notes in Artificial Intelligence*, pages 245–266. Springer, 2006.

[73] H. M. Edwards. *Galois Theory*. Springer, Berlin-New York, 1984.

[74] V. Efremovič. The geometry of proximity. *Matematiceskij Sbornik (New Series)*, 31:189–200, 1952.

[75] M. Erné, J. Koslowski, A. Melton, and G. E. Strecker. A primer on Galois connections. *Annals of the New York Academy of Science*, 704:103–125, 1993.

[76] L. Esakia. Topological Kripke models. *Soviet Mathematics – Doklady*, 15:147–151, 1974.

[77] L. Esakia. The modalized Heyting calculus: a conservative modal extension of the intuitionistic logic. *Journal of Applied Non–classical Logics*, 16(3–4):349–366, 2006.

[78] F. Esteva and L. Godo. Monoidal t–norm based logic: towards a logic for left–continuous t–norms. *Fuzzy Sets and Systems*, 124:271–288, 2001.

[79] F. Esteva, L. Godo, and A. Garcia-Cerdaña. On the hierarchy of t-norm based residuated fuzzy logics. In M. Fitting and E. Orłowska, editors, *Beyond Two: Theory and Applications of Multiple-Valued Logic*, volume 114 of *Studies in Fuzziness and Soft Computing*, pages 251–272. Physica Verlag, 2003.

[80] W. B. Ewald. Intuitionistic tense and modal logic. *The Journal of Symbolic Logic*, 51(1):39–70, 1986.

[81] R. Fagin, J. Y. Halpern, Y. Moses, and M. Y. Vardi. *Reasoning about Knowledge*. MIT Press, Cambridge, Mass., 1995.

[82] G. Fischer-Servi. Axiomatization of some intuitionistic modal logic. *Rendiconti del Seminario Matematico dell Universit'a Politecnica di Torino*, 42:179–194, 1984.

[83] G. Fisher-Servi. On modal logics with an intuitionistic base. *Studia Logica*, 36:141–149, 1977.

[84] G. Fisher-Servi. Semantics for a class of intuitionistic modal calculi. In M. Dalla Chiara, editor, *Italian Studies in the Philosophy of Science*, pages 59–72. Reidel, 1980.

[85] D. J. Foulis and C. H. Randall. Lexicographic orthogonality. *Journal of Combinatorial Theory*, 11:157–162, 1971.

[86] L. Fuchs. *Partially-Ordered Algebraic Systems*. Pergamon Press, Oxford, 1963.

[87] H. Furusawa, K. Nishizawa, and N. Tsumagari. Multirelational models of lazy, monodic tree, and probabilistic Kleene algebras. *Bulletin of Informatics and Cybernetics*, 41:11–24, 2009.

[88] H. Furusawa and N. Nishizawa. Relational and multirelational representation theorems for complete idempotent semirings. In H. de Swart, editor, *Relational and Algebraic Methods in Computer Science. 12th International Conference, RAMICS–2011, Rotterdam, The Netherlands, May 30–June 3, 2011, Proceedings*, volume 6663 of *Lecture Notes in Computer Science*, pages 148–163. Springer, 2011.

[89] H. Furusawa, N. Tsumagari, and K. Nishizawa. A non-probabilistic relational model of probabilistic Kleene algebras. In R. Berghammer, B. Möller, and G. Struth, editors, *Relations and Kleene Algebra in Computer Science. 10th International Conference on Relational Methods in Computer Science and 5th International Conference on Applications of Kleene Algebra RelMiCS/AKA 2008, Frauenwörth, Germany, April 7–11, 2008. Proceedings*, volume 4988 of *Lecture Notes in Computer Science*, pages 110–122. Springer, 2008.

[90] D. Gabbay and H. J. Ohlbach. Quantifier elimination in second order predicate logic. *South African Computer Journal*, 7:35–43, 1992.

[91] D. M. Gabbay, R. A. Schmidt, and A. Szalas. Second-order quantifier elimination: Foundations, computational aspects and applications. *Studies in Logic: Mathematical Logic and Foundations*, 12, 2008.

[92] D. M. Gabbay and H. Wansing, editors. *What is negation?*, number 13 in Applied Logic Series. Kluwer Academic Publishers, Dordrecht, 1999.

[93] N. Galatos. *Varieties of Residuated Lattices.* PhD thesis, Vanderbilt University, 2003.

[94] N. Galatos, P. Jipsen, T. Kowalski, and H.Ono. *Residuated Lattices: An Algebraic Glimpse at Substructural Logics.* Studies in Logic and Foundations of Mathematics vol. 151. Elsevier, 2007.

[95] E. Galois. Memoire sur les conditions de résolubilité des équestions par radicaux. *Journal de Mathematiques Pures et Appliques*, 1846.

[96] B. Ganter and R. Wille. *Formal Concept Analysis: Mathematical Foundations.* Springer, Heidelberg, 1999.

[97] G. Gargov and S. Passy. A note on Boolean modal logic. In F. Petkov, editor, *Mathematical Logic. Proceedings of the Summer School and Conference dedicated to Arend Heyting*, pages 311–321, New York, 1988. Plenum Press.

[98] G. Gargov, S. Passy, and T. Tinchev. Modal environment for Boolean speculations. In D. Skordev, editor, *Mathematical Logic and Its Applications*, pages 253–263. Plenum Press, New York, 1987.

[99] G. Gediga and I. Düntsch. Modal-style operators in qualitative data analysis. In *Proceedings of the 2002 IEEE International Conference on Data Mining (ICDM'02)*, pages 155–162, 2002.

[100] M. Gehrke, J. M. Dunn, and A. Palmigiano. Canonical extensions and relational completeness of some substructural logics. *Journal of Symbolic Logic*, 70(3):713–740, 2005.

[101] M. Gehrke and J. Harding. Bounded lattice expansions. *Journal of Algebra*, 238(1):345–371, 2001.

[102] M. Gehrke, J. Harding, and Y. Venema. MacNeille completions and canonical extensions. *Transactions of the American Mathematical Society*, 358(2):573–590, 2005.

[103] M. Gehrke and B. Jónsson. Bounded distributive lattices with operators. *Mathematica Japonica*, 40(2):207–215, 1994.

[104] M. Gehrke and B. Jónsson. Monotone distributive lattice expansions. *Mathematica Japonica*, 52(2):197–213, 2000.

[105] M. Gehrke and B. Jónsson. Bounded distributive lattice expansions. *Mathematica Scandinavica*, 94:13–45, 2004.

[106] M. Gehrke, H. Nagahashi, and Y. Venema. A Sahlqvist theorem for distributive modal logic. *Annals of Pure and Applied Logic*, 131(1-3):65–102, 2005.

[107] M. Gehrke and H. A. Priestley. Canonical extensions and completions of posets and lattices. *Reports on Mathematical Logic*, 43:133–152, 2008.

[108] S. Ghilardi and G. Meloni. Constructive canonicity in non-classical logics. *Annals of Pure and Applied Logics*, 86:1–32, 1997.

[109] R. Goldbaltt. On closure under canonical embedding algebras. In H. Andreka, J. D. Monk, and I. Nemeti, editors, *Algebraic Logic*, volume 54 of *Colloquia Mathematica Societatis Janos Bolyai*, pages 217–229. North–Holland Publishing Co,, Amsterdam, 1991.

[110] R. Goldbaltt. Mathematic of Modality. CSLI Lecture Notes No 43, CSLI Publications, Stanford, CA, 1993.

[111] R. Goldblatt. Semantic analysis of orthologic. *Journal of Philosophical Logic*, 3:19–35, 1974.

[112] R. Goldblatt. The Stone space of an ortholattice. *Bulletin of the London Mathematical Society*, 7:45–48, 1975.

[113] R. Goldblatt. Varieties of complex algebras. *Annals of Pure and Applied Logic*, 44:303–316, 1989.

[114] V. Goranko. Modal definability in enriched language. *Notre Dame Journal of Formal Logic*, 31:81–105, 1990.

[115] G. Grätzer and E. Schmidt. On a problem of M. H. Stone. *Acta Mathematica Academiae Scientiarum Hungaricae*, 8:455–460, 1957.

[116] G. A. Grätzer. *Universal Algebra.* Springer–Verlag, New York, 1979.

[117] G. A. Grätzer. *General Lattice Theory.* Birkhäuser, Basel, 2 edition, 2000.

[118] P. Hájek. *Metamathematics of Fuzzy Logic.* Kluwer, Dordrecht, 1998.

[119] P. Hájek. Observations on the monoidal t-norm logic. *Fuzzy Sets and Systems*, 132:107–112, 2002.

[120] H. H. Hansen. Monotonic Modal Logic. Master's thesis, University of Amsterdam, 2003. Preprint 2003–2004 ILLC.

[121] J. Harding and G. Bezhanishvili. MacNeille completions of Heyting algebras. *Houston Journal of Mathematics*, 30(4):937–952, 2004.

[122] C. Hartonas and J. M. Dunn. Stone duality for lattices. *Algebra Universalis*, 37:391–401, 1997.

[123] G. Hartung. A topological representation of lattices. *Algebra Universalis*, 29:273–299, 1992.

[124] G. Hartung. An extended duality for lattices. In K. Denecke and H. J. Vogel, editors, *General Algebra and Applications*, pages 126–142. Heldermann-Verlag, Berlin, 1993.

[125] M. Haviar and H. Priestley. Canonical extensions of Stone and double Stone algebras. *Mathematica Slovaka*, 56:53–78, 2006.

[126] L. Henkin, J. D. Monk, and A. Tarski. *Cylindric Algebras, Part I*, volume 64 of *Studies in Logic and the Foundations of Mathematics.* North-Holland Publishing Co., Amsterdam-London, 1971.

[127] L. Henkin, J. D. Monk, and A. Tarski. *Cylindric Algebras, Part II*, volume 115 of *Studies in Logic and the Foundations of Mathematics.* North-Holland Publishing Co., Amsterdam, New York, 1985.

[128] D. Hilbert. Über das Unendliche. *Mathematische Annalen*, 95:161–190, 1926.

[129] J. Hintikka. *Knowledge and Belief.* Cornell University Press, London, 1962.

[130] U. Höhle. Commutative, residuated l–monoids. In U. Höhle and U. P. Klement, editors, *Non–Classical Logics and their Applications to Fuzzy Subsets*, pages 53–106. Kluwer Academic Publishers, Dordrecht, 1996.

[131] R. Horčik, C. Noguera, and M. Petrik. On n-contractive fuzzy logics. *Mathematical Logic Quarterly*, 53(3):268–288, 2007.

[132] I. Humberstone. Inaccessible words. *Notre Dame Journal of Formal Logic*, 24:346–352, 1983.

[133] L. Iturrioz. Symmetrical Heyting algebras with operators. *Zeitschrift für Mathematische Logik und Grundlagen der Mathematik*, 29:33–70, 1983.

[134] L. Iturrioz and E. Orłowska. A Kripke-style and relational semantics for logics based on Łukasiewicz algebras. *Multiple Valued Logic and Soft Computing*, 12(1-2):131–147, 2006.

[135] N. Ivanov and D. Vakarelov. A system of relational syllogistic incorporating full Boolean reasoning. *Journal of Logic, Language and Information*, 21:433–459, 2012.

[136] J. Järvinen and E. Orłowska. Relational correspondences for lattices with operators. *Lecture Notes in Computer Science*, 3929:134–146, 2006.

[137] S. Jenei and F. Montagna. A proof of standard completeness for Esteva-Godo logic MTL. *Studia Logica*, 70:183–192, 2002.

[138] I. Johansson. Minimalkalkül, ein Reduzierte Intuitionistischer Formalismus. *Compositio Mathematica*, 4:119–136, 1936.

[139] B. Jónsson and A. Tarski. Boolean algebras with operators. Part I. *American Journal of Mathematics*, 73:891–939, 1951.

[140] B. Jónsson and A. Tarski. Boolean algebras with operators. Part II. *American Journal of Mathematics*, 74:127–162, 1952.

[141] A. Jung, M. Kegelmann, and M.A. Moshier. Stable compact spaces and closed relations. In S. Brookes and M. Mislove, editors, *17th Conference on Mathematical Foundations of Programming Semantics, Electronic Notes in Theoretical Computer Science Volume 45*, pages 1–24, 2001.

[142] A. Jung and M.A. Moshier. On the bitopological nature of Stone duality. Schoof of Computer Science Research Report, 110 pages, 2006.

[143] A. Jung and Ph. Sünderhauf. On the duality of compact vs. open. In S. Andima, R.C. Flagg, G. Itzkowitz, P. Misra, Y. Kong, and R. Kopperman, editors, *11th Summer Conference at the University of Southern Maine, Annals of the New York Academy of Sciences Volume 806*, pages 214–230, 1996.

[144] S. Kanger, editor. *Completeness and Correspondence in the First and Second Order Semantics for Modal Logics*. North-Holland, Amsterdam, 1975.

[145] J. P. Katoen. *Quantitative and Qualitative Extensions of Event Structures*. PhD thesis, University of Twente, 1996.

[146] T. Katriňák. The structure of distributive double p-algebras. *Algebra Universalis*, 3:238–246, 1973.

[147] T. Katriňák. Construction of regular double p–algebras. *Bull. Soc. Roy. Sci. Liège*, 43:294–301, 1974.

[148] J.L. Kelley. *General Topology.* Van Nostrand, Princeton, 1955.

[149] T. Kowalski and T. Litak. Completions of GBL-algebras: negative results. *Algebra Universalis*, 58(4):373–384, 2008.

[150] W. Krull. Axiomatische Begründung der algemeinen Idealtheorie. *Sitzungsberichte der physikalischmedizinischen Societät zu Erlangen*, 56:47–63, 1924.

[151] J. Lambek. The mathematics of sentence structure. *American Mathematical Monthly*, 65(3):157–170, 1958.

[152] J. Lambek. Logic without structural rules (another look at cut elimination). In K. Dosen and P. Schroeder-Heister, editors, *Substructural Logics, Studies in Logic and Computation vol. 2*, pages 179–206. Oxford University Press, 1993.

[153] W. Lenzen. Recent work in epistemic logic. *Acta Philosophica Fennica*, 30:1–129, 1978.

[154] S. Leśniewski. Grundzüge eines neuen Systems der Grundlagen der Mathematik. *Fundamenta Mathematicae*, 14:1–81, 1929. English translation in: S. J. Surma, J. T. Srzednicki, D. I. Bernett, V. Rickey (eds), Stanisław Leśniewski: Collected Works, 1992.

[155] J. Łukasiewicz. *Aristotles Syllogistic from the standpoint of modern formal logic.* Clarendon Press, Oxford, 2nd edition edition, 1957.

[156] W. MacCaul and D. Vakarelov. Lattice-based paraconsistent logic. In W. MacCaul, M.Winter, and I. Düntsch, editors, *Relational Methods in Computer Science, 8th International Seminar on Relational Methods in Computer Science, 3rd International Workshop on Applications of Kleene Algebra, and Workshop of COST Action 274: TARSKI, St. Catharines, ON, Canada, February 22-26, 2005*, volume 3929 of *Lecture Notes in Computer Science*, pages 173–187. Springer, 2006.

[157] S. MacLane. Mathematical models: A sketch for the philosophy of mathematics. *The American Mathematical Monthly*, 88:462–472, 1981.

[158] S. MacLane. *Categories for the Working Mathematician.* Springer, 1997.

[159] H.M. MacNeille. Partially ordered sets. *Transactions of the American Mathematical Society*, 42(3):416460, 1937.

[160] R. Maddux. Some varieties containing relation algebras. *Transactions of the American Mathematical Society*, 272:501–526, 1982.

[161] R. Maddux. Finite integral relation algebras. *Lecture Notes in Mathematics*, 1149:175–197, 1985.

[162] R. Maddux. Introductory course on relation algebras, finite-dimensional cylindric algebras, and their interconnections. In H. Andreka, J. D. Monk, and I. Nemeti, editors, *Algebraic Logic*, volume 54 of *Colloq. Math. Soc. J. Bolyai*, pages 361–392. North Holland Publishing Co., Amsterdam, 1991.

[163] R. Maddux. Relation algebras. In C. Brink, W. Kahl, and G. Schmidt, editors, *Relational Methods in Computer Science*, Advances in Computer Science, pages 22–38. Springer, Wien, New York, 1997.

[164] R. Maddux. Relation algebras. In A. Abramsky, S. Artemov, and D. M. Gabbay, editors, *Studies in Logic and the Foundations of Mathematics*, volume 150. Elsevier, Amsterdam, 2006.

[165] J. J. Ch. Meyer and W. van der Hoek. *Epistemic Logic for AI and Computer Science*, volume 41 of *Cambridge Tracks in Theoretical Computer Science*. Cambridge University Press, Cambridge, 1995.

[166] G. C. Moisil. Recherches sur les logiques non chrysippiennes. *Annales Scientifiques de l'Université de Jassy*, 26:431–466, 1940.

[167] G. C. Moisil. *Essais sur les Logiques Non Chrysippiennes.* Editions de l'Academie de la Republique Socialiste de Roumanie, Bucharest, 1972.

[168] S. A. Naimpally and B. D. Warrack. *Proximity Spaces.* Cambridge University Press, Cambridge, 1972.

[169] A. Nonnengart, H.J. Ohlbach, and A. Szałas. Elimination of predicate quantifiers. In H. J. Ohlbach and U. Reyle, editors, *Logic, Language, and Reasoning, Trends in Logic vol. 5*, pages 149–171. Kluwer Academic Publishers, 1999.

[170] H. Ono. Semantics for substructural logics. In K. Došen and P. Schroeder-Heister, editors, *Substructural Logics*, pages 259–291. Oxford University Press, 1993.

[171] E. Orłowska. Representation of vague information. *ICS PAS Report*, 503, 1983.

[172] E. Orłowska. Semantics of vague concepts. In G. Dorn and P. Weingartner, editors, *Foundations of Logic and Linguistics. Problems and Their Solutions. Selected Contributions to the 70th Intenational Congress of Logic Methodology and Philosophy of Science, Salzburg, 1983*, pages 465–482. Plenum Press, New York and London, 1985.

[173] E. Orłowska. Logic for reasoning about knowledge. *Zeitschrift für Mathematische Logik und Grundlagen der Mathematik*, 35:556–568, 1989.

[174] E. Orłowska and J. Golińska-Pilarek. *Dual Tablaux: Foundations, Methodology, Case Studies*, volume 33 of *Trends in Logic*. Springer, 2011.

[175] E. Orlowska and A. Radzikowska. Discrete duality for some axiomatic extensions of MTL algebras. In P. Cintula, Z. Hanikova, and V. Svejdar, editors, *Witnessed Years: Essays in Honour of Petr Hajek*, pages 329–344. King's College Publications, 2009.

[176] E. Orłowska and A. M. Radzikowska. Relational representability for algebras of substructural logics. In W. MacCaull, M. Winter, and I. Düntsch, editors, *Relational Methods in Computer Science*, volume 3929 of *Lecture Notes in Computer Science*, pages 212–226. Springer–Verlag, 2006.

[177] E. Orłowska and A. M. Radzikowska. Representation theorems for some fuzzy logics based on residuated non–distributive lattices. *Fuzzy Sets and Systems*, 159:1247–1259, 2008.

[178] E. Orłowska and A. M. Radzikowska. Knowledge algebras and their discrete duality. In A. Skowron and Z. Suraj, editors, *Rough Sets and Intelligent Systems – Professor Pawlak in Memorium*, volume 43 of *Intelligent Systems Reference Library*, chapter 2, pages 7–20. Springer–Verlag Berlin Heidelberg, 2013.

[179] E. Orłowska and I. Rewitzky. Duality via truth: Semantic frameworks for lattice–based logics. *Logic Journal of the IGPL*, 13:467–490, 2005.

[180] E. Orłowska and I. Rewitzky. Context algebras, context frames and their discrete duality. In J. Peters, A. Skowron, and H. Rybiński, editors, *Transactions on Rough Sets IX*, volume 5390 of *Lecture Notes in Computer Science*, pages 212–220. Springer-Verlag, Berlin, Heidelberg, 2008.

[181] E. Orłowska and I. Rewitzky. Discrete duality for relation algebras and cylindric algebras. In B. Möller R. Berghammer, A. Jaoua, editor, *Relations and Kleene Algebra in Computer Science, Proceedings of the 11th International Conference on Relational Methods in Computer Science and the 6th International Conference on Applications of Kleene Algebra, Doha, Qatar, 2009*, volume 5827 of *Lecture Notes in Computer Science*, pages 291–305. Springer-Verlag, Berlin, Heidelberg, 2009.

[182] E. Orłowska and I. Rewitzky. Algebras for Galois–style connections and their discrete duality. *Fuzzy Sets and Systems*, 161(9):1325–1342, 2010.

[183] E. Orłowska, I. Rewitzky, and I. Düntsch. Relational semantics through duality. In W. MacCaull, M. Winter, and I. Düntsch, editors, *Relational Methods in Computer Science*, volume 3929 of *Lecture Notes in Computer Science*, pages 17–32. Springer–Verlag, 2006.

[184] E. Orłowska and D. Vakarelov. Lattice–based modal algebras and modal logics. In P. Hájek, L. Valdes, and D. Westerstahl, editors, *Logic, Methodology and Philosophy of Science*, pages 147–170. College Publications, King's College London, 2005.

[185] P. Pagliani. Rough sets and Nelson algebras. *Fundamenta Informaticae*, 27(2–3):205–219, 1996.

[186] P. Pagliani and M. Chakraborty. *A Geometry of Approximation*, volume 27 of *Trends in Logic*. Springer, 2008.

[187] A. Palmigiano. Dualities for intuitionistic modal logic. Preprint, 2007.

[188] R. Parikh. Knowledge and the problem of logical omniscience. In Z. W. Raś and M. Zemankova, editors, *Methodologies for Intelligent Systems, Proceeding of the second international conference, Charlotte, North Carolina, USA October 14-17 1987*, pages 432–439. North–Holland/Elsevier, 1987.

[189] R. Parikh. Logical omniscience. In D. Leivant, editor, *Logic and Computational Complexity*, volume 960 of *Lecture Notes in Computer Science*, pages 22–29. Springer–Verlag, 1995.

[190] M. Pauly. Formal methods and the theory of social choice (invited talk). In R. Berghammer, B. Möller, and G. Struth, editors, *Relations and Kleene Algebra in Computer Science. 10th International Conference on Relational Methods in Computer Science and 5th International Conference on Applications of Kleene Algebra, RelMiCS/AKA 2008 Frauenwörth, Germany, April 7–11, 2008. Proceedings*, volume 4988 of *Lecture Notes in Computer Science*, pages 1–2. Springer, Heidelberg, 2008.

[191] Z. Pawlak. Information systems – theoretical foundations. *Information Systems*, 6:205–218, 1981.

[192] Z. Pawlak. *Rough relations*. ICS PAS Reports 45. Institute of Computer Science, Polish Academy of Sciences, 1981.

[193] Z. Pawlak. *Rough sets*. ICS PAS Reports 431. Institute of Computer Science, Polish Academy of Sciences, 1981.

[194] Z. Pawlak. Rough Sets. *International Journal of Computer and Information Sciences*, 11(5):341–356, 1982.

[195] Z. Pawlak. *Rough Sets - Theoretical Aspects of Reasoning about Data*. Kluwer Academic Publishers, Dordrecht, 1991.

[196] W. Penczek. A temporal logic for event structures. *Fundamenta Informaticae*, XI:297–326, 1988.

[197] C. S. Pierce. Note B: The logic of relatives. In C. S. Pierce, editor, *Studies in Logic by Members of the John Hopkins University*, pages 187–203. Little, Brown and Co., Boston, 1983.

[198] G. Plotkin and C. Stirling. A framework for intuitionistic modal logic. In J. Y. Halpern, editor, *Theorical Aspects of Reasoning and Knowledge*, pages 399–406. Morgan–Kaufmann, 1986.

[199] J. Pomykała and J. A. Pomykała. The Stone algebra of rough sets. *Bulletin of the Polish Academy of Sciences, Mathematics*, 36:495–508, 1988.

[200] H. Priestley. Stone lattices: a topological approach. *Fundamenta Mathematicae*, 84:127–143, 1974.

[201] H. Priestley. The construction of spaces dual to pseudocomplemented distributive lattices. *Quarterly Journal of Mathematics*, 26(1):215–228, 1975.

[202] H. A. Priestley. Representation of distributive lattices by means of ordered Stone spaces. *Bulletin of the London Mathematical Society*, 2:186–190, 1970.

[203] H. A. Priestley. Ordered topological spaces and the representation of distributive lattices. *Proceedings of the London Mathematical Society*, 24:507–530, 1972.

[204] A. M. Radzikowska. Relational representation theorems for some modal algebras based on non-distributive lattices. submitted, 2014.

[205] D. A. Randell, Z. Cui, and A. G. Cohn. A spatial logic based on regions and connection. In B. Nebel, W. Swartout, and C. Rich, editors, *Proceedings of the 3rd International Conference on Principles of Knowledge Representation and Reasoning*, pages 165–176. Morgan Kaufmann, Cambridge, 1992.

[206] H. Rasiowa. *An Algebraic Approach to Non–Classical Logics*, volume 78 of *Studies in Logic and the Foundations of Mathematics*. Polish Scientific Publishers, Warsaw and North Holland Publishing Company, Amsterdam, 1974.

[207] H. Rasiowa and R. Sikorski. Algebraic treatment of the notion of satisfiability. *Fundamenta Mathematicae*, 73:193–200, 1950.

[208] H. Rasiowa and R. Sikorski. *Mathematics of Metamathematics*. PWN, Warsaw, 1963.

[209] C. Rauszer. Semi-Boolean algebras and their applications to intuitionistic logic with dual operations. *Fundamentas Mathematicae*, 83:219–249, 1974.

[210] S. Read. *Thinking about Logic*. Oxford University Press, 1994.

[211] I. Rewitzky. Binary multirelations. In H. de Swart, E. Orłowska, G. Schmidt, and M. Roubens, editors, *Theory and Applications of Relational Structures as Knowledge Instruments*, number 2929 in Lecture Notes in Computer Science, pages 259–274. Springer, 2003.

[212] F. Richman. Removing inequality. Unpublished, 2002.

[213] M. De Rijke and Y. Venema. Salqvists theorem for Boolean algebras with operators with applications to cylindric algebras. *Studia Logica*, 54:61–78, 1995.

[214] A. Romanowska. On some equational classes of distributive double p-algebras. *Demonstratio Mathematicae*, 9:593–607, 1976.

[215] M. Roubens and Ph. Vincke. *Preference Modelling*, volume 250 of *Lecture Notes in Economics and Mathematical Science*. Springer, 1985.

[216] J. J. M. M. Rutten. Universal coalgebra: A theory of systems. *Theoretical Computer Science*, 249:3–80, 2000.

[217] C. Martin S. Curtis and I. Rewitzky. Modelling angelic and demonic nondeterminism using multirelations. *Science of Computer Programming*, 65:140–158, 2007.

[218] S. Salbany. Bitopological spaces, compactifications and completions. Mathematical Monographs of the University of Cape Town, No 1, 1974.

[219] J. C. Shepherdson. On the interpretation of Aristotelian syllogistic. *The Journal of Symbolic Logic*, 21:137–147, 1956.

[220] A. K. Simpson. *The Proof Theory and Semantics of Intuitionistic Modal Logic*. PhD thesis, University of Edinburg, 1994.

[221] M.B. Smyth. Topology. In S. Abramsky, D.M. Gabbay, and T.S.E. Maibaum, editors, *Handbook of Logic in Computer Science, Volume 1*. Oxford University Press, Oxford, 1992.

[222] V. Sofronie-Stokkermans. Duality and canonical extensions of bounded distributive lattices with operators, and applications to the semantics of non-classical logics. Part I. *Studia Logica*, 64:93–122, 2000.

[223] V. Sofronie-Stokkermans. Duality and canonical extensions of bounded distributive lattices with operators, and applications to the semantics of non-classical logics. Part II. *Studia Logica*, 64:151–172, 2000.

[224] V. Sofronie-Stokkermans. Representation theorems and the semantics of non-classical logics, and applications to automated theorem proving. In M. Fitting and E. Orłowska, editors, *Beyond Two: Theory and Applications of Multiple Valued Logic*, volume 114 of *Studies in Fuzziness and Soft Computing*, chapter 3, pages 59–100. Springer, Berlin, January 2003.

[225] V. Sotirov. Modal theories with intuitionistic logic. In *Proceedings of the Conference on Mathematical Logic*, pages 139–171. Bulgarian Academy of Science, 1984.

[226] J. Stell. Boolean connection algebras: A new approach to the region connection calculus. *Artificial Intelligence*, 122:111–136, 2000.

[227] M. H. Stone. The theory of representations of Boolean algebras. *Transactions of the American Mathematical Society*, 40(1):37–111, 1936.

[228] M. H. Stone. Topological representation of distributive lattices and Brouwerian logics. *Časopis Pro Pestování Matematiky*, 67:1–25, 1937.

[229] T. Suzuki. Canonicity results of substructural and lattice-based logics. *The Review of Symbolic Logic*, 4(1):1–42, 2011.

[230] T. Suzuki. A Sahlqvist theorem for substructural logics. *The Review of Symbolic Logic*, 6(2):229–253, 2013.

[231] A. Szałas. On correspondence between modal and classical logic: An automated approach. Technical report, Max Planck Institut für Informatik, Saarbrüken, 1992.

[232] A. Szałas. On correspondence between modal and classical logic: An automated approach. *Journal of Logic and Computation*, 3:605–620, 1993.

[233] A. Tarski. On the calculus of relations. *The Journal of Symbolic Logic*, 6(3):73–89, 1941.

[234] M. Theunissen and Y. Venema. MacNeille completions of lattice expansions. *Algebra Universalis*, 57:143–193, 2007.

[235] A. Urquhart. A topological representation theorem for lattices. *Algebra Universalis*, 8:45–58, 1978.

[236] A. Urquhart. Equational classes of distributive double p-algebras. *Algebra Universalis*, 14:235–243, 1982.

[237] A. Urquhart. Duality for algebras of relevant logics. *Studia Logica*, 56:263–276, 1996.

[238] D. Vakarelov. Consistency, completeness and negation. In G. Priest, R. Routley, and J. Norman, editors, *Paraconsistent Logic: Essays on the Inconsistent*, pages 328–363. Philosophia Verlag, Munich, 1989.

[239] D. Vakarelov. A modal approach to dynamic ontology: modal mereotopology. *Logic and Logical Philosophy*, 17:167–187, 2008.

[240] D. Vakarelov, G. Dimov, I. Düntsch, and B. Bennet. A proximity approach to some region-based theories of space. *Journal of Applied Non-Classical Logics*, 12(3–4):527–559, 2002.

[241] C.J. van Alten. Implicational subreducts of n-potent commutative residuated lattices. *Algebra Universalis*, 57(1):47–62, 2007.

[242] J. van Benthem. Minimal deontic logic. *Bulletin of the Section of Logic*, 8(1):36–42, 1979.

[243] J. van Benthem. Correspondence theory. In D. Gabbay and F. Guenthner, editors, *Handbook of Philosophical Logic*, volume 2, pages 167–247. D. Reidel, Dordrecht, 1984.

[244] J. van Benthem and A. ter Meulen (eds). *Handbook of Logic and Language*. Elsevier, 1997.

[245] W. van der Hoek and J. J. Meyer. A complete epistemic logic for multiple agents. In M. Bacharach, L. A. Gérard-Varet, P. Mongin, and H. Shin, editors, *Epistemic Logic and the Theory of Games and Decisions*, pages 35–68. Kluwer Academic Publishers, Dordrecht, 1999.

[246] J. Varlet. A regular variety of type $< 2, 2, 1, 1, 0, 0 >$. *Algebra Universalis*, 2:218–223, 1972.

[247] L. Vita. On complements of sets and the Efremovič condition in pre–apartness spaces. *J. UCS*, 11(12):2159–2164, 2005.

[248] G. H. von Wright. *An Essay in Modal Logic.* North-Holland Publishing Company, Amsterdam, 1952.

[249] H. Wansing. A general possible worlds framework for reasoning about knowledge and belief. *Studia Logica*, 49(4):523–539, 1990.

[250] H. Wansing, editor. *Negation: A Notion in Focus*, volume 7 of *Perspectives in Analytical Philosophy.* Walter de Gruyter, Berlin, New York, 1996.

[251] M. Ward and R. P. Dilworth. Residuated lattices. *Transactions of the American Mathematical Society*, 45:335–354, 1939.

[252] A. Wedberg. The Aristotelian theory of classes. *Ajatus*, 15:299–314, 1948.

[253] D. Westerstähl. Aristotelian syllogisms and generalized quantifiers. *Studia Logica*, 48:577–585, 1989.

[254] A. N. Whitehead. *Process and Reality.* MacMillan, London, 1929.

[255] S. Willard. *General Topology.* Addison-Wesley, Reading, Mass, 1970.

[256] R. Wille. Restructuring lattice theory: An approach based on hierarchies of concepts. In I. Rival, editor, *Ordered Sets*, volume 82 of *NATO Advanced Studies Institute*, pages 445–470. NATO Advanced Studies Institute, 1982.

[257] G. Winskel. *Events in Computation.* PhD thesis, University of Edinburgh, 1980. Technical Report CST-1-80.

[258] G. Winskel and M. Nielsen. Models for concurrency. In A. Abramsky, D. Gabbay, and T. S. Maibaum, editors, *Handbook of Logic in Computer Science: Semantic Modelling*, volume 4, pages 2–148. Oxford University Press, 1995.

Index

www.ingramcontent.com/pod-product-compliance
Lightning Source LLC
Chambersburg PA
CBHW060115200326
41518CB00008B/829